OWL Information Menu Bar

At the top of each question page, you can click **Chemical Formulas** or **Scientific Notation** on the information menu bar.

Chemical Formulas	Scientific Notation	Periodic Table	Tables

Periodic Table and **Tables** on the left menu take you to additional resources in **OWL**. You can also access these resources by choosing **Chemistry Tables** from the left navigation menu on most **OWL** screens.

Answer Formatting in OWL

Like most online homework systems, when you answer questions in **OWL**, your responses must be correctly formatted in order to be graded. For some answers, you must enter superscripts and subscripts in your answer. Proper answer formatting is especially important when entering chemical formulas, mathematical equations, and units of measurement. Improper formatting will lead to otherwise correct answers being marked wrong.

Be sure to complete the **Intro to OWL** assignments to learn the basics of answer formatting and setting up your browser correctly. Answer formatting basics are summarized in the **Chemical Formulas** screen below (accessible from the information menu bar on any question page):

To Write a Chemical Formula in OWL

Enclose **subscripts** with *underscores* _.	Enclose **superscripts** with *carats* ^.
The *underscore* key is next to the number zero on the keyboard.	The *carat* key is the number six on the keyboard.
$H\_2\_O = H_2O$	$Cr\string^3+\string^ = Cr^{3+}$
Combined: $SO\_4\_\string^2-\string^ = SO_4{}^{2-}$	

Ions	Unit Charge Ions
Write the **number first** and then the charge.	Do **not** include the number one in unit charge ions.
$N\string^3-\string^ = N^{3-}$ $Ca\string^2+\string^ = Ca^{2+}$	$Na\string^+\string^ = Na^+$ $Cl\string^-\string^ = Cl^-$

Using the Chemical Formula Input

The chemical formula input box displays the superscripts and subscripts as you enter the formula. There are 3 ways to use the input box.

- **Keyboard:** Use the keyboard to enter underscores and carats on your own.
- **Buttons after:** Enter the formula without underscores or carats, then highlight each superscript and/or subscript, click the appropriate subscript or superscript button, and the underscores or carats will be filled in automatically.
- **Button during:** Use the subscript or superscript buttons to enter the underscores and carats while you type the formula.

Using the Chemical Formula Input Tool

Use the Chemical Formula Input Tool to help correctly format your answers with subscripts and superscripts.*

Click A_2 to subscript text and click A^2 to superscript text.

1. For example, to write the chemical formula for water, type a capital "**H**" in the input box.
2. Click A_2 and then type "**2**."
3. Click A_2 again to set the text format back to normal font. Note that you press the button before AND after you enter the character to be formatted.
4. Type a capital "**O**."

Check the Preview Area to see your entry before submitting it.

*If you are familiar with previous versions of **OWL**, you can still format your answers with underscores *(Shift-minus)* for subscripts and carats *(Shift-6)* for superscripts.

Scientific Notation in OWL

The basics of formatting scientific notation in **OWL** appear in the Scientific Notation screen below *(accessible from the information menu bar on any question page):*

SCIENTIFIC NOTATION IN OWL

OWL uses a similar format to the "EE" or "EXP" format found on most calculators.

6.02×10^{23} should be entered as **6.02E23**

Where "**E23**" represents "$\times 10^{23}$"

DO NOT confuse **SCIENTIFIC NOTATION** with entering **SUPERSCRIPTS!**

Use "**E**" notation for **NUMERICAL VALUES**.   6.02E23 **This is a numerical value.**

Use "**carat**" notation for **TEXT** input.   Cu^2+^ **This is text.**

Examples:

Number to Enter	OWL Format	Calculator Format
6.02×10^{23}	6.02E23	6.02 EE 23 or 6.02 EXP 23
0.000000454	4.54E-7	4.54 EE -7 or 4.54 EXP -7

Note:
Also do not confuse this notation with the irrational number "e" = 2.718282 which forms the base for the natural log, ln. On calculators this is usually the e^x or **inv ln** key.

For more info, go to **www.cengage.com/owl** and choose For Students >Getting Started.
For Technical Support for **OWL**, please visit **www.cengage.com/owl**.

Study Card to Accompany Zumdahl's *Introductory Chemistry* Series

Measurements and Calculations

Table 2.2 — The Commonly Used Prefixes in the Metric System

p. 19

Prefix	Symbol	Meaning	Power of 10 for Scientific Notation
mega	M	1,000,000.000000001	10^6
kilo	k	1000	10^3
deci	d	0.1	10^{-1}
centi	c	0.01	10^{-2}
milli	m	0.001	10^{-3}
micro	μ	0.000001	10^{-6}
nano	n	0.000000001	10^{-9}

Table 2.6 — Some Examples of Commonly Used Units

p. 22

length	A dime is 1 mm thick. A quarter is 2.5 cm in diameter. The average height of an adult man is 1.8 m.
mass	A nickel has a mass of about 5 g. A 120-lb woman has a mass of about 55 kg.
volume	A 12-oz can of soda has a volume of about 360 mL. A half gallon of milk is equal to about 2 L of milk.

$1 \text{ cm}^3 = 1 \text{ mL}$
density of $H_2O(l) = 1.0 \text{ g/mL}$
density = mass/volume
Avogadro's number = 6.022×10^{23}

Energy

Heat Required = Q = specific heat capacity $\times$ mass $\times \Delta T$
Specific heat capacity of $H_2O(l) = 4.184 \text{ J/g } °C$
Kinetic energy = $mv^2/2$
Exothermic reactions produce heat
Endothermic reactions absorb heat

Kinds of Chemical Reactions

p. 191

Figure 7.12 — Summary of classes of reactions.

Atomic Structure

Mass number (A) (number of protons and neutrons)
$^{23}_{11}\text{Na}$ ← Element symbol
Atomic number (Z) (number of protons)
$A - Z = \#n^0$ $\#p^+ - \#e^- = \text{charge}$

Gases

STP: 0 °C, 1 atm
Volume of 1 mole of ideal gas at STP = 22.4 L
$PV = nRT$ (Ideal Gas Law)
$R = 0.08206 \text{ L atm/K mol}$
Process at constant n and T: $P_1V_1 = P_2V_2$ (Boyle's law)
Process at constant n and P: $V_1/T_1 = V_2/T_2$ (Charles's law)
Process at constant T and P: $V_1/n_1 = V_2/n_2$
(Avogadro's law)

Types of Crystalline Solids

p. 459

Crystalline solids

Ionic solids	Molecular solids	Atomic solids
Components are ions.	Components are molecules.	Components are atoms.

Figure 14.13 — The classes of crystalline solids.

Solutions

$$\text{Mass percent} = \frac{\text{mass of solute}}{\text{mass of solution}} \times 100\% \quad \text{(p. 481)}$$

$$M = \text{molarity} = \frac{\text{moles of solute}}{\text{liters of solution}} = \frac{\text{mol}}{\text{L}} \quad \text{(p. 483)}$$

Mass of solute = (molar mass of solute) $\times$ (L of solution) (Molarity)

$$\text{Normality} = N = \frac{\text{number of equivalents}}{1 \text{ liter of solution}}$$

$$= \frac{\text{equivalents}}{\text{liter}} = \frac{\text{equiv}}{\text{L}} \quad \text{(p. 499)}$$

Acids and Bases

Common Strong Acids: HCl, HNO_3, H_2SO_4, $HClO_4$, HI
Common Weak Acids: HSO_4^-, CH_3COOH (often written $HC_2H_3O_2$), HF
Common Strong Bases: $NaOH$, KOH
Common Weak Bases: NH_3

$K_w = 10^{-14} = [H^+][OH^-]$ (ion-product constant for water) (p. 523)

$pH = -\log[H^+]$ (p. 526)

$pOH = -\log[OH^-]$ (p. 527)

$pH + pOH = 14.00$ (p. 529)

Equilibrium Constants

$a\text{A} + b\text{B} \rightleftharpoons c\text{C} + d\text{D}, \; K = [C]^c[D]^d/[A]^a[B]^b$ (p. 553)
$A_2B_3 \text{ (s)} \rightleftharpoons 2A^{3+}(aq) + 3B^{2-}(aq), \; K_{sp} = [A^{3+}]^2[B^{2-}]^3$
$[X]$ = Molarity of X

Study Card to Accompany Zumdahl's *Introductory Chemistry* Series

Chemical Bonding

Table 12.4 Arrangements of Electron Pairs and the Resulting Molecular Structures for Two, Three, and Four Electron Pairs *p. 388*

Number of Electron Pairs	Bonds	Electron Pair Arrangement	Ball-and-Stick Model	Molecular Structure	Partial Lewis Structure	Ball-and-Stick Model
2	2	Linear	180°	Linear	A—B—A	
3	3	Trigonal planar (triangular)	120°	Trigonal planar (triangular)	A—B—A (with A below)	
4	4	Tetrahedral	109.5°	Tetrahedral	A—B—A (A above and below)	
4	3	Tetrahedral	109.5°	Trigonal pyramid	A—B—A (A below)	
4	2	Tetrahedral	109.5°	Bent or V-shaped	A—B—A	

Common Lewis Dot Fragments

Oxidation–Reduction Reactions

Oxidation is loss of electrons (OIL)
Reduction is gain of electrons (RIG)

Rules for Assigning Oxidation States *p. 586*

1. The oxidation state of an atom in an uncombined element is 0.
2. The oxidation state of a monatomic ion is the same as its charge.
3. Oxygen is assigned an oxidation state of −2 in most of its covalent compounds. Important exception: peroxides (compounds containing the O_2^{2-} group), in which each oxygen is assigned an oxidation state of −1.
4. In its covalent compounds with nonmetals, hydrogen is assigned an oxidation state of +1.
5. In binary compounds, the element with the greater electronegativity is assigned a negative oxidation state equal to its charge as an anion in its ionic compounds.
6. For an electrically neutral compound, the sum of the oxidation states must be zero.
7. For an ionic species, the sum of the oxidation states must equal the overall charge.

Periodic Table of Elements

1 1A	Alkaline earth metals																Noble gases 18 8A
1 H	2 2A											13 3A	14 4A	15 5A	16 6A	17 7A	2 He
3 Li	4 Be											5 B	6 C	7 N	8 O	9 F	10 Ne
11 Na	12 Mg	3	4	5	6	7	8	9	10	11	12	13 Al	14 Si	15 P	16 S	17 Cl	18 Ar
19 K	20 Ca	21 Sc	22 Ti	23 V	24 Cr	25 Mn	26 Fe	27 Co	28 Ni	29 Cu	30 Zn	31 Ga	32 Ge	33 As	34 Se	35 Br	36 Kr
37 Rb	38 Sr	39 Y	40 Zr	41 Nb	42 Mo	43 Tc	44 Ru	45 Rh	46 Pd	47 Ag	48 Cd	49 In	50 Sn	51 Sb	52 Te	53 I	54 Xe
55 Cs	56 Ba	57 La*	72 Hf	73 Ta	74 W	75 Re	76 Os	77 Ir	78 Pt	79 Au	80 Hg	81 Tl	82 Pb	83 Bi	84 Po	85 At	86 Rn
87 Fr	88 Ra	89 Ac†	104 Rf	105 Db	106 Sg	107 Bh	108 Hs	109 Mt	110 Ds	111 Rg	112 Uub	113 Uut	114 Uuq	115 Uup			118 Uuo

Alkali metals

Transition metals

Halogens

*Lanthanides	58 Ce	59 Pr	60 Nd	61 Pm	62 Sm	63 Eu	64 Gd	65 Tb	66 Dy	67 Ho	68 Er	69 Tm	70 Yb	71 Lu
†Actinides	90 Th	91 Pa	92 U	93 Np	94 Pu	95 Am	96 Cm	97 Bk	98 Cf	99 Es	100 Fm	101 Md	102 No	103 Lr

Table of Atomic Masses*

Element	Symbol	Atomic Number	Atomic Mass	Element	Symbol	Atomic Number	Atomic Mass	Element	Symbol	Atomic Number	Atomic Mass
Actinium	Ac	89	[227]§	Gold	Au	79	197.0	Praseodymium	Pr	59	140.9
Aluminum	Al	13	26.98	Hafnium	Hf	72	178.5	Promethium	Pm	61	[145]
Americium	Am	95	[243]	Hassium	Hs	108	[265]	Protactinium	Pa	91	[231]
Antimony	Sb	51	121.8	Helium	He	2	4.003	Radium	Ra	88	226
Argon	Ar	18	39.95	Holmium	Ho	67	164.9	Radon	Rn	86	[222]
Arsenic	As	33	74.92	Hydrogen	H	1	1.008	Rhenium	Re	75	186.2
Astatine	At	85	[210]	Indium	In	49	114.8	Rhodium	Rh	45	102.9
Barium	Ba	56	137.3	Iodine	I	53	126.9	Roentgenium	Rg	111	[272]
Berkelium	Bk	97	[247]	Iridium	Ir	77	192.2	Rubidium	Rb	37	85.47
Beryllium	Be	4	9.012	Iron	Fe	26	55.85	Ruthenium	Ru	44	101.1
Bismuth	Bi	83	209.0	Krypton	Kr	36	83.80	Rutherfordium	Rf	104	[261]
Bohrium	Bh	107	[264]	Lanthanum	La	57	138.9	Samarium	Sm	62	150.4
Boron	B	5	10.81	Lawrencium	Lr	103	[260]	Scandium	Sc	21	44.96
Bromine	Br	35	79.90	Lead	Pb	82	207.2	Seaborgium	Sg	106	[263]
Cadmium	Cd	48	112.4	Lithium	Li	3	6.9419	Selenium	Se	34	78.96
Calcium	Ca	20	40.08	Lutetium	Lu	71	175.0	Silicon	Si	14	28.09
Californium	Cf	98	[251]	Magnesium	Mg	12	24.31	Silver	Ag	47	107.9
Carbon	C	6	12.01	Manganese	Mn	25	54.94	Sodium	Na	11	22.99
Cerium	Ce	58	140.1	Meitnerium	Mt	109	[268]	Strontium	Sr	38	87.62
Cesium	Cs	55	132.90	Mendelevium	Md	101	[258]	Sulfur	S	16	32.07
Chlorine	Cl	17	35.45	Mercury	Hg	80	200.6	Tantalum	Ta	73	180.9
Chromium	Cr	24	52.00	Molybdenum	Mo	42	95.94	Technetium	Tc	43	[98]
Cobalt	Co	27	58.93	Neodymium	Nd	60	144.2	Tellurium	Te	52	127.6
Copper	Cu	29	63.55	Neon	Ne	10	20.18	Terbium	Tb	65	158.9
Curium	Cm	96	[247]	Neptunium	Np	93	[237]	Thallium	Tl	81	204.4
Darmstadtium	Ds	110	[271]	Nickel	Ni	28	58.69	Thorium	Th	90	232.0
Dubnium	Db	105	[262]	Niobium	Nb	41	92.91	Thulium	Tm	69	168.9
Dysprosium	Dy	66	162.5	Nitrogen	N	7	14.01	Tin	Sn	50	118.7
Einsteinium	Es	99	[252]	Nobelium	No	102	[259]	Titanium	Ti	22	47.88
Erbium	Er	68	167.3	Osmium	Os	76	190.2	Tungsten	W	74	183.9
Europium	Eu	63	152.0	Oxygen	O	8	16.00	Uranium	U	92	238.0
Fermium	Fm	100	[257]	Palladium	Pd	46	106.4	Vanadium	V	23	50.94
Fluorine	F	9	19.00	Phosphorus	P	15	30.97	Xenon	Xe	54	131.3
Francium	Fr	87	[223]	Platinum	Pt	78	195.1	Ytterbium	Yb	70	173.0
Gadolinium	Gd	64	157.3	Plutonium	Pu	94	[244]	Yttrium	Y	39	88.91
Gallium	Ga	31	69.72	Polonium	Po	84	[209]	Zinc	Zn	30	65.38
Germanium	Ge	32	72.59	Potassium	K	19	39.10	Zirconium	Zr	40	91.22

*The values given here are to four significant figures where possible.   §A value given in parentheses denotes the mass of the longest-lived isotope.

Table 5.1 Common Simple Cations and Anions

Cation	Name	Anion	Name*
H^+	hydrogen	H^-	hydride
Li^+	lithium	F^-	fluoride
Na^+	sodium	Cl^-	chloride
K^+	potassium	Br^-	bromide
Cs^+	cesium	I^-	iodide
Be^{2+}	beryllium	O^{2-}	oxide
Mg^{2+}	magnesium	S^{2-}	sulfide
Ca^{2+}	calcium		
Ba^{2+}	barium		
Al^{3+}	aluminum		
Ag^+	silver		
Zn^{2+}	zinc		

*The root is given in color.

Table 5.2 Common Type II Cations

Ion	Systematic Name	Older Name
Fe^{3+}	iron(III)	ferric
Fe^{2+}	iron(II)	ferrous
Cu^{2+}	copper(II)	cupric
Cu^+	copper(I)	cuprous
Co^{3+}	cobalt(III)	cobaltic
Co^{2+}	cobalt(II)	cobaltous
Sn^{4+}	tin(IV)	stannic
Sn^{2+}	tin(II)	stannous
Pb^{4+}	lead(IV)	plumbic
Pb^{2+}	lead(II)	plumbous
Hg^{2+}	mercury(II)	mercuric
Hg_2^{2+}*	mercury(I)	mercurous

*Mercury(I) ions always occur bound together in pairs to form Hg_2^{2+}.

Table 5.4 Names of Common Polyatomic Ions

Ion	Name	Ion	Name
NH_4^+	ammonium	CO_3^{2-}	carbonate
NO_2^-	nitrite	HCO_3^-	hydrogen carbonate (bicarbonate is a widely used common name)
NO_3^-	nitrate		
SO_3^{2-}	sulfite		
SO_4^{2-}	sulfate	ClO^-	hypochlorite
HSO_4^-	hydrogen sulfate (bisulfate is a widely used common name)	ClO_2^-	chlorite
		ClO_3^-	chlorate
		ClO_4^-	perchlorate
OH^-	hydroxide	$C_2H_3O_2^-$	acetate
CN^-	cyanide	MnO_4^-	permanganate
PO_4^{3-}	phosphate	$Cr_2O_7^{2-}$	dichromate
HPO_4^{2-}	hydrogen phosphate	CrO_4^{2-}	chromate
$H_2PO_4^-$	dihydrogen phosphate	O_2^{2-}	peroxide

Table 7.1 General Rules for Solubility of Ionic Compounds (Salts) in Water at 25 °C

1. Most nitrate (NO_3^-) salts are soluble.
2. Most salts of Na^+, K^+, and NH_4^+ are soluble.
3. Most chloride salts are soluble. Notable exceptions are $AgCl$, $PbCl_2$, and Hg_2Cl_2.
4. Most sulfate salts are soluble. Notable exceptions are $BaSO_4$, $PbSO_4$, and $CaSO_4$.
5. Most hydroxide compounds are only slightly soluble.* The important exceptions are $NaOH$ and KOH. $Ba(OH)_2$ and $Ca(OH)_2$ are only moderately soluble.
6. Most sulfide (S^{2-}), carbonate (CO_3^{2-}), and phosphate (PO_4^{3-}) salts are only slightly soluble.*

*The terms *insoluble* and *slightly soluble* really mean the same thing: such a tiny amount dissolves that it is not possible to detect it with the naked eye.

SI Units and Conversion Factors*

Length SI Unit: Meter (m)

1 meter	= 1.0936 yards
1 centimeter	= 0.39370 inch
1 inch	= 2.54 centimeters (exactly)
1 kilometer	= 0.62137 mile
1 mile	= 5280. feet
	= 1.6093 kilometers

Mass SI Unit: Kilogram (kg)

1 kilogram	= 1000 grams
	= 2.2046 pounds
1 pound	= 453.59 grams
	= 0.45359 kilogram
	= 16 ounces
1 atomic mass unit	= 1.66057×10^{-27} kilograms

Volume SI Unit: Cubic Meter (m^3)

1 liter	= 10^{-3} m^3
	= 1 dm^3
	= 1.0567 quarts
1 gallon	= 4 quarts
	= 8 pints
	= 3.7854 liters
1 quart	= 32 fluid ounces
	= 0.94635 liter

Pressure SI Unit: Pascal (Pa)

1 atmosphere	= 101.325 kilopascals
	= 760. torr (mm Hg)
	= 14.70 pounds per square inch

Energy SI Unit: Joule (J)

1 joule	= 0.23901 calorie
1 calorie	= 4.184 joules

Introductory Chemistry

A FOUNDATION

HYBRID EDITION

Introductory Chemistry

A FOUNDATION

HYBRID EDITION

Steven S. Zumdahl
University of Illinois

Donald J. DeCoste
University of Illinois

BROOKS/COLE
CENGAGE Learning™

Australia • Brazil • Japan • Korea • Mexico • Singapore • Spain • United Kingdom • United States

Introductory Chemistry: A Foundation,
Seventh Edition, Hybrid Edition
Steven S. Zumdahl and Donald J. DeCoste

Publisher: Charles Hartford

Development Editor: Alyssa White

Senior Content Project Manager: Cathy Brooks

Art Directors: Jill Haber, Cate Barr

Senior Marketing Manager: Nicole Hamm

Senior Technology Project Manager: Rebecca Berardy-Schwartz

Assistant Editor: Stephanie VanCamp

Editorial Assistant: Jon Olafsson

Photo Researcher: Sharon Donahue

Senior Marketing Communications Manager: Linda Yip

Print Buyer: Judy Inouye

Permissions Editor: Roberta Broyer

Text Designer: Janet Theurer

Production Service: Graphic World Inc.

Production Project Manager, Graphic World: Megan Greiner

Copyeditor: Graphic World Inc.

Illustrators: Jessica Broekman, Rossi Art, and Graphic World Inc.

Cover Designer: Brian Salisbury

Cover Image: Radius Images/Photo Library

OWL Producers: Stephen Battisti, Cindy Stein, David Hart (Center for Educational Software Development; University of Massachusetts, Amherst).

Compositor: Graphic World Inc.

For product information and technology assistance, contact us at
Cengage Learning Customer & Sales Support, 1-800-354-9706
For permission to use material from this text or product,
submit all requests online at **www.cengage.com/permissions**
Further permissions questions can be e-mailed to
permissionrequest@cengage.com

Library of Congress Control Number: 2009934046

ISBN-13: 978-0-538-75708-9

ISBN-10: 0-538-75708-6

Brooks/Cole
20 Davis Drive
Belmont, CA 94002-3098
USA

Cengage Learning is a leading provider of customized learning solutions with office locations around the globe, including Singapore, the United Kingdom, Australia, Mexico, Brazil, and Japan. Locate your local office at **international.cengage.com/region**.

Cengage Learning products are represented in Canada by Nelson Education, Ltd.

For your course and learning solutions, visit **academic.cengage.com**.

Purchase any of our products at your local college store or at our preferred online store **www.CengageBrain.com**.

Printed in Canada
1 2 3 4 5 6 7 14 13 12 11 10

BRIEF CONTENTS

CONTENTS

The Hybrid Edition of *Introductory Chemistry: A Foundation* continues toward the goals we have pursued in all editions: to make chemistry interesting, accessible, and understandable to the beginning student. For this edition, we have included additional support for instructors and students to help achieve these goals.

Learning chemistry can be very rewarding. And even the novice, we believe, can relate the macroscopic world of chemistry—the observation of color changes and precipitate formation—to the microscopic world of ions and molecules. To achieve that goal, instructors are making a sincere attempt to provide more interesting and more effective ways to learn chemistry, and we hope that *Introductory Chemistry: A Foundation: Hybrid Edition* will be perceived as a part of that effort. In this text we have presented concepts in a clear and sensible manner using language and analogies that students can relate to. We have also written the book in a way that supports active learning. In particular, the Active Learning Questions, found at the end of each chapter, provide excellent material for collaborative work by students. In addition, we have connected chemistry to real-life experience at every opportunity, from chapter opening discussions of chemical applications to "Chemistry in Focus" features throughout the book. We are convinced that this approach will foster enthusiasm and real understanding as the student uses this text. Highlights of the *Introductory Chemistry: A Foundation: Hybrid Edition* program are described below.

In This Edition

Building on the success of *Introductory Chemistry: A Foundation,* the following changes have been made to further enhance the text:

Updates to the Student Text and Instructor's Annotated Edition

Changes to the student text and the accompanying Instructor's Annotated Edition are outlined below:

Instructor's Annotated Edition The marginal annotations have been revised and expanded to include point-of-use references to print and new media.

Section 8.4: Learning to Solve Problems We have added a new section that emphasizes the importance of conceptual problem solving in which students are shown how to think their way through a problem. The students will learn that this "big picture approach" produces more long-term, meaningful learning rather than simply memorizing specific steps that are soon forgotten.

Problem-Solving Approach in *Examples* Using the general conceptual problem-solving approach outlined in the new Section 8.4, we have introduced a series of questions into the in-chapter *Examples*. This more active approach helps students think their way through the solution to the problem. We use this approach for most of the quantitative *Examples* beginning in Section 8.4.

Section 9.4: The Concept of Limiting Reactants We have added a new section that helps students better understand the concept of a limiting reactant. Familiar examples such as making sandwiches and lemonade are used, as well as molecular-level illustrations of chemistry reactions. Students will learn how to think about limiting reactants before being asked to make mass–mass calculations to determine which reactant is limiting for a given reaction.

New Active Learning Questions We have written new Active Learning Questions for each chapter (over 170 new problems in all). In addition, many of the new problems include visual components such as graphs or molecular-level illustrations.

New End-of-Chapter Questions Using the "Chemistry in Focus" Boxes We have written new end-of-chapter questions that specifically address topics covered in the "Chemistry in Focus" boxes (over 40 new questions in all).

Art Program We have revised almost every figure in the textbook to better serve visual learners. Most of the glassware, orbitals, graphs, and flowcharts have been redrawn.

"Chemistry in Focus" Boxes Approximately 20% of the "Chemistry in Focus" boxes in the seventh edition are new, and many more have been revised, with up-to-date topics such as hybrid cars, artificial sweeteners, and positron emission tomography (PET).

End-of-Chapter Exercises We have replaced 20% of the end-of-chapter exercises and problems and cumulative review exercises. As before, the margins of the Annotated Instructor's Edition include answers to all of the Self-Check and end-of-chapter exercises, along with additional examples for all Example problems. In the student edition, answers to Self-Check Exercises are provided at the back of the book.

Enhanced Teaching Resources for the Instructor

OWL: Online Web-based Learning by Roberta Day and Beatrice Botch of the University of Massachusetts, Amherst, and William Vining of the State University of New York at Oneonta. Developed at the University of Massachusetts, Amherst, and class-tested by tens of thousands of chemistry students, OWL is a fully customizable and flexible web-based learning system. OWL supports mastery learning and offers numerical, chemical, and contextual parameterization to produce thousands of problems correlated to this text. The OWL system also features a database of simulations, tutorials, and exercises, as well as end-of-chapter problems from the text. In addition, OWL now includes *ChemWork* assignments, which help students learn key chemical concepts while guiding them in the process to become problem solvers. (See description below.) With OWL, you get the most widely used online learning system available for chemistry with unsurpassed reliability and dedicated training and support. Also new in OWL is *Go Chemistry*™—27 mini-video lectures covering key chemistry concepts that students can view onscreen or download to their portable video player to study on the go! The **eBook in OWL** includes the complete electronic version of the text, including the original end-of-chapter homework problems. Most e-Books in OWL are interactive and offer highlighting, notetaking, and bookmarking features that can all be saved. In addition, the eBook includes links to Thinkwell® mini-video lectures. To view an OWL demo and for more information, visit **www.cengage.com/owl** or contact your Cengage Learning, Brooks/Cole representative.

ChemWork Offered in OWL, *ChemWork* assignments offer students another opportunity to practice. These problems are designed for students to use in one of two ways: to *learn* the problem-solving process (while doing actual homework problems) or as a *capstone* assignment to determine whether they understand how to solve problems (perhaps in final preparation for an exam). *ChemWork* assignments test students' understanding of core concepts from each chapter. Students who solve a particular problem with no assistance can proceed directly to the answer and receive congratulations. However, students who need help can get assistance through a series of hints. The procedure for assisting students is modeled after the way a teacher would help with a homework problem in his or her office. The hints are usually in the form of interactive questions that guide students through the problem-solving process. Students cannot receive the right answer from the system; rather it encourages them to continue working on the problem through this system of multiple hints. *ChemWork* is chemically and numerically parameterized so that each student in the course receives a unique set of problems.

PowerLecture with ExamView® and JoinIn™ Instructor's DVD (ISBN-10: 0-538-73643-7; ISBN-13: 978-0-538-73643-5): PowerLecture is a one-stop digital library and presentation tool that includes:

- **Prepared Microsoft® PowerPoint® Lecture Slides** that cover all key points from the text in a convenient format that you can enhance with your own materials or with additional interactive video and animations from the CD-ROM for personalized, media-enhanced lectures.

- **Image Libraries** in PowerPoint® and JPEG formats that contain electronic files for all text art, most photographs, and all numbered tables in the text. These files can be used to print transparencies or to enhance PowerPoint® lectures.

- **JoinIn™ "Clicker" Slides** include questions that are written specifically for the use of *Introductory Chemistry* with the classroom response system of your choice, and allows you to seamlessly display student answers.

- **ExamView®** testing software, with all the test items from the Online Test Bank in electronic format. The electronic test bank by Steven S. Zumdahl and Donald J. DeCoste provides over 1600 multiple-choice, true-false, short-answer, matching, and completion questions. Approximately 200 questions from the previous edition have been made into algorithms, which enable you to create even more customized tests.

Enhanced Learning Resources for the Student

Student Companion Website Accessible from **www.cengage.com/chemistry/zumdahl,** this site provides online study tools, including practice tests and flashcards.

GoChemistry™ **for General Chemistry** (27-Module Set) (ISBN-10: 0-495-38228-0; ISBN-13: 978-0-495-38228-7) *GoChemistry*™ is a set of easy-to-use essential videos that can be downloaded to your video iPod or portable video player—ideal for the student on the go! Developed by award-winning chemists, these new electronic tools are designed to help students quickly review essential chemistry topics. Mini-video lectures include animations and problems for a quick summary of key concepts. Selected *GoChemistry* modules have e-flashcards to briefly introduce a key concept and then test student understanding of the basics with a series of questions. *GoChemistry* also plays on QuickTime, iTunes, and iPhones. Modules are also available separately. To purchase, enter ISBN 0-495-38228-0 at **www.CengageBrain.com.**

ⓊWL **OWL for General Chemistry** by Roberta Day and Beatrice Botch of the University of Massachusetts, Amherst, and William Vining of the State University of New York at Oneonta [OWL Instant Access (4 Semesters) ISBN-10: 0-495-05099-7; ISBN-13: 978-0-495-05099-5]. See the previous description in the instructor support materials.

Emphasis on Reaction Chemistry

We continue to emphasize chemical reactions early in the book, leaving the more abstract material on orbitals for later chapters. In a course in which many students encounter chemistry for the first time, it seems especially important that we present the chemical nature of matter before we discuss the theoretical intricacies of atoms and orbitals. Reactions are inherently interesting to students and can help us draw them to chemistry. In particular, reactions can form the basis for fascinating classroom demonstrations and laboratory experiments.

We have therefore chosen to emphasize reactions before going on to the details of atomic structure. Relying only on very simple ideas about the atom, Chapters 6 and 7 represent a thorough treatment of chemical reactions, including how to recognize a chemical change and what a chemical equation means. The properties of aqueous solutions are discussed in detail, and careful attention is given to precipitation and acid–base reactions. In addition, a simple treatment of oxidation–reduction reactions is given. These chapters should provide a solid foundation, relatively early in the course, for reaction-based laboratory experiments.

For instructors who feel that it is desirable to introduce orbitals early in the course, prior to chemical reactions, the chapters on atomic theory and bonding (Chapters 11 and 12) can be covered directly after Chapter 4. Chapter 5 deals solely with nomenclature and can be used wherever it is needed in a particular course.

Development of Problem-Solving Skills

Problem solving is a high priority in chemical education. We all want our students to acquire problem-solving skills. Fostering the development of such skills has been a central focus of the earlier editions of this text and we have maintained this approach in this edition.

In the first chapters we spend considerable time guiding students to an understanding of the importance of learning chemistry. At the same time, we explain that the complexities that can make chemistry frustrating at times can also provide the opportunity to develop the problem-solving skills that are beneficial in any profession. Learning to think like a chemist is useful to everyone. To emphasize this idea, we apply scientific thinking to some real-life problems in Chapter 1.

One reason chemistry can be challenging for beginning students is that they often do not possess the required mathematical skills. Thus we have paid careful attention to such fundamental mathematical skills as using scientific notation, rounding off to the correct number of significant figures, and rearranging equations to solve for a particular quantity. And we have meticulously followed the rules we have set down, so as not to confuse students.

Attitude plays a crucial role in achieving success in problem solving. Students must learn that a systematic, thoughtful approach to problems is better than brute force memorization. We foster this attitude early in the book, using temperature conversions as a vehicle in Chapter 2. Throughout the book we encourage an approach that starts with trying to represent the essence of the problem using symbols and/or diagrams, and ends with thinking about whether the answer makes sense. We approach new concepts by carefully working through the material before we give mathematical formulas or overall strategies. We encourage a thoughtful step-by-step approach rather than the premature use of algorithms. Once we have provided the necessary foundation, we highlight important rules and processes in skill development boxes so that students can locate them easily.

We have written a new section (Section 8.4: Learning to Solve Problems) so that students will better understand how to think their way through a problem. We discuss how to solve problems in a flexible, creative way based on understanding the fundamental ideas of chemistry and asking and answering key questions. We model this approach in the in-text *Examples* throughout the text.

Many of the worked examples are followed by Self-Check Exercises, which provide additional practice. The Self-Check Exercises are keyed to end-of-chapter exercises to offer another opportunity for students to practice a particular problem-solving skill or understand a particular concept.

End-of-chapter exercises for this special *Hybrid Edition* appear in OWL. Because of OWL's unique parameterization functionality where numeric chemical variables change each time the student attempts a question, not all "matched pair" problems needed to be coded; however, care was taken to ensure proper content coverage. Active Learning Questions, which are designed to be open-ended, remain in the text itself. The OWL eBook contains the original exercise sets, as does the AIE.

Handling the Language of Chemistry and Applications

We have gone to great lengths to make this book "student friendly" and have received enthusiastic feedback from students who have used it.

As in the earlier editions, we present a systematic and thorough treatment of chemical nomenclature. Once this framework is established, students can progress through the book comfortably.

Along with chemical reactions, applications form an important part of descriptive chemistry. Because students are interested in chemistry's impact on their lives, we have included many new "Chemistry in Focus" boxes, which describe current applications of chemistry. These special interest boxes cover such topics as new technology to replace the incandescent lightbulb, using bees to detect drugs and bombs at airports, and analyzing isotopes in human hair to identify disaster victims' country of origin.

Visual Impact of Chemistry

In response to instructors' requests to include graphic illustrations of chemical reactions, phenomena, and processes, we use a full-color design that enables color to be used functionally, thoughtfully, and consistently to help students understand chemistry and to make the subject more inviting to them. We have included only those photos that illustrate a chemical reaction or phenomenon or that make a connection between chemistry and the real world. Many new photos enhance the seventh edition.

Supplements for the Text

A main focus of this revision is to provide instructors and students with an unparalleled level of support. In addition to the media components described previously, we offer the following materials.

For the Student

Student Companion Website Accessible from **www.cengage.com/chemistry/zumdahl,** this site provides online study tools, including practice tests and flashcards.

Study Guide by Donald J. DeCoste of the University of Illinois contains Chapter Discussions and Learning Review (practice chapter tests) (ISBN-10: 0-538-73640-2; ISBN-13: 978-0-538-73640-4).

Solutions Guide by James F. Hall, University of Massachusetts, Lowell, contains detailed solutions for the even-numbered end-of-chapter questions and exercises and cumulative review exercises (ISBN-10: 0-538-73641-0; ISBN-13: 978-0-538-73641-1).

Introductory Chemistry in the Laboratory by James F. Hall contains experiments organized according to the topical presentation in the text. Annotations in the Annotated Instructor's Edition indicate where the experiments from this manual are relevant to chapter content. The lab manual has been updated and revised for this edition (ISBN-10: 0-538-73642-9; ISBN-13: 978-0-538-73642-8).

For the Instructor

Annotated Instructor's Edition The Annotated Instructor's Edition gathers a wealth of teaching support in one convenient package. The AIE contains all 21 chapters (the full contents of *Introductory Chemistry: A Foundation,* Seventh Edition). Annotations in the wrap-around margins of the AIE include:

- Answers to Self-Check Exercises, at point-of-use.

- Answers to all end-of-chapter questions and exercises, at point-of-use.

- Additional Examples with answers to supplemental worked-out Examples in the text.

- Technology Information about incorporating animations and video clips from the electronic support materials in lecture.

- Teaching Support Suggestions for specific lecture/instruction methods, activities, and in-class demonstrations to help convey concepts.

- An Overview of the chapter's learning objectives.

- Teaching Tips: Guidelines for highlighting critical information in the chapter.

- Misconceptions: Tips on where students may have trouble or be confused with a topic.

- Demonstrations: Detailed instructions for in-class demonstrations and activities. (These are similar to material in Teaching Support, and may be referenced in Teaching Support annotations.)

- Laboratory Experiments: Information on which labs in the Laboratory Manual are relevant to chapter content.

- Background Information: Explanations of conventions used in the text.

- Icons mark material correlations between the main text and the electronic support materials, the Test Bank, and the Laboratory Manual.

- Historical Notes: Biographical or other historical information about science and scientists.

PowerLecture with ExamView® and JoinIn™ Instructor's DVD (ISBN-10: 0-538-73643-7; ISBN-13: 978-0-538-73643-5): PowerLecture is a one-stop digital library and presentation tool that includes:

- **Prepared Microsoft® PowerPoint® Lecture Slides** that cover all key points from the text in a convenient format that you can enhance with your own materials or with additional interactive video and animations from the CD-ROM for personalized, media-enhanced lectures.

- **Image Libraries** in PowerPoint® and JPEG formats that contain electronic files for all text art, most photographs, and all numbered tables in the text. These files can be used to print transparencies or to enhance PowerPoint® lectures.

- **JoinIn™ "Clicker" Slides** include questions that are written specifically for the use of *Introductory Chemistry* with the classroom response system of your choice, and allows you to seamlessly display student answers.

- *Complete Solutions Manual* (James F. Hall, University of Massachusetts, Lowell) The *Complete Solutions Manual* contains detailed solutions to all of the end-of chapter problems, problems, and cumulative review exercises.

- **Answers to Active Learning Questions** from the end-of-chapter questions, written by Donald J. DeCoste.

- *Instructor's Guide for Introductory Chemistry in the Laboratory* by James F. Hall includes general notes about each experiment, estimated completion time, materials required, and answers to both pre- and post-laboratory questions. Annotations in the AIE indicate where experiments from this manual are relevant to chapter content. The lab manual has been updated and revised for this edition.

- **Sample Chapters** from the *Student Solutions Manual* and *Study Guide*.

- **ExamView**® testing software, with all the test items from the Online Test Bank in electronic format. The electronic test bank by Steven S. Zumdahl and Donald J. DeCoste provides over 1600 multiple-choice, true-false, short-answer, matching, and completion questions. Approximately 200 questions from the previous edition have been made into algorithms, which enable you to create even more customized tests.

Acknowledgments

This book represents the collaborative efforts of many talented and dedicated people to whom we are greatly indebted. Charles Hartford, Publisher, was extremely supportive of the revision. Charlie asked good questions and provided helpful and creative ideas. We also wish to thank Cathy Brooks, Content Project Manager, who has an eye for detail and an uncanny ability to do everything at once and all of it well. We appreciate the efforts of Alyssa White, Development Editor, who, along with her other tasks, was instrumental in making sure the art was appealing and correct. We are grateful to have worked with Sharon Donahue, Photo Researcher, who once again displayed her remarkable ability for finding outstanding photos.

Jim Hall of the University of Massachusetts, Lowell, contributed in many different ways to the success of this project. He has been a tremendous help with the end-of-chapter questions and problems and the cumulative review exercises, along with writing the *Solutions Guides, Introductory Chemistry in the Laboratory,* and the *Instructor's Guide for Introductory Chemistry in the Laboratory.*

We especially appreciate the efforts of Gretchen Adams of the University of Illinois for her work on revising the PowerPoint® media component, Richard Triplett of Des Moines Area Community College for reviewing the ancillaries, and Linda Bush for revising the test bank.

Thanks to others who provided valuable assistance on this revision: Stephanie VanCamp, Assistant Editor for the ancillaries; Rebecca Berardy-Schwartz, Technology Project Manager; Jon Olaffson, Editorial Assistant; Nicole Hamm, Marketing Manager; Megan Greiner, Project Manager (Graphic World); Jill Haber and Cate Barr, Art Directors; Betty Litt, Copyeditor; and David Shinn, who checked the textbook and solutions for accuracy.

Our sincerest appreciation goes to all of the reviewers whose feedback and suggestions contributed to the success of this project.

Angela Bickford
Northwest Missouri State University

Simon Bott
University of Houston

Jabe Breland
St. Petersburg College

Frank Calvagna
Rock Valley College

Jing-Yi Chin
Suffolk County Community College

Carl David
University of Connecticut

Cory DiCarlo
Grand Valley State University

Cathie Keenan
Chaffey College

Pamela Kimbrough
Crafton Hills College

Wendy Lewis
Stark State College of Technology

Guillermo Muhlmann
Capital Community College

Lydia Martinez Rivera
University of Texas at San Antonio

Sharadha Sambasivan
Suffolk County Community College

Perminder Sandhu
Bellevue Community College

Lois Schadewald
Normandale Community College

Marie Villarba
Seattle Central Community College

1

Chemistry: An Introduction

● Chemistry deals with the natural world. *(Dr. John Brackenbury/Science Photo Library/Photo Researchers, Inc.)*

Did you ever see a fireworks display on July Fourth and wonder how it's possible to produce those beautiful, intricate designs in the air? Have you read about dinosaurs—how they ruled the earth for millions of years and then suddenly disappeared? Although the extinction happened 65 million years ago and may seem unimportant, could the same thing happen to us? Have you ever wondered why an ice cube (pure water) floats in a glass of water (also pure water)? Did you know that the "lead" in your pencil is made of the same substance (carbon) as the diamond in an engagement ring? Did you ever wonder how a corn plant or a palm tree grows seemingly by magic, or why leaves turn beautiful colors in autumn? Do you know how the battery works to start your car or run your calculator? Surely some of these things and many others in the world around you have intrigued you. The fact is that we can explain all of these things in convincing ways using the models of chemistry and the related physical and life sciences.

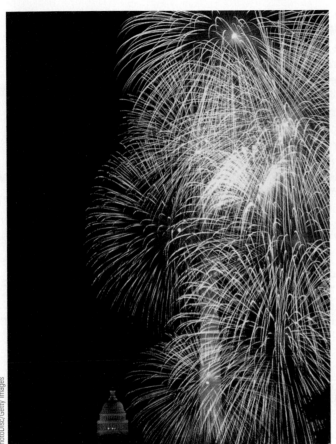

PhotoDisc/Getty Images

Fireworks are a beautiful illustration of chemistry in action.

1.1 Chemistry: An Introduction

OBJECTIVE: To understand the importance of learning chemistry.

Although chemistry might seem to have little to do with dinosaurs, knowledge of chemistry was the tool that enabled paleontologist Luis W. Alvarez and his coworkers from the University of California at Berkeley to "crack the case" of the disappearing dinosaurs. The key was the relatively high level of iridium found in the sediment that represents the boundary between the earth's Cretaceous (K) and Tertiary (T) periods—the time when the dinosaurs disappeared virtually overnight (on the geological scale). The Berkeley researchers knew that meteorites also have unusually high iridium content (relative to the earth's composition), which led them to suggest that a large meteorite impacted the earth 65 million years ago, causing the climatic changes that wiped out the dinosaurs.

A knowledge of chemistry is useful to almost everyone—chemistry occurs all around us all of the time, and an understanding of chemistry is useful to doctors, lawyers, mechanics, business people, firefighters, and poets among others. Chemistry is important—there is no doubt about that. It lies at the heart of our efforts to produce new materials that make our lives safer and easier, to produce new sources of energy that are abundant and nonpolluting, and to understand and control the many diseases that threaten us and our food supplies. Even if your future career does not require the daily use of chemical principles, your life will be greatly influenced by chemistry.

A strong case can be made that the use of chemistry has greatly enriched all of our lives. However, it is important to understand that the principles of chemistry are inherently neither good nor bad—it's what we do with this knowledge that really matters. Although humans are clever, resourceful, and concerned about others, they also can be greedy, selfish, and ignorant. In addition, we tend to be shortsighted; we concentrate too much on the present and do not think enough about the long-range implications of our actions. This type of thinking has already caused us a great deal of trouble—severe environmental damage has occurred on many fronts. We cannot place all the responsibility on the chemical companies, because everyone has contributed to these problems. However, it is less important to lay blame than to figure out how to solve these problems. An important part of the answer must rely on chemistry.

One of the "hottest" fields in the chemical sciences is environmental chemistry—an area that involves studying our environmental ills and finding creative ways to address them. For example, meet Bart Eklund, who works in the atmospheric chemistry field for Radian Corporation in Austin, Texas. Bart's interest in a career in environmental science was fostered by two environmental chemistry courses and two ecology courses he took as an undergraduate. His original plan to gain several years of industrial experience and then to return to school for a graduate degree changed when he discovered that professional advancement with a B.S. degree was possible in the environmental research field. The multidisciplinary nature of environmental problems has allowed Bart to pursue his interest in several fields at the same time. You might say that he specializes in being a generalist.

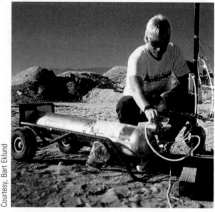

Bart Eklund checking air quality at a hazardous waste site.

The environmental consulting field appeals to Bart for a number of reasons: the chance to define and solve a number of research problems; the simultaneous work on a number of diverse projects; the mix of desk, field, and laboratory work; the travel; and the opportunity to perform rewarding work that has a positive effect on people's lives.

Among his career highlights are the following:

- Spending a winter month doing air sampling in the Grand Tetons, where he also met his wife and learned to ski;

- Driving sampling pipes by hand into the rocky ground of Death Valley Monument in California;

- Working regularly with experts in their fields and with people who enjoy what they do;

- Doing vigorous work in 100 °F weather while wearing a rubberized suit, double gloves, and a respirator; and

- Getting to work in and see Alaska, Yosemite Park, Niagara Falls, Hong Kong, the People's Republic of China, Mesa Verde, New York City, and dozens of other interesting places.

Bart Eklund's career demonstrates how chemists are helping to solve our environmental problems. It is how we use our chemical knowledge that makes all the difference.

An example that shows how technical knowledge can be a "double-edged sword" is the case of chlorofluorocarbons (CFCs). When the compound CCl_2F_2 (originally called Freon-12) was first synthesized, it was hailed as a near-miracle substance. Because of its noncorrosive nature and its unusual ability to resist decomposition, Freon-12 was rapidly applied in refrigeration and air-conditioning systems, cleaning applications, the blowing of foams used for insulation and packing materials, and many other ways. For years everything seemed fine—the CFCs actually replaced more dangerous materials, such as the ammonia formerly used in refrigeration systems. The CFCs were definitely viewed as "good guys." But then a problem was discovered—the ozone in the upper atmosphere that protects us from the high-energy radiation of the sun began to decline. What was happening to cause the destruction of the vital ozone?

Much to everyone's amazement, the culprits turned out to be the seemingly beneficial CFCs. Inevitably, large quantities of CFCs had leaked into the atmosphere but nobody was very worried about this development because these compounds seemed totally benign. In fact, the great stability of the CFCs (a tremendous advantage for their various applications) was in the end a great disadvantage when they were released into the environment. Professor F. S. Rowland and his colleagues at the University of California at Irvine demonstrated that the CFCs eventually drifted to high altitudes in the atmosphere, where the energy of the sun stripped off chlorine atoms. These chlorine atoms in turn promoted the decomposition of the ozone in the upper atmosphere. (We will discuss this in more detail in Chapter 13.) Thus a substance that possessed many advantages in earth-bound applications turned against us in the atmosphere. Who could have guessed it would turn out this way?

The good news is that the U.S. chemical industry is leading the way to find environmentally safe alternatives to CFCs, and the levels of CFCs in the atmosphere are already dropping.

The saga of the CFCs demonstrates that we can respond relatively quickly to a serious environmental problem if we decide to do so. Also, it is important to understand that chemical manufacturers have a new attitude about the environment—they are now among the leaders in finding ways to address our environmental ills. The industries that apply the chemical sciences are now determined to be part of the solution rather than part of the problem.

As you can see, learning chemistry is both interesting and important. A chemistry course can do more than simply help you learn the principles of chemistry, however. A major by-product of your study of chemistry is that you will become a better problem solver. One reason chemistry has the reputation of being "tough" is that it often deals with rather complicated systems that require some effort to figure out. Although this might at first seem like a disadvantage, you can turn it to your advantage if you have the right attitude. Recruiters for companies of all types maintain that one of the first things they look for in a prospective employee is the ability to solve problems. We will spend a good deal of time solving various types of problems in this book by using a systematic, logical approach that will serve you well in solving any kind of problem in any field. Keep this broader goal in mind as you learn to solve the specific problems connected with chemistry.

Although learning chemistry is often not easy, it's never impossible. In fact, anyone who is interested, patient, and willing to work can learn

A chemist in the laboratory.

Dr. Ruth—Cotton Hero

Ruth Benerito, the inventor of easy-care cotton.

Dr. Ruth Rogan Benerito may have saved the cotton industry in the United States. In the 1960s, synthetic fibers posed a serious competitive threat to cotton, primarily because of wrinkling. Synthetic fibers such as polyester can be formulated to be highly resistant to wrinkles both in the laundering process and in wearing. On the other hand, 1960s' cotton fabrics wrinkled easily—white cotton shirts had to be ironed to look good. This requirement put cotton at a serious disadvantage and endangered an industry very important to the economic health of the South.

During the 1960s Ruth Benerito worked as a scientist for the Department of Agriculture, where she was instrumental in developing the chemical treatment of cotton to make it wrinkle resistant. In so doing she enabled cotton to remain a preeminent fiber in the market—a place it continues to hold today. She was honored with the Lemelson–MIT Lifetime Achievement Award for Inventions in 2002 when she was 86 years old.

Dr. Benerito, who holds 55 patents, including the one for wrinkle-free cotton awarded in 1969, began her career when women were not expected to enter scientific fields. However, her mother, who was an artist, adamantly encouraged her to be anything she wanted to be.

Dr. Benerito graduated from high school at 14 and attended Newcomb College, the women's college associated with Tulane University. She majored in chemistry with minors in physics and math. At that time she was one of only two women allowed to take the physical chemistry course at Tulane. She earned her B.S. degree in 1935 at age 19 and subsequently earned a master's degree at Tulane and a Ph.D. at the University of Chicago.

In 1953 Dr. Benerito began working in the Agriculture Department's Southern Regional Research Center in New Orleans, where she mainly worked on cotton and cotton-related products. She also invented a special method for intravenous feeding in long-term medical patients.

Since her retirement in 1986, she has continued to tutor science students to keep busy. Everyone who knows Dr. Benerito describes her as a class act.

---

the fundamentals of chemistry. In this book we will try very hard to help you understand what chemistry is and how it works and to point out how chemistry applies to the things going on in your life.

Our sincere hope is that this text will motivate you to learn chemistry, make its concepts understandable to you, and demonstrate how interesting and vital the study of chemistry is.

1.2 What Is Chemistry?

OBJECTIVE: To define chemistry.

> Chemical and physical changes will be discussed in Chapter 3.

Chemistry can be defined as *the science that deals with the materials of the universe and the changes that these materials undergo.* Chemists are involved in activities as diverse as examining the fundamental particles of matter, looking for molecules in space, synthesizing and formulating new materials of all types, using bacteria to produce such chemicals as insulin, and inventing new diagnostic methods for early detection of disease.

Chemistry is often called the central science—and with good reason. Most of the phenomena that occur in the world around us involve chemical changes, changes where one or more substances become different substances. Here are some examples of chemical changes:

Wood burns in air, forming water, carbon dioxide, and other substances.

A plant grows by assembling simple substances into more complex substances.

The steel in a car rusts.

Eggs, flour, sugar, and baking powder are mixed and baked to yield a cake.

The definition of the term *chemistry* is learned and stored in the brain.

Emissions from a power plant lead to the formation of acid rain.

As we proceed, you will see how the concepts of chemistry allow us to understand the nature of these and other changes and thus help us manipulate natural materials to our benefit.

The launch of the space shuttle gives clear indications that chemical reactions are occurring.

1.3   Solving Problems Using a Scientific Approach

OBJECTIVE: To understand scientific thinking.

One of the most important things we do in everyday life is solve problems. In fact, most of the decisions you make each day can be described as solving problems.

It's 8:30 A.M. on Friday. Which is the best way to drive to school to avoid traffic congestion?

You have two tests on Monday. Should you divide your study time equally or allot more time to one than to the other?

Your car stalls at a busy intersection and your little brother is with you. What should you do next?

These are everyday problems of the type we all face. What process do we use to solve them? You may not have thought about it before, but there are several steps that almost everyone uses to solve problems:

1. Recognize the problem and state it clearly. Some information becomes known, or something happens that requires action. In science we call this step *making an observation*.

2. Propose *possible* solutions to the problem or *possible* explanations for the observation. In scientific language, suggesting such a possibility is called *formulating a hypothesis*.

A Mystifying Problem

To illustrate how science helps us solve problems, consider a true story about two people, David and Susan (not their real names). Several years ago David and Susan were healthy 40-year-olds living in California, where David was serving in the Air Force. Gradually Susan became quite ill, showing flu-like symptoms including nausea and severe muscle pains. Even her personality changed: she became uncharacteristically grumpy. She seemed like a totally different person from the healthy, happy woman of a few months earlier. Following her doctor's orders, she rested and drank a lot of fluids, including large quantities of coffee and orange juice from her favorite mug, part of a 200-piece set of pottery dishes recently purchased in Italy. However, she just got sicker, developing extreme abdominal cramps and severe anemia.

During this time David also became ill and exhibited symptoms much like Susan's: weight loss, excruciating pain in his back and arms, and uncharacteristic fits of temper. The disease became so debilitating that he retired early from the Air Force and the couple moved to Seattle. For a short time their health improved, but after they unpacked all their belongings (including those pottery dishes), their health began to deteriorate again. Susan's body became so sensitive that she could not tolerate the weight of a blanket. She was near death. What was wrong? The doctors didn't know, but one suggested she might have porphyria, a rare blood disease.

Desperate, David began to search the medical literature himself. One day while he was reading about porphyria, a phrase jumped off the page: "Lead poisoning can sometimes be confused with porphyria." Could the problem be lead poisoning?

We have described a very serious problem with life-or-death implications. What should David do next? Overlooking for a moment the obvious response of calling the couple's doctor immediately to discuss the possibility of lead poisoning, could David solve the problem via scientific thinking? Let's use the three steps described in Section 1.3 to attack the problem one part at a time. This is important: usually we solve complex problems by breaking them down into manageable parts. We can then assemble the solution to the overall problem from the answers we have found "piecemeal."

In this case there are many parts to the overall problem:

What is the disease?

Where is it coming from?

Can it be cured?

Let's attack "What is the disease?" first.

Observation: David and Susan are ill with the symptoms described. Is the disease lead poisoning?

Hypothesis: The disease is lead poisoning.

Experiment: If the disease is lead poisoning, the symptoms must match those known to characterize lead poisoning. Look up the symptoms of lead poisoning. David did this and found that they matched the couple's symptoms almost exactly.

This discovery points to lead poisoning as the source of their problem, but David needed more evidence.

Observation: Lead poisoning results from high levels of lead in the bloodstream.

Hypothesis: The couple have high levels of lead in their blood.

Experiment: Perform a blood analysis. Susan arranged for such an analysis, and the results showed high lead levels for both David and Susan.

3. Decide which of the solutions is the best or decide whether the explanation proposed is reasonable. To do this we search our memory for any pertinent information or we seek new information. In science we call searching for new information *performing an experiment.*

Italian pottery.

This confirms that lead poisoning is probably the cause of the trouble, but the overall problem is still not solved. David and Susan are likely to die unless they find out where the lead is coming from.

Observation: There is lead in the couple's blood.

Hypothesis: The lead is in their food or drink when they buy it.

Experiment: Find out whether anyone else who shopped at the same store was getting sick (no one was). Also note that moving to a new area did not solve the problem.

Observation: The food they buy is free of lead.

Hypothesis: The dishes they use are the source of the lead poisoning.

Experiment: Find out whether their dishes contain lead. David and Susan learned that lead compounds are often used to put a shiny finish on pottery objects. And laboratory analysis of their Italian pottery dishes showed that lead was present in the glaze.

Observation: Lead is present in their dishes, so the dishes are a possible source of their lead poisoning.

Hypothesis: The lead is leaching into their food.

Experiment: Place a beverage, such as orange juice, in one of the cups and then analyze the beverage for lead. The results showed high levels of lead in drinks that had been in contact with the pottery cups.

After many applications of the scientific method, the problem is solved. We can summarize the answer to the problem (David and Susan's illness) as follows: the Italian pottery they used for everyday dishes contained a lead glaze that contaminated their food and drink with lead. This lead accumulated in their bodies to the point where it interfered seriously with normal functions and produced severe symptoms. This overall explanation, which summarizes the hypotheses that agree with the experimental results, is called a *theory* in science. This explanation accounts for the results of all the experiments performed.*

We could continue to use the scientific method to study other aspects of this problem, such as

What types of food or drink leach the most lead from the dishes?

Do all pottery dishes with lead glazes produce lead poisoning?

As we answer questions using the scientific method, other questions naturally arise. By repeating the three steps over and over, we can come to understand a given phenomenon thoroughly.

*"David" and "Susan" recovered from their lead poisoning and are now publicizing the dangers of using lead-glazed pottery. This happy outcome is the answer to the third part of their overall problem, "Can the disease be cured?" They simply stopped eating from that pottery!

As we will discover in the next section, scientists use these same procedures to study what happens in the world around us. The important point here is that scientific thinking can help you in all parts of your life. It's worthwhile to learn how to think scientifically—whether you want to be a scientist, an auto mechanic, a doctor, a politician, or a poet!

1.4 The Scientific Method

OBJECTIVE: To describe the method scientists use to study nature.

In the last section we began to see how the methods of science are used to solve problems. In this section we will further examine this approach.

Science is a framework for gaining and organizing knowledge. Science is not simply a set of facts but also a plan of action—a *procedure* for processing and understanding certain types of information. Although scientific thinking is useful in all aspects of life, in this text we will use it to understand how the natural world operates. The process that lies at the center of scientific inquiry is called the **scientific method.** As we saw in the previous section, it consists of the following steps:

Steps in the Scientific Method

1. *State the problem and collect data (make observations).* Observations may be *qualitative* (the sky is blue; water is a liquid) or *quantitative* (water boils at 100 °C; a certain chemistry book weighs 4.5 pounds). A qualitative observation does not involve a number. A quantitative observation is called a **measurement** and does involve a number (and a unit, such as pounds or inches). We will discuss measurements in detail in Chapter 2.
2. *Formulate hypotheses.* A hypothesis is a *possible* explanation for the observation.
3. *Perform experiments.* An experiment is something we do to test the hypothesis. We gather new information that allows us to decide whether the hypothesis is supported by the new information we have learned from the experiment. Experiments always produce new observations, and this brings us back to the beginning of the process again.

Quantitative observations involve a number. Qualitative ones do not.

To explain the behavior of a given part of nature, we repeat these steps many times. Gradually we accumulate the knowledge necessary to understand what is going on.

Once we have a set of hypotheses that agrees with our various observations, we assemble them into a theory that is often called a *model*. A **theory** (model) is a set of tested hypotheses that gives an overall explanation of some part of nature (see Figure 1.1).

It is important to distinguish between observations and theories. An observation is something that is witnessed and can be recorded. A theory is an *interpretation*—a possible explanation of *why* nature behaves in a particular way. Theories inevitably change as more information becomes available. For example, the motions of the sun and stars have remained virtually the same over the thousands of years during which humans have been observing them, but our explanations—our theories—have changed greatly since ancient times.

The point is that we don't stop asking questions just because we have devised a theory that seems to account satisfactorily for some aspect of natural behavior. We continue doing experiments to refine our theories. We generally do this by using the theory to make a prediction and then doing an experiment (making a new observation) to see whether the results bear out this prediction.

Figure 1.1

The various parts of the scientific method.

Always remember that theories (models) are human inventions. They represent our attempts to explain observed natural behavior in terms of our human experiences. We must continue to do experiments and refine our theories to be consistent with new knowledge if we hope to approach a more nearly complete understanding of nature.

As we observe nature, we often see that the same observation applies to many different systems. For example, studies of innumerable chemical changes have shown that the total mass of the materials involved is the same before and after the change. We often formulate such generally observed behavior into a statement called a **natural law.** The observation that the total mass of materials is not affected by a chemical change in those materials is called the law of conservation of mass.

You must recognize the difference between a law and a theory. A law is a summary of observed (measurable) behavior, whereas a theory is an explanation of behavior. *A law tells what happens; a theory (model) is our attempt to explain why it happens.*

In this section, we have described the scientific method (which is summarized in Figure 1.1) as it might ideally be applied. However, it is important to remember that science does not always progress smoothly and efficiently. Scientists are human. They have prejudices; they misinterpret data; they can become emotionally attached to their theories and thus lose objectivity; and they play politics. Science is affected by profit motives, budgets, fads, wars, and religious beliefs. Galileo, for example, was forced to recant his astronomical observations in the face of strong religious resistance. Lavoisier, the father of modern chemistry, was beheaded because of his political affiliations. And great progress in the chemistry of nitrogen fertilizers resulted from the desire to produce explosives to fight wars. The progress of science is often slowed more by the frailties of humans and their institutions than by the limitations of scientific measuring devices. The scientific method is only as effective as the humans using it. It does not automatically lead to progress.

> Law: A summary of observed behavior.

> Theory: An explanation of behavior.

1.5 Learning Chemistry

OBJECTIVE: To develop successful strategies for learning chemistry.

Chemistry courses have a universal reputation for being difficult. There are some good reasons for this. For one thing, the language of chemistry is unfamiliar in the beginning; many terms and definitions need to be memorized. As with any language, *you must know the vocabulary* before you can communicate effectively. We will try to help you by pointing out those things that need to be memorized.

But memorization is only the beginning. Don't stop there or your experience with chemistry will be frustrating. Be willing to do some thinking, and learn to trust yourself to figure things out. To solve a typical chemistry problem, you must sort through the given information and decide what is really crucial.

It is important to realize that chemical systems tend to be complicated—there are typically many components—and we must make approximations in describing them. Therefore, trial and error play a major role in solving chemical problems. In tackling a complicated system, a practicing chemist really does not expect to be right the first time he or she analyzes the problem. The usual practice is to make several simplifying assumptions and then give it a try. If the answer obtained doesn't make sense, the chemist adjusts the assumptions, using feedback from the first attempt, and tries

Chemistry: An Important Component of Your Education

What is the purpose of education? Because you are spending considerable time, energy, and money to pursue an education, this is an important question.

Some people seem to equate education with the storage of facts in the brain. These people apparently believe that education simply means memorizing the answers to all of life's present and future problems. Although this is clearly unreasonable, many students seem to behave as though this were their guiding principle. These students want to memorize lists of facts and to reproduce them on tests. They regard as unfair any exam questions that require some original thought or some processing of information. Indeed, it might be tempting to reduce education to a simple filling up with facts, because that approach can produce short-term satisfaction for both student and teacher. And of course, storing facts in the brain *is* important. You cannot function without knowing that red means stop, electricity is hazardous, ice is slippery, and so on.

However, mere recall of abstract information, without the ability to process it, makes you little better than a talking encyclopedia. Former students always seem to bring the same message

Students pondering the structure of a molecule.

Claude Charlier/Photo Researchers, Inc.

when they return to campus. The characteristics that are most important to their success are a knowledge of the fundamentals of their fields, the ability to recognize and solve problems, and the ability to communicate effectively. They also emphasize the importance of a high level of motivation.

How does studying chemistry help you achieve these characteristics? The fact that chemical systems are complicated is really a blessing, though one that is well disguised. Studying chemistry will not by itself make you a good problem solver, but it can help you develop a positive, aggressive attitude toward problem solving and can help boost your confidence. Learning to "think like a chemist" can be valuable to anyone in any field. In fact, the chemical industry is heavily populated at all levels and in all areas by chemists and chemical engineers. People who were trained as chemical professionals often excel not only in chemical research and production but also in the areas of personnel, marketing, sales, development, finance, and management. The point is that much of what you learn in this course can be applied to any field of endeavor. So be careful not to take too narrow a view of this course. Try to look beyond short-term frustration to long-term benefits. It may not be easy to learn to be a good problem solver, but it's well worth the effort.

again. The point is this: in dealing with chemical systems, do not expect to understand immediately everything that is going on. In fact, it is typical (even for an experienced chemist) *not* to understand at first. Make an attempt to solve the problem and then analyze the feedback. *It is no disaster to make a mistake as long as you learn from it.*

The only way to develop your confidence as a problem solver is to practice solving problems. To help you, this book contains examples worked out in detail. Follow these through carefully, making sure you understand each step. These examples are usually followed by a similar exercise (called a self-check exercise) that you should try on your own (detailed solutions of the self-check exercises are given at the end of each chapter). Use the self-check exercises to test whether you are understanding the material as you go along.

There are questions and problems at the end of each chapter. The questions review the basic concepts of the chapter and give you an opportunity to check whether you properly understand the vocabulary introduced. Some of the problems are really just exercises that are very similar to examples done in the chapter. If you understand the material in the chapter, you should be able to do these exercises in a straightforward way. Other problems require more creativity. These contain a knowledge gap—some unfamiliar territory that you must cross—and call for thought and patience on your part. For this course to be really useful to you, it is important to go beyond the questions and exercises. Life offers us many exercises, routine events that we deal with rather automatically, but the real challenges in life are true problems. This course can help you become a more creative problem solver.

As you do homework, be sure to use the problems correctly. If you cannot do a particular problem, do not immediately look at the solution. Review the relevant material in the text and then try the problem again. Don't be afraid to struggle with a problem. Looking at the solution as soon as you get stuck short-circuits the learning process.

Learning chemistry takes time. Use all the resources available to you and study on a regular basis. Don't expect too much of yourself too soon. You may not understand everything at first, and you may not be able to do many of the problems the first time you try them. This is normal. It doesn't mean you can't learn chemistry. Just remember to keep working and to keep learning from your mistakes, and you will make steady progress.

CHAPTER 1 REVIEW

Key Terms

chemistry (1.2)
scientific method (1.4)
theory (1.4)
natural law (1.4)

Active Learning Questions

These questions are designed to be considered by groups of students in class. Often these questions work well for introducing a particular topic in class.

1. Discuss how a hypothesis can become a theory. Can a theory become a law? Explain.

2. Make five qualitative and five quantitative observations about the room in which you now sit.

3. List as many chemical reactions you can think of that are part of your everyday life. Explain.

4. Differentiate between a "theory" and a "scientific theory."

5. Describe three situations when you used the scientific method (outside of school) in the past month.

6. Scientific models do not describe reality. They are simplifications and therefore incorrect at some level. So why are models useful?

7. Theories should inspire questions. Discuss a scientific theory you know and the questions it brings up.

8. Describe how you would set up an experiment to test the relationship between completion of assigned homework and the final grade you receive in the course.

9. If all scientists use the scientific method to try to arrive at a better understanding of the world, why do so many debates arise among scientists?

10. As stated in the text, there is no one scientific method. However, making observations, formulating hypotheses, and performing experiments are generally components of "doing science." Read the following passage, and list any observations, hypotheses, and experiments. Support your answer.

    Joyce and Frank are eating raisins and drinking ginger ale. Frank accidentally drops a raisin into his ginger ale. They both notice that the raisin falls to the bottom of the glass. Soon, the raisin rises to the surface of the ginger ale, and then sinks. Within a couple of minutes, it rises and sinks again. Joyce asks, "I wonder why that happened?" Frank says, "I don't know, but let's see if it works in water." Joyce fills a glass with water and drops the raisin into the glass. After a few minutes, Frank says, "No, it doesn't go up and down in the water." Joyce closely observes the raisins in the two glasses and states, "Look, there are bubbles on the raisins in the ginger ale but not on the raisins in the water." Frank says, "It must be the bubbles that make the raisin rise." Joyce asks, "OK, but then why do they sink again?"

11. In Section 1.3 the statement is made that it is worthwhile for scientists, auto mechanics, doctors, politicians, and poets to take a scientific approach to their professions. Discuss how each of these people could use a scientific approach in his or her profession.

12. As part of a science project, you study traffic patterns in your city at an intersection in the middle of downtown. You set up a device that counts the cars passing through this intersection for a 24-hour period during a weekday. The graph of hourly traffic looks like this.

a. At what time(s) does the highest number of cars pass through the intersection?
b. At what time(s) does the lowest number of cars pass through the intersection?
c. Briefly describe the trend in numbers of cars over the course of the day.
d. Provide a hypothesis explaining the trend in numbers of cars over the course of the day.
e. Provide a possible experiment that could test your hypothesis.

13. Confronted with the box shown in the diagram, you wish to discover something about its internal workings. You have no tools and cannot open the box. You pull on rope B, and it moves rather freely. When you pull on rope A, rope C appears to be pulled slightly into the box. When you pull on rope C, rope A almost disappears into the box.*

a. Based on these observations, construct a model for the interior mechanism of the box.
b. What further experiments could you do to refine your model?

*From Yoder, Suydam, and Snavely, *Chemistry* (New York: Harcourt Brace Jovanovich, 1975), pp. 9–11.

ⓦWL Questions and problems for this chapter are available in OWL. Use the OWL access card that came with this textbook to access assigned questions and problems for this chapter.

2 Measurements and Calculations

● An enlarged view of a graduated cylinder (*Masterfile*)

⊌WL Sign in to OWL at
www.cengage.com/owl to view
tutoritals and simulations, develop
problem-solving skills, and complete
online homework assigned by your
professor.

As we pointed out in Chapter 1, making observations is a key part of the scientific process. Sometimes observations are *qualitative* ("the substance is a yellow solid") and sometimes they are *quantitative* ("the substance weighs 4.3 grams"). A quantitative observation is called a **measurement.** Measurements are very important in our daily lives. For example, we pay for gasoline by the gallon, so the gas pump must accurately measure the gas delivered to our fuel tank. The efficiency of the modern automobile engine depends on various measurements, including the amount of oxygen in the exhaust gases, the temperature of the coolant, and the pressure of the lubricating oil. In addition, cars with traction control systems have devices to measure and compare the rates of rotation of all four wheels. As we will see in the "Chemistry in Focus" discussion in this chapter, measuring devices have become very sophisticated in dealing with our fast-moving and complicated society.

As we will discuss in this chapter, a measurement always consists of two parts: a number and a unit. Both parts are necessary to make the measurement meaningful. For example, suppose a friend tells you that she saw a bug 5 long. This statement is meaningless as it stands. Five what? If it's 5 millimeters, the bug is quite small. If it's 5 centimeters, the bug is quite large. If it's 5 meters, run for cover!

The point is that for a measurement to be meaningful, it must consist of both a number and a unit that tells us the scale being used.

In this chapter we will consider the characteristics of measurements and the calculations that involve measurements.

A gas pump measures the amount of gasoline delivered.

2.1

Scientific Notation

OBJECTIVE: To show how very large or very small numbers can be expressed as the product of a number between 1 and 10 and a power of 10.

> A measurement must always consist of a number *and* a unit.

The numbers associated with scientific measurements are often very large or very small. For example, the distance from the earth to the sun is approximately 93,000,000 (93 million) miles. Written out, this number is rather bulky. Scientific notation is a method for making very large or very small numbers more compact and easier to write.

To see how this is done, consider the number 125, which can be written as the product

$$125 = 1.25 \times 100$$

Because $100 = 10 \times 10 = 10^2$, we can write

$$125 = 1.25 \times 100 = 1.25 \times 10^2$$

Similarly, the number 1700 can be written

$$1700 = 1.7 \times 1000$$

and because $1000 = 10 \times 10 \times 10 = 10^3$, we can write

$$1700 = 1.7 \times 1000 = 1.7 \times 10^3$$

Ray Simons/Photo Researchers, Inc.

When describing very small distances, such as the diameter of a swine flu virus (shown here magnified 16,537 times), it is convenient to use scientific notation.

MATH SKILL BUILDER
Keep one digit to the left of the decimal point.

Scientific notation simply expresses a number as *a product of a number between 1 and 10 and the appropriate power of 10*. For example, the number 93,000,000 can be expressed as

$$93{,}000{,}000 = 9.3 \times 10{,}000{,}000 = 9.3 \times 10^7$$

Number between 1 and 10   Appropriate power of 10 ($10{,}000{,}000 = 10^7$)

The easiest way to determine the appropriate power of 10 for scientific notation is to start with the number being represented and count the number of places the decimal point must be moved to obtain a number between 1 and 10. For example, for the number

$$9\ 3\ 0\ 0\ 0\ 0\ 0\ 0$$
$$7\ 6\ 5\ 4\ 3\ 2\ 1$$

we must move the decimal point seven places to the left to get 9.3 (a number between 1 and 10). To compensate for every move of the decimal point to the left, we must multiply by 10. That is, each time we move the decimal point to the left, we make the number smaller by one power of 10. So for each move of the decimal point to the left, we must multiply by 10 to restore the number to its original magnitude. Thus moving the decimal point seven places to the left means we must multiply 9.3 by 10 seven times, which equals 10^7:

$$93{,}000{,}000 = 9.3 \times 10^7$$

We moved the decimal point seven places to the left, so we need 10^7 to compensate.

MATH SKILL BUILDER
Moving the decimal point to the left requires a positive exponent.

Remember: whenever the decimal point is moved to the *left*, the exponent of 10 is *positive*.

We can represent numbers smaller than 1 by using the same convention, but in this case the power of 10 is negative. For example, for the number 0.010 we must move the decimal point two places to the right to obtain a number between 1 and 10:

$$0\ .\ 0\ 1\ 0$$
$$1\ 2$$

MATH SKILL BUILDER
Moving the decimal point to the right requires a negative exponent.

This requires an exponent of -2, so $0.010 = 1.0 \times 10^{-2}$. Remember: whenever the decimal point is moved to the *right*, the exponent of 10 is *negative*.

Next consider the number 0.000167. In this case we must move the decimal point four places to the right to obtain 1.67 (a number between 1 and 10):

$$0\ .\ 0\ 0\ 0\ 1\ 6\ 7$$
$$1\ 2\ 3\ 4$$

MATH SKILL BUILDER
Read the Appendix if you need a further discussion of exponents and scientific notation.

Moving the decimal point four places to the right requires an exponent of -4. Therefore,

$$0.000167 = 1.67 \times 10^{-4}$$

We moved the decimal point four places to the right.

We summarize these procedures below.

Using Scientific Notation

MATH SKILL BUILDER
$100 = 1.0 \times 10^2$
$0.010 = 1.0 \times 10^{-2}$

MATH SKILL BUILDER
Left Is Positive; remember LIP.

- Any number can be represented as the product of a number between 1 and 10 and a power of 10 (either positive or negative).
- The power of 10 depends on the number of places the decimal point is moved and in which direction. The *number of places* the decimal point is moved determines the *power of 10*. The *direction* of the move determines whether the power of 10 is *positive* or *negative.* If the decimal point is moved to the left, the power of 10 is positive; if the decimal point is moved to the right, the power of 10 is negative.

EXAMPLE 2.1 | Scientific Notation: Powers of 10 (Positive)

MATH SKILL BUILDER
A number that is greater than 1 will always have a positive exponent when written in scientific notation.

Represent the following numbers in scientific notation.

   a. 238,000

   b. 1,500,000

SOLUTION

   a. First we move the decimal point until we have a number between 1 and 10, in this case 2.38.

$$2\,3\,8\,0\,0\,0$$
$$5\,4\,3\,2\,1$$ The decimal point was moved five places to the left.

Because we moved the decimal point five places to the left, the power of 10 is positive 5. Thus $238{,}000 = 2.38 \times 10^5$.

   b. $$1\,5\,0\,0\,0\,0\,0$$
$$6\,5\,4\,3\,2\,1$$ The decimal point was moved six places to the left, so the power of 10 is 6.

Thus $1{,}500{,}000 = 1.5 \times 10^6$. ∎

EXAMPLE 2.2 | Scientific Notation: Powers of 10 (Negative)

MATH SKILL BUILDER
A number that is less than 1 will always have a negative exponent when written in scientific notation.

Represent the following numbers in scientific notation.

   a. 0.00043

   b. 0.089

SOLUTION

   a. First we move the decimal point until we have a number between 1 and 10, in this case 4.3.

$$0\,.\,0\,0\,0\,4\,3$$
$$1\,2\,3\,4$$ The decimal point was moved four places to the right.

Because we moved the decimal point four places to the right, the power of 10 is negative 4. Thus $0.00043 = 4.3 \times 10^{-4}$.

b. 0.0 8 9

  1 2   The power of 10 is negative 2 because the decimal point was moved two places to the right.

Thus $0.089 = 8.9 \times 10^{-2}$.

Self-Check   **EXERCISE 2.1**   Write the numbers 357 and 0.0055 in scientific notation. If you are having difficulty with scientific notation at this point, reread the Appendix.

See Problems 2.5 through 2.14. ■

2.2 Units

OBJECTIVE: To learn the English, metric, and SI systems of measurement.

The **units** part of a measurement tells us what *scale* or *standard* is being used to represent the results of the measurement. From the earliest days of civilization, trade has required common units. For example, if a farmer from one region wanted to trade some of his grain for the gold of a miner who lived in another region, the two people had to have common standards (units) for measuring the amount of the grain and the weight of the gold.

The need for common units also applies to scientists, who measure quantities such as mass, length, time, and temperature. If every scientist had her or his own personal set of units, complete chaos would result. Unfortunately, although standard systems of units did arise, different systems were adopted in different parts of the world. The two most widely used systems are the **English system** used in the United States and the **metric system** used in most of the rest of the industrialized world.

The metric system has long been preferred for most scientific work. In 1960 an international agreement set up a comprehensive system of units called the **International System** (*le Système Internationale* in French), or **SI**. The SI units are based on the metric system and units derived from the metric system. The most important fundamental SI units are listed in Table 2.1. Later in this chapter we will discuss how to manipulate some of these units.

Because the fundamental units are not always a convenient size, the SI system uses prefixes to change the size of the unit. The most commonly used prefixes are listed in Table 2.2. Although the fundamental unit for length is the meter (m), we can also use the decimeter (dm), which represents one-tenth (0.1) of a meter; the centimeter (cm), which represents one one-hundredth (0.01) of a meter; the millimeter (mm), which represents one one-thousandth (0.001) of a meter; and so on. For example, it's much more convenient to specify the diameter of a certain contact lens as 1.0 cm than as 1.0×10^{-2} m.

Table 2.1 Some Fundamental SI Units

Physical Quantity	Name of Unit	Abbreviation
mass	kilogram	kg
length	meter	m
time	second	s
temperature	kelvin	K

Critical Units!

How important are conversions from one unit to another? If you ask the National Aeronautics and Space Administration (NASA), very important! In 1999 NASA lost a $125 million Mars Climate Orbiter because of a failure to convert from English to metric units.

The problem arose because two teams working on the Mars mission were using different sets of units. NASA's scientists at the Jet Propulsion Laboratory in Pasadena, California, assumed that the thrust data for the rockets on the Orbiter they received from Lockheed Martin Astronautics in Denver, which built the spacecraft, were in metric units. In reality, the units were English. As a result the Orbiter dipped 100 kilometers lower into the Mars atmosphere than planned and the friction from the atmosphere caused the craft to burn up.

NASA's mistake refueled the controversy over whether Congress should require the United States to switch to the metric system. About 95% of the world now uses the metric sys-tem, and the United States is slowly switching from English to metric. For example, the automobile industry has adopted metric fasteners and we buy our soda in two-liter bottles.

Units can be very important. In fact, they can mean the difference between life and death on some occasions. In 1983, for example, a Canadian jetliner almost ran out of fuel when someone pumped 22,300 pounds of fuel into the aircraft instead of 22,300 kilograms. Remember to watch your units!

Artist's conception of the lost Mars Climate Orbiter.

Table 2.2	The Commonly Used Prefixes in the Metric System		
Prefix	Symbol	Meaning	Power of 10 for Scientific Notation
mega	M	1,000,000	10^6
kilo	k	1000	10^3
deci	d	0.1	10^{-1}
centi	c	0.01	10^{-2}
milli	m	0.001	10^{-3}
micro	μ	0.000001	10^{-6}
nano	n	0.000000001	10^{-9}

Measurements consist of both a number and a unit, and both are crucial. Just as you would not report a measurement without a numerical value, you would not report a measurement without a unit. You already use units in your daily life, whether you tell somebody, "Let's meet in one hour" (hour is the unit), or you and your friends order two pizzas for dinner (pizza is the unit).

2.3 Measurements of Length, Volume, and Mass

OBJECTIVE: To understand the metric system for measuring length, volume, and mass.

The fundamental SI unit of length is the **meter,** which is a little longer than a yard (1 meter = 39.37 inches). In the metric system fractions of a meter or multiples of a meter can be expressed by powers of 10, as summarized in Table 2.3.

The English and metric systems are compared on the ruler shown in Figure 2.1. Note that

$$1 \text{ inch} = 2.54 \text{ centimeters}$$

Other English–metric equivalences are given in Section 2.6.

Volume is the amount of three-dimensional space occupied by a substance. The fundamental unit of volume in the SI system is based on the volume of a cube that measures 1 meter in each of the three directions. That is, each edge of the cube is 1 meter in length. The volume of this cube is

$$1 \text{ m} \times 1 \text{ m} \times 1 \text{ m} = (1 \text{ m})^3 = 1 \text{ m}^3$$

or, in words, one cubic meter.

In Figure 2.2 this cube is divided into 1000 smaller cubes. Each of these small cubes represents a volume of 1 dm^3, which is commonly called the **liter** (rhymes with "meter" and is slightly larger than a quart) and abbreviated L.

The meter was originally defined, in the eighteenth century, as one ten-millionth of the distance from the equator to the North Pole and then, in the late nineteenth century, as the distance between two parallel marks on a special metal bar stored in a vault in Paris. More recently, for accuracy and convenience, a definition expressed in terms of light waves has been adopted.

Table 2.3   The Metric System for Measuring Length

Unit	Symbol	Meter Equivalent
kilometer	km	1000 m or 10^3 m
meter	m	1 m
decimeter	dm	0.1 m or 10^{-1} m
centimeter	cm	0.01 m or 10^{-2} m
millimeter	mm	0.001 m or 10^{-3} m
micrometer	μm	0.000001 m or 10^{-6} m
nanometer	nm	0.000000001 m or 10^{-9} m

Figure 2.1

Comparison of English and metric units for length on a ruler.

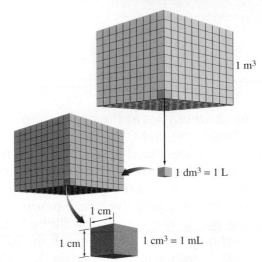

Figure 2.2

The largest drawing represents a cube that has sides 1 m in length and a volume of 1 m³. The middle-size cube has sides 1 dm in length and a volume of 1 dm³, or 1 L. The smallest cube has sides 1 cm in length and a volume of 1 cm³, or 1 mL.

Figure 2.3

A 100-mL graduated cylinder.

Figure 2.4

An electronic analytical balance used in chemistry labs.

The cube with a volume of 1 dm³ (1 liter) can in turn be broken into 1000 smaller cubes, each representing a volume of 1 cm³. This means that each liter contains 1000 cm³. One cubic centimeter is called a **milliliter** (abbreviated mL), a unit of volume used very commonly in chemistry. This relationship is summarized in Table 2.4.

The *graduated cylinder* (see Figure 2.3), commonly used in chemical laboratories for measuring the volumes of liquids, is marked off in convenient units of volume (usually milliliters). The graduated cylinder is filled to the desired volume with the liquid, which then can be poured out.

Another important measurable quantity is **mass,** which can be defined as the quantity of matter present in an object. The fundamental SI unit of mass is the **kilogram.** Because the metric system, which existed before the SI system, used the gram as the fundamental unit, the prefixes for the various mass units are based on the **gram,** as shown in Table 2.5.

In the laboratory we determine the mass of an object by using a balance. A balance compares the mass of the object to a set of standard masses ("weights"). For example, the mass of an object can be determined by using a single-pan balance (Figure 2.4).

To help you get a feeling for the common units of length, volume, and mass, some familiar objects are described in Table 2.6.

Table 2.4   The Relationship of the Liter and Milliliter

Unit	Symbol	Equivalence
liter	L	1 L = 1000 mL
milliliter	mL	$\frac{1}{1000}$ L = 10^{-3} L = 1 mL

Table 2.5   The Most Commonly Used Metric Units for Mass

Unit	Symbol	Gram Equivalent
kilogram	kg	1000 g = 10^3 g = 1 kg
gram	g	1 g
milligram	mg	0.001 g = 10^{-3} g = 1 mg

CHEMISTRY *IN* FOCUS

Measurement: Past, Present, and Future

Measurement lies at the heart of doing science. We obtain the data for formulating laws and testing theories by doing measurements. Measurements also have very practical importance; they tell us if our drinking water is safe, whether we are anemic, and the exact amount of gasoline we put in our cars at the filling station.

Although the fundamental measuring devices we consider in this chapter are still widely used, new measuring techniques are being developed every day to meet the challenges of our increasingly sophisticated world. For example, engines in modern automobiles have oxygen sensors that analyze the oxygen content in the exhaust gases. This information is sent to the computer that controls the engine functions so that instantaneous adjustments can be made in spark timing and air–fuel mixtures to provide efficient power with minimum air pollution.

As another example, consider airline safety: How do we rapidly, conveniently, and accurately determine whether a given piece of baggage contains an explosive device? A thorough hand-search of each piece of luggage is out of the question. Scientists are now developing a screening procedure that bombards the luggage with high-

A pollution control officer measuring the oxygen content of river water.

energy particles that cause any substance present to emit radiation characteristic of that substance. This radiation is monitored to identify luggage with unusually large quantities of nitrogen, because most chemical explosives are based on compounds containing nitrogen.

Scientists are also examining the natural world to find supersensitive detectors because many organisms are sensitive to tiny amounts of chemicals in their environments—recall, for example, the sensitive noses of bloodhounds. One of these natural measuring devices uses the sensory hairs from Hawaiian red swimming crabs, which are connected to electrical analyzers and used to detect hormones down to levels of 10^{-8} g/L. Likewise, tissues from pineapple cores can be used to detect tiny amounts of hydrogen peroxide.

These types of advances in measuring devices have led to an unexpected problem: detecting all kinds of substances in our food and drinking water scares us. Although these substances were always there, we didn't worry so much when we couldn't detect them. Now that we know they are present what should we do about them? How can we assess whether these trace substances are harmful or benign? Risk assessment has become much more complicated as our sophistication in taking measurements has increased.

Table 2.6	Some Examples of Commonly Used Units
length	A dime is 1 mm thick.
	A quarter is 2.5 cm in diameter.
	The average height of an adult man is 1.8 m.
mass	A nickel has a mass of about 5 g.
	A 120-lb woman has a mass of about 55 kg.
volume	A 12-oz can of soda has a volume of about 360 mL.
	A half gallon of milk is equal to about 2 L of milk.

2.4 Uncertainty in Measurement

OBJECTIVES: To understand how uncertainty in a measurement arises. • To learn to indicate a measurement's uncertainty by using significant figures.

When you measure the amount of something by counting, the measurement is exact. For example, if you asked your friend to buy four apples from the store and she came back with three or five apples, you would be surprised. However, measurements are not always exact. For example, whenever a measurement is made with a device such as a ruler or a graduated cylinder, an estimate is required. We can illustrate this by measuring the pin shown in Figure 2.5a. We can see from the ruler that the pin is a little longer than 2.8 cm and a little shorter than 2.9 cm. Because there are no graduations on the ruler between 2.8 and 2.9, we must estimate the pin's length between 2.8 and 2.9 cm. We do this by *imagining* that the distance between 2.8 and 2.9 is broken into 10 equal divisions (Figure 2.5b) and estimating to which division the end of the pin reaches. The end of the pin appears to come about halfway between 2.8 and 2.9, which corresponds to 5 of our 10 imaginary divisions. So we estimate the pin's length as 2.85 cm. The result of our measurement is that the pin is approximately 2.85 cm in length, but we had to rely on a visual estimate, so it might actually be 2.84 or 2.86 cm.

A student performing a titration in the laboratory.

Because the last number is based on a visual estimate, it may be different when another person makes the same measurement. For example, if five different people measured the pin, the results might be

Person	Result of Measurement
1	2.85 cm
2	2.84 cm
3	2.86 cm
4	2.85 cm
5	2.86 cm

Note that the first two digits in each measurement are the same regardless of who made the measurement; these are called the *certain* numbers of the measurement. However, the third digit is estimated and can vary; it is called an *uncertain* number. When one is making a measurement, the custom is to record all of the certain numbers plus the *first* uncertain number. It would not make any sense to try to measure the pin to the third decimal place (thousandths of a centimeter), because this ruler requires an estimate of even the second decimal place (hundredths of a centimeter).

It is very important to realize that *a measurement always has some degree of uncertainty*. The uncertainty of a measurement depends on the

Every measurement has some degree of uncertainty.

The length is between 2.8 cm and 2.9 cm.

Imagine that the distance between 2.8 and 2.9 is divided into 10 equal parts. The end of the pin occurs after about 5 of these divisions.

Figure 2.5

Measuring a pin.

measuring device. For example, if the ruler in Figure 2.5 had marks indicating hundredths of a centimeter, the uncertainty in the measurement of the pin would occur in the thousandths place rather than the hundredths place, but some uncertainty would still exist.

The numbers recorded in a measurement (all the certain numbers plus the first uncertain number) are called **significant figures.** The number of significant figures for a given measurement is determined by the inherent uncertainty of the measuring device. For example, the ruler used to measure the pin can give results only to hundredths of a centimeter. Thus, when we record the significant figures for a measurement, we automatically give information about the uncertainty in a measurement. The uncertainty in the last number (the estimated number) is usually assumed to be ±1 unless otherwise indicated. For example, the measurement 1.86 kilograms can be interpreted as 1.86 ± 0.01 kilograms, where the symbol ± means plus or minus. That is, it could be 1.86 kg − 0.01 kg = 1.85 kg or 1.86 kg + 0.01 kg = 1.87 kg.

2.5 Significant Figures

OBJECTIVE: To learn to determine the number of significant figures in a calculated result.

We have seen that any measurement involves an estimate and thus is uncertain to some extent. We signify the degree of certainty for a particular measurement by the number of significant figures we record.

Because doing chemistry requires many types of calculations, we must consider what happens when we do arithmetic with numbers that contain uncertainties. It is important that we know the degree of uncertainty in the final result. Although we will not discuss the process here, mathematicians have studied how uncertainty accumulates and have designed a set of rules to determine how many significant figures the result of a calculation should have. You should follow these rules whenever you carry out a calculation. The first thing we need to do is learn how to count the significant figures in a given number. To do this we use the following rules:

Rules for Counting Significant Figures

1. *Nonzero integers.* Nonzero integers *always* count as significant figures. For example, the number 1457 has four nonzero integers, all of which count as significant figures.
2. *Zeros.* There are three classes of zeros:
    a. *Leading zeros* are zeros that *precede* all of the nonzero digits. They *never* count as significant figures. For example, in the number 0.0025, the three zeros simply indicate the position of the decimal point. The number has only two significant figures, the 2 and the 5.
    b. *Captive zeros* are zeros that fall *between* nonzero digits. They *always* count as significant figures. For example, the number 1.008 has four significant figures.
    c. *Trailing zeros* are zeros at the *right end* of the number. They are significant only if the number is written with a decimal point. The number one hundred written as 100 has only one significant figure, but written as 100., it has three significant figures.
3. *Exact numbers.* Often calculations involve numbers that were not obtained using measuring devices but were determined by counting: 10 experiments, 3 apples, 8 molecules. Such numbers are called *exact numbers*. They can be assumed to have an unlimited number of significant figures. Exact numbers can also arise from definitions. For example, 1 inch is defined as *exactly* 2.54 centimeters. Thus in the statement 1 in. = 2.54 cm, neither 2.54 nor 1 limits the number of significant figures when it is used in a calculation.

MATH SKILL BUILDER
Leading zeros are never significant figures.

MATH SKILL BUILDER
Captive zeros are always significant figures.

MATH SKILL BUILDER
Trailing zeros are sometimes significant figures.

MATH SKILL BUILDER
Exact numbers never limit the number of significant figures in a calculation.

MATH SKILL BUILDER
Significant figures are easily indicated by scientific notation.

Rules for counting significant figures also apply to numbers written in scientific notation. For example, the number 100. can also be written as 1.00×10^2, and both versions have three significant figures. Scientific notation offers two major advantages: the number of significant figures can be indicated easily, and fewer zeros are needed to write a very large or a very small number. For example, the number 0.000060 is much more conveniently represented as 6.0×10^{-5}, and the number has two significant figures, written in either form.

EXAMPLE 2.3 | Counting Significant Figures

Give the number of significant figures for each of the following measurements.

a. A sample of orange juice contains 0.0108 g of vitamin C.

b. A forensic chemist in a crime lab weighs a single hair and records its mass as 0.0050060 g.

c. The distance between two points was found to be 5.030×10^3 ft.

d. In yesterday's bicycle race, 110 riders started but only 60 finished.

SOLUTION

a. The number contains three significant figures. The zeros to the left of the 1 are leading zeros and are not significant, but the remaining zero (a captive zero) is significant.

b. The number contains five significant figures. The leading zeros (to the left of the 5) are not significant. The captive zeros between the 5 and the 6 are significant, and the trailing zero to the right of the 6 is significant because the number contains a decimal point.

c. This number has four significant figures. Both zeros in 5.030 are significant.

d. Both numbers are exact (they were obtained by counting the riders). Thus these numbers have an unlimited number of significant figures.

Self-Check   EXERCISE 2.2   Give the number of significant figures for each of the following measurements.

a. 0.00100 m

b. 2.0800×10^2 L

c. 480 Corvettes

See Problems 2.33 and 2.34. ■

▶ Rounding Off Numbers

When you perform a calculation on your calculator, the number of digits displayed is usually greater than the number of significant figures that the result should possess. So you must "round off" the number (reduce it to fewer digits). The rules for **rounding off** follow.

These rules reflect the way calculators round off.

Rules for Rounding Off

1. If the digit to be removed
   a. is less than 5, the preceding digit stays the same. For example, 1.33 rounds to 1.3.
   b. is equal to or greater than 5, the preceding digit is increased by 1. For example, 1.36 rounds to 1.4, and 3.15 rounds to 3.2.
2. In a series of calculations, carry the extra digits through to the final result and *then* round off.* This means that you should carry all of the digits that show on your calculator until you arrive at the final number (the answer) and then round off, using the procedures in Rule 1.

We need to make one more point about rounding off to the correct number of significant figures. Suppose the number 4.348 needs to be

---

*This practice will not be followed in the worked-out examples in this text, because we want to show the correct number of significant figures in each step of the example.

rounded to two significant figures. In doing this, we look *only* at the *first number* to the right of the 3:

$$4.348$$

$$\uparrow$$

Look at this
number to round off
to two significant figures.

The number is rounded to 4.3 because 4 is less than 5. It is incorrect to round sequentially. For example, do *not* round the 4 to 5 to give 4.35 and then round the 3 to 4 to give 4.4.

When rounding off, *use only the first number to the right of the last significant figure.*

▶ Determining Significant Figures in Calculations

Next we will learn how to determine the correct number of significant figures in the result of a calculation. To do this we will use the following rules.

Rules for Using Significant Figures in Calculations

1. For *multiplication* or *division*, the number of significant figures in the result is the same as that in the measurement with the *smallest number* of significant figures. We say this measurement is *limiting*, because it limits the number of significant figures in the result. For example, consider this calculation:

$$4.56 \times 1.4 = 6.384 \quad \boxed{\text{Round off}} \quad 6.4$$

| Three significant figures | Limiting (two significant figures) | Two significant figures |

Because 1.4 has only two significant figures, it limits the result to two significant figures. Thus the product is correctly written as 6.4, which has two significant figures. Consider another example. In the division $\dfrac{8.315}{298}$, how many significant figures should appear in the answer? Because 8.315 has four significant figures, the number 298 (with three significant figures) limits the result. The calculation is correctly represented as

Four significant figures

$$\frac{8.315}{298} = 0.0279027 \quad \boxed{\text{Round off}} \quad 2.79 \times 10^{-2}$$

| Limiting (three significant figures) | Result shown on calculator | Three significant figures |

(continued)

MATH SKILL BUILDER
If you need help in using your calculator, see the Appendix.

2. For *addition* or *subtraction*, the limiting term is the one with the smallest number of decimal places. For example, consider the following sum:

$$
\begin{array}{r}
12.11 \\
18.0 \quad \text{Limiting term (has one decimal place)} \\
\underline{1.013} \\
31.123 \quad \boxed{\text{Round off}} \Rightarrow 31.1 \\
\uparrow \\
\text{One decimal place}
\end{array}
$$

Why is the answer limited by the term with the smallest number of decimal places? Recall that the last digit reported in a measurement is actually an uncertain number. Although 18, 18.0, and 18.00 are treated as the same quantities by your calculator, they are different to a scientist. The problem above can be thought of as follows:

$$
\begin{array}{r}
12.11? \text{ mL} \\
18.0?? \text{ mL} \\
\underline{1.013 \text{ mL}} \\
31.1?? \text{ mL}
\end{array}
$$

Because the term 18.0 is reported only to the tenths place, our answer must be reported this way as well.

The correct result is 31.1 (it is limited to one decimal place because 18.0 has only one decimal place). Consider another example:

$$
\begin{array}{r}
0.6875 \\
\underline{-0.1} \quad \text{Limiting term (one decimal place)} \\
0.5875
\end{array}
$$

$$\boxed{\text{Round off}} \Rightarrow 0.6$$

Note that *for multiplication and division, significant figures are counted. For addition and subtraction, the decimal places are counted.*

Now we will put together the things you have learned about significant figures by considering some mathematical operations in the following examples.

EXAMPLE 2.4 | Counting Significant Figures in Calculations

Without performing the calculations, tell how many significant figures each answer should contain.

a. 5.19
    1.9
    +0.842

b. 1081 − 7.25

c. 2.3 × 3.14

d. the total cost of 3 boxes of candy at $2.50 a box

SOLUTION

a. The answer will have one digit after the decimal place. The limiting number is 1.9, which has one decimal place, so the answer has two significant figures.

b. The answer will have no digits after the decimal point. The number 1081 has no digits to the right of the decimal point and limits the result, so the answer has four significant figures.

c. The answer will have two significant figures because the number 2.3 has only two significant figures (3.14 has three).

d. The answer will have three significant figures. The limiting factor is 2.50 because 3 (boxes of candy) is an exact number. ∎

> ### EXAMPLE 2.5 | Calculations Using Significant Figures
>
> Carry out the following mathematical operations and give each result to the correct number of significant figures.
>
> a. 5.18×0.0208 d. $116.8 - 0.33$
>
> b. $(3.60 \times 10^{-3}) \times (8.123) \div 4.3$ e. $(1.33 \times 2.8) + 8.41$
>
> c. $21 + 13.8 + 130.36$
>
> #### SOLUTION
>
> Limiting terms   Round to this digit.
>   ↓
>
> a. $5.18 \times 0.0208 = 0.107744$ ⟹ 0.108
>
> The answer should contain three significant figures because each number being multiplied has three significant figures (Rule 1). The 7 is rounded to 8 because the following digit is greater than 5.
>
>   Round to this digit.
>   ↓
>
> b. $\dfrac{(3.60 \times 10^{-3})(8.123)}{4.3} = 6.8006 \times 10^{-3}$ ⟹ 6.8×10^{-3}
>
>   ↑
> Limiting term
>
> Because 4.3 has the least number of significant figures (two), the result should have two significant figures (Rule 1).
>
> c.   21   In this case 21 is limiting (there are no digits
>   13.8   after the decimal point). Thus the answer must
>   +130.36   have no digits after the decimal point, in ac-
>   165.16 ⟹ 165   cordance with the rule for addition (Rule 2).
>
> d.   116.8   Because 116.8 has only one decimal place, the
>   − 0.33   answer must have only one decimal place
>   116.47 ⟹ 116.5   (Rule 2). The 4 is rounded up to 5 because the
>   digit to the right (7) is greater than 5.
>
> e. $1.33 \times 2.8 = 3.724$ ⟹ 3.7   3.7 ← Limiting term
>   + 8.41
>   12.11 ⟹ 12.1
>
> Note that in this case we multiplied and then rounded the result to the correct number of significant figures before we performed the addition so that we would know the correct number of decimal places.

MATH SKILL BUILDER
When we multiply and divide in a problem, perform all calculations before rounding the answer to the correct number of significant figures.

MATH SKILL BUILDER
When we multiply (or divide) and then add (or subtract) in a problem, round the first answer from the first operation (in this case, multiplication) before performing the next operation (in this case, addition). We need to know the correct number of decimal places.

Self-Check EXERCISE 2.3   Give the answer for each calculation to the correct number of significant figures.

a. 12.6×0.53

b. $(12.6 \times 0.53) - 4.59$

c. $(25.36 - 4.15) \div 2.317$ See Problems 2.47 through 2.52. ∎

2.6 Problem Solving and Dimensional Analysis

OBJECTIVE: To learn how dimensional analysis can be used to solve various types of problems.

Suppose that the boss at the store where you work on weekends asks you to pick up 2 dozen doughnuts on the way to work. However, you find that the doughnut shop sells by the doughnut. How many doughnuts do you need?

This "problem" is an example of something you encounter all the time: converting from one unit of measurement to another. Examples of this occur in cooking (The recipe calls for 3 cups of cream, which is sold in pints. How many pints do I buy?); traveling (The purse costs 250 pesos. How much is that in dollars?); sports (A recent Tour de France bicycle race was 3215 kilometers long. How many miles is that?); and many other areas.

How do we convert from one unit of measurement to another? Let's explore this process by using the doughnut problem.

$$2 \text{ dozen doughnuts} = ? \text{ individual doughnuts}$$

where ? represents a number you don't know yet. The essential information you must have is the definition of a dozen:

$$1 \text{ dozen} = 12$$

You can use this information to make the needed conversion as follows:

$$2 \text{ dozen doughnuts} \times \frac{12}{1 \text{ dozen}} = 24 \text{ doughnuts}$$

You need to buy 24 doughnuts.

Note two important things about this process.

1. The factor $\frac{12}{1 \text{ dozen}}$ is a conversion factor based on the definition of the term *dozen*. This conversion factor is a ratio of the two parts of the definition of a dozen given above.

2. The unit "dozen" itself cancels.

Now let's generalize a bit. To change from one unit to another we will use a conversion factor.

$$\text{Unit}_1 \times \text{conversion factor} = \text{Unit}_2$$

The **conversion factor** is a ratio of the two parts of the statement that relates the two units. We will see this in more detail on the following pages.

Earlier in this chapter we considered a pin that measured 2.85 cm in length. What is the length of the pin in inches? We can represent this problem as

$$2.85 \text{ cm} \rightarrow ? \text{ in.}$$

The question mark stands for the number we want to find. To solve this problem, we must know the relationship between inches and centimeters. In Table 2.7, which gives several equivalents between the English and metric systems, we find the relationship

$$2.54 \text{ cm} = 1 \text{ in.}$$

MATH SKILL BUILDER
Since 1 dozen = 12, when we multiply by $\frac{12}{1 \text{ dozen}}$, we are multiplying by 1. The unit "dozen" cancels.

Table 2.7 English–Metric and English–English Equivalents

Length	1 m = 1.094 yd
	2.54 cm = 1 in.
	1 mi = 5280. ft
	1 mi = 1760. yd
Mass	1 kg = 2.205 lb
	453.6 g = 1 lb
Volume	1 L = 1.06 qt
	1 ft³ = 28.32 L

This is called an **equivalence statement.** In other words, 2.54 cm and 1 in. stand for *exactly the same distance.* (See Figure 2.1.) The respective numbers are different because they refer to different *scales* (*units*) of distance.

The equivalence statement 2.54 cm = 1 in. can lead to either of two conversion factors:

$$\frac{2.54 \text{ cm}}{1 \text{ in.}} \quad \text{or} \quad \frac{1 \text{ in.}}{2.54 \text{ cm}}$$

Note that these *conversion factors* are *ratios of the two parts of the equivalence statement* that relates the two units. Which of the two possible conversion factors do we need? Recall our problem:

$$2.85 \text{ cm} = ? \text{ in.}$$

That is, we want to convert from units of centimeters to inches:

$$2.85 \text{ cm} \times \text{conversion factor} = ? \text{ in.}$$

We choose a conversion factor that cancels the units we want to discard and leaves the units we want in the result. Thus we do the conversion as follows:

MATH SKILL BUILDER
Units cancel just as numbers do.

$$2.85 \text{ cm} \times \frac{1 \text{ in.}}{2.54 \text{ cm}} = \frac{2.85 \text{ in.}}{2.54} = 1.12 \text{ in.}$$

Note two important facts about this conversion:

1. The centimeter units cancel to give inches for the result. This is exactly what we had wanted to accomplish. Using the other conversion factor $\left(2.85 \text{ cm} \times \dfrac{2.54 \text{ cm}}{1 \text{ in.}} \right)$ would not work because the units would not cancel to give inches in the result.

MATH SKILL BUILDER
When you finish a calculation, always check to make sure that the answer makes sense.

2. As the units changed from centimeters to inches, the number changed from 2.85 to 1.12. Thus 2.85 cm has exactly the same value (is the same length) as 1.12 in. Notice that in this conversion, the number decreased from 2.85 to 1.12. This makes sense because the inch is a larger unit of length than the centimeter is. That is, it takes fewer inches to make the same length in centimeters.

The result in the foregoing conversion has three significant figures as required. Caution: Noting that the term 1 appears in the conversion, you might think that because this number appears to have only one significant figure, the result should have only one significant figure. That is, the answer should be given as 1 in. rather than 1.12 in. However, in the equivalence statement 1 in. = 2.54 cm, the 1 is an exact number (by definition). In other words, exactly 1 in. equals 2.54 cm. Therefore, the 1 does not limit the number of significant digits in the result.

MATH SKILL BUILDER
When exact numbers are used in a calculation, they never limit the number of significant digits.

We have seen how to convert from centimeters to inches. What about the reverse conversion? For example, if a pencil is 7.00 in. long, what is its length in centimeters? In this case, the conversion we want to make is

$$7.00 \text{ in.} \rightarrow ? \text{ cm}$$

What conversion factor do we need to make this conversion?

Remember that two conversion factors can be derived from each equivalence statement. In this case, the equivalence statement 2.54 cm = 1 in. gives

$$\frac{2.54 \text{ cm}}{1 \text{ in.}} \quad \text{or} \quad \frac{1 \text{ in.}}{2.54 \text{ cm}}$$

Again, we choose which factor to use by looking at the *direction* of the required change. For us to change from inches to centimeters, the inches must cancel. Thus the factor

$$\frac{2.54 \text{ cm}}{1 \text{ in.}}$$

is used, and the conversion is done as follows:

$$7.00 \text{ in.} \times \frac{2.54 \text{ cm}}{1 \text{ in.}} = (7.00)(2.54) \text{ cm} = 17.8 \text{ cm}$$

Here the inch units cancel, leaving centimeters as required.

Note that in this conversion, the number increased (from 7.00 to 17.8). This makes sense because the centimeter is a smaller unit of length than the inch. That is, it takes more centimeters to make the same length in inches. *Always take a moment to think about whether your answer makes sense.* This will help you avoid errors.

Changing from one unit to another via conversion factors (based on the equivalence statements between the units) is often called **dimensional analysis.** We will use this method throughout our study of chemistry.

We can now state some general steps for doing conversions by dimensional analysis.

Converting from One Unit to Another

Step 1 To convert from one unit to another, use the equivalence statement that relates the two units. The conversion factor needed is a ratio of the two parts of the equivalence statement.

Step 2 Choose the appropriate conversion factor by looking at the direction of the required change (make sure the unwanted units cancel).

Step 3 Multiply the quantity to be converted by the conversion factor to give the quantity with the desired units.

Step 4 Check that you have the correct number of significant figures.

Step 5 Ask whether your answer makes sense.

We will now illustrate this procedure in Example 2.6.

EXAMPLE 2.6 | Conversion Factors: One-Step Problems

An Italian bicycle has its frame size given as 62 cm. What is the frame size in inches?

SOLUTION

We can represent the problem as

$$62 \text{ cm} = ? \text{ in.}$$

In this problem we want to convert from centimeters to inches.

$$62 \text{ cm} \times \text{conversion factor} = ? \text{ in.}$$

Step 1 To convert from centimeters to inches, we need the equivalence statement 1 in. = 2.54 cm. This leads to two conversion factors:

$$\frac{1 \text{ in.}}{2.54 \text{ cm}} \quad \text{and} \quad \frac{2.54 \text{ cm}}{1 \text{ in.}}$$

Step 2 In this case, the direction we want is

$$\text{Centimeters} \rightarrow \text{inches}$$

so we need the conversion factor $\dfrac{1 \text{ in.}}{2.54 \text{ cm}}$. We know this is the one we want because using it will make the units of centimeters cancel, leaving units of inches.

Step 3 The conversion is carried out as follows:

$$62 \text{ cm} \times \frac{1 \text{ in.}}{2.54 \text{ cm}} = 24 \text{ in.}$$

Step 4 The result is limited to two significant figures by the number 62. The centimeters cancel, leaving inches as required.

Step 5 Note that the number decreased in this conversion. This makes sense; the inch is a larger unit of length than the centimeter.

Self-Check **EXERCISE 2.4**   Wine is often bottled in 0.750-L containers. Using the appropriate equivalence statement from Table 2.7, calculate the volume of such a wine bottle in quarts.

See Problems 2.59 and 2.60. ■

Next we will consider a conversion that requires several steps.

EXAMPLE 2.7 | **Conversion Factors: Multiple-Step Problems** |

The length of the marathon race is approximately 26.2 mi. What is this distance in kilometers?

SOLUTION

The problem before us can be represented as follows:

$$26.2 \text{ mi} = ? \text{ km}$$

We could accomplish this conversion in several different ways, but because Table 2.7 gives the equivalence statements 1 mi = 1760 yd and 1 m = 1.094 yd, we will proceed as follows:

$$\text{Miles} \rightarrow \text{yards} \rightarrow \text{meters} \rightarrow \text{kilometers}$$

This process will be carried out one conversion at a time to make sure everything is clear.

MILES → YARDS: We convert from miles to yards using the conversion factor $\dfrac{1760 \text{ yd}}{1 \text{ mi}}$.

$$26.2 \text{ mi} \times \frac{1760 \text{ yd}}{1 \text{ mi}} = 46{,}112 \text{ yd}$$

Result shown
on calculator

46,112 yd ⟩ Round off ⟩ 46,100 yd = 4.61×10^4 yd

YARDS → METERS: The conversion factor used to convert yards to meters is $\dfrac{1 \text{ m}}{1.094 \text{ yd}}$.

$$4.61 \times 10^4 \text{ yd} \times \dfrac{1 \text{ m}}{1.094 \text{ yd}} = 4.213894 \times 10^4 \text{ m}$$

Result shown on calculator

$$4.213894 \times 10^4 \text{ m} \quad \boxed{\text{Round off}} \Rightarrow 4.21 \times 10^4 \text{ m}$$

METERS → KILOMETERS: Because 1000 m = 1 km, or 10^3 m = 1 km, we convert from meters to kilometers as follows:

$$4.21 \times 10^4 \text{ m} \times \dfrac{1 \text{ km}}{10^3 \text{ m}} = 4.21 \times 10^1 \text{ km}$$
$$= 42.1 \text{ km}$$

Thus the marathon (26.2 mi) is 42.1 km.

Once you feel comfortable with the conversion process, you can combine the steps. For the above conversion, the combined expression is

miles → yards → meters → kilometers

$$26.2 \text{ mi} \times \dfrac{1760 \text{ yd}}{1 \text{ mi}} \times \dfrac{1 \text{ m}}{1.094 \text{ yd}} \times \dfrac{1 \text{ km}}{10^3 \text{ m}} = 42.1 \text{ km}$$

Note that the units cancel to give the required kilometers and that the result has three significant figures.

MATH SKILL BUILDER

Remember that we are rounding off at the end of each step to show the correct number of significant figures. However, in doing a multi-step calculation, *you* should retain the extra numbers that show on your calculator and round off only at the end of the calculation.

Self-Check **EXERCISE 2.5** Racing cars at the Indianapolis Motor Speedway now routinely travel around the track at an average speed of 225 mi/h. What is this speed in kilometers per hour?

See Problems 2.65 and 2.66. ■

Units provide a very valuable check on the validity of your solution. Always use them.

Recap: Whenever you work problems, remember the following points:

1. Always include the units (a measurement always has two parts: a number *and* a unit).

2. Cancel units as you carry out the calculations.

3. Check that your final answer has the correct units. If it doesn't, you have done something wrong.

4. Check that your final answer has the correct number of significant figures.

5. Think about whether your answer makes sense.

2.7 Temperature Conversions: An Approach to Problem Solving

OBJECTIVES: To learn the three temperature scales. • To learn to convert from one scale to another. • To continue to develop problem-solving skills.

When the doctor tells you your temperature is 102 degrees and the weatherperson on TV says it will be 75 degrees tomorrow, they are using the **Fahrenheit scale.** Water boils at 212 °F and freezes at 32 °F, and normal body temperature is 98.6 °F (where °F signifies "Fahrenheit degrees"). This temperature scale is widely used in the United States and Great Britain, and it is the scale employed in most of the engineering sciences. Another temperature scale, used in Canada and Europe and in the physical and life sciences in most countries, is the **Celsius scale.** In keeping with the metric system, which is based on powers of 10, the freezing and boiling points of water on the Celsius scale are assigned as 0 °C and 100 °C, respectively. On both the Fahrenheit and the Celsius scales, the unit of temperature is called a degree, and the symbol for it is followed by the capital letter representing the scale on which the units are measured: °C or °F.

Still another temperature scale used in the sciences is the **absolute** or **Kelvin scale.** On this scale water freezes at 273 K and boils at 373 K. On the Kelvin scale, the unit of temperature is called a kelvin and is symbolized by K. Thus, on the three scales, the boiling point of water is stated as 212 Fahrenheit degrees (212 °F), 100 Celsius degrees (100 °C), and 373 kelvins (373 K).

The three temperature scales are compared in Figures 2.6 and 2.7. There are several important facts you should note.

> Although 373 K is often stated as 373 degrees Kelvin, it is more correct to say 373 kelvins.

1. The size of each temperature unit (each degree) is the same for the Celsius and Kelvin scales. This follows from the fact that the *difference* between the boiling and freezing points of water is 100 units on both of these scales.

2. The Fahrenheit degree is smaller than the Celsius and Kelvin units. Note that on the Fahrenheit scale there are 180 Fahrenheit degrees between the boiling and freezing points of water, as compared with 100 units on the other two scales.

3. The zero points are different on all three scales.

Figure 2.6

Thermometers based on the three temperature scales in **a** ice water and **b** boiling water.

In your study of chemistry, you will sometimes need to convert from one temperature scale to another. We will consider in some detail how this is done. In addition to learning how to change temperature scales, you should also use this section as an opportunity to further develop your skills in problem solving.

▶ Converting Between the Kelvin and Celsius Scales

It is relatively simple to convert between the Celsius and Kelvin scales because the temperature unit is the same size; only the zero points are different. Because 0 °C corresponds to 273 K, converting from Celsius to Kelvin requires that we add 273 to the Celsius temperature. We will illustrate this procedure in Example 2.8.

EXAMPLE 2.8 | Temperature Conversion: Celsius to Kelvin

Boiling points will be discussed further in Chapter 14.

The boiling point of water at the top of Mt. Everest is 70. °C. Convert this temperature to the Kelvin scale. (The decimal point after the temperature reading indicates that the trailing zero is significant.)

SOLUTION

This problem asks us to find 70. °C in units of kelvins. We can represent this problem simply as

$$70. \ °C = ? \ K$$

In solving problems, it is often helpful to draw a diagram that depicts what the words are telling you.

In doing problems, it is often helpful to draw a diagram in which we try to represent the words in the problem with a picture. This problem can be diagramed as shown in Figure 2.8a.

In this picture we have shown what we want to find: "What temperature (in kelvins) is the same as 70. °C?" We also know from Figure 2.7 that 0 °C represents the same temperature as 273 K. How many degrees above 0 °C is 70. °C? The answer, of course, is 70. Thus we must add 70. to 0 °C to reach 70. °C. Because degrees are the *same size* on both the Celsius scale

Converting 70. °C to units measured on the Kelvin scale.

a

We know 0 °C = 273 K. We want to know 70. °C = ? K.

b

There are 70 degrees on the Celsius scale between 0 °C and 70. °C. Because units on these scales are the same size, there are also 70 kelvins in this same distance on the Kelvin scale.

and the Kelvin scale (see Figure 2.8b), we must also add 70. to 273 K (same temperature as 0 °C) to reach ? K. That is,

$$? \text{ K} = 273 + 70. = 343 \text{ K}$$

Thus 70. °C corresponds to 343 K.

Note that to convert from the Celsius to the Kelvin scale, we simply add the temperature in °C to 273. That is,

$$T_{°C} \quad + \quad 273 \quad = \quad T_K$$

Temperature   Temperature
in Celsius   in kelvins
degrees

Using this formula to solve the present problem gives

$$70. + 273 = 343$$

(with units of kelvins, K), which is the correct answer. ∎

We can summarize what we learned in Example 2.8 as follows: to convert from the Celsius to the Kelvin scale, we can use the formula

$$T_{°C} \quad + \quad 273 \quad = \quad T_K$$

Temperature   Temperature
in Celsius   in kelvins
degrees

EXAMPLE 2.9 | **Temperature Conversion: Kelvin to Celsius**

Liquid nitrogen boils at 77 K. What is the boiling point of nitrogen on the Celsius scale?

SOLUTION

The problem to be solved here is 77 K = ? °C. Let's explore this question by examining the picture on the following page representing the two

Tiny Thermometers

Can you imagine a thermometer that has a diameter equal to one one-hundredth of a human hair? Such a device has actually been produced by scientists Yihica Gao and Yoshio Bando of the National Institute for Materials Science in Tsukuba, Japan. The thermometer they constructed is so tiny that it must be read using a powerful electron microscope.

It turns out that the tiny thermometers were produced by accident. The Japanese scientists were actually trying to make tiny (nanoscale) gallium nitride wires. However, when they examined the results of their experiment, they discovered tiny tubes of carbon atoms that were filled with elemental gallium. Because gallium is a liquid over an unusually large temperature range, it makes a perfect working liquid for a thermometer. Just as in mercury thermometers, which have mostly been phased out because of the toxicity of mercury, the gallium expands as the temperature increases. Therefore, gallium moves up the tube as the temperature increases.

These minuscule thermometers are not useful in the normal macroscopic world—they can't even be seen with the naked eye. However, they should be valuable for monitoring temperatures from 50 °C to 500 °C in materials in the nanoscale world.

Liquid gallium expands within a carbon nanotube as the temperature increases (left to right).

Dr. Yoshio Bando/National Institute for Materials Sciences

temperature scales. One key point is to recognize that 0 °C = 273 K. Also note that the difference between 273 K and 77 K is 196 kelvins ($273 - 77 = 196$). That is, 77 K is 196 kelvins below 273 K. The degree size is the same on these two temperature scales, so 77 K must correspond to 196 Celsius degrees below zero or −196 °C. Thus 77 K = ? °C = −196 °C.

We can also solve this problem by using the formula

$$T_{°C} + 273 = T_K$$

However, in this case we want to solve for the Celsius temperature, $T_{°C}$. That is, we want to isolate $T_{°C}$ on one side of the equals sign. To do this we use an important general principle: doing *the same thing on both sides of the equals sign* preserves the equality. In other words, it's always okay to perform the same operation on both sides of the equals sign.

To isolate $T_{°C}$ we need to subtract 273 from both sides:

$$T_{°C} + 273 - 273 = T_K - 273$$

↑   ↑
Sum is zero

to give

$$T_{°C} = T_K - 273$$

Using this equation to solve the problem, we have

$$T_{°C} = T_K - 273 = 77 - 273 = -196$$

So, as before, we have shown that

$$77 \text{ K} = -196 \text{ °C}$$

Self-Check   **EXERCISE 2.6**   Which temperature is colder, 172 K or -75 °C?

See Problems 2.73 and 2.74. ■

In summary, because the Kelvin and Celsius scales have the same size unit, to switch from one scale to the other we must simply account for the different zero points. We must add 273 to the Celsius temperature to obtain the temperature on the Kelvin scale:

$$T_K = T_{°C} + 273$$

To convert from the Kelvin scale to the Celsius scale, we must subtract 273 from the Kelvin temperature:

$$T_{°C} = T_K - 273$$

▶ Converting Between the Fahrenheit and Celsius Scales

The conversion between the Fahrenheit and Celsius temperature scales requires two adjustments:

1. For the different size units

2. For the different zero points

To see how to adjust for the different unit sizes, consider the diagram in Figure 2.9. Note that because 212 °F = 100 °C and 32 °F = 0 °C,

$$212 - 32 = 180 \text{ Fahrenheit degrees} = 100 - 0 = 100 \text{ Celsius degrees}$$

Thus

$$180. \text{ Fahrenheit degrees} = 100. \text{ Celsius degrees}$$

MATH SKILL BUILDER
Remember, it's okay to do the same thing to both sides of the equation.

Dividing both sides of this equation by 100. gives

$$\frac{180.}{100.} \text{ Fahrenheit degrees} = \frac{\cancel{100.}}{\cancel{100.}} \text{ Celsius degrees}$$

or

$$1.80 \text{ Fahrenheit degrees} = 1.00 \text{ Celsius degree}$$

The factor 1.80 is used to convert from one degree size to the other.

Figure 2.9

Comparison of the Celsius and Fahrenheit scales.

Next we have to account for the fact that 0 °C is *not* the same as 0 °F. In fact, 32 °F = 0 °C. Although we will not show how to derive it, the equation to convert a temperature in Celsius degrees to the Fahrenheit scale is

$$T_{°F} = 1.80(T_{°C}) + 32$$

Temperature   Temperature
in °F   in °C

In this equation the term $1.80(T_{°C})$ adjusts for the difference in degree size between the two scales. The 32 in the equation accounts for the different zero points. We will now show how to use this equation.

EXAMPLE 2.10   Temperature Conversion: Celsius to Fahrenheit

On a summer day the temperature in the laboratory, as measured on a lab thermometer, is 28 °C. Express this temperature on the Fahrenheit scale.

SOLUTION

This problem can be represented as 28 °C = ? °F. We will solve it using the formula

$$T_{°F} = 1.80 (T_{°C}) + 32$$

In this case,

$$T_{°C}$$
$$\downarrow$$
$$T_{°F} = ? \ °F = 1.80(28) + 32 = 50.4 + 32$$

Rounds
off to 50

$$= 50. + 32 = 82$$

Thus 28 °C = 82 °F. ∎

Note that 28 °C is approximately equal to 82 °F. Because the numbers are just reversed, this is an easy reference point to remember for the two scales.

EXAMPLE 2.11 | Temperature Conversion: Celsius to Fahrenheit

Express the temperature $-40.$ °C on the Fahrenheit scale.

SOLUTION

We can express this problem as $-40.$ °C = ? °F. To solve it we will use the formula

$$T_{°F} = 1.80 \ (T_{°C}) + 32$$

In this case,

$$T_{°C}$$
$$\downarrow$$
$$T_{°F} = ? \ °F = 1.80(-40.) + 32$$
$$= -72 + 32 = -40$$

So -40 °C $= -40$ °F. This is a very interesting result and is another useful reference point.

Self-Check **EXERCISE 2.7** Hot tubs are often maintained at 41 °C. What is this temperature in Fahrenheit degrees?

See Problems 2.75 through 2.78. ∎

To convert from Celsius to Fahrenheit, we have used the equation

$$T_{°F} = 1.80 \ (T_{°C}) + 32$$

To convert a Fahrenheit temperature to Celsius, we need to rearrange this equation to isolate Celsius degrees ($T_{°C}$). Remember, we can always do the same operation to both sides of the equation. First subtract 32 from each side:

$$T_{°F} - 32 = 1.80 \ (T_{°C}) + 32 - 32$$
$$\uparrow \qquad \uparrow$$
$$\text{Sum is zero}$$

to give

$$T_{°F} - 32 = 1.80(T_{°C})$$

Next divide both sides by 1.80

$$\frac{T_{°F} - 32}{1.80} = \frac{1.80(T_{°C})}{1.80}$$

to give

$$\frac{T_{°F} - 32}{1.80} = T_{°C}$$

or

$$T_{°C} = \frac{T_{°F} - 32}{1.80}$$

Temperature in °F

Temperature in °C

$$T_{°C} = \frac{T_{°F} - 32}{1.80}$$

EXAMPLE 2.12 | **Temperature Conversion: Fahrenheit to Celsius**

One of the body's responses to an infection or injury is to elevate its temperature. A certain flu victim has a body temperature of 101 °F. What is this temperature on the Celsius scale?

SOLUTION

The problem is 101 °F = ? °C. Using the formula

$$T_{°C} = \frac{T_{°F} - 32}{1.80}$$

yields

$$T_{°C} = ? °C = \frac{\overset{T_{°F}}{101} - 32}{1.80} = \frac{69}{1.80} = 38$$

That is, 101 °F = 38 °C.

Self-Check **EXERCISE 2.8** An antifreeze solution in a car's radiator boils at 239 °F. What is this temperature on the Celsius scale?

See Problems 2.75 through 2.78. ■

In doing temperature conversions, you will need the following formulas.

Temperature Conversion Formulas	
• Celsius to Kelvin	$T_K = T_{°C} + 273$
• Kelvin to Celsius	$T_{°C} = T_K - 273$
• Celsius to Fahrenheit	$T_{°F} = 1.80(T_{°C}) + 32$
• Fahrenheit to Celsius	$T_{°C} = \dfrac{T_{°F} - 32}{1.80}$

2.8 Density

OBJECTIVE: To define density and its units.

Lead has a greater density than feathers.

When you were in elementary school, you may have been embarrassed by your answer to the question "Which is heavier, a pound of lead or a pound of feathers?" If you said lead, you were undoubtedly thinking about density, not mass. **Density** can be defined as the amount of matter present *in a given volume* of substance. That is, density is mass per unit volume, the ratio of the mass of an object to its volume:

$$\text{Density} = \frac{\text{mass}}{\text{volume}}$$

It takes a much bigger volume to make a pound of feathers than to make a pound of lead. This is because lead has a much greater mass per unit volume—a greater density.

The density of a liquid can be determined easily by weighing a known volume of the substance as illustrated in Example 2.13.

EXAMPLE 2.13

Calculating Density

Suppose a student finds that 23.50 mL of a certain liquid weighs 35.062 g. What is the density of this liquid?

SOLUTION

We can calculate the density of this liquid simply by applying the definition

$$\text{Density} = \frac{\text{mass}}{\text{volume}} = \frac{35.062 \text{ g}}{23.50 \text{ mL}} = 1.492 \text{ g/mL}$$

This result could also be expressed as 1.492 g/cm³ because 1 mL = 1 cm³. ■

The volume of a solid object is often determined indirectly by submerging it in water and measuring the volume of water displaced. In fact, this is the most accurate method for measuring a person's percent body fat. The person is submerged momentarily in a tank of water, and the increase in volume is measured (see Figure 2.10). It is possible to calculate the body density by using the person's weight (mass) and the volume of the person's body determined by submersion. Fat, muscle, and bone have different densities (fat is less dense than muscle tissue, for example), so the fraction of the person's body that is fat can be calculated. The more muscle and the less fat a person has, the higher his or her body density. For example, a muscular person weighing 150 lb has a smaller body volume (and thus a higher density) than a fat person weighing 150 lb.

EXAMPLE 2.14

Determining Density

The most common units for density are g/mL = g/cm³.

At a local pawn shop a student finds a medallion that the shop owner insists is pure platinum. However, the student suspects that the medallion may actually be silver and thus much less valuable. The student buys the medallion only after the shop owner agrees to refund the price if the medallion is

Figure 2.10

a *Tank of water*

b *Person submerged in the tank, raising the level of the water* returned within two days. The student, a chemistry major, then takes the medallion to her lab and measures its density as follows. She first weighs the medallion and finds its mass to be 55.64 g. She then places some water in a graduated cylinder and reads the volume as 75.2 mL. Next she drops the medallion into the cylinder and reads the new volume as 77.8 mL. Is the medallion platinum (density = 21.4 g/cm^3) or silver (density = 10.5 g/cm^3)?

SOLUTION

The densities of platinum and silver differ so much that the measured density of the medallion will show which metal is present. Because by definition

$$\text{Density} = \frac{\text{mass}}{\text{volume}}$$

to calculate the density of the medallion, we need its mass and its volume. The mass of the medallion is 55.64 g. The volume of the medallion can be obtained by taking the difference between the volume readings of the water in the graduated cylinder before and after the medallion was added.

$$\text{Volume of medallion} = 77.8 \text{ mL} - 75.2 \text{ mL} = 2.6 \text{ mL}$$

The volume appeared to increase by 2.6 mL when the medallion was added, so 2.6 mL represents the volume of the medallion. Now we can use the measured mass and volume of the medallion to determine its density:

$$\text{Density of medallion} = \frac{\text{mass}}{\text{volume}} = \frac{55.64 \text{ g}}{2.6 \text{ mL}} = 21 \text{ g/mL}$$

$$\text{or}$$

$$= 21 \text{ g/cm}^3$$

The medallion is really platinum.

Self-Check   **EXERCISE 2.9**   A student wants to identify the main component in a commercial liquid cleaner. He finds that 35.8 mL of the cleaner weighs 28.1 g. Of the following possibilities, which is the main component of the cleaner?

Substance	Density, g/cm^3
chloroform	1.483
diethyl ether	0.714
isopropyl alcohol	0.785
toluene	0.867

See Problems 2.89 and 2.90. ■

EXAMPLE 2.15 | Using Density in Calculations

Mercury has a density of 13.6 g/mL. What volume of mercury must be taken to obtain 225 g of the metal?

SOLUTION

To solve this problem, start with the definition of density,

$$\text{Density} = \frac{\text{mass}}{\text{volume}}$$

and then rearrange this equation to isolate the required quantity. In this case we want to find the volume. Remember that we maintain an equality

Spherical droplets of mercury, a very dense liquid.

when we do the same thing to both sides. For example, if we multiply *both sides* of the density definition by volume,

$$\text{Volume} \times \text{density} = \frac{\text{mass}}{\cancel{\text{volume}}} \times \cancel{\text{volume}}$$

volume cancels on the right, leaving

$$\text{Volume} \times \text{density} = \text{mass}$$

We want the volume, so we now divide both sides by density,

$$\frac{\text{Volume} \times \cancel{\text{density}}}{\cancel{\text{density}}} = \frac{\text{mass}}{\text{density}}$$

to give

$$\text{Volume} = \frac{\text{mass}}{\text{density}}$$

Now we can solve the problem by substituting the given numbers:

$$\text{Volume} = \frac{225 \text{ g}}{13.6 \text{ g/mL}} = 16.5 \text{ mL}$$

We must take 16.5 mL of mercury to obtain an amount that has a mass of 225 g. ■

The densities of various common substances are given in Table 2.8.

Besides being a tool for the identification of substances, density has many other uses. For example, the liquid in your car's lead storage battery (a solution of sulfuric acid) changes density because the sulfuric acid is consumed as the battery discharges. In a fully charged battery, the density of the solution is about 1.30 g/cm^3. When the density falls below 1.20 g/cm^3, the battery has to be recharged. Density measurement is also used to determine the amount of antifreeze, and thus the level of protection against freezing, in the cooling system of a car. Water and antifreeze have different densities, so the measured density of the mixture tells us how much of each is present. The device used to test the density of the solution—a hydrometer—is shown in Figure 2.11.

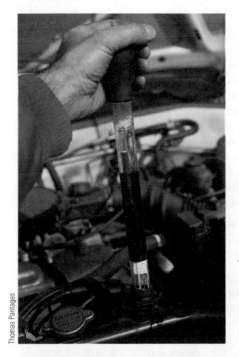

Figure 2.11

A hydrometer being used to determine the density of the antifreeze solution in a car's radiator.

Table 2.8	Densities of Various Common Substances at 20 °C	
Substance	Physical State	Density (g/cm^3)
oxygen	gas	0.00133*
hydrogen	gas	0.000084*
ethanol	liquid	0.785
benzene	liquid	0.880
water	liquid	1.000
magnesium	solid	1.74
salt (sodium chloride)	solid	2.16
aluminum	solid	2.70
iron	solid	7.87
copper	solid	8.96
silver	solid	10.5
lead	solid	11.34
mercury	liquid	13.6
gold	solid	19.32

*At 1 atmosphere pressure

In certain situations, the term *specific gravity* is used to describe the density of a liquid. **Specific gravity** is defined as the ratio of the density of a given liquid to the density of water at 4 °C. Because it is a ratio of densities, specific gravity has no units.

CHAPTER 2 REVIEW

Key Terms

measurement (p. 15)

scientific notation (2.1)

units (2.2)

English system (2.2)

metric system (2.2)

SI units (2.2)

volume (2.3)

mass (2.3)

significant figures (2.4)

rounding off (2.5)

conversion factor (2.6)

equivalence statement (2.6)

dimensional analysis (2.6)

Fahrenheit scale (2.7)

Celsius scale (2.7)

Kelvin (absolute) scale (2.7)

density (2.8)

specific gravity (2.8)

Summary

1. A quantitative observation is called a measurement and always consists of a number and a unit.

2. We can conveniently express very large or very small numbers using scientific notation, which represents the number as a number between 1 and 10 multiplied by 10 raised to a power.

3. Units give a scale on which to represent the results of a measurement. The three systems discussed are the English, metric, and SI systems. The metric and SI systems use prefixes (Table 2.2) to change the size of the units.

4. The mass of an object represents the quantity of matter in that object.

5. All measurements have a degree of uncertainty, which is reflected in the number of significant figures used to express them. Various rules are used to round off to the correct number of significant figures in a calculated result.

6. We can convert from one system of units to another by a method called dimensional analysis, in which conversion factors are used.

7. Temperature can be measured on three different scales: Fahrenheit, Celsius, and Kelvin. We can readily convert among these scales.

8. Density is the amount of matter present in a given volume (mass per unit volume). That is,

$$\text{Density} = \frac{\text{mass}}{\text{volume}}$$

Active Learning Questions

These questions are designed to be considered by groups of students in class. Often these questions work well for introducing a particular topic in class.

1. a. There are 365 days/year, 24 hours/day, 12 months/year, and 60 minutes/hour. How many minutes are there in one month?
   b. There are 24 hours/day, 60 minutes/hour, 7 days/week, and 4 weeks/month. How many minutes are there in one month?
   c. Why are these answers different? Which (if either) is more correct and why?

2. You go to a convenience store to buy candy and find the owner to be rather odd. He allows you to buy pieces only in multiples of four, and to buy four, you need $0.23. He allows you only to use 3 pennies and 2 dimes. You have a bunch of pennies and dimes, and instead of counting them, you decide to weigh them. You have 636.3 g of pennies, and each penny weighs an average of 3.03 g. Each dime weighs an average of 2.29 g. Each piece of candy weighs an average of 10.23 g.

   a. How many pennies do you have?
   b. How many dimes do you need to buy as much candy as possible?
   c. How much would all of your dimes weigh?
   d. How many pieces of candy could you buy (based on the number of dimes from part b)?
   e. How much would this candy weigh?
   f. How many pieces of candy could you buy with twice as many dimes?

3. When a marble is dropped into a beaker of water, it sinks to the bottom. Which of the following is the best explanation?

   a. The surface area of the marble is not large enough for the marble to be held up by the surface tension of the water.
   b. The mass of the marble is greater than that of the water.
   c. The marble weighs more than an equivalent volume of the water.
   d. The force from dropping the marble breaks the surface tension of the water.
   e. The marble has greater mass and volume than the water.

Explain each choice. That is, for choices you did not pick, explain why you feel they are wrong, and justify the choice you did pick.

4. Consider water in each graduated cylinder as shown:

You add both samples of water to a beaker. How would you write the number describing the total volume? What limits the precision of this number?

5. What is the numerical value of a conversion factor? Why must this be true?

6. For each of the following numbers, indicate which zeros are significant and explain. Do not merely cite the rule that applies, but explain the rule.

    a. 10.020   b. 0.002050   c. 190   d. 270

7. Consider the addition of "15.4" to "28." What would a mathematician say the answer is? What would a scientist say? Justify the scientist's answer, not merely citing the rule, but explaining it.

8. Consider multiplying "26.2" by "16.43." What would a mathematician say the answer is? What would a scientist say? Justify the scientist's answer, not merely citing the rule, but explaining it.

9. In lab you report a measured volume of 128.7 mL of water. Using significant figures as a measure of the error, what range of answers does your reported volume imply? Explain.

10. Sketch two pieces of glassware: one that can measure volume to the thousandths place, and one that can measure volume only to the ones place.

11. Oil floats on water but is "thicker" than water. Why do you think this fact is true?

12. Show how converting numbers to scientific notation can help you decide which digits are significant.

13. You are driving 65 mph and take your eyes off the road "just for a second." How many feet do you travel in this time?

14. You have a 1.0-cm^3 sample of lead and a 1.0-cm^3 sample of glass. You drop each in a separate beaker of water. How do the volumes of water that are displaced by the samples compare? Explain.

15. The beakers shown below have different precisions.

    a. Label the amount of water in each of the three beakers to the correct number of significant figures.
    b. Is it possible for each of the three beakers to contain the exact same amount of water? If no, why not? If yes, did you report the volumes as the same in part a? Explain.
    c. Suppose you pour the water from these three beakers into one container. What should be the volume in the container reported to the correct number of significant figures?

16. True or False? For any mathematical operation performed on two measurements, the number of significant figures in the answer is the same as the least number of significant figures in either of the measurements. Explain your answer.

17. Complete the following and explain each in your own words: leading zeros are (never/sometimes/always) significant; captive zeros are (never/sometimes/always) significant; and trailing zeros are (never/sometimes/always) significant.

    For any statement with an answer of "sometimes," give examples of when the zero is significant and when it is not, and explain.

18. For each of the following figures, a through d, decide which block is more dense: the orange block, the blue block, or it cannot be determined. Explain your answers.

19. For the pin shown below, why is the third digit determined for the length of the pin uncertain? Considering that the third digit is uncertain, explain why the length of the pin is indicated as 2.85 cm rather than, for example, 2.83 or 2.87 cm.

20. Why can the length of the pin shown below not be recorded as 2.850 cm?

21. Use the figure below to answer the following questions.

a. Derive the relationship between °C and °X.
b. If the temperature outside is 22.0 °C, what is the temperature in units of °X?
c. Convert 58.0 °X to units of °C, K, and °F.

OWL Questions and problems for this chapter are available in OWL. Use the OWL access card that came with this textbook to access assigned questions and problems for this chapter.

3

Matter

● An iceberg in Greenland.
(Frank Krahmer/Masterfile)

☊WL Sign in to OWL at
www.cengage.com/owl to view
tutoritals and simulations, develop
problem-solving skills, and complete
online homework assigned by your
professor.

As you look around you, you must wonder about the properties of matter. How do plants grow and why are they green? Why is the sun hot? Why does a hot dog get hot in a microwave oven? Why does wood burn whereas rocks do not? What is a flame? How does soap work? Why does soda fizz when you open the bottle? When iron rusts, what's happening? And why doesn't aluminum rust? How does a cold pack for an athletic injury, which is stored for weeks or months at room temperature, suddenly get cold when you need it? How does a hair permanent work?

© Cengage Learning

Why does soda fizz when you open the bottle?

The answers to these and endless other questions lie in the domain of chemistry. In this chapter we begin to explore the nature of matter: how it is organized and how and why it changes.

3.1 Matter

OBJECTIVE: To learn about matter and its three states.

Matter, the "stuff" of which the universe is composed, has two characteristics: it has mass and it occupies space. Matter comes in a great variety of forms: the stars, the air that you are breathing, the gasoline that you put in your car, the chair on which you are sitting, the turkey in the sandwich you may have had for lunch, the tissues in your brain that enable you to read and comprehend this sentence, and so on.

To try to understand the nature of matter, we classify it in various ways. For example, wood, bone, and steel share certain characteristics. These things are all rigid; they have definite shapes that are difficult to change. On the other hand, water and gasoline, for example, take the shape of any container into which they are poured (see Figure 3.1). Even so, 1 L of water has a volume of 1 L whether it is in a pail or a beaker. In contrast, air takes the shape of its container and fills any container uniformly.

The substances we have just described illustrate the three **states of matter: solid, liquid,** and **gas.** These are defined and illustrated in Table 3.1. The state of a given sample of matter depends on the strength of the forces among the particles contained in the matter; the stronger these forces, the more rigid the matter. We will discuss this in more detail in the next section.

Richard Megna/Fundamental Photographs

Figure 3.1

Liquid water takes the shape of its container.

Table 3.1	The Three States of Matter	
State	Definition	Examples
solid	rigid; has a fixed shape and volume	ice cube, diamond, iron bar
liquid	has a definite volume but takes the shape of its container	gasoline, water, alcohol, blood
gas	has no fixed volume or shape; takes the shape and volume of its container	air, helium, oxygen

3.2 Physical and Chemical Properties and Changes

OBJECTIVES: To learn to distinguish between physical and chemical properties. • To learn to distinguish between physical and chemical changes.

How does this lush vegetation grow in a tropical rain forest, and why is it green?

Brian Parker/Tom Stack & Associates

When you see a friend, you immediately respond and call him or her by name. We can recognize a friend because each person has unique characteristics or properties. The person may be thin and tall, may have blonde hair and blue eyes, and so on. The characteristics just mentioned are examples of **physical properties.** Substances also have physical properties. Typical physical properties of a substance include odor, color, volume, state (gas, liquid, or solid), density, melting point, and boiling point. We can also describe a pure substance in terms of its **chemical properties,** which refer to its ability to form new substances. An example of a chemical change is wood burning in a fireplace, giving off heat and gases and leaving a residue of ashes. In this process, the wood is changed to several new substances. Other examples of chemical changes include the rusting of the steel in our cars, the digestion of food in our stomachs, and the growth of grass in our yards. In a chemical change a given substance changes to a fundamentally different substance or substances.

EXAMPLE 3.1 | Identifying Physical and Chemical Properties

Classify each of the following as a physical or a chemical property.

   a. The boiling point of a certain alcohol is 78 °C.

   b. Diamond is very hard.

   c. Sugar ferments to form alcohol.

   d. A metal wire conducts an electric current.

SOLUTION

Items (a), (b), and (d) are physical properties; they describe inherent characteristics of each substance, and no change in composition occurs. A metal wire has the same composition before and after an electric current has passed through it. Item (c) is a chemical property of sugar. Fermentation of sugars involves the formation of a new substance (alcohol).

Self-Check **EXERCISE 3.1**

Gallium metal has such a low melting point (30 °C) that it melts from the heat of a hand.

© Cengage Learning

Which of the following are physical properties and which are chemical properties?

   a. Gallium metal melts in your hand.

   b. Platinum does not react with oxygen at room temperature.

   c. This page is white.

   d. The copper sheets that form the "skin" of the Statue of Liberty have acquired a greenish coating over the years.

See Problems 3.11 through 3.14. ∎

Matter can undergo changes in both its physical and its chemical properties. To illustrate the fundamental differences between physical and chemical changes, we will consider water. As we will see in much more detail in later chapters, a sample of water contains a very large number of individual

The letters indicate atoms and the lines indicate attachments (bonds) between atoms.

units (called molecules), each made up of two atoms of hydrogen and one atom of oxygen—the familiar H_2O. This molecule can be represented as

where the letters stand for atoms and the lines show attachments (called bonds) between atoms, and the molecular model (on the right) represents water in a more three-dimensional fashion.

The purpose here is to give an overview. Don't worry about the precise definitions of *atom* and *molecule* now. We will explore these concepts more fully in Chapter 4.

What is really occurring when water undergoes the following changes?

Solid (ice) ⟹ Liquid (water) ⟹ Gas (steam)

Melting   Boiling

We will describe these changes of state precisely in Chapter 14, but you already know something about these processes because you have observed them many times.

When ice melts, the rigid solid becomes a mobile liquid that takes the shape of its container. Continued heating brings the liquid to a boil, and the water becomes a gas or vapor that seems to disappear into "thin air." The changes that occur as the substance goes from solid to liquid to gas are represented in Figure 3.2. In ice the water molecules are locked into fixed positions (although they are vibrating). In the liquid the molecules are still very close together, but some motion is occurring; the positions of the molecules are no longer fixed as they are in ice. In the gaseous state the molecules are much farther apart and move randomly, hitting each other and the walls of the container.

An iron pyrite crystal (gold color) on a white quartz crystal.

Solid (Ice)   **Liquid (Water)**   **Gas (Steam)**

Figure 3.2

The three states of water (where red spheres represent oxygen atoms and blue spheres represent hydrogen atoms).

a — *Solid: the water molecules are locked into rigid positions and are close together.*

b — *Liquid: the water molecules are still close together but can move around to some extent.*

c — *Gas: the water molecules are far apart and move randomly.*

Chip Clark

Figure 3.3

Electrolysis, the decomposition of water by an electric current, is a chemical process.

The most important thing about all these changes is that the water molecules are still intact. The motions of individual molecules and the distances between them change, but *H_2O molecules are still present*. These changes of state are **physical changes** because they do not affect the composition of the substance. In each state we still have water (H_2O), not some other substance.

Now suppose we run an electric current through water as illustrated in Figure 3.3. Something very different happens. The water disappears and is replaced by two new gaseous substances, hydrogen and oxygen. An electric current actually causes the water molecules to come apart—the water *decomposes* to hydrogen and oxygen. We can represent this process as follows:

This is a **chemical change** because water (consisting of H_2O molecules) has changed into different substances: hydrogen (containing H_2 molecules) and oxygen (containing O_2 molecules). Thus in this process, the H_2O molecules have been replaced by O_2 and H_2 molecules. Let us summarize:

Physical and Chemical Changes

1. A *physical change* involves a change in one or more physical properties, but no change in the fundamental components that make up the substance. The most common physical changes are changes of state: solid ⇔ liquid ⇔ gas.
2. A *chemical change* involves a change in the fundamental components of the substance; a given substance changes into a different substance or substances. Chemical changes are called **reactions**: silver tarnishes by reacting with substances in the air; a plant forms a leaf by combining various substances from the air and soil; and so on.

EXAMPLE 3.2 | Identifying Physical and Chemical Changes

Classify each of the following as a physical or a chemical change.

a. Iron metal is melted.

b. Iron combines with oxygen to form rust.

c. Wood burns in air.

d. A rock is broken into small pieces.

SOLUTION

a. Melted iron is just liquid iron and could cool again to the solid state. This is a physical change.

b. When iron combines with oxygen, it forms a different substance (rust) that contains iron and oxygen. This is a chemical change because a different substance forms.

c. Wood burns to form different substances (as we will see later, they include carbon dioxide and water). After the fire, the wood is no longer in its original form. This is a chemical change.

d. When the rock is broken up, all the smaller pieces have the same composition as the whole rock. Each new piece differs from the original only in size and shape. This is a physical change.

Oxygen combines with the chemicals in wood to produce flames. Is a physical or chemical change taking place?

Jim Pickerell/Stock Connection

Self-Check **EXERCISE 3.2**   Classify each of the following as a chemical change, a physical change, or a combination of the two.

a. Milk turns sour.

b. Wax is melted over a flame and then catches fire and burns.

See Problems 3.17 and 3.18. ■

3.3 Elements and Compounds

OBJECTIVE: To understand the definitions of elements and compounds.

Element: a substance that cannot be broken down into other substances by chemical methods.

As we examine the chemical changes of matter, we encounter a series of fundamental substances called **elements.** Elements cannot be broken down into other substances by chemical means. Examples of elements are iron, aluminum, oxygen, and hydrogen. All of the matter in the world around us contains elements. The elements sometimes are found in an isolated state, but more often they are combined with other elements. Most substances contain several elements combined together.

The atoms of certain elements have special affinities for each other. They bind together in special ways to form **compounds**, substances that have the same composition no matter where we find them. Because compounds are made of elements, they can be broken down into elements through chemical changes:

> Compound: a substance composed of a given combination of elements that can be broken down into those elements by chemical methods.

$$\text{Compounds} \quad \Longrightarrow \quad \text{Elements}$$
Chemical changes

Water is an example of a compound. Pure water always has the same composition (the same relative amounts of hydrogen and oxygen) because it consists of H_2O molecules. Water can be broken down into the elements hydrogen and oxygen by chemical means, such as by the use of an electric current (see Figure 3.3).

As we will discuss in more detail in Chapter 4, each element is made up of a particular kind of atom: a pure sample of the element aluminum contains only aluminum atoms, elemental copper contains only copper atoms, and so on. Thus an element contains only one kind of atom; a sample of iron contains many atoms, but they are all iron atoms. Samples of certain pure elements do contain molecules; for example, hydrogen gas contains H—H (usually written H_2) molecules, and oxygen gas contains O—O (O_2) molecules. However, any pure sample of an element contains only atoms of that element, *never* any atoms of any other element.

A compound *always* contains atoms of *different* elements. For example, water contains hydrogen atoms and oxygen atoms, and there are always exactly twice as many hydrogen atoms as oxygen atoms because water consists of H—O—H molecules. A different compound, carbon dioxide, consists of CO_2 molecules and so contains carbon atoms and oxygen atoms (always in the ratio 1:2).

A compound, although it contains more than one type of atom, *always has the same composition*—that is, the same combination of atoms. The properties of a compound are typically very different from those of the elements it contains. For example, the properties of water are quite different from the properties of pure hydrogen and pure oxygen.

3.4 Mixtures and Pure Substances

OBJECTIVE: To learn to distinguish between mixtures and pure substances.

Virtually all of the matter around us consists of mixtures of substances. For example, if you closely observe a sample of soil, you will see that it has many types of components, including tiny grains of sand and remnants of plants. The air we breathe is a complex mixture of such gases as oxygen, nitrogen, carbon dioxide, and water vapor. Even the sparkling water from a drinking fountain contains many substances besides water.

A **mixture** can be defined as something that has variable composition. For example, wood is a mixture (its composition varies greatly depending on the tree from which it originates); wine is a mixture (it can be red or pale yellow, sweet or dry); coffee is a mixture (it can be strong, weak, or bitter); and, although it looks very pure, water pumped from deep in the earth is a mixture (it contains dissolved minerals and gases).

Concrete—An Ancient Material Made New

Concrete, which was invented more than 2000 years ago by the ancient Romans, is being transformed into a high-tech building material through the use of our knowledge of chemistry. There is little doubt that concrete is the world's most important material. It is used to construct highways, bridges, buildings, floors, countertops, and countless other objects. In its simplest form concrete consists of about 70% sand and gravel, 15% water, and 15% cement (a mixture prepared by heating and grinding limestone, clay, shale, and gypsum). Because concrete forms the skeleton of much of our society, improvements to make it last longer and perform better are crucial.

One new type of concrete is Ductal, which was developed by the French company Lafarge. Unlike traditional concrete, which is brittle and can rupture suddenly under a heavy load, Ductal can bend. Even better, Ductal is five times stronger than traditional concrete. The secret behind Ductal's near-magical properties lies in the addition of small steel or polymeric fibers, which are dispersed throughout the structure. The fibers eliminate the need for steel reinforcing bars (rebar) for structures such as bridges. Bridges built of Ductal are lighter, thinner, and much more corrosion resistant than bridges built with traditional concrete containing rebar.

In another innovation, the Hungarian company Litracon has developed a translucent concrete material by incorporating optical fibers of various diameters into the concrete. With this light-transmitting concrete, architects can design buildings with translucent concrete walls and concrete floors that can be lighted from below.

Another type of concrete being developed by the Italian company Italcementi Group has a self-cleaning surface. This new material is made by mixing titanium oxide particles into the concrete. Titanium oxide can absorb ultraviolet light and promote the decomposition of pollutants that would otherwise darken the surface of the building. This material has already been used for several buildings in Italy. One additional bonus of using this material for buildings and roads in cities is that it may actually act to reduce air pollution very significantly.

Concrete is an ancient material, but one that is showing the flexibility to be a high-tech material. Its adaptability will ensure that it finds valuable uses far into the future.

An object made of translucent concrete.

A **pure substance,** on the other hand, will always have the same composition. Pure substances are either elements or compounds. For example, pure water is a compound containing individual H_2O molecules. However, as we find it in nature, liquid water always contains other substances in addition to pure water—it is a mixture. This is obvious from the different tastes, smells, and colors of water samples obtained from various locations. However, if we take great pains to purify samples of water from various sources (such as oceans, lakes, rivers, and the earth's interior), we always end up with the same pure substance—water, which is made up only of H_2O molecules. Pure water always has the same physical and chemical properties and is always made of molecules containing hydrogen and oxygen in exactly the same proportions, regardless of the original source of the water. The properties of a pure substance make it possible to identify that substance conclusively.

Mixtures can be separated into pure substances: elements and/or compounds.

> Although we say we can separate mixtures into pure substances, it is virtually impossible to separate mixtures into totally pure substances. No matter how hard we try, some impurities (components of the original mixture) remain in each of the "pure substances."

Mixtures ⟹ Two or more pure substances

For example, the mixture known as air can be separated into oxygen (element), nitrogen (element), water (compound), carbon dioxide (compound), argon (element), and other pure substances.

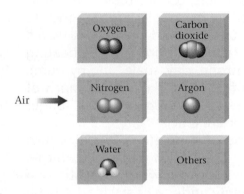

> A solution is a homogeneous mixture.

Mixtures can be classified as either homogeneous or heterogeneous. A **homogeneous mixture** is *the same throughout.* For example, when we dissolve some salt in water and stir well, all regions of the resulting mixture have the same properties. A homogeneous mixture is also called a **solution.** Of course, different amounts of salt and water can be mixed to form various solutions, but a homogeneous mixture (a solution) does not vary in composition from one region to another (see Figure 3.4).

Figure 3.4

When table salt is stirred into water *(left),* a homogeneous mixture called a solution forms *(right).*

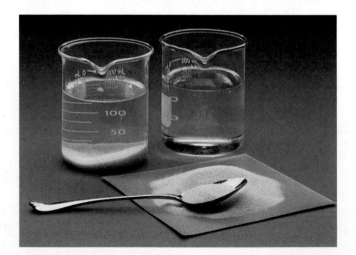

Figure 3.5

Sand and water do not mix to form a uniform mixture. After the mixture is stirred, the sand settles back to the bottom.

Coffee is a solution that has variable composition. It can be strong or weak.

The air around you is a solution—it is a homogeneous mixture of gases. Solid solutions also exist. Brass is a homogeneous mixture of the metals copper and zinc.

A **heterogeneous mixture** contains regions that have different properties from those of other regions. For example, when we pour sand into water, the resulting mixture has one region containing water and another, very different region containing mostly sand (see Figure 3.5).

EXAMPLE 3.3 | Distinguishing Between Mixtures and Pure Substances

Identify each of the following as a pure substance, a homogeneous mixture, or a heterogeneous mixture.

a. gasoline

b. a stream with gravel at the bottom

c. air

d. brass

e. copper metal

SOLUTION

a. Gasoline is a homogeneous mixture containing many compounds.

b. A stream with gravel on the bottom is a heterogeneous mixture.

c. Air is a homogeneous mixture of elements and compounds.

d. Brass is a homogeneous mixture containing the elements copper and zinc. Brass is not a pure substance because the relative amounts of copper and zinc are different in different brass samples.

e. Copper metal is a pure substance (an element).

Self-Check **EXERCISE 3.3**   Classify each of the following as a pure substance, a homogeneous mixture, or a heterogeneous mixture.

a. maple syrup

b. the oxygen and helium in a scuba tank

c. oil and vinegar salad dressing

d. common salt (sodium chloride)

See Problems 3.29 through 3.31. ∎

3.5 Separation of Mixtures

OBJECTIVE: To learn two methods of separating mixtures.

We have seen that the matter found in nature is typically a mixture of pure substances. For example, seawater is water containing dissolved minerals. We can separate the water from the minerals by boiling, which changes the water to steam (gaseous water) and leaves the minerals behind as solids. If we collect and cool the steam, it condenses to pure water. This separation process, called **distillation,** is shown in Figure 3.6.

The separation of a mixture sometimes occurs in the natural environment and can be to our benefit (see photo on p. 66).

When we carry out the distillation of salt water, water is changed from the liquid state to the gaseous state and then back to the liquid state. These

When the solution is boiled, steam (gaseous water) is driven off. If this steam is collected and cooled, it condenses to form pure water, which drips into the collection flask as shown.

After all of the water has been boiled off, the salt remains in the original flask and the water is in the collection flask.

Figure 3.6

Distillation of a solution consisting of salt dissolved in water.

changes of state are examples of physical changes. We are separating a mixture of substances, but we are not changing the composition of the individual substances. We can represent this as shown in Figure 3.7.

Suppose we scooped up some sand with our sample of seawater. This sample is a heterogeneous mixture, because it contains an undissolved solid as well as the saltwater solution. We can separate out the sand by simple **filtration.** We pour the mixture onto a mesh, such as a filter paper, which allows the liquid to pass through and leaves the solid behind (see Figure 3.8). The salt can then be separated from the water by distillation. The total separation process is represented in Figure 3.9. All the changes involved are physical changes.

We can summarize the description of matter given in this chapter with the diagram shown in Figure 3.10. Note that a given sample of matter can be a pure substance (either an element or a compound) or, more commonly, a mixture (homogeneous or heterogeneous). We have seen that all matter exists as elements or can be broken down into elements, the most fundamental substances we have encountered up to this point. We will have more to say about the nature of elements in the next chapter.

Figure 3.7

No chemical change occurs when salt water is distilled.

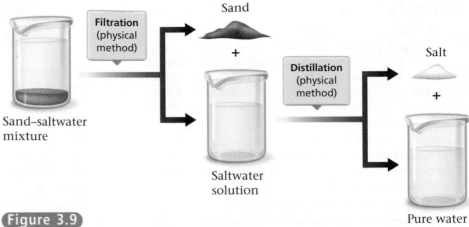

Figure 3.9

Separation of a sand–saltwater mixture.

Figure 3.8

Filtration separates a liquid from a solid. The liquid passes through the filter paper, but the solid particles are trapped.

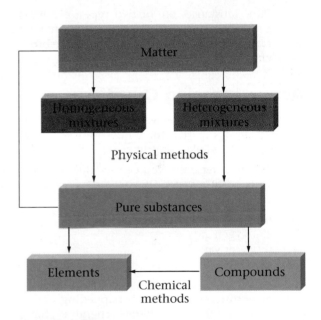

Figure 3.10

The organization of matter.

CHAPTER 3 REVIEW

Key Terms

matter (3.1)
states of matter (3.1)
solid (3.1)
liquid (3.1)
gas (3.1)
physical properties (3.2)

chemical properties (3.2)
physical change (3.2)
chemical change (3.2)
reaction (3.2)
element (3.3)
compound (3.3)

mixture (3.4)
pure substance (3.4)
homogeneous
  mixture (3.4)
solution (3.4)

heterogeneous
  mixture (3.4)
distillation (3.5)
filtration (3.5)

Summary

1. Matter can exist in three states—solid, liquid, and gas—and can be described in terms of its physical and chemical properties. Chemical properties describe a substance's ability to undergo a change to a different substance. Physical properties are the characteristics a substance exhibits as long as no chemical change occurs.

2. A physical change involves a change in one or more physical properties, but no change in composition. A chemical change transforms a substance into a new substance or substances.

3. A mixture has variable composition. A homogeneous mixture has the same properties throughout; a heterogeneous mixture does not. A pure substance always has the same composition. We can physically separate mixtures of pure substances by distillation and filtration.

4. Pure substances are of two types: elements, which cannot be broken down chemically into simpler substances, and compounds, which can be broken down chemically into elements.

Active Learning Questions

These questions are designed to be considered by groups of students in class. Often these questions work well for introducing a particular topic in class.

1. When water boils, you can see bubbles rising to the surface of the water. Of what are these bubbles made?

   a. air
   b. hydrogen and oxygen gas
   c. oxygen gas
   d. water vapor
   e. carbon dioxide gas

2. If you place a glass rod over a burning candle, the glass turns black. What is happening to each of the following (physical change, chemical change, both, or neither) as the candle burns? Explain.

   a. the wax
   b. the wick
   c. the glass rod

3. The boiling of water is a

   a. physical change because the water disappears.
   b. physical change because the gaseous water is chemically the same as the liquid.
   c. chemical change because heat is needed for the process to occur.
   d. chemical change because hydrogen and oxygen gases are formed from water.
   e. chemical and physical change.

   Explain your answer.

4. Is there a difference between a homogeneous mixture of hydrogen and oxygen in a 2:1 ratio and a sample of water vapor? Explain.

5. Sketch a magnified view (showing atoms and/or molecules) of each of the following and explain why the specified type of mixture is

   a. a heterogeneous mixture of two different compounds
   b. a homogeneous mixture of an element and a compound

6. Are all physical changes accompanied by chemical changes? Are all chemical changes accompanied by physical changes? Explain.

7. Why would a chemist find fault with the phrase "pure orange juice"?

8. Are separations of mixtures physical or chemical changes? Explain.

9. Explain the terms *element, atom,* and *compound.* Provide an example and microscopic drawing of each.

10. Mixtures can be classified as either homogeneous or heterogeneous. Compounds cannot be classified in this way. Why not? In your answer, explain what is meant by heterogeneous and homogeneous.

11. Provide microscopic drawings down to the atoms for Figure 3.10 in your text.

12. Look at Table 2.8 in your text. How do the densities of gases, liquids, and solids compare to each other? Use microscopic pictures to explain why this is true.

13. Label each of the following as an atomic element, a molecular element, or a compound.

14. Match each description below with the following microscopic pictures. More than one picture may fit each description. A picture may be used more than once or not used at all.

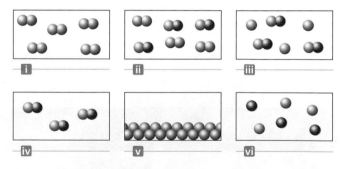

   a. a gaseous compound
   b. a mixture of two gaseous elements
   c. a solid element
   d. a mixture of gaseous element and a gaseous compound

4

Chemical Foundations: Elements, Atoms, and Ions

● Gold leafing on the ceiling of the Library of Congress in Washington, D.C. *(John Zoiner)*

© Cengage Learning

Lithium is administered in the form of lithium carbonate pills.

The Granger Collection, New York

Robert Boyle at 62 years of age.

The chemical elements are very important to each of us in our daily lives. Although certain elements are present in our bodies in tiny amounts, they can have a profound impact on our health and behavior. As we will see in this chapter, lithium can be a miracle treatment for someone with bipolar disorder, and our cobalt levels can have a remarkable impact on whether we behave violently.

Since ancient times, humans have used chemical changes to their advantage. The processing of ores to produce metals for ornaments and tools and the use of embalming fluids are two applications of chemistry that were used before 1000 B.C.

The Greeks were the first to try to explain why chemical changes occur. By about 400 B.C. they had proposed that all matter was composed of four fundamental substances: fire, earth, water, and air.

The next 2000 years of chemical history were dominated by alchemy. Some alchemists were mystics and fakes who were obsessed with the idea of turning cheap metals into gold. However, many alchemists were sincere scientists, and this period saw important events: the elements mercury, sulfur, and antimony were discovered, and alchemists learned how to prepare acids.

The first scientist to recognize the importance of careful measurements was the Irishman Robert Boyle (1627–1691). Boyle is best known for his pioneering work on the properties of gases, but his most important contribution to science was probably his insistence that science should be firmly grounded in experiments. For example, Boyle held no preconceived notions about how many elements there might be. His definition of the term *element* was based on experiments: a substance was an element unless it could be broken down into two or more simpler substances. For example, air could not be an element as the Greeks believed, because it could be broken down into many pure substances.

As Boyle's experimental definition of an element became generally accepted, the list of known elements grew, and the Greek system of four elements died. But although Boyle was an excellent scientist, he was not always right. For some reason he ignored his own definition of an element and clung to the alchemists' views that metals were not true elements and that a way would be found eventually to change one metal into another.

4.1 The Elements

OBJECTIVES: To learn about the relative abundances of the elements. • To learn the names of some elements.

In studying the materials of the earth (and other parts of the universe), scientists have found that all matter can be broken down chemically into about 100 different elements. At first it might seem amazing that the millions of known substances are composed of so few fundamental elements. Fortunately for those trying to understand and systematize it, nature often uses a relatively small number of fundamental units to assemble even extremely complex materials. For example, proteins, a group of substances that serve the human body in almost uncountable ways, are all made by linking together a few fundamental units to form huge molecules. A nonchemical example is the English language, where hundreds of thousands of words are constructed from only 26 letters. If you take apart the thousands of words in an English dictionary, you will find only these 26 fundamental components. In much the same way, when we take apart all of the substances in the world around us, we find only about 100 fundamental building blocks—the elements. Compounds are made by combining atoms of the various elements, just as words are constructed from the 26 letters of the alphabet. And just as you had to learn the letters of the alphabet before you learned to read and write, you need to learn the names and symbols of the chemical elements before you can read and write chemistry.

Presently about 116 different elements are known,* 88 of which occur naturally. (The rest have been made in laboratories.) The elements vary tremendously in abundance. In fact, only 9 elements account for most of the compounds found in the earth's crust. In Table 4.1, the elements are listed in order of their abundance (mass percent) in the earth's crust, oceans, and atmosphere. Note that nearly half of the mass is accounted for by oxygen alone. Also note that the 9 most abundant elements account for over 98% of the total mass.

Oxygen, in addition to accounting for about 20% of the earth's atmosphere (where it occurs as O_2 molecules), is found in virtually all the rocks, sand, and soil on the earth's crust. In these latter materials, oxygen is not present as O_2 molecules but exists in compounds that usually contain silicon and aluminum atoms. The familiar substances of the geological world, such as rocks and sand, contain large groups of silicon and oxygen atoms bound together to form huge clusters.

The list of elements found in living matter is very different from the list of elements found in the earth's crust. Table 4.2 shows the distribution of elements in the human body. Oxygen, carbon, hydrogen, and nitrogen form the basis for all biologically important molecules. Some elements found in the body (called trace elements) are crucial for life, even though they are present in relatively small amounts. For example, chromium helps the body use sugars to provide energy.

Footprints in the sand of the Namib Desert in Namibia.

Table 4.1	Distribution (Mass Percent) of the 18 Most Abundant Elements in the Earth's Crust, Oceans, and Atmosphere		
Element	Mass Percent	Element	Mass Percent
oxygen	49.2	titanium	0.58
silicon	25.7	chlorine	0.19
aluminum	7.50	phosphorus	0.11
iron	4.71	manganese	0.09
calcium	3.39	carbon	0.08
sodium	2.63	sulfur	0.06
potassium	2.40	barium	0.04
magnesium	1.93	nitrogen	0.03
hydrogen	0.87	fluorine	0.03
		all others	0.49

*This number changes as new elements are made in particle accelerators.

Table 4-2	Abundance of Elements in the Human Body	
Major Elements	Mass Percent	Trace Elements (in alphabetical order)
oxygen	65.0	arsenic
carbon	18.0	chromium
hydrogen	10.0	cobalt
nitrogen	3.0	copper
calcium	1.4	fluorine
phosphorus	1.0	iodine
magnesium	0.50	manganese
potassium	0.34	molybdenum
sulfur	0.26	nickel
sodium	0.14	selenium
chlorine	0.14	silicon
iron	0.004	vanadium
zinc	0.003	

One more general comment is important at this point. As we have seen, elements are fundamental to understanding chemistry. However, students are often confused by the many different ways that chemists use the term *element.* Sometimes when we say *element,* we mean a single atom of that element. We might call this the microscopic form of an element. Other times when we use the term *element,* we mean a sample of the element large enough to weigh on a balance. Such a sample contains many, many atoms of the element, and we might call this the macroscopic form of the element. There is yet a further complication. As we will see in more detail in Section 4.9 the macroscopic forms of several elements contain molecules rather than individual atoms as the fundamental components. For example, chemists know that oxygen gas consists of molecules with two oxygen atoms connected together (represented as O—O or more commonly as O_2). Thus when we refer to the element oxygen we might mean a single atom of oxygen, a single O_2 molecule, or a macroscopic sample containing many O_2 molecules. Finally, we often use the term *element* in a generic fashion. When we say the human body contains the element sodium or lithium, we do not mean that free elemental sodium or lithium is present. Rather, we mean that atoms of these elements are present in some form. In this text we will try to make clear what we mean when we use the term *element* in a particular case.

4.2   Symbols for the Elements

OBJECTIVE: To learn the symbols of some elements.

The names of the chemical elements have come from many sources. Often an element's name is derived from a Greek, Latin, or German word that describes some property of the element. For example, gold was originally called *aurum,* a Latin word meaning "shining dawn," and lead was known as *plumbum,* which means "heavy." The names for chlorine and iodine come

Trace Elements: Small but Crucial

We all know that certain chemical elements, such as calcium, carbon, nitrogen, phosphorus, and iron, are essential for humans to live. However, many other elements that are present in tiny amounts in the human body are also essential to life. Examples are chromium, cobalt, iodine, manganese, and copper. Chromium assists in the metabolism of sugars, cobalt is present in vitamin B_{12}, iodine is necessary for the proper functioning of the thyroid gland, manganese appears to play a role in maintaining the proper calcium levels in bones, and copper is involved in the production of red blood cells.

It is becoming clear that certain trace elements are very important in determining human behavior. For example, lithium (administered as lithium carbonate) has been a miracle drug for some people afflicted with bipolar disorder, a disease that produces oscillatory behavior between inappropriate "highs" and the blackest of depressions. Although its exact function remains unknown, lithium seems to moderate the levels of neurotransmitters (compounds that are essential to nerve function), thus relieving some of the extreme emotions in sufferers of bipolar disorder.

In addition, a chemist named William Walsh has done some very interesting studies on the inmates of Stateville Prison in Illinois. By analyzing the trace elements in the hair of prisoners, he has found intriguing relationships between the behavior of the inmates and their trace element profile. For example, Walsh found an inverse relationship between the level of cobalt in the prisoner's body and the degree of violence in his behavior.

Besides the levels of trace elements in our bodies, the various substances in the water, the food we consume, and the air we breathe also are of great importance to our health. For example, many scientists are concerned about our exposure to aluminum, through aluminum compounds used in water purification, baked goods and cheese (sodium aluminum phosphate acts as a leavening agent and also is added to cheese to make it softer and easier to melt), and the aluminum that dissolves from our cookware and utensils. The effects of exposure to low levels of aluminum on humans are not presently clear, but there are some indications that we should limit our intake of this element.

Another example of low-level exposure to an element is the fluoride placed in many water supplies and toothpastes to control tooth decay by making tooth enamel more resistant to dissolving. However, the exposure of large numbers of people to fluoride is quite controversial—many people think it is harmful.

The chemistry of trace elements is fascinating and important. Keep your eye on the news for further developments.

---

from Greek words describing their colors, and the name for bromine comes from a Greek word meaning "stench." In addition, it is very common for an element to be named for the place where it was discovered. You can guess where the elements francium, germanium, californium,* and americium* were first found. Some of the heaviest elements are named after famous scientists—for example, einsteinium* and nobelium.*

We often use abbreviations to simplify the written word. For example, it is much easier to put MA on an envelope than to write out Massachusetts, and we often write USA instead of United States of America. Likewise, chemists have invented a set of abbreviations or **element symbols** for the chemical elements. These symbols usually consist of the first letter or the

---

*These elements are made artificially. They do not occur naturally.

first two letters of the element names. The first letter is always capitalized, and the second is not. Examples include

fluorine	F	neon	Ne
oxygen	O	silicon	Si
carbon	C		

Sometimes, however, the two letters used are not the first two letters in the name. For example,

| zinc | Zn | cadmium | Cd |
| chlorine | Cl | platinum | Pt |

The symbols for some other elements are based on the original Latin or Greek name.

Current Name	Original Name	Symbol
gold	aurum	Au
lead	plumbum	Pb
sodium	natrium	Na
iron	ferrum	Fe

A list of the most common elements and their symbols is given in Table 4.3. You can also see the elements represented on a table in the inside front cover of this text. We will explain the form of this table (which is called the periodic table) in later chapters.

Walter Urie/West Light/Corbis

Various forms of the element gold.

Table 4.3 The Names and Symbols of the Most Common Elements

Element	Symbol	Element	Symbol
aluminum	Al	lithium	Li
antimony (stibium)*	Sb	magnesium	Mg
argon	Ar	manganese	Mn
arsenic	As	mercury (hydrargyrum)	Hg
barium	Ba	neon	Ne
bismuth	Bi	nickel	Ni
boron	B	nitrogen	N
bromine	Br	oxygen	O
cadmium	Cd	phosphorus	P
calcium	Ca	platinum	Pt
carbon	C	potassium (kalium)	K
chlorine	Cl	radium	Ra
chromium	Cr	silicon	Si
cobalt	Co	silver (argentium)	Ag
copper (cuprum)	Cu	sodium (natrium)	Na
fluorine	F	strontium	Sr
gold (aurum)	Au	sulfur	S
helium	He	tin (stannum)	Sn
hydrogen	H	titanium	Ti
iodine	I	tungsten (wolfram)	W
iron (ferrum)	Fe	uranium	U
lead (plumbum)	Pb	zinc	Zn

*Where appropriate, the original name is shown in parentheses so that you can see where some of the symbols came from.

4.3 Dalton's Atomic Theory

OBJECTIVES: To learn about Dalton's theory of atoms. • To understand and illustrate the law of constant composition.

Figure 4.1

John Dalton (1766–1844) was an English scientist who made his living as a teacher in Manchester. Although Dalton is best known for his atomic theory, he made contributions in many other areas, including meteorology (he recorded daily weather conditions for 46 years, producing a total of 200,000 data entries). A rather shy man, Dalton was colorblind to red (a special handicap for a chemist) and suffered from lead poisoning contracted from drinking stout (strong beer or ale) that had been drawn through lead pipes.

Figure 4.2

Dalton pictured compounds as collections of atoms. Here NO, NO_2, and N_2O are represented. Note that the number of atoms of each type in a molecule is given by a subscript, except that the number 1 is always assumed and never written.

As scientists of the eighteenth century studied the nature of materials, several things became clear:

1. Most natural materials are mixtures of pure substances.

2. Pure substances are either elements or combinations of elements called compounds.

3. A given compound always contains the same proportions (by mass) of the elements. For example, water *always* contains 8 g of oxygen for every 1 g of hydrogen, and carbon dioxide *always* contains 2.7 g of oxygen for every 1 g of carbon. This principle became known as the **law of constant composition.** It means that a given compound always has the same composition, regardless of where it comes from.

John Dalton (Figure 4.1), an English scientist and teacher, was aware of these observations, and in about 1808 he offered an explanation for them that became known as **Dalton's atomic theory.** The main ideas of this theory (model) can be stated as follows:

Dalton's Atomic Theory

1. Elements are made of tiny particles called **atoms.**
2. All atoms of a given element are identical.
3. The atoms of a given element are different from those of any other element.
4. Atoms of one element can combine with atoms of other elements to form compounds. A given compound always has the same relative numbers and types of atoms.
5. Atoms are indivisible in chemical processes. That is, atoms are not created or destroyed in chemical reactions. A chemical reaction simply changes the way the atoms are grouped together.

Dalton's model successfully explained important observations such as the law of constant composition. This law makes sense because if a compound always contains the same relative numbers of atoms, it will always contain the same proportions by mass of the various elements.

Like most new ideas, Dalton's model was not accepted immediately. However, Dalton was convinced he was right and *used his model to predict* how a given pair of elements might combine to form more than one compound. For example, nitrogen and oxygen might form a compound containing one atom of nitrogen and one atom of oxygen (written NO), a compound containing two atoms of nitrogen and one atom of oxygen (written N_2O), a compound containing one atom of nitrogen and two atoms of oxygen (written NO_2), and so on (Figure 4.2). When the existence of these substances was verified, it was a triumph for Dalton's model. Because Dalton was able to predict correctly the formation of multiple compounds between two elements, his atomic theory became widely accepted.

No Laughing Matter

Sometimes solving one problem leads to another. One such example involves the catalytic converters now required on all automobiles sold around much of the world. The purpose of these converters is to remove harmful pollutants such as CO and NO_2 from automobile exhausts. The good news is that these devices are quite effective and have led to much cleaner air in congested areas. The bad news is that these devices produce significant amounts of nitrous oxide, N_2O, commonly known as laughing gas because when inhaled it produces relaxation and mild inebriation. It was long used by dentists to make their patients more tolerant of some painful dental procedures.

The problem with N_2O is not that it is an air pollutant but that it is a "greenhouse gas." Certain molecules, such as CO_2, CH_4, N_2O, and others, strongly absorb infrared light ("heat radiation"), which causes the earth's atmosphere to retain more of its heat energy. Human activities have significantly increased the concentrations of these gases in the atmosphere. Mounting evidence suggests that the earth is warming as a result, leading to possible dramatic climatic changes.

A recent study by the Environmental Protection Agency (EPA) indicates that N_2O now accounts for over 7% of the greenhouse gases in the atmosphere and that automobiles equipped with catalytic converters produce nearly half of this N_2O. Ironically, N_2O is not regulated, because the Clean Air Act of 1970 was written to control smog—not greenhouse gases. The United States and other industrialized nations are now negotiating to find ways to control global warming, but no agreement is in place.

The N_2O situation illustrates just how complex environmental issues are. Clean may not necessarily be "green."

4.4 Formulas of Compounds

OBJECTIVE: To learn how a formula describes a compound's composition.

A **compound** is a distinct substance that is composed of the atoms of two or more elements and always contains exactly the same relative masses of those elements. In light of Dalton's atomic theory, this simply means that a compound always contains the same relative *numbers* of atoms of each element. For example, water always contains two hydrogen atoms for each oxygen atom.

> Here, *relative* refers to ratios.

The types of atoms and the number of each type in each unit (molecule) of a given compound are conveniently expressed by a **chemical formula.** In a chemical formula the atoms are indicated by the element symbols, and the number of each type of atom is indicated by a subscript, a number that appears to the right of and below the symbol for the element. The formula for water is written H_2O, indicating that each molecule of water contains two atoms of hydrogen and one atom of oxygen (the subscript 1 is always understood and not written). Following are some general rules for writing formulas:

Rules for Writing Formulas

1. Each atom present is represented by its element symbol.
2. The number of each type of atom is indicated by a subscript written to the right of the element symbol.
3. When only one atom of a given type is present, the subscript 1 is not written.

EXAMPLE 4.1 | **Writing Formulas of Compounds**

Write the formula for each of the following compounds, listing the elements in the order given.

a. Each molecule of a compound that has been implicated in the formation of acid rain contains one atom of sulfur and three atoms of oxygen.

b. Each molecule of a certain compound contains two atoms of nitrogen and five atoms of oxygen.

c. Each molecule of glucose, a type of sugar, contains six atoms of carbon, twelve atoms of hydrogen, and six atoms of oxygen.

SOLUTION

a. Symbol for sulfur ——————— Symbol for oxygen

$$SO_3$$

One atom of sulfur ——— Three atoms of oxygen

b. Symbol for nitrogen ——————— Symbol for oxygen

$$N_2O_5$$

Two atoms of nitrogen ——— Five atoms of oxygen

c.   Symbol for hydrogen

Symbol for carbon ——————— Symbol for oxygen

$$C_6H_{12}O_6$$

Six atoms of carbon ——— Six atoms of oxygen

Twelve atoms of hydrogen

Self-Check **EXERCISE 4.1** Write the formula for each of the following compounds, listing the elements in the order given.

a. A molecule contains four phosphorus atoms and ten oxygen atoms.

b. A molecule contains one uranium atom and six fluorine atoms.

c. A molecule contains one aluminum atom and three chlorine atoms.

See Problems 4.19 and 4.20. ∎

4.5 The Structure of the Atom

OBJECTIVES: To learn about the internal parts of an atom. • To understand Rutherford's experiment to characterize the atom's structure.

Dalton's atomic theory, proposed in about 1808, provided such a convincing explanation for the composition of compounds that it became generally accepted. Scientists came to believe that *elements consist of atoms* and that *compounds are a specific collection of atoms* bound together in some way. But what is an atom like? It might be a tiny ball of matter that is the same throughout with no internal structure—like a ball bearing. Or the atom

Figure 4.3

One of the early models of the atom was the plum pudding model, in which the electrons were pictured as embedded in a positively charged spherical cloud, much as raisins are distributed in an old-fashioned plum pudding.

Some historians credit J. J. Thomson for the plum pudding model.

might be composed of parts—it might be made up of a number of subatomic particles. But if the atom contains parts, there should be some way to break up the atom into its components.

Many scientists pondered the nature of the atom during the 1800s, but it was not until almost 1900 that convincing evidence became available that the atom has a number of different parts.

A physicist in England named J. J. Thomson showed in the late 1890s that the atoms of any element can be made to emit tiny negative particles. (He knew the particles had a negative charge because he could show that they were repelled by the negative part of an electric field.) Thus he concluded that all types of atoms must contain these negative particles, which are now called **electrons.**

On the basis of his results, Thomson wondered what an atom must be like. Although he knew that atoms contain these tiny negative particles, he also knew that whole atoms are not negatively *or* positively charged. Thus he concluded that the atom must also contain positive particles that balance exactly the negative charge carried by the electrons, giving the atom a zero overall charge.

Another scientist pondering the structure of the atom was William Thomson (better known as Lord Kelvin and no relation to J. J. Thomson). Lord Kelvin got the idea (which might have occurred to him during dinner) that the atom might be something like plum pudding (a pudding with raisins randomly distributed throughout). Kelvin reasoned that the atom might be thought of as a uniform "pudding" of positive charge with enough negative electrons scattered within to counterbalance that positive charge (see Figure 4.3). Thus the plum pudding model of the atom came into being.

If you had taken this course in 1910, the plum pudding model would have been the only picture of the atom described. However, our ideas about the atom were changed dramatically in 1911 by a physicist named Ernest Rutherford (Figure 4.4), who learned physics in J. J. Thomson's laboratory in the late 1890s. By 1911 Rutherford had become a distinguished scientist with many important discoveries to his credit. One of his main areas of interest involved alpha particles (α particles), positively charged particles with a mass approximately 7500 times that of an electron. In studying the flight of these particles through air, Rutherford found that some of the α particles were deflected by something in the air. Puzzled by this, he designed an experiment that involved directing α particles toward a thin metal foil. Surrounding the foil was a detector coated with a substance that produced tiny

Figure 4.4

Ernest Rutherford (1871–1937) was born on a farm in New Zealand. In 1895 he placed second in a scholarship competition to attend Cambridge University but was awarded the scholarship when the winner decided to stay home and get married. Rutherford was an intense, hard-driving person who became a master at designing just the right experiment to test a given idea. He was awarded the Nobel Prize in chemistry in 1908.

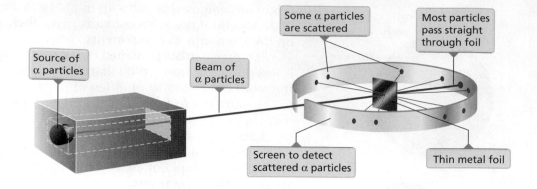

Figure 4.5

Rutherford's experiment on α-particle bombardment of metal foil.

flashes wherever it was hit by an α particle (Figure 4.5). The results of the experiment were very different from those Rutherford anticipated. Although most of the α particles passed straight through the foil, some of the particles were deflected at large angles, as shown in Figure 4.5, and some were reflected backward.

This outcome was a great surprise to Rutherford. (He described this result as comparable to shooting a gun at a piece of paper and having the bullet bounce back.) Rutherford knew that if the plum pudding model of the atom was correct, the massive α particles would crash through the thin foil like cannonballs through paper (as shown in Figure 4.6a). So he expected the α particles to travel through the foil experiencing, at most, very minor deflections of their paths.

Rutherford concluded from these results that the plum pudding model for the atom could not be correct. The large deflections of the α particles could be caused only by a center of concentrated positive charge that would repel the positively charged α particles, as illustrated in Figure 4.6b. Most of the α particles passed directly through the foil because the atom is mostly open space. The deflected α particles were those that had a "close encounter" with the positive center of the atom, and the few reflected α particles were those that scored a "direct hit" on the positive center. In Rutherford's mind these results could be explained only in terms of a **nuclear atom**—an atom with a dense center of positive charge (the **nucleus**) around which tiny electrons moved in a space that was otherwise empty.

He concluded that the nucleus must have a positive charge to balance the negative charge of the electrons and that it must be small and dense.

One of Rutherford's coworkers in this experiment was an undergraduate named Ernest Marsden who, like Rutherford, was from New Zealand.

Figure 4.6

(a) The results that the metal foil experiment would have yielded if the plum pudding model had been correct.

(b) Actual results of Rutherford's experiment.

What was it made of? By 1919 Rutherford concluded that the nucleus of an atom contained what he called protons. A **proton** has the same magnitude (size) of charge as the electron, but its charge is *positive*. We say that the proton has a charge of $1+$ and the electron a charge of $1-$.

Rutherford reasoned that the hydrogen atom has a single proton at its center and one electron moving through space at a relatively large distance from the proton (the hydrogen nucleus). He also reasoned that other atoms must have nuclei (the plural of *nucleus*) composed of many protons bound together somehow. In addition, Rutherford and a coworker, James Chadwick, were able to show in 1932 that most nuclei also contain a neutral particle that they named the **neutron.** A neutron is slightly more massive than a proton but has no charge.

> If the atom were expanded to the size of a huge stadium, the nucleus would be only about as big as a fly at the center.

4.6 Introduction to the Modern Concept of Atomic Structure

OBJECTIVE: To understand some important features of subatomic particles.

> In this model the atom is called a nuclear atom because the positive charge is localized in a small, compact structure (the nucleus) and not spread out uniformly, as in the plum pudding view.

In the years since Thomson and Rutherford, a great deal has been learned about atomic structure. The simplest view of the atom is that it consists of a tiny nucleus (about 10^{-13} cm in diameter) and electrons that move about the nucleus at an average distance of about 10^{-8} cm from it (Figure 4.7). To visualize how small the nucleus is compared with the size of the atom, consider that if the nucleus were the size of a grape, the electrons would be about one *mile* away on average. The nucleus contains protons, which have a positive charge equal in magnitude to the electron's negative charge, and neutrons, which have almost the same mass as a proton but no charge. The neutrons' function in the nucleus is not obvious. They may help hold the protons (which repel each other) together to form the nucleus, but we will not be concerned with that here. The relative masses and charges of the electron, proton, and neutron are shown in Table 4.4.

> The *chemistry* of an atom arises from its electrons.

An important question arises at this point: *"If all atoms are composed of these same components, why do different atoms have different chemical properties?"* The answer lies in the number and arrangement of the electrons. The space in which the electrons move accounts for most of the atomic volume. The electrons are the parts of atoms that "intermingle" when atoms combine to form molecules. Therefore, the number of electrons a given atom possesses greatly affects the way it can interact with other atoms. As a result, atoms of different elements, which have different numbers of electrons, show different chemical behavior. Although the atoms of different elements also differ in their numbers of protons, it is the number of electrons that really determines chemical behavior. We will discuss how this happens in later chapters.

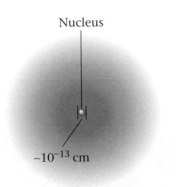

Nucleus

$\sim 10^{-13}$ cm

$\longleftarrow \sim 10^{-8}$ cm $\longrightarrow$

Figure 4.7

A nuclear atom viewed in cross section. (The symbol $\sim$ means approximately.) This drawing does not show the actual scale. The nucleus is actually *much* smaller compared with the size of an atom.

Table 4.4	The Mass and Charge of the Electron, Proton, and Neutron	
Particle	Relative Mass*	Relative Charge
electron	1	$1-$
proton	1836	$1+$
neutron	1839	none

*The electron is arbitrarily assigned a mass of 1 for comparison.

4.7   Isotopes

OBJECTIVES: To learn about the terms isotope, atomic number, and mass number. • To understand the use of the symbol $^A_Z X$ to describe a given atom.

We have seen that an atom has a nucleus with a positive charge due to its protons and has electrons in the space surrounding the nucleus at relatively large distances from it.

As an example, consider a sodium atom, which has 11 protons in its nucleus. Because an atom has no overall charge, the number of electrons must equal the number of protons. Therefore, a sodium atom has 11 electrons in the space around its nucleus. It is *always* true that a sodium atom has 11 protons and 11 electrons. However, each sodium atom also has neutrons in its nucleus, and different types of sodium atoms exist that have different numbers of neutrons.

When Dalton stated his atomic theory in the early 1800s, he assumed all of the atoms of a given element were identical. This idea persisted for over a hundred years, until James Chadwick discovered that the nuclei of most atoms contain neutrons as well as protons. (This is a good example of how a theory changes as new observations are made.) After the discovery of the neutron, Dalton's statement that all atoms of a given element are identical had to be changed to "All atoms of the same element contain the same number of protons and electrons, but atoms of a given element may have different numbers of neutrons."

To illustrate this idea, consider the sodium atoms represented in Figure 4.8. These atoms are **isotopes,** or *atoms with the same number of protons but different numbers of neutrons.* The number of protons in a nucleus is called the atom's **atomic number.** The *sum* of the number of neutrons and the number of protons in a given nucleus is called the atom's **mass number.** To specify which of the isotopes of an element we are talking about, we use the symbol

$$^A_Z X$$

where

X = the symbol of the element
A = the mass number (number of protons and neutrons)
Z = the atomic number (number of protons)

> All atoms of the same element have the same number of protons (the element's atomic number) and the same number of electrons.

> In a free atom, the positive and negative charges always balance to yield a net zero charge.

> Atomic number: the number of protons. Mass number: the sum of protons and neutrons.

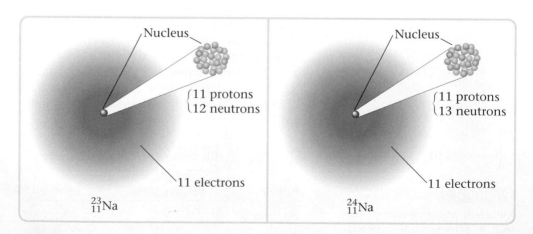

Nucleus
11 protons
12 neutrons
11 electrons
$^{23}_{11}\text{Na}$

Nucleus
11 protons
13 neutrons
11 electrons
$^{24}_{11}\text{Na}$

Figure 4.8

Two isotopes of sodium. Both have 11 protons and 11 electrons, but they differ in the number of neutrons in their nuclei.

"Whair" Do You Live?

Picture a person who has been the victim of a crime in a large city in the eastern United States. The person has been hit in the head and, as a result, has total amnesia. The person's ID has been stolen, but the authorities suspect he may not be from the local area. Is there any way to find out where the person might be from? The answer is yes. Recent research indicates that the relative amounts of the isotopes of hydrogen and oxygen in a person's hair indicate in which part of the United States a person lives.

Support for this idea has come from a recent study by James Ehleringer, a chemist at the University of Utah in Salt Lake City. Noting that the concentrations of hydrogen-2 (deuterium) and oxygen-18 in drinking water vary significantly from region to region in the United States (see accompanying illustration), Ehleringer and his colleagues collected hair samples from barbershops in 65 cities and 18 states. Their analyses showed that 86% of the variations in the hydrogen and oxygen isotopes in the hair samples result from the isotopic composition of the local water. Based on their results, the group was able to develop estimates of the isotopic signature of peoples' hair from various regions of the country. Although this method cannot be used to pinpoint a person's place of residence, it can give a general region. This method might be helpful for the amnesia victim described above by showing where to look for his family. His picture could be shown on TV in the region indicated by analysis of his hair. Another possible use of this technique is identifying the country of origin of victims of a natural disaster in a tourist region with visitors from all over the world. In fact, a similar technique was used to specify the countries of origin of the victims of the tsunami that devastated southern Asia in December 2004.

An interesting verification of this technique occurred when the researchers examined a strand of hair from a person who had recently moved from Beijing, China, to Salt Lake City. Analysis of various parts of the hair showed a distinct change in isotopic distribution corresponding to his change of residence. Thus the isotopes of elements can provide useful information in unexpected ways.

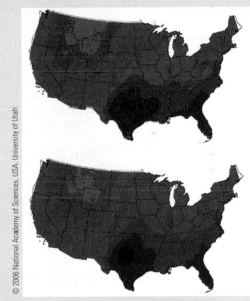

© 2006 National Academy of Sciences, USA, University of Utah

Maps of predicted concentrations of hydrogen-2 (top) and oxygen-18 (bottom). Red represents the highest concentration, and blue represents the lowest concentration of each isotope.

For example, the symbol for one particular type of sodium atom is written

$$^{23}_{11}\text{Na}$$

Mass number (number of protons and neutrons) ↓

← Element symbol

↑ Atomic number (number of protons)

The particular atom represented here is called sodium-23, because it has a mass number of 23. Let's specify the number of each type of subatomic particle. From the atomic number 11 we know that the nucleus contains

11 protons. And because the number of electrons is equal to the number of protons, we know that this atom contains 11 electrons. How many neutrons are present? We can calculate the number of neutrons from the definition of the mass number

$$\text{Mass number} = \text{number of protons} + \text{number of neutrons}$$

or, in symbols,

$$A = Z + \text{number of neutrons}$$

We can isolate (solve for) the number of neutrons by subtracting Z from both sides of the equation

$$A - Z = Z - Z + \text{number of neutrons}$$
$$A - Z = \text{number of neutrons}$$

This is a general result. You can always determine the number of neutrons present in a given atom by subtracting the atomic number from the mass number. In this case ($^{23}_{11}\text{Na}$), we know that $A = 23$ and $Z = 11$. Thus

$$A - Z = 23 - 11 = 12 = \text{number of neutrons}$$

In summary, sodium-23 has 11 electrons, 11 protons, and 12 neutrons.

EXAMPLE 4.2 | **Interpreting Symbols for Isotopes**

In nature, elements are usually found as a mixture of isotopes. Three isotopes of elemental carbon are $^{12}_{6}\text{C}$ (carbon-12), $^{13}_{6}\text{C}$ (carbon-13), and $^{14}_{6}\text{C}$ (carbon-14). Determine the number of each of the three types of subatomic particles in each of these carbon atoms.

SOLUTION

The number of protons and electrons is the same in each of the isotopes and is given by the atomic number of carbon, 6. The number of neutrons can be determined by subtracting the atomic number (Z) from the mass number (A):

$$A - Z = \text{number of neutrons}$$

The numbers of neutrons in the three isotopes of carbon are

$$^{12}_{6}\text{C: number of neutrons} = A - Z = 12 - 6 = 6$$
$$^{13}_{6}\text{C: number of neutrons} = 13 - 6 = 7$$
$$^{14}_{6}\text{C: number of neutrons} = 14 - 6 = 8$$

In summary,

Symbol	Number of Protons	Number of Electrons	Number of Neutrons
$^{12}_{6}\text{C}$	6	6	6
$^{13}_{6}\text{C}$	6	6	7
$^{14}_{6}\text{C}$	6	6	8

Self-Check **EXERCISE 4.2** Give the number of protons, neutrons, and electrons in the atom symbolized by $^{90}_{38}\text{Sr}$. Strontium-90 occurs in fallout from nuclear testing. It can accumulate in bone marrow and may cause leukemia and bone cancer.

See Problems 4.39 and 4.42. ■

Isotope Tales

The atoms of a given element typically consist of several isotopes—atoms with the same number of protons but different numbers of neutrons. It turns out that the ratio of isotopes found in nature can be very useful in natural detective work. One reason is that the ratio of isotopes of elements found in living animals and humans reflects their diets. For example, African elephants that feed on grasses have a different $^{13}C/^{12}C$ ratio in their tissues than elephants that primarily eat tree leaves. This difference arises because grasses have a different growth pattern than leaves do, resulting in different amounts of ^{13}C and ^{12}C being incorporated from the CO_2 in the air. Because leaf-eating and grass-eating elephants live in different areas of Africa, the observed differences in the $^{13}C/^{12}C$ isotope ratios in elephant ivory samples have enabled authorities to identify the sources of illegal samples of ivory.

Another case of isotope detective work involves the tomb of King Midas, who ruled the kingdom Phyrgia in the eighth century B.C. Analysis of nitrogen isotopes in the king's decayed casket has revealed details about the king's diet. Scientists have learned that the $^{15}N/^{14}N$ ratios of carnivores are higher than those of herbivores, which in turn are higher than those of plants. It turns out that the organism responsible for decay of the king's wooden casket has an unusually large requirement for nitrogen. The source of this nitrogen was the body of the dead king. Because the decayed wood under his now-decomposed body showed a high $^{15}N/^{14}N$ ratio, researchers feel sure that the king's diet was rich in meat.

A third case of historical isotope detective work concerns the Pueblo ancestor people (commonly called the Anasazi), who lived in what is now northwestern New Mexico between A.D. 900 and 1150. The center of their civilization, Chaco Canyon, was a thriving cultural center boasting dwellings made of hand-hewn sandstone and more than 200,000 logs. The sources of the logs have always been controversial. Many theories have been advanced concerning the distances over which the logs were hauled. Recent research by Nathan B. English, a geochemist at the University of Arizona in Tucson, has used the distribution of strontium isotopes in the wood to identify the probable sources of the logs. This effort has enabled scientists to understand more clearly the Anasazi building practices.

These stories illustrate how isotopes can serve as valuable sources of biologic and historical information.

Ancient Anasazi Indian cliff dwellings.

Paul Chesley/National Geographic/Getty Images

Self-Check **EXERCISE 4.3**   Give the number of protons, neutrons, and electrons in the atom symbolized by $^{201}_{80}Hg$.

See Problems 4.39 and 4.42. ■

EXAMPLE 4.3   Writing Symbols for Isotopes

Write the symbol for the magnesium atom (atomic number 12) with a mass number of 24. How many electrons and how many neutrons does this atom have?

Magnesium burns in air to give a bright white flame.

SOLUTION

The atomic number 12 means the atom has 12 protons. The element magnesium is symbolized by Mg. The atom is represented as

$$^{24}_{12}\text{Mg}$$

and is called magnesium-24. Because the atom has 12 protons, it must also have 12 electrons. The mass number gives the total number of protons and neutrons, which means that this atom has 12 neutrons ($24 - 12 = 12$). ■

EXAMPLE 4.4 | Calculating Mass Number

Write the symbol for the silver atom ($Z = 47$) that has 61 neutrons.

SOLUTION

The element symbol is ^A_ZAg, where we know that $Z = 47$. We can find A from its definition, $A = Z +$ number of neutrons. In this case,

$$A = 47 + 61 = 108$$

The complete symbol for this atom is $^{108}_{47}\text{Ag}$.

Self-Check   **EXERCISE 4.4**   Give the symbol for the phosphorus atom ($Z = 15$) that contains 17 neutrons.

See Problem 4.42. ■

4.8 Introduction to the Periodic Table

OBJECTIVES: To learn about various features of the periodic table. • To learn some of the properties of metals, nonmetals, and metalloids.

In any room where chemistry is taught or practiced, you are almost certain to find a chart called the **periodic table** hanging on the wall. This chart shows all of the known elements and gives a good deal of information about each. As our study of chemistry progresses, the usefulness of the periodic table will become more obvious. This section will simply introduce it.

A simple version of the periodic table is shown in Figure 4.9. Note that each box of this table contains a number written over one, two, or three letters. The letters are the symbols for the elements. The number shown above each symbol is the atomic number (the number of protons and also the number of electrons) for that element. For example, carbon (C) has atomic number 6:

| 6 |
| C |

Figure 4.9

The periodic table.

Lead (Pb) has atomic number 82:

82
Pb

Notice that elements 112 through 115 and 118 have unusual three-letter designations beginning with U. These are abbreviations for the systematic names of the atomic numbers of these elements. "Regular" names for these elements will be chosen eventually by the scientific community.

Note that the elements are listed on the periodic table in order of increasing atomic number. They are also arranged in specific horizontal rows and vertical columns. The elements were first arranged in this way in 1869 by Dmitri Mendeleev, a Russian scientist. Mendeleev arranged the elements in this way because of similarities in the chemical properties of various "families" of elements. For example, fluorine and chlorine are reactive gases that form similar compounds. It was also known that sodium and potassium behave very similarly. Thus the name *periodic table* refers to the fact that as we increase the atomic numbers, every so often an element occurs with

Mendeleev actually arranged the elements in order of increasing atomic mass rather than atomic number.

properties similar to those of an earlier (lower-atomic-number) element. For example, the elements

9 F
17 Cl
35 Br
53 I
85 At

all show similar chemical behavior and so are listed vertically, as a "family" of elements.

These families of elements with similar chemical properties that lie in the same vertical column on the periodic table are called **groups.** Groups are often referred to by the number over the column (see Figure 4.9). Note that the group numbers are accompanied by the letter A on the periodic table in Figure 4.9 and the one inside the front cover of the text. For simplicity we will delete the A's when we refer to groups in the text. Many of the groups have special names. For example, the first column of elements (Group 1) has the name **alkali metals.** The Group 2 elements are called the **alkaline earth metals,** the Group 7 elements are the **halogens,** and the elements in Group 8 are called the **noble gases.** A large collection of elements that spans many vertical columns consists of the **transition metals.**

Most of the elements are **metals.** Metals have the following characteristic physical properties:

Physical Properties of Metals

1. Efficient conduction of heat and electricity
2. Malleability (they can be hammered into thin sheets)
3. Ductility (they can be pulled into wires)
4. A lustrous (shiny) appearance

For example, copper is a typical metal. It is lustrous (although it tarnishes readily); it is an excellent conductor of electricity (it is widely used in electrical wires); and it is readily formed into various shapes, such as pipes for water systems. Copper is one of the transition metals—the metals shown in the center of the periodic table. Iron, aluminum, and gold are other familiar elements that have metallic properties. All of the elements shown to the left of and below the heavy "stair-step" black line in Figure 4.9 are classified as metals, except for hydrogen (Figure 4.10).

The relatively small number of elements that appear in the upper-right corner of the periodic table (to the right of the heavy line in Figures 4.9 and 4.10) are called **nonmetals.** Nonmetals generally lack those properties that characterize metals and show much more variation in their properties than metals do. Whereas almost all metals are solids at normal temperatures,

Figure 4.10

The elements classified as metals and as nonmetals.

many nonmetals (such as nitrogen, oxygen, chlorine, and neon) are gaseous and one (bromine) is a liquid. Several nonmetals (such as carbon, phosphorus, and sulfur) are also solids.

The elements that lie close to the "stair-step" line as shown in blue in Figure 4.10 often show a mixture of metallic and nonmetallic properties. These elements, which are called **metalloids** or **semimetals,** include silicon, germanium, arsenic, antimony, and tellurium.

As we continue our study of chemistry, we will see that the periodic table is a valuable tool for organizing accumulated knowledge and that it helps us predict the properties we expect a given element to exhibit. We will also develop a model for atomic structure that will explain why there are groups of elements with similar chemical properties.

EXAMPLE 4.5 | Interpreting the Periodic Table

For each of the following elements, use the periodic table in the front of the book to give the symbol and atomic number and to specify whether the element is a metal or a nonmetal. Also give the named family to which the element belongs (if any).

a. iodine   b. magnesium   c. gold   d. lithium

SOLUTION

Indonesian men carrying chunks of elemental sulfur in baskets.

a. Iodine (symbol I) is element 53 (its atomic number is 53). Iodine lies to the right of the stair-step line in Figure 4.10 and thus is a nonmetal. Iodine is a member of Group 7, the family of halogens.

b. Magnesium (symbol Mg) is element 12 (atomic number 12). Magnesium is a metal and is a member of the alkaline earth metal family (Group 2).

c. Gold (symbol Au) is element 79 (atomic number 79). Gold is a metal and is not a member of a named vertical family. It is classed as a transition metal.

d. Lithium (symbol Li) is element 3 (atomic number 3). Lithium is a metal in the alkali metal family (Group 1).

Self-Check   **EXERCISE 4.5**   Give the symbol and atomic number for each of the following elements. Also indicate whether each element is a metal or a nonmetal and whether it is a member of a named family.

a. argon   b. chlorine   c. barium   d. cesium

See Problems 4.53 and 4.54. ∎

Putting the Brakes on Arsenic

The toxicity of arsenic is well known. Indeed, arsenic has often been the poison of choice in classic plays and films—rent *Arsenic and Old Lace* sometime. Contrary to its treatment in the aforementioned movie, arsenic poisoning is a serious, contemporary problem. For example, the World Health Organization estimates that 77 million people in Bangladesh are at risk from drinking water that contains large amounts of naturally occurring arsenic. Recently, the Environmental Protection Agency announced more stringent standards for arsenic in U.S. public drinking water supplies. Studies show that prolonged exposure to arsenic can lead to a higher risk of bladder, lung, and skin cancers as well as other ailments, although the levels of arsenic that induce these symptoms remain in dispute in the scientific community.

Cleaning up arsenic-contaminated soil and water poses a significant problem. One approach is to find plants that will leach arsenic from the soil. Such a plant, the brake fern, recently has been shown to have a voracious appetite for arsenic. Research led by Lenna Ma, a chemist at the University of Florida in Gainesville, has shown that the brake fern accumulates arsenic at a rate 200 times that of the average plant. The arsenic, which becomes concentrated in fronds that grow up to 5 feet long, can be easily harvested and hauled away. Researchers are now investigating the best way to dispose of the plants so the arsenic can be isolated. The fern (*Pteris vittata*) looks promising for putting the brakes on arsenic pollution.

Tara Piasio/IFAS/University of Florida

Lenna Ma and Pteris vittata—*called the brake fern.*

4.9 Natural States of the Elements

OBJECTIVE: To learn the natures of the common elements.

As we have noted, the matter around us consists mainly of mixtures. Most often these mixtures contain compounds, in which atoms from different elements are bound together. Most elements are quite reactive: their atoms tend to combine with those of other elements to form compounds. Thus we do not often find elements in nature in pure form—uncombined with other elements. However, there are notable exceptions. The gold nuggets found at Sutter's Mill in California that launched the Gold Rush in 1849 are virtually pure elemental gold. And platinum and silver are often found in nearly pure form.

Gold, silver, and platinum are members of a class of metals called *noble metals* because they are relatively unreactive. (The term *noble* implies a class set apart.)

Other elements that appear in nature in the uncombined state are the elements in Group 8: helium, neon, argon, krypton, xenon, and radon. Because the atoms of these elements do not combine readily with those of other elements, we call them the *noble gases*. For example, helium gas is found in uncombined form in underground deposits with natural gas.

A gold nugget weighing 13 lb, 7 oz, which came to be called Tom's Baby, was found by Tom Grove near Breckenridge, Colorado, on July 23, 1887.

Recall that a molecule is a collection of atoms that behaves as a unit. Molecules are always electrically neutral (zero charge).

When we take a sample of air (the mixture of gases that constitute the earth's atmosphere) and separate it into its components, we find several pure elements present. One of these is argon. Argon gas consists of a collection of separate argon atoms, as shown in Figure 4.11.

Air also contains nitrogen gas and oxygen gas. When we examine these two gases, however, we find that they do not contain single atoms, as argon does, but instead contain **diatomic molecules:** molecules made up of *two atoms,* as represented in Figure 4.12. In fact, any sample of elemental oxygen gas at normal temperatures contains O_2 molecules. Likewise, nitrogen gas contains N_2 molecules.

Hydrogen is another element that forms diatomic molecules. Though virtually all of the hydrogen found on earth is present in compounds with other elements (such as with oxygen in water), when hydrogen is prepared as a free element it contains diatomic H_2 molecules. For example, an electric current can be used to decompose water (see Figure 4.13 and Figure 3.3 on p. 60) into elemental hydrogen and oxygen containing H_2 and O_2 molecules, respectively.

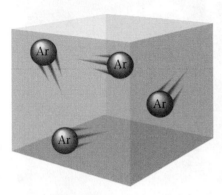

Figure 4.11

Argon gas consists of a collection of separate argon atoms.

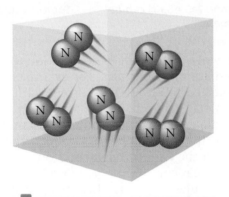

a

Nitrogen gas contains N—N (N_2) molecules.

b

Oxygen gas contains O—O (O_2) molecules.

Figure 4.12

Gaseous nitrogen and oxygen contain diatomic (two-atom) molecules.

Figure 4.13

The decomposition of two water molecules (H_2O) to form two hydrogen molecules (H_2) and one oxygen molecule (O_2). Note that only the grouping of the atoms changes in this process; no atoms are created or destroyed. There must be the same number of H atoms and O atoms before and after the process. Thus the decomposition of two H_2O molecules (containing four H atoms and two O atoms) yields one O_2 molecule (containing two O atoms) and two H_2 molecules (containing a total of four H atoms).

a *Sodium chloride (common table salt) can be decomposed to its elements.*

b *Sodium metal (on the left) and chlorine gas.*

Figure 4.14

Several other elements, in addition to hydrogen, nitrogen, and oxygen, exist as diatomic molecules. For example, when sodium chloride is melted and subjected to an electric current, chlorine gas is produced (along with sodium metal). This chemical change is represented in Figure 4.14. Chlorine gas is a pale green gas that contains Cl_2 molecules.

Chlorine is a member of Group 7, the halogen family. All the elemental forms of the Group 7 elements contain diatomic molecules. Fluorine is a pale yellow gas containing F_2 molecules. Bromine is a brown liquid made up of Br_2 molecules. Iodine is a lustrous, purple solid that contains I_2 molecules.

Table 4.5 lists the elements that contain diatomic molecules in their pure, elemental forms.

So far we have seen that several elements are gaseous in their elemental forms at normal temperatures (~25 °C). The noble gases (the Group 8 elements) contain individual atoms, whereas several other gaseous elements contain diatomic molecules (H_2, N_2, O_2, F_2, and Cl_2).

Only two elements are liquids in their elemental forms at 25 °C: the nonmetal bromine (containing Br_2 molecules) and the metal mercury. The metals gallium and cesium almost qualify in this category; they are solids at 25 °C, but both melt at ~30 °C.

> The only elemental hydrogen found naturally on earth occurs in the exhaust gases of volcanoes.

> ~ means "approximately."

Platinum is a noble metal used in jewelry and in many industrial processes.

Table 4.5	Elements That Exist as Diatomic Molecules in Their Elemental Forms	
Element Present	Elemental State at 25 °C	Molecule
hydrogen	colorless gas	H_2
nitrogen	colorless gas	N_2
oxygen	pale blue gas	O_2
fluorine	pale yellow gas	F_2
chlorine	pale green gas	Cl_2
bromine	reddish brown liquid	Br_2
iodine	lustrous, dark purple solid	I_2

Liquid bromine in a flask with bromine vapor.

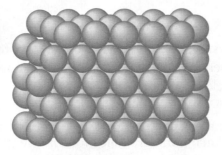

Figure 4.15

In solid metals, the spherical atoms are packed closely together.

Cut diamond, held over coal.

The other elements are solids in their elemental forms at 25 °C. For metals these solids contain large numbers of atoms packed together much like marbles in a jar (see Figure 4.15).

The structures of solid nonmetallic elements are more varied than those of metals. In fact, different forms of the same element often occur. For example, solid carbon occurs in three forms. Different forms of a given element are called *allotropes*. The three allotropes of carbon are the familiar diamond and graphite forms plus a form that has only recently been discovered called *buckminsterfullerene*. These elemental forms have very different properties because of their different structures (see Figure 4.16). Diamond is the hardest natural substance known and is often used for industrial cutting tools. Diamonds are also valued as gemstones. Graphite, by contrast, is a rather soft material useful for writing (pencil "lead" is really graphite) and (in the form of a powder) for lubricating locks. The rather odd name given to buckminsterfullerene comes from the structure of the C_{60} molecules that form this allotrope. The soccer-ball-like structure contains five- and six-member rings reminiscent of the structure of geodesic domes suggested by the late industrial designer Buckminster Fuller. Other "fullerenes" containing molecules with more than 60 carbon atoms have also been discovered, leading to a new area of chemistry.

| Diamond | Graphite | Buckminsterfullerene |

Figure 4.16

The three solid elemental forms of carbon (allotropes): diamond, graphite, and buckminsterfullerene. The representations of diamond and graphite are fragments of much larger structures that extend in all directions from the parts shown here. Buckminsterfullerene contains C_{60} molecules, one of which is shown.

4.10 Ions

OBJECTIVES: To understand the formation of ions from their parent atoms, and learn to name them. • To learn how the periodic table can help predict which ion a given element forms.

goChemistry Module 2: Predicting Ion Charges covers concepts in this section.

We have seen that an atom has a certain number of protons in its nucleus and an equal number of electrons in the space around the nucleus. This results in an exact balance of positive and negative charges. We say that an atom is a neutral entity—it has *zero net charge*.

We can produce a charged entity, called an **ion,** by taking a neutral atom and adding or removing one or more electrons. For example, a sodium atom ($Z = 11$) has eleven protons in its nucleus and eleven electrons outside its nucleus.

11 electrons
(11–)

11+

Neutral sodium
atom (Na)

> An ion has a net positive or negative charge.

If one of the electrons is lost, there will be eleven positive charges but only ten negative charges. This gives an ion with a net positive one (1+) charge: $(11+) + (10-) = 1+$. We can represent this process as follows:

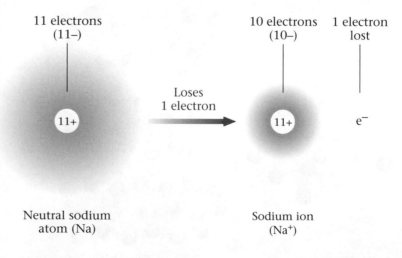

11 electrons
(11–)

11+

Neutral sodium
atom (Na)

Loses
1 electron

10 electrons
(10–)

1 electron
lost

11+   e⁻

Sodium ion
(Na⁺)

or, in shorthand form, as

$$Na \rightarrow Na^+ + e^-$$

where Na represents the neutral sodium atom, Na⁺ represents the 1+ ion formed, and e⁻ represents an electron.

A positive ion, called a **cation** (pronounced *cat' eye on*), is produced when one or more electrons are *lost* from a neutral atom. We have seen that sodium loses one electron to become a 1+ cation. Some atoms lose more than one electron. For example, a magnesium atom typically loses two electrons to form a 2+ cation:

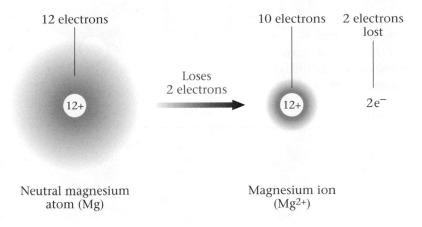

We usually represent this process as follows:

$$Mg \rightarrow Mg^{2+} + 2e^-$$

Aluminum forms a 3+ cation by losing three electrons:

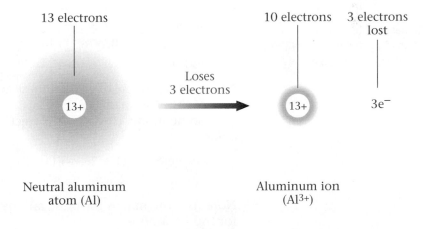

Note the size decreases dramatically when an atom loses one or more electrons to form a positive ion.

or

$$Al \rightarrow Al^{3+} + 3e^-$$

A cation is named using the name of the parent atom. Thus Na^+ is called the sodium ion (or sodium cation), Mg^{2+} is called the magnesium ion (or magnesium cation), and Al^{3+} is called the aluminum ion (or aluminum cation).

When electrons are *gained* by a neutral atom, an ion with a negative charge is formed. A negatively charged ion is called an **anion** (pronounced *an' ion*). An atom that gains one extra electron forms an anion with a

1– charge. An example of an atom that forms a 1– anion is the chlorine atom, which has seventeen protons and seventeen electrons:

Neutral chlorine
atom (Cl)

Chloride ion
(Cl⁻)

> Note the size increases dramatically when an atom gains one or more electrons to form a negative ion.

or

$$Cl + e^- \rightarrow Cl^-$$

Note that the anion formed by chlorine has eighteen electrons but only seventeen protons, so the net charge is (18−) + (17+) = 1−. Unlike a cation, which is named for the parent atom, an anion is named by taking the root name of the atom and changing the ending. For example, the Cl⁻ anion produced from the Cl (chlorine) atom is called the *chloride* ion (or chloride anion). Notice that the word *chloride* is obtained from the root of the atom name (*chlor-*) plus the suffix *-ide*. Other atoms that add one electron to form 1– ions include

> The name of an anion is obtained by adding *-ide* to the root of the atom name.

fluorine	$F + e^- \rightarrow F^-$	(*fluor*ide ion)
bromine	$Br + e^- \rightarrow Br^-$	(*brom*ide ion)
iodine	$I + e^- \rightarrow I^-$	(*iod*ide ion)

Note that the name of each of these anions is obtained by adding *-ide* to the root of the atom name.

Some atoms can add two electrons to form 2– anions. Examples include

oxygen	$O + 2e^- \rightarrow O^{2-}$	(*ox*ide ion)
sulfur	$S + 2e^- \rightarrow S^{2-}$	(*sulf*ide ion)

Note that the names for these anions are derived in the same way as those for the 1– anions.

It is important to recognize that ions are always formed by removing electrons from an atom (to form cations) or adding electrons to an atom (to form anions). *Ions are never formed by changing the number of protons in an atom's nucleus.*

It is essential to understand that isolated atoms do not form ions on their own. Most commonly, ions are formed when metallic elements combine with nonmetallic elements. As we will discuss in detail in Chapter 7, when metals and nonmetals react, the metal atoms tend to lose one or more electrons, which are in turn gained by the atoms of the nonmetal. Thus reactions between metals and nonmetals tend to form compounds that contain metal cations and nonmetal anions. We will have more to say about these compounds in Section 4.11.

Figure 4.17

The ions formed by selected members of Groups 1, 2, 3, 6, and 7.

▶ Ion Charges and the Periodic Table

We find the periodic table very useful when we want to know what type of ion is formed by a given atom. Figure 4.17 shows the types of ions formed by atoms in several of the groups on the periodic table. Note that the Group 1 metals all form 1+ ions (M^+), the Group 2 metals all form 2+ ions (M^{2+}), and the Group 3 metals form 3+ ions (M^{3+}). Thus for Groups 1 through 3 the charges of the cations formed are identical to the group numbers.

> For Groups 1, 2, and 3, the charges of the cations equal the group numbers.

In contrast to the Group 1, 2, and 3 metals, most of the many *transition metals* form cations with various positive charges. For these elements there is no easy way to predict the charge of the cation that will be formed.

Note that metals always form positive ions. This tendency to lose electrons is a fundamental characteristic of metals. Nonmetals, on the other hand, form negative ions by gaining electrons. Note that the Group 7 atoms all gain one electron to form 1− ions and that all the nonmetals in Group 6 gain two electrons to form 2− ions.

At this point you should memorize the relationships between the group number and the type of ion formed, as shown in Figure 4.17. You will understand why these relationships exist after we further discuss the theory of the atom in Chapter 11.

4.11 Compounds That Contain Ions

OBJECTIVE: To learn how ions combine to form neutral compounds.

Chemists have good reasons to believe that many chemical compounds contain ions. For instance, consider some of the properties of common table salt, sodium chloride (NaCl). It must be heated to about 800 °C to melt and to almost 1500 °C to boil (compare to water, which boils at 100 °C). As a solid, salt will not conduct an electric current, but when melted it is a very good conductor. Pure water does not conduct electricity (does not allow an electric current to flow), but when salt is dissolved in water, the resulting solution readily conducts electricity (see Figure 4.18).

> Melting means that the solid, where the ions are locked into place, is changed to a liquid, where the ions can move.

Chemists have come to realize that we can best explain these properties of sodium chloride (NaCl) by picturing it as containing Na^+ ions and Cl^- ions packed together as shown in Figure 4.19. Because the positive and negative charges attract each other very strongly, it must be heated to a very high temperature (800 °C) before it melts.

To explore further the significance of the electrical conductivity results, we need to discuss briefly the nature of electric currents. An electric current

a

Pure water does not conduct a current, so the circuit is not complete and the bulb does not light.

b

Water containing dissolved salt conducts electricity and the bulb lights.

Figure 4.18

A substance containing ions that can move can conduct an electric current.

can travel along a metal wire because *electrons are free to move* through the wire; the moving electrons carry the current. In ionic substances the ions carry the current. Thus substances that contain ions can conduct an electric current *only if the ions can move*—the current travels by the movement of the charged ions. In solid NaCl the ions are tightly held and cannot move, but when the solid is melted and changed to a liquid, the structure is disrupted and the ions can move. As a result, an electric current can travel through the melted salt.

Dissolving NaCl causes the ions to be randomly dispersed in the water, allowing them to move freely. Dissolving is not the same as melting, but both processes free the ions to move.

The same reasoning applies to NaCl dissolved in water. When the solid dissolves, the ions are dispersed throughout the water and can move around in the water, allowing it to conduct a current.

Thus, we recognize substances that contain ions by their characteristic properties. They often have very high melting points, and they conduct an electric current when melted or when dissolved in water.

An ionic compound cannot contain only anions or only cations, because the net charge of a compound must be zero.

Many substances contain ions. In fact, whenever a compound forms between a metal and a nonmetal, it can be expected to contain ions. We call these substances **ionic compounds.**

a

The arrangement of sodium ions (Na$^+$) and chloride ions (Cl$^-$) in the ionic compound sodium chloride.

b

Solid sodium chloride highly magnified.

E. R. Degginger/Color-Pic, Inc.

Figure 4.19

One fact very important to remember is that *a chemical compound must have a net charge of zero*. This means that if a compound contains ions, then

1. Both positive ions (cations) and negative ions (anions) must be present.

2. The numbers of cations and anions must be such that the net charge is zero.

> The net charge of a compound (zero) is the sum of the positive and negative charges.

For example, note that the formula for sodium chloride is written NaCl, indicating one of each type of these elements. This makes sense because sodium chloride contains Na^+ ions and Cl^- ions. Each sodium ion has a 1+ charge and each chloride ion has a 1− charge, so they must occur in equal numbers to give a net charge of zero.

Charge: 1+   Charge: 1−   Net charge: 0

And for *any* ionic compound,

$$\begin{array}{c} \text{Total charge} \\ \text{of cations} \end{array} + \begin{array}{c} \text{Total charge} \\ \text{of anions} \end{array} = \begin{array}{c} \text{Zero} \\ \text{net charge} \end{array}$$

Consider an ionic compound that contains the ions Mg^{2+} and Cl^-. What combination of these ions will give a net charge of zero? To balance the 2+ charge on Mg^{2+}, we will need two Cl^- ions to give a net charge of zero.

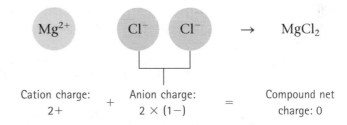

Cation charge: Anion charge: Compound net
2+   +   2 × (1−)   =   charge: 0

This means that the formula of the compound must be $MgCl_2$. Remember that subscripts are used to give the relative numbers of atoms (or ions).

Now consider an ionic compound that contains the ions Ba^{2+} and O^{2-}. What is the correct formula? These ions have charges of the same size (but opposite sign), so they must occur in equal numbers to give a net charge of zero. The formula of the compound is BaO, because (2+) + (2−) = 0.

Similarly, the formula of a compound that contains the ions Li^+ and N^{3-} is Li_3N, because three Li^+ cations are needed to balance the charge of the N^{3-} anion.

Positive charge: Negative charge: Net charge:
3 × (1+)   +   (3−)   =   0

EXAMPLE 4.6 Writing Formulas for Ionic Compounds

The pairs of ions contained in several ionic compounds are listed below. Give the formula for each compound.

a. Ca^{2+} and Cl^- b. Na^+ and S^{2-} c. Ca^{2+} and P^{3-}

> The subscript 1 in a formula is not written.

SOLUTION

a. Ca^{2+} has a 2+ charge, so two Cl^- ions (each with the charge 1−) will be needed.

where   2+   +   2(1−)   = 0

The formula is $CaCl_2$.

b. In this case S^{2-}, with its 2− charge, requires two Na^+ ions to produce a zero net charge.

where   2(1+)   +   2−   = 0

The formula is Na_2S.

c. We have the ions Ca^{2+} (charge 2+) and P^{3-} (charge 3−). We must figure out how many of each are needed to balance exactly the positive and negative charges. Let's try two Ca^{2+} and one P^{3-}.

The resulting net charge is 2(2+) + (3−) = (4+) + (3−) = 1−. This doesn't work because the net charge is not zero. We can obtain the same total positive and total negative charges by having three Ca^{2+} ions and two P^{3-} ions.

where   3(2+)   +   2(3−)   = 0

Thus the formula must be Ca_3P_2.

Self-Check **EXERCISE 4.6** Give the formulas for the compounds that contain the following pairs of ions.

a. K^+ and I^- b. Mg^{2+} and N^{3-} c. Al^{3+} and O^{2-}

See Problems 4.83 and 4.84. ■

CHAPTER 4 REVIEW

Key Terms

element symbols (4.2)	periodic table (4.8)
law of constant composition (4.3)	groups (4.8)
	alkali metals (4.8)
Dalton's atomic theory (4.3)	alkaline earth metals (4.8)
	halogens (4.8)
atom (4.3)	noble gases (4.8)
compound (4.4)	transition metals (4.8)
chemical formula (4.4)	metals (4.8)
electron (4.5)	nonmetals (4.8)
nuclear atom (4.5)	metalloids
nucleus (4.5)	(semimetals) (4.8)
proton (4.5)	diatomic molecule (4.9)
neutron (4.5)	ion (4.10)
isotopes (4.7)	cation (4.10)
atomic number, Z (4.7)	anion (4.10)
mass number, A (4.7)	ionic compound (4.11)

Summary

1. Of the more than 100 different elements now known, only 9 account for about 98% of the total mass of the earth's crust, oceans, and atmosphere. In the human body, oxygen, carbon, hydrogen, and nitrogen are the most abundant elements.

2. Elements are represented by symbols that usually consist of the first one or two letters of the element's name. Sometimes, however, the symbol is taken from the element's original Latin or Greek name.

3. The law of constant composition states that a given compound always contains the same proportions by mass of the elements of which it is composed.

4. Dalton accounted for this law with his atomic theory. He postulated that all elements are composed of atoms; that all atoms of a given element are identical, but that atoms of different elements are different; that chemical compounds are formed when atoms combine; and that atoms are not created or destroyed in chemical reactions.

5. A compound can be represented by a chemical formula that uses the symbol for each type of atom and gives the number of each type of atom that appears in a molecule of the compound.

6. Atoms consist of a nucleus containing protons and neutrons, surrounded by electrons that occupy a large volume relative to the size of the nucleus. Electrons have a relatively small mass (1/1836 of the proton mass) and a negative charge. Protons have a positive charge equal in magnitude (but opposite in sign) to that of the electron. A neutron has a slightly greater mass than the proton but no charge.

7. Isotopes are atoms with the same number of protons but different numbers of neutrons.

8. The periodic table displays the elements in rows and columns in order of increasing atomic number. Elements that have similar chemical properties fall into vertical columns called groups. Most of the elements are metals. These occur on the left-hand side of the periodic table; the nonmetals appear on the right-hand side.

9. Each chemical element is composed of a given type of atom. These elements may exist as individual atoms or as groups of like atoms. For example, the noble gases contain single, separated atoms. However, elements such as oxygen, nitrogen, and chlorine exist as diatomic (two-atom) molecules.

10. When an atom loses one or more electrons, it forms a positive ion called a cation. This behavior is characteristic of metals. When an atom gains one or more electrons, it becomes a negatively charged ion called an anion. This behavior is characteristic of non-metals. Oppositely charged ions form ionic compounds. A compound is always neutral overall—it has zero net charge.

11. The elements in Groups 1 and 2 on the periodic table form 1+ and 2+ cations, respectively. Group 7 atoms can gain one electron to form 1− ions. Group 6 atoms form 2− ions.

Active Learning Questions

These questions are designed to be considered by groups of students in class. Often these questions work well for introducing a particular topic in class.

1. Knowing the number of protons in the atom of a neutral element enables you to determine which of the following?

   a. the number of neutrons in the atom of the neutral element
   b. the number of electrons in the atom of the neutral element
   c. the name of the element
   d. two of the above
   e. none of the above

   Explain.

2. The average mass of a carbon atom is 12.011. Assuming you could pick up one carbon atom, what is the chance that you would randomly get one with a mass of 12.011?

a. 0%	d. 12.011%
b. 0.011%	e. greater than 50%
c. about 12%	f. none of the above

   Explain.

3. How is an ion formed?

   a. by either adding or subtracting protons from the atom
   b. by either adding or subtracting neutrons from the atom
   c. by either adding or subtracting electrons from the atom
   d. all of the above
   e. two of the above

   Explain.

4. The formula of water, H_2O, suggests which of the following?

   a. There is twice as much mass of hydrogen as oxygen in each molecule.
   b. There are two hydrogen atoms and one oxygen atom per water molecule.
   c. There is twice as much mass of oxygen as hydrogen in each molecule.
   d. There are two oxygen atoms and one hydrogen atom per water molecule.
   e. Two of the above.

   Explain.

5. The vitamin niacin (nicotinic acid, $C_6H_5NO_2$) can be isolated from a variety of natural sources, such as liver, yeast, milk, and whole grain. It also can be synthesized from commercially available materials. Which source of nicotinic acid, from a nutritional view, is best for use in a multivitamin tablet? Why?

6. One of the best indications of a useful theory is that it raises more questions for further experimentation than it originally answered. How does this apply to Dalton's atomic theory? Give examples.

7. Dalton assumed that all atoms of the same element are identical in all their properties. Explain why this assumption is not valid.

8. How does Dalton's atomic theory account for the law of constant composition?

9. Which of the following is true about the state of an individual atom?

   a. An individual atom should be considered to be a solid.
   b. An individual atom should be considered to be a liquid.
   c. An individual atom should be considered to be a gas.
   d. The state of the atom depends on which element it is.
   e. An individual atom cannot be considered to be a solid, liquid, or gas.

   For choices you did not pick, explain what you feel is wrong with them, and justify the choice you did pick.

10. These questions concern the work of J. J. Thomson:

    a. From Thomson's work, which particles do you think he would feel are most important in the formation of compounds (chemical changes) and why?
    b. Of the remaining two subatomic particles, which do you place second in importance for forming compounds and why?
    c. Come up with three models that explain Thomson's findings and evaluate them. To be complete you should include Thomson's findings.

11. Heat is applied to an ice cube until only steam is present. Draw a sketch of this process, assuming you can see it at an extremely high level of magnification. What happens to the size of the molecules? What happens to the total mass of the sample?

12. What makes a carbon atom different from a nitrogen atom? How are they alike?

13. Hundreds of years ago, alchemists tried to turn lead into gold. Is this possible? If not, why not? If yes, how would you do it?

14. Chlorine has two prominent isotopes, ^{37}Cl and ^{35}Cl. Which is more abundant? How do you know?

15. Differentiate between an atomic element and a molecular element. Provide an example and microscopic drawing of each.

16. Science often develops by using the known theories and expanding, refining, and perhaps changing these theories. As discussed in Section 4.5, Rutherford used Thompson's ideas when thinking about his model of the atom. What if Rutherford had not known about Thompson's work? How might Rutherford's model of the atom have been different?

17. Rutherford was surprised when some of the α-particles bounced back. He was surprised because he was thinking of Thompson's model of the atom. What if Rutherford believed atoms were as Dalton envisioned them? What do you suppose Rutherford would have expected, and what would have surprised him?

18. It is good practice to actively read the textbook and to try to verify claims that are made when you can. The following claim is made in your textbook: " . . . if the nucleus were the size of a grape, the electrons would be about one mile away on average."

    Provide mathematical support for this statement.

19. Why is the term "sodium chloride molecule" incorrect but the term "carbon dioxide molecule" is correct?

20. Both atomic elements and molecular elements exist. Are there such entities as atomic compounds and molecular compounds? If so, provide an example and microscopic drawing. If not, explain why not.

21. Now that you have gone through Chapter 4, go back to Section 4.3 and review Dalton's Atomic Theory. Which of the premises are no longer accepted? Explain your answer.

22. Write the formula for each of the following substances, listing the elements in the order given.

    a.

    List the phosphorus atom first.
    b. a molecule containing two boron atoms and six hydrogen atoms
    c. a compound containing one calcium atom for every two chlorine atoms
    d.

    List the carbon atom first.
    e. a compound containing two iron atoms for every three oxygen atoms
    f. a molecule containing three hydrogen atoms, one phosphorus atom, and four oxygen atoms

23. Use the following figures to identify the element or ion. Write the symbol for each, using the $^A_Z X$

    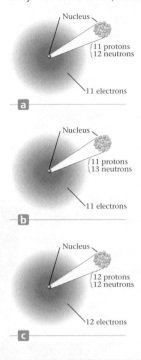

OWL Questions and problems for this chapter are available in OWL. Use the OWL access card that came with this textbook to access assigned questions and problems for this chapter.

5

Nomenclature

● Clouds over tufa towers in Mono Lake, California. *(Fred Hirschmann/Science Faction)*

When chemistry was an infant science, there was no system for naming compounds. Names such as sugar of lead, blue vitriol, quicklime, Epsom salts, milk of magnesia, gypsum, and laughing gas were coined by early chemists. Such names are called *common names.* As our knowledge of chemistry grew, it became clear that using common names for compounds was not practical. More than four million chemical compounds are currently known. Memorizing common names for all these compounds would be impossible.

The solution, of course, is a *system* for naming compounds in which the name tells something about the composition of the compound. After learning the system, you should be able to name a compound when you are given its formula. And, conversely, you should be able to construct a compound's formula, given its name. In the next few sections we will specify the most important rules for naming compounds other than organic compounds (those based on chains of carbon atoms).

An artist using plaster of Paris, a gypsum plaster.

Bob Daemmrich/The Image Works

Naming Compounds

OBJECTIVE: To understand why it is necessary to have a system for naming compounds.

We will begin by discussing the system for naming **binary compounds**— compounds composed of two elements. We can divide binary compounds into two broad classes:

1. Compounds that contain a metal and a nonmetal

2. Compounds that contain two nonmetals

We will describe how to name compounds in each of these classes in the next several sections. Then, in succeeding sections, we will describe the systems used for naming more complex compounds.

Sugar of Lead

In ancient Roman society it was common to boil wine in a lead-lined vessel, driving off much of the water to produce a very sweet, viscous syrup called *sapa*. This syrup was commonly used as a sweetener for many types of food and drink.

We now realize that a major component of this syrup was lead acetate, $Pb(C_2H_3O_2)_2$. This compound has a very sweet taste—hence its original name, sugar of lead.

Many historians believe that the fall of the Roman Empire was due at least in part to lead poisoning, which causes lethargy and mental malfunctions. One major source of this lead was the sapa syrup. In addition, the Romans' highly advanced plumbing system employed lead water pipes, which allowed lead to be leached into their drinking water.

Sadly, this story is more relevant to today's society than you might think. Lead-based solder was widely used for many years to connect the copper pipes in water systems in homes and commercial buildings. There is evidence that dangerous amounts of lead can be leached from these soldered joints into drinking water. In fact, large quantities of lead have been found in the water that some drinking fountains and water coolers dispense. In response to these problems, the U.S. Congress has passed a law banning lead from the solder used in plumbing systems for drinking water.

An ancient painting showing Romans drinking wine.

5.2 Naming Binary Compounds That Contain a Metal and a Nonmetal (Types I and II)

OBJECTIVE: To learn to name binary compounds of a metal and a nonmetal.

As we saw in Section 4.11, when a metal such as sodium combines with a nonmetal such as chlorine, the resulting compound contains ions. The metal loses one or more electrons to become a cation, and the nonmetal gains one or more electrons to form an anion. The resulting substance is called a **binary ionic compound.** Binary ionic compounds contain a positive ion (cation), which is always written first in the formula, and a negative ion (anion). *To name these compounds we simply name the ions.*

In this section we will consider binary ionic compounds of two types based on the cations they contain. Certain metal atoms form only one cation. For example, the Na atom always forms Na^+, *never* Na^{2+} or Na^{3+}. Likewise, Cs always forms Cs^+, Ca always forms Ca^{2+}, and Al always forms Al^{3+}. We will call compounds that contain this type of metal atom Type I binary compounds and the cations they contain Type I cations. Examples of Type I cations are Na^+, Ca^{2+}, Cs^+, and Al^{3+}.

go Chemistry **Module 3: Names to Formulas of Ionic Compounds** covers concepts in this section.

Table 5.1 Common Simple Cations and Anions

Cation	Name	Anion	Name*
H^+	hydrogen	H^-	hydride
Li^+	lithium	F^-	fluoride
Na^+	sodium	Cl^-	chloride
K^+	potassium	Br^-	bromide
Cs^+	cesium	I^-	iodide
Be^{2+}	beryllium	O^{2-}	oxide
Mg^{2+}	magnesium	S^{2-}	sulfide
Ca^{2+}	calcium		
Ba^{2+}	barium		
Al^{3+}	aluminum		
Ag^+	silver		
Zn^{2+}	zinc		

*The root is given in color.

Other metal atoms can form two or more cations. For example, Cr can form Cr^{2+} and Cr^{3+} and Cu can form Cu^+ and Cu^{2+}. We will call such ions Type II cations and their compounds Type II binary compounds.

In summary:

Type I compounds: The metal present forms only one type of cation.

Type II compounds: The metal present can form two (or more) cations that have different charges.

Some common cations and anions and their names are listed in Table 5.1. You should memorize these. They are an essential part of your chemical vocabulary.

▶ Type I Binary Ionic Compounds

The following rules apply for Type I ionic compounds:

Rules for Naming Type I Ionic Compounds

1. The cation is always named first and the anion second.
2. A simple cation (obtained from a single atom) takes its name from the name of the element. For example, Na^+ is called sodium in the names of compounds containing this ion.
3. A simple anion (obtained from a single atom) is named by taking the first part of the element name (the root) and adding *-ide*. Thus the Cl^- ion is called chloride.

A simple cation has the same name as its parent element.

We will illustrate these rules by naming a few compounds. For example, the compound NaI is called sodium iodide. It contains Na^+ (the sodium cation, named for the parent metal) and I^- (iodide: the root of *iodine* plus *-ide*). Similarly, the compound CaO is called calcium oxide because it contains Ca^{2+} (the calcium cation) and O^{2-} (the oxide anion).

The rules for naming binary compounds are also illustrated by the following examples:

Compound	Ions Present	Name
NaCl	Na^+, Cl^-	sodium chloride
KI	K^+, I^-	potassium iodide
CaS	Ca^{2+}, S^{2-}	calcium sulfide
CsBr	Cs^+, Br^-	cesium bromide
MgO	Mg^{2+}, O^{2-}	magnesium oxide

It is important to note that in the *formulas* of ionic compounds, simple ions are represented by the element symbol: Cl means Cl^-, Na means Na^+, and so on. However, when *individual ions* are shown, the charge is always included. Thus the formula of potassium bromide is written KBr, but when the potassium and bromide ions are shown individually, they are written K^+ and Br^-.

EXAMPLE 5.1   **Naming Type I Binary Compounds**

Name each binary compound.

　　a. CsF 　　　b. $AlCl_3$ 　　　c. MgI_2

SOLUTION

We will name these compounds by systematically following the rules given above.

　　a. CsF

Step 1 Identify the cation and anion. Cs is in Group 1, so we know it will form the 1+ ion Cs^+. Because F is in Group 7, it forms the 1− ion F^-.

Step 2 Name the cation. Cs^+ is simply called cesium, the same as the element name.

Step 3 Name the anion. F^- is called fluoride: we use the root name of the element plus *-ide*.

Step 4 Name the compound by combining the names of the individual ions. The name for CsF is cesium fluoride. (Remember that the name of the cation is always given first.)

b.

Compound	Ions Present	Ion Names	Comments
$AlCl_3$ — Cation →	Al^{3+}	aluminum	Al (Group 3) always forms Al^{3+}.
$AlCl_3$ — Anion →	Cl^-	chloride	Cl (Group 7) always forms Cl^-.

The name of $AlCl_3$ is aluminum chloride.

c.

Compound	Ions Present	Ion Names	Comments
MgI_2 — Cation →	Mg^{2+}	magnesium	Mg (Group 2) always forms Mg^{2+}.
MgI_2 — Anion →	I^-	iodide	I (Group 7) gains one electron to form I^-.

The name of MgI_2 is magnesium iodide.

Self-Check **EXERCISE 5.1** Name the following compounds.

a. Rb_2O b. SrI_2 c. K_2S

See Problems 5.9 and 5.10. ■

Example 5.1 reminds us of three things:

1. Compounds formed from metals and nonmetals are ionic.

2. In an ionic compound the cation is always named first.

3. The *net* charge on an ionic compound is always zero. Thus, in CsF, one of each type of ion (Cs^+ and F^-) is required: ⓛ⁺ + ⓛ⁻ = 0 charge. In $AlCl_3$, however, three Cl^- ions are needed to balance the charge of Al^{3+}: ③⁺ + 3 ⓛ⁻ = 0 charge. In MgI_2, two I^- ions are needed for each Mg^{2+} ion: ②⁺ + 2 ⓛ⁻ = 0 charge.

▶ Type II Binary Ionic Compounds

> Type II binary ionic compounds contain a metal that can form more than one type of cation.

So far we have considered binary ionic compounds (Type I) containing metals that always give the same cation. For example, sodium always forms the Na^+ ion, calcium always forms the Ca^{2+} ion, and aluminum always forms the Al^{3+} ion. As we said in the previous section, we can predict with certainty that each Group 1 metal will give a 1+ cation and each Group 2 metal will give a 2+ cation. Aluminum always forms Al^{3+}.

However, there are many metals that can form more than one type of cation. For example, lead (Pb) can form Pb^{2+} or Pb^{4+} in ionic compounds. Also, iron (Fe) can produce Fe^{2+} or Fe^{3+}, chromium (Cr) can produce Cr^{2+} or Cr^{3+}, gold (Au) can produce Au^+ or Au^{3+}, and so on. This means that if we saw the name gold chloride, we wouldn't know whether it referred to the compound AuCl (containing Au^+ and Cl^-) or the compound $AuCl_3$ (containing Au^{3+} and three Cl^- ions). Therefore, we need a way of specifying which cation is present in compounds containing metals that can form more than one type of cation.

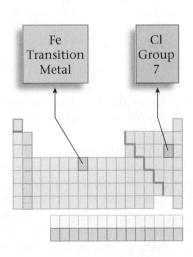

Chemists have decided to deal with this situation by using a Roman numeral to specify the charge on the cation. To see how this works, consider the compound $FeCl_2$. Iron can form Fe^{2+} or Fe^{3+}, so we must first decide which of these cations is present. We can determine the charge on the iron cation, because we know it must just balance the charge on the two 1− anions (the chloride ions). Thus if we represent the charges as

$$\underset{\substack{\uparrow \\ \text{Charge} \\ \text{on iron} \\ \text{cation}}}{?⁺} + 2 \underset{\substack{\uparrow \\ \text{Charge} \\ \text{on Cl}^-}}{1⁻} = \underset{\substack{\uparrow \\ \text{Net} \\ \text{charge}}}{0}$$

we know that ? must represent 2 because

$$(2+) + 2(1-) = 0$$

> $FeCl_3$ must contain Fe^{3+} to balance the charge of three Cl^- ions.

The compound $FeCl_2$, then, contains one Fe^{2+} ion and two Cl^- ions. We call this compound iron(II) chloride, where the II tells the charge of the iron cation. That is, Fe^{2+} is called iron(II). Likewise, Fe^{3+} is called iron(III). And $FeCl_3$, which contains one Fe^{3+} ion and three Cl^- ions, is called iron(III) chloride. Remember that the Roman numeral tells the *charge* on the ion, not the number of ions present in the compound.

Copper(II) sulfate crystals.

Table 5.2	Common Type II Cations	
Ion	Systematic Name	Older Name
Fe^{3+}	iron(III)	ferric
Fe^{2+}	iron(II)	ferrous
Cu^{2+}	copper(II)	cupric
Cu^{+}	copper(I)	cuprous
Co^{3+}	cobalt(III)	cobaltic
Co^{2+}	cobalt(II)	cobaltous
Sn^{4+}	tin(IV)	stannic
Sn^{2+}	tin(II)	stannous
Pb^{4+}	lead(IV)	plumbic
Pb^{2+}	lead(II)	plumbous
Hg^{2+}	mercury(II)	mercuric
$Hg_2^{2+}*$	mercury(I)	mercurous

*Mercury(I) ions always occur bound together in pairs to form Hg_2^{2+}.

Note that in the preceding examples the Roman numeral for the cation turned out to be the same as the subscript needed for the anion (to balance the charge). This is often not the case. For example, consider the compound PbO_2. Since the oxide ion is O^{2-}, for PbO_2 we have

$$\underset{\substack{\uparrow \\ \text{Charge on} \\ \text{lead ion}}}{(?+)} \quad + \quad \underset{\substack{\uparrow \\ (4-) \\ \text{Charge on} \\ \text{two } O^{2-} \text{ ions}}}{2\,(2-)} \quad = \quad \underset{\substack{\uparrow \\ \text{Net} \\ \text{charge}}}{0}$$

Thus the charge on the lead ion must be 4+ to balance the 4− charge of the two oxide ions. The name of PbO_2 is therefore lead(IV) oxide, where the IV indicates the presence of the Pb^{4+} cation.

There is another system for naming ionic compounds containing metals that form two cations. *The ion with the higher charge has a name ending in* -ic, *and the one with the lower charge has a name ending in* -ous. In this system, for example, Fe^{3+} is called the ferric ion, and Fe^{2+} is called the ferrous ion. The names for $FeCl_3$ and $FeCl_2$, in this system, are ferric chloride and ferrous chloride, respectively. Table 5.2 gives both names for many Type II cations. We will use the system of Roman numerals exclusively in this text; the other system is falling into disuse.

To help distinguish between Type I and Type II cations, remember that Group 1 and 2 metals are always Type I. On the other hand, transition metals are almost always Type II.

Rules for Naming Type II Ionic Compounds

1. The cation is always named first and the anion second.
2. Because the cation can assume more than one charge, the charge is specified by a Roman numeral in parentheses.

EXAMPLE 5.2 | Naming Type II Binary Compounds

Give the systematic name of each of the following compounds.

a. CuCl   b. HgO   c. Fe_2O_3 d. MnO_2 e. $PbCl_4$

SOLUTION

All these compounds include a metal that can form more than one type of cation; thus we must first determine the charge on each cation. We do this by recognizing that a compound must be electrically neutral; that is, the positive and negative charges must balance exactly. We will use the known charge on the anion to determine the charge of the cation.

a. In CuCl we recognize the anion as Cl^-. To determine the charge on the copper cation, we invoke the principle of charge balance.

$$\underset{\substack{\uparrow \\ \text{Charge} \\ \text{on copper} \\ \text{ion}}}{\boxed{?+}} + \underset{\substack{\uparrow \\ \text{Charge} \\ \text{on } Cl^-}}{\boxed{1-}} = \underset{\substack{\uparrow \\ \text{Net charge} \\ \text{(must be zero)}}}{0}$$

In this case, ?+ must be 1+ because $(1+) + (1-) = 0$. Thus the copper cation must be Cu^+. Now we can name the compound by using the regular steps.

Compound	Ions Present	Ion Names	Comments
CuCl — Cation →	Cu^+	copper(I)	Copper forms other cations (it is a transition metal), so we must include the I to specify its charge.
CuCl — Anion →	Cl^-	chloride	

The name of CuCl is copper(I) chloride.

b. In HgO we recognize the O^{2-} anion. To yield zero net charge, the cation must be Hg^{2+}.

Compound	Ions Present	Ion Names	Comments
HgO — Cation →	Hg^{2+}	mercury(II)	The II is necessary to specify the charge.
HgO — Anion →	O^{2-}	oxide	

The name of HgO is mercury(II) oxide.

c. Because Fe_2O_3 contains three O^{2-} anions, the charge on the iron cation must be 3+.

$$\underset{\substack{\uparrow \\ Fe^{3+}}}{2(3+)} + \underset{\substack{\uparrow \\ O^{2-}}}{3(2-)} = \underset{\substack{\uparrow \\ \text{Net charge}}}{0}$$

Compound	Ions Present	Ion Names	Comments
Fe_2O_3 — Cation →	Fe^{3+}	iron(III)	Iron is a transition metal and requires a III to specify the charge on the cation.
Fe_2O_3 — Anion →	O^{2-}	oxide	

The name of Fe_2O_3 is iron(III) oxide.

d. MnO_2 contains two O^{2-} anions, so the charge on the manganese cation is 4+.

$$(4+) + 2(2-) = \quad 0$$
$$\uparrow \qquad\quad \uparrow \qquad\quad \uparrow$$
$$Mn^{4+} \qquad O^{2-} \qquad \text{Net charge}$$

Compound	Ions Present	Ion Names	Comments
	Mn^{4+}	manganese(IV)	Manganese is a transition metal and requires a IV to specify the charge on the cation.
MnO_2 — Cation / Anion			
	O^{2-}	oxide	

The name of MnO_2 is manganese(IV) oxide.

e. Because $PbCl_4$ contains four Cl^- anions, the charge on the lead cation is 4+.

$$(4+) + 4(1-) = \quad 0$$
$$\uparrow \qquad\quad \uparrow \qquad\quad \uparrow$$
$$Pb^{4+} \qquad Cl^- \qquad \text{Net charge}$$

Compound	Ions Present	Ion Names	Comments
	Pb^{4+}	lead(IV)	Lead forms both Pb^{2+} and Pb^{4+}, so a Roman numeral is required.
$PbCl_4$ — Cation / Anion			
	Cl^-	chloride	

The name for $PbCl_4$ is lead(IV) chloride. ∎

> Sometimes transition metals form only one ion, such as silver, which forms Ag^+; zinc, which forms Zn^{2+}; and cadmium, which forms Cd^{2+}. In these cases, chemists do not use a Roman numeral, although it is not "wrong" to do so.

The use of a Roman numeral in a systematic name for a compound is required only in cases where more than one ionic compound forms between a given pair of elements. This occurs most often for compounds that contain transition metals, which frequently form more than one cation. *Metals that form only one cation do not need to be identified by a Roman numeral.* Common metals that do not require Roman numerals are the Group 1 elements, which form only 1+ ions; the Group 2 elements, which form only 2+ ions; and such Group 3 metals as aluminum and gallium, which form only 3+ ions.

As shown in Example 5.2, when a metal ion that forms more than one type of cation is present, the charge on the metal ion must be determined by balancing the positive and negative charges of the compound. To do this, you must be able to recognize the common anions and you must know their charges (see Table 5.1).

EXAMPLE 5.3 | **Naming Binary Ionic Compounds: Summary**

Give the systematic name of each of the following compounds.

a. $CoBr_2$

b. $CaCl_2$

c. Al_2O_3

d. $CrCl_3$

SOLUTION

	Compound	Ions and Names	Compound Name	Comments
a.	CoBr$_2$	Co^{2+} cobalt(II) Br$^-$ bromide	cobalt(II) bromide	Cobalt is a transition metal; the name of the compound must have a Roman numeral. The two Br$^-$ ions must be balanced by a Co^{2+} cation.
b.	CaCl$_2$	Ca^{2+} calcium Cl$^-$ chloride	calcium chloride	Calcium, a Group 2 metal, forms only the Ca^{2+} ion. A Roman numeral is not necessary.
c.	Al$_2$O$_3$	Al^{3+} aluminum O^{2-} oxide	aluminum oxide	Aluminum forms only Al^{3+}. A Roman numeral is not necessary.
d.	CrCl$_3$	Cr^{3+} chromium(III) Cl$^-$ chloride	chromium(III) chloride	Chromium is a transition metal. The name of the compound must have a Roman numeral. CrCl$_3$ contains Cr^{3+}.

───

Self-Check   **EXERCISE 5.2**   Give the names of the following compounds.

a. PbBr$_2$ and PbBr$_4$ b. FeS and Fe$_2$S$_3$ c. AlBr$_3$ d. Na$_2$S   e. CoCl$_3$

See Problems 5.9, 5.10, and 5.13 through 5.16. ∎

The following flow chart is useful when you are naming binary ionic compounds:

5.3

OBJECTIVE:

Naming Binary Compounds That Contain Only Nonmetals (Type III)

To learn how to name binary compounds containing only nonmetals.

Table 5.3 Prefixes Used to Indicate Numbers in Chemical Names

Prefix	Number Indicated
mono-	1
di-	2
tri-	3
tetra-	4
penta-	5
hexa-	6
hepta-	7
octa-	8

Binary compounds that contain only nonmetals are named in accordance with a system similar in some ways to the rules for naming binary ionic compounds, but there are important differences. *Type III binary compounds contain only nonmetals.* The following rules cover the naming of these compounds.

Rules for Naming Type III Binary Compounds

1. The first element in the formula is named first, and the full element name is used.
2. The second element is named as though it were an anion.
3. Prefixes are used to denote the numbers of atoms present. These prefixes are given in Table 5.3.
4. The prefix *mono-* is never used for naming the first element. For example, CO is called carbon monoxide, *not* monocarbon monoxide.

We will illustrate the application of these rules in Example 5.4.

EXAMPLE 5.4 | Naming Type III Binary Compounds

Name the following binary compounds, which contain two nonmetals (Type III).

    a. BF_3 b. NO   c. N_2O_5

SOLUTION

  a. BF_3

     Rule 1   Name the first element, using the full element name: boron.

     Rule 2   Name the second element as though it were an anion: fluoride.

     Rules 3 and 4   Use prefixes to denote numbers of atoms. One boron atom: do not use *mono-* in first position. Three fluorine atoms: use the prefix *tri-*.

     The name of BF_3 is boron trifluoride.

  b.

Compound	Individual Names	Prefixes	Comments
NO	nitrogen	none	*Mono-* is not used
	oxide	*mono-*	for the first element.

     The name for NO is nitrogen monoxide. Note that the second *o* in *mono-* has been dropped for easier pronunciation. The *common* name for NO, which is often used by chemists, is nitric oxide.

  c.

Compound	Individual Names	Prefixes	Comments
N_2O_5	nitrogen	*di-*	two N atoms
	oxide	*penta-*	five O atoms

     The name for N_2O_5 is dinitrogen pentoxide. The *a* in *penta-* has been dropped for easier pronunciation.

© Cengage Learning

A piece of copper metal about to be placed in nitric acid (*left*). Copper reacts with nitric acid to produce colorless NO, which immediately reacts with the oxygen in the air to form reddish-brown NO_2 gas and Cu^{2+} ions in solution (which produce the green color) (*right*).

Self-Check **EXERCISE 5.3**   Name the following compounds.

a. CCl_4 b. NO_2 c. IF_5 See Problems 5.17 and 5.18. ∎

The previous examples illustrate that, to avoid awkward pronunciation, we often drop the final *o* or *a* of the prefix when the second element is oxygen. For example, N_2O_4 is called dinitrogen tetroxide, *not* dinitrogen tetr*a*oxide, and CO is called carbon monoxide, *not* carbon mon*o*oxide.

> Water and ammonia are always referred to by their common names.

Some compounds are always referred to by their common names. The two best examples are water and ammonia. The systematic names for H_2O and NH_3 are never used.

To make sure you understand the procedures for naming binary non-metallic compounds (Type III), study Example 5.5 and then do Self-Check Exercise 5.4.

EXAMPLE 5.5 | Naming Type III Binary Compounds: Summary

Name each of the following compounds.

a. PCl_5 c. SF_6 d. SO_2

b. P_4O_6 d. SO_3 f. N_2O_3

SOLUTION

Compound	Name
a. PCl_5	phosphorus pentachloride
b. P_4O_6	tetraphosphorus hexoxide
c. SF_6	sulfur hexafluoride
d. SO_3	sulfur trioxide
e. SO_2	sulfur dioxide
f. N_2O_3	dinitrogen trioxide

Self-Check   **EXERCISE 5.4**   Name the following compounds.

a. SiO_2 b. O_2F_2 c. XeF_6

See Problems 5.17 and 5.18. ■

5.4 Naming Binary Compounds: A Review

OBJECTIVE: To review the naming of Type I, Type II, and Type III binary compounds.

Because different rules apply for naming various types of binary compounds, we will now consider an overall strategy to use for these compounds. We have considered three types of binary compounds, and naming each of them requires different procedures.

Type I: Ionic compounds with metals that always form a cation with the same charge

Type II: Ionic compounds with metals (usually transition metals) that form cations with various charges

Type III: Compounds that contain only nonmetals

In trying to determine which type of compound you are naming, use the periodic table to help identify metals and nonmetals and to determine which elements are transition metals.

The flow chart given in Figure 5.1 should help you as you name binary compounds of the various types.

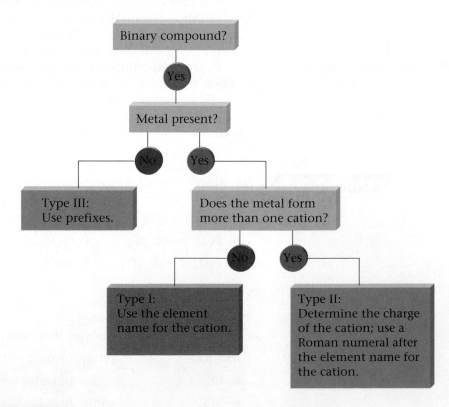

Figure 5.1

A flow chart for naming binary compounds.

Chemophilately

Philately is the study of postage stamps. Chemophilately, a term coined by the Israeli chemist Zvi Rappoport, refers to the study of stamps that have some sort of chemical connection. Collectors estimate that more than 2000 chemical-related stamps have been printed throughout the world. Relatively few of these stamps have been produced in the United States. One example is a 29¢ stamp honoring minerals that shows a copper nugget.

Courtesy, Daniel Rabinovich

Chemists also have been honored on U.S. postage stamps. One example is a 29¢ stamp printed in 1993 honoring Percy L. Julian, an African-American chemist who was the grandson of slaves. Julian is noted for his synthesis of steroids used to treat glaucoma and rheumatoid arthritis. As a holder of more than 100 patents, he was inducted into the National Inventors Hall of Fame in 1990.

Courtesy, Daniel Rabinovich

In 1983 the United States issued a stamp honoring Joseph Priestley, whose experiments led to the discovery of oxygen.

Courtesy, Whitman Publishing LLC

A stamp from 2005 pictures J. Willard Gibbs, a Yale professor who was instrumental in developing thermodynamics—the study of energy and its transformations.

In 2008, a 41¢ stamp was issued that honors Linus C. Pauling, who pioneered the concept of the chemical bond. Pauling received two Nobel Prizes: one for his work on chemical bonds and the other for his work championing world peace. His stamp includes drawings of red blood cells to commemorate his work on the study of hemoglobin, which led to the classification of sickle cell anemia as a molecular disease.

Courtesy, Daniel Rabinovich

Postal chemistry also shows up in postmarks from places in the United States with chemical names. Examples include Radium, KS, Neon, KY, Boron, CA, Bromide, OK, and Telluride, CO.

Courtesy, Daniel Rabinovich

Chemophilately—further proof that chemistry is everywhere!

EXAMPLE 5.6   Naming Binary Compounds: Summary

Name the following binary compounds.

a. CuO   c. B_2O_3 e. K_2S g. NH_3

b. SrO   d. $TiCl_4$ f. OF_2

SOLUTION

a.

The name of CuO is copper(II) oxide.

b.

The name of SrO is strontium oxide.

c.

The name of B_2O_3 is diboron trioxide.

d.

The name of $TiCl_4$ is titanium(IV) chloride.

e.

The name of K_2S is potassium sulfide.

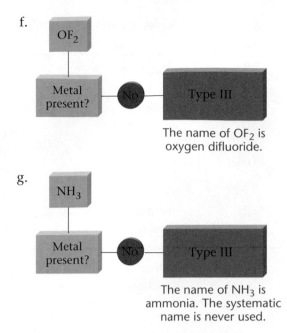

f.

OF$_2$

Metal present? — No — Type III

The name of OF$_2$ is oxygen difluoride.

g.

NH$_3$

Metal present? — No — Type III

The name of NH$_3$ is ammonia. The systematic name is never used.

Self-Check **EXERCISE 5.5** Name the following binary compounds.

a. ClF_3 d. MnO_2

b. VF_5 e. MgO

c. $CuCl$ f. H_2O

See Problems 5.19 through 5.22. ■

5.5 Naming Compounds That Contain Polyatomic Ions

OBJECTIVE: To learn the names of common polyatomic ions and how to use them in naming compounds.

Ionic compounds containing polyatomic ions are not binary compounds, because they contain more than two elements.

The names and charges of polyatomic ions must be memorized. They are an important part of the vocabulary of chemistry.

A type of ionic compound that we have not yet considered is exemplified by ammonium nitrate, NH_4NO_3, which contains the **polyatomic ions** NH_4^+ and NO_3^-. As their name suggests, polyatomic ions are charged entities composed of several atoms bound together. Polyatomic ions are assigned special names that you *must memorize* to name the compounds containing them. The most important polyatomic ions and their names are listed in Table 5.4.

Note in Table 5.4 that several series of polyatomic anions exist that contain an atom of a given element and different numbers of oxygen atoms. These anions are called **oxyanions.** When there are two members in such a series, the name of the one with the smaller number of oxygen atoms ends in *-ite,* and the name of the one with the larger number ends in *-ate.* For example, SO_3^{2-} is sulfite and SO_4^{2-} is sulfate. When more than two oxyanions make up a series, *hypo-* (less than) and *per-* (more than) are used as prefixes to name the members of the series with the fewest and the most oxygen

Table 5.4   Names of Common Polyatomic Ions

Ion	Name	Ion	Name
NH_4^+	ammonium	CO_3^{2-}	carbonate
NO_2^-	nitrite	HCO_3^-	hydrogen carbonate (bicarbonate is a widely used common name)
NO_3^-	nitrate		
SO_3^{2-}	sulfite		
SO_4^{2-}	sulfate	ClO^-	hypochlorite
HSO_4^-	hydrogen sulfate (bisulfate is a widely used common name)	ClO_2^-	chlorite
		ClO_3^-	chlorate
		ClO_4^-	perchlorate
OH^-	hydroxide	$C_2H_3O_2^-$	acetate
CN^-	cyanide	MnO_4^-	permanganate
PO_4^{3-}	phosphate	$Cr_2O_7^{2-}$	dichromate
HPO_4^{2-}	hydrogen phosphate	CrO_4^{2-}	chromate
$H_2PO_4^-$	dihydrogen phosphate	O_2^{2-}	peroxide

atoms, respectively. The best example involves the oxyanions containing chlorine:

ClO^-	*hypo*chlor*ite*
ClO_2^-	chlor*ite*
ClO_3^-	chlor*ate*
ClO_4^-	*per*chlor*ate*

Naming ionic compounds that contain polyatomic ions is very similar to naming binary ionic compounds. For example, the compound NaOH is called sodium hydroxide, because it contains the Na^+ (sodium) cation and the OH^- (hydroxide) anion. To name these compounds, *you must learn to recognize the common polyatomic ions.* That is, you must learn the *composition* and *charge* of each of the ions in Table 5.4. Then when you see the formula $NH_4C_2H_3O_2$, you should immediately recognize its two "parts":

The correct name is ammonium acetate.

Remember that when a metal is present that forms more than one cation, a Roman numeral is required to specify the cation charge, just as in naming Type II binary ionic compounds. For example, the compound $FeSO_4$ is called iron(II) sulfate, because it contains Fe^{2+} (to balance the 2− charge on SO_4^{2-}). Note that to determine the charge on the iron cation, you must know that sulfate has a 2− charge.

EXAMPLE 5.7   **Naming Compounds That Contain Polyatomic Ions**

Give the systematic name of each of the following compounds.

a. Na_2SO_4 c. $Fe(NO_3)_3$ e. Na_2SO_3

b. KH_2PO_4 d. $Mn(OH)_2$ f. NH_4ClO_3

SOLUTION

Compound	Ions Present	Ion Names	Compound Name
a. Na_2SO_4	two Na^+ SO_4^{2-}	sodium sulfate	sodium sulfate
b. KH_2PO_4	K^+ $H_2PO_4^-$	potassium dihydrogen phosphate	potassium dihydrogen phosphate
c. $Fe(NO_3)_3$	Fe^{3+} three NO_3^-	iron(III) nitrate	iron(III) nitrate
d. $Mn(OH)_2$	Mn^{2+} two OH^-	manganese(II) hydroxide	manganese(II) hydroxide
e. Na_2SO_3	two Na^+ SO_3^{2-}	sodium sulfite	sodium sulfite
f. NH_4ClO_3	NH_4^+ ClO_3^-	ammonium chlorate	ammonium chlorate

---

Self-Check **EXERCISE 5.6** Name each of the following compounds.

a. $Ca(OH)_2$ d. $(NH_4)_2Cr_2O_7$ g. $Cu(NO_2)_2$

b. Na_3PO_4 e. $Co(ClO_4)_2$

c. $KMnO_4$ f. $KClO_3$

See Problems 5.35 and 5.36. ■

Example 5.7 illustrates that when more than one polyatomic ion appears in a chemical formula, parentheses are used to enclose the ion and a subscript is written after the closing parenthesis. Other examples are $(NH_4)_2SO_4$ and $Fe_3(PO_4)_2$.

In naming chemical compounds, use the strategy summarized in Figure 5.2. If the compound being considered is binary, use the procedure summarized in Figure 5.1. If the compound has more than two elements, ask yourself whether it has any polyatomic ions. Use Table 5.4 to help you recognize these ions until you have committed them to memory. If a polyatomic ion is present, name the compound using procedures very similar to those for naming binary ionic compounds.

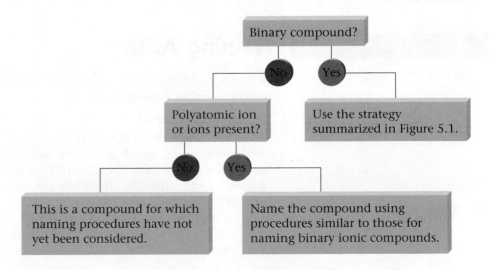

Figure 5.2

Overall strategy for naming chemical compounds.

EXAMPLE 5.8   Summary of Naming Binary Compounds and Compounds That Contain Polyatomic Ions

Name the following compounds.

a. Na_2CO_3

b. $FeBr_3$

c. $CsClO_4$

d. PCl_3

e. $CuSO_4$

SOLUTION

Compound	Name	Comments
a. Na_2CO_3	sodium carbonate	Contains $2Na^+$ and CO_3^{2-}.
b. $FeBr_3$	iron(III) bromide	Contains Fe^{3+} and $3Br^-$.
c. $CsClO_4$	cesium perchlorate	Contains Cs^+ and ClO_4^-.
d. PCl_3	phosphorus trichloride	Type III binary compound (both P and Cl are non-metals).
e. $CuSO_4$	copper(II) sulfate	Contains Cu^{2+} and SO_4^{2-}.

Self-Check EXERCISE 5.7   Name the following compounds.

a. $NaHCO_3$

b. $BaSO_4$

c. $CsClO_4$

d. BrF_5

e. NaBr

f. KOCl

g. $Zn_3(PO_4)_2$

Although we have emphasized that a Roman numeral is required in the name of a compound that contains a transition metal ion, certain transition metals form only one ion. Common examples are zinc (forms only Zn^{2+}) and silver (forms only Ag^+). For these cases the Roman numeral is omitted from the name.

See Problems 5.29 through 5.36. ■

5.6   Naming Acids

OBJECTIVES: To learn how the anion composition determines the acid's name. • To learn names for common acids.

When dissolved in water, certain molecules produce H^+ ions (protons). These substances, which are called **acids,** were first recognized by the sour taste of their solutions. For example, citric acid is responsible for the tartness of lemons and limes. Acids will be discussed in detail later. Here we simply present the rules for naming acids.

An acid can be viewed as a molecule with one or more H^+ ions attached to an anion. The rules for naming acids depend on whether the anion contains oxygen.

Rules for Naming Acids

1. If the *anion does not contain oxygen*, the acid is named with the prefix *hydro-* and the suffix *-ic* attached to the root name for the element. For example, when gaseous HCl (hydrogen chloride) is dissolved in water, it forms hydrochloric acid. Similarly, hydrogen cyanide (HCN) and dihydrogen sulfide (H_2S) dissolved in water are called hydrocyanic acid and hydrosulfuric acid, respectively.

2. When the *anion contains oxygen*, the acid name is formed from the root name of the central element of the anion or the anion name, with a suffix of *-ic* or *-ous*. When the anion name ends in *-ate*, the suffix *-ic* is used. For example,

Acid	Anion	Name
H_2SO_4	SO_4^{2-} (sulfate)	sulfuric acid
H_3PO_4	PO_4^{3-} (phosphate)	phosphoric acid
$HC_2H_3O_2$	$C_2H_3O_2^{-}$ (acetate)	acetic acid

When the anion name ends in *-ite*, the suffix *-ous* is used in the acid name. For example,

Acid	Anion	Name
H_2SO_3	SO_3^{2-} (sulfite)	sulfurous acid
HNO_2	NO_2^{-} (nitrite)	nitrous acid

Table 5.5 Names of Acids That Do Not Contain Oxygen

Acid	Name
HF	hydrofluoric acid
HCl	hydrochloric acid
HBr	hydrobromic acid
HI	hydroiodic acid
HCN	hydrocyanic acid
H_2S	hydrosulfuric acid

Table 5.6 Names of Some Oxygen-Containing Acids

Acid	Name
HNO_3	nitric acid
HNO_2	nitrous acid
H_2SO_4	sulfuric acid
H_2SO_3	sulfurous acid
H_3PO_4	phosphoric acid
$HC_2H_3O_2$	acetic acid

The application of Rule 2 can be seen in the names of the acids of the oxyanions of chlorine below.

Acid	Anion	Name
$HClO_4$	perchlor*ate*	perchlor*ic* acid
$HClO_3$	chlor*ate*	chlor*ic* acid
$HClO_2$	chlor*ite*	chlor*ous* acid
$HClO$	hypochlor*ite*	hypochlor*ous* acid

The rules for naming acids are given in schematic form in Figure 5.3. The names of the most important acids are given in Table 5.5 and Table 5.6. These should be memorized.

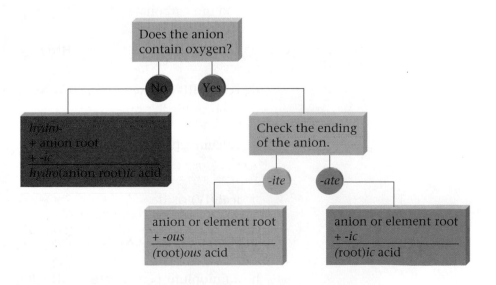

Figure 5.3

A flow chart for naming acids. The acid is considered as one or more H^+ ions attached to an anion.

5.7 Writing Formulas from Names

OBJECTIVE: To learn to write the formula of a compound, given its name.

So far we have started with the chemical formula of a compound and decided on its systematic name. Being able to reverse the process is also important. Often a laboratory procedure describes a compound by name, but the label on the bottle in the lab shows only the formula of the chemical it contains. It is essential that you are able to get the formula of a compound from its name. In fact, you already know enough about compounds to do this. For example, given the name calcium hydroxide, you can write the formula as $Ca(OH)_2$ because you know that calcium forms only Ca^{2+} ions and that, since hydroxide is OH^-, two of these anions are required to give a neutral compound. Similarly, the name iron(II) oxide implies the formula FeO, because the Roman numeral II indicates the presence of the cation Fe^{2+} and the oxide ion is O^{2-}.

We emphasize at this point that it is essential to learn the name, composition, and charge of each of the common polyatomic anions (and the NH_4^+ cation). If you do not recognize these ions by formula and by name, you will not be able to write the compound's name given its formula or the compound's formula given its name. You must also learn the names of the common acids.

EXAMPLE 5.9 | Writing Formulas from Names

Give the formula for each of the following compounds.

a. potassium hydroxide   e. calcium chloride

b. sodium carbonate   f. lead(IV) oxide

c. nitric acid   g. dinitrogen pentoxide

d. cobalt(III) nitrate   h. ammonium perchlorate

SOLUTION

Name	Formula	Comments
a. potassium hydroxide	KOH	Contains K^+ and OH^-.
b. sodium carbonate	Na_2CO_3	We need two Na^+ to balance CO_3^{2-}.
c. nitric acid	HNO_3	Common strong acid; memorize.
d. cobalt(III) nitrate	$Co(NO_3)_3$	Cobalt(III) means Co^{3+}; we need three NO_3^- to balance Co^{3+}.
e. calcium chloride	$CaCl_2$	We need two Cl^- to balance Ca^{2+}; Ca (Group 2) always forms Ca^{2+}.
f. lead(IV) oxide	PbO_2	Lead(IV) means Pb^{4+}; we need two O^{2-} to balance Pb^{4+}.
g. dinitrogen pentoxide	N_2O_5	*Di-* means two; *pent(a)-* means five.
h. ammonium perchlorate	NH_4ClO_4	Contains NH_4^+ and ClO_4^-.

Self-Check **EXERCISE 5.8** Write the formula for each of the following compounds.

a. ammonium sulfate

b. vanadium(V) fluoride

c. disulfur dichloride

d. rubidium peroxide

e. aluminum oxide

See Problems 5.41 through 5.46. ■

CHAPTER 5 REVIEW

Key Terms

binary compound (5.1)
binary ionic
  compound (5.2)

polyatomic ion (5.5)
oxyanion (5.5)
acid (5.6)

Summary

1. Binary compounds can be named systematically by following a set of relatively simple rules. For compounds containing both a metal and a nonmetal, the metal is always named first, followed by a name derived from the root name for the nonmetal. For compounds containing a metal that can form more than one cation (Type II), we use a Roman numeral to specify the cation's charge. In binary compounds containing only nonmetals (Type III), prefixes are used to specify the numbers of atoms.

2. Polyatomic ions are charged entities composed of several atoms bound together. These have special names that must be memorized. Naming ionic compounds that contain polyatomic ions is very similar to naming binary ionic compounds.

3. The names of acids (molecules with one or more H^+ ions attached to an anion) depend on whether the anion contains oxygen.

Active Learning Questions

These questions are designed to be considered by groups of students in class. Often these questions work well for introducing a particular topic in class.

1. In some cases the Roman numeral in a name is the same as a subscript in the formula, and in some cases it is not. Provide an example (formula and name) for each of these cases. Explain why the Roman numeral is not necessarily the same as the subscript.

2. The formulas $CaCl_2$ and $CoCl_2$ look very similar. What is the name for each compound? Why do we name them differently?

3. The formulas MgO and CO look very similar. What is the name for each compound? Why do we name them differently?

4. Explain how to use the periodic table to determine that there are two chloride ions for every magnesium ion in magnesium chloride and one chloride ion for every sodium ion in sodium chloride. Then write the formulas for calcium oxide and potassium oxide and explain how you got them.

5. What is the general formula for an ionic compound formed by elements in the following groups? Explain your reasoning and provide an example for each (name and formula).

   a. Group 1 with group 7
   b. Group 2 with group 7
   c. Group 1 with group 6
   d. Group 2 with group 6

6. An element forms an ionic compound with chlorine, leading to a compound having the formula XCl_2. The ion of element X has mass number 89 and 36 electrons. Identify the element X, tell how many neutrons it has, and name the compound.

7. Name each of the following compounds.

   a.

   b.

   c. SO_s
   d. P_2S_5

8. Why do we call $Ba(NO_3)_2$ barium nitrate but call $Fe(NO_3)_2$ iron(II) nitrate?

9. What is the difference between sulfuric acid and hydrosulfuric acid?

OWL Questions and problems for this chapter are available in OWL. Use the OWL access card that came with this textbook to access assigned questions and problems for this chapter.

6 Chemical Reactions: An Introduction

● Lightning over Seattle, Washington. *(Chuck Pefley/Tips Images)*

Chemistry is about change. Grass grows. Steel rusts. Hair is bleached, dyed, "permed," or straightened. Natural gas burns to heat houses. Nylon is produced for jackets, swimsuits, and pantyhose. Water is decomposed to hydrogen and oxygen gas by an electric current. Grape juice ferments in the production of wine. The bombardier beetle concocts a toxic spray to shoot at its enemies (see "Chemistry in Focus," p. 153).

These are just a few examples of chemical changes that affect each of us. Chemical reactions are the heart and soul of chemistry, and in this chapter we will discuss the fundamental ideas about chemical reactions.

Nylon is a strong material that makes sturdy parasails.

Production of plastic film for use in containers such as soft drink bottles *(left)*. Nylon being drawn from the boundary between two solutions containing different reactants *(right)*.

6.1 Evidence for a Chemical Reaction

OBJECTIVE: To learn the signals that show a chemical reaction has occurred.

Energy and chemical reactions will be discussed in more detail in Chapter 7.

How do we know when a chemical reaction has occurred? That is, what are the clues that a chemical change has taken place? A glance back at the processes in the introduction suggests that *chemical reactions often give a*

Table 6.1 Some Clues That a Chemical Reaction Has Occurred

1. The color changes.
2. A solid forms.
3. Bubbles form.
4. Heat and/or a flame is produced, or heat is absorbed.

visual signal. Steel changes from a smooth, shiny material to a reddish-brown, flaky substance when it rusts. Hair changes color when it is bleached. Solid nylon is formed when two particular liquid solutions are brought into contact. A blue flame appears when natural gas reacts with oxygen. Chemical reactions, then, often give *visual* clues: a color changes, a solid forms, bubbles are produced (see Figure 6.1), a flame occurs, and so on. However, reactions are not always visible. Sometimes the only signal that a reaction is occurring is a change in temperature as heat is produced or absorbed (see Figure 6.2).

Table 6.1 summarizes common clues to the occurrence of a chemical reaction, and Figure 6.3 gives some examples of reactions that show these clues.

Oxygen gas

Hydrogen gas

© Cengage Learning

Figure 6.1

Bubbles of hydrogen and oxygen gas form when an electric current is used to decompose water.

Spencer Grant/PhotoEdit

© Cengage Learning

a

An injured girl wearing a cold pack to help prevent swelling. The pack is activated by breaking an ampule; this initiates a chemical reaction that absorbs heat rapidly, lowering the temperature of the area to which the pack is applied.

b

A hot pack used to warm hands and feet in winter. When the package is opened, oxygen from the air penetrates a bag containing solid chemicals. The resulting reaction produces heat for several hours.

Figure 6.2

a

When colorless hydrochloric acid is added to a red solution of cobalt(II) nitrate, the solution turns blue, a sign that a chemical reaction has taken place.

b

A solid forms when a solution of sodium dichromate is added to a solution of lead nitrate.

c

Bubbles of hydrogen gas form when calcium metal reacts with water.

d

Methane gas reacts with oxygen to produce a flame in a Bunsen burner.

Figure 6.3

6.2 Chemical Equations

OBJECTIVE: To learn to identify the characteristics of a chemical reaction and the information given by a chemical equation.

Chemists have learned that a chemical change always involves a rearrangement of the ways in which the atoms are grouped. For example, when the methane, CH_4, in natural gas combines with oxygen, O_2, in the air and burns, carbon dioxide, CO_2, and water, H_2O, are formed. A chemical change such as this is called a **chemical reaction.** We represent a chemical reaction by writing a **chemical equation** in which the chemicals present before the reaction (the **reactants**) are shown to the left of an arrow and the chemicals formed by the reaction (the **products**) are shown to the right of an arrow. The arrow indicates the direction of the change and is read as "yields" or "produces":

$$\text{Reactants} \rightarrow \text{Products}$$

For the reaction of methane with oxygen, we have

$$\underset{\text{Methane}}{CH_4} \; + \; \underset{\text{Oxygen}}{O_2} \; \rightarrow \; \underset{\substack{\text{Carbon} \\ \text{dioxide}}}{CO_2} \; + \; \underset{\text{Water}}{H_2O}$$

Reactants   Products

Note from this equation that the products contain the same atoms as the reactants but that the atoms are associated in different ways. That is, a *chemical reaction involves changing the ways the atoms are grouped.*

It is important to recognize that **in a chemical reaction, atoms are neither created nor destroyed.** *All atoms present in the reactants must be accounted for among the products.* In other words, there must be the same number of each type of atom on the product side as on the reactant side of the arrow. Making sure that the equation for a reaction obeys this rule is called **balancing the chemical equation** for a reaction.

= C   = O   = H

Figure 6.4

The reaction between methane and oxygen to give water and carbon dioxide. Note that there are four oxygen atoms in the products *and* in the reactants; none has been gained or lost in the reaction. Similarly, there are four hydrogen atoms and one carbon atom in the reactants *and* in the products. The reaction simply changes the way the atoms are grouped.

The equation that we have shown for the reaction between CH_4 and O_2 is not balanced. We can see that it is not balanced by taking the reactants and products apart.

The reaction cannot happen this way because, as it stands, this equation states that one oxygen atom is created and two hydrogen atoms are destroyed. A reaction is only rearrangement of the way the atoms are grouped; atoms are not created or destroyed. The total number of each type of atom must be the same on both sides of the arrow. We can fix the imbalance in this equation by involving one more O_2 molecule on the left and by showing the production of one more H_2O molecule on the right.

This *balanced chemical equation* shows the actual numbers of molecules involved in this reaction (see Figure 6.4).

When we write the balanced equation for a reaction, we group like molecules together. Thus

$$CH_4 + \boxed{O_2 + O_2} \longrightarrow CO_2 + \boxed{H_2O + H_2O}$$

is written

$$CH_4 + \boxed{2O_2} \longrightarrow CO_2 + \boxed{2H_2O}$$

The chemical equation for a reaction provides us with two important types of information:

1. The identities of the reactants and products

2. The relative numbers of each

H
Group
1

O
Group
6

▶ **Physical States**

Besides specifying the compounds involved in the reaction, we often indicate in the equation the *physical states* of the reactants and products by using the following symbols:

Symbol	State
(s)	solid
(l)	liquid
(g)	gas
(aq)	dissolved in water (in aqueous solution)

a

The reactant potassium metal
(stored in mineral oil to prevent
oxidation).

Figure 6.5

b

The reactant water.

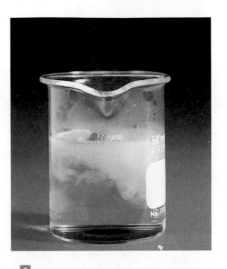

c

The reaction of potassium with
water. The flame occurs because
hydrogen gas, $H_2(g)$, produced by
the reaction burns in air [reacts with
$O_2(g)$] at the high temperatures
caused by the reaction.

For example, when solid potassium reacts with liquid water, the products are hydrogen gas and potassium hydroxide; the latter remains dissolved in the water. From this information about the reactants and products, we can write the equation for the reaction. Solid potassium is represented by $K(s)$; liquid water is written as $H_2O(l)$; hydrogen gas contains diatomic molecules and is represented as $H_2(g)$; potassium hydroxide dissolved in water is written as $KOH(aq)$. So the *unbalanced* equation for the reaction is

Solid		Hydrogen	Potassium hydroxide
potassium	Water	gas	dissolved in water

$$K(s) + H_2O(l) \rightarrow H_2(g) + KOH(aq)$$

This reaction is shown in Figure 6.5.

The hydrogen gas produced in this reaction then reacts with the oxygen gas in the air, producing gaseous water and a flame. The *unbalanced* equation for this second reaction is

$$H_2(g) + O_2(g) \rightarrow H_2O(g)$$

Both of these reactions produce a great deal of heat. In Example 6.1 we will practice writing the unbalanced equations for reactions. Then, in the next section, we will discuss systematic procedures for balancing equations.

EXAMPLE 6.1 | Chemical Equations: Recognizing Reactants and Products

Write the *unbalanced* chemical equation for each of the following reactions.

a. Solid mercury(II) oxide decomposes to produce liquid mercury metal and gaseous oxygen.

b. Solid carbon reacts with gaseous oxygen to form gaseous carbon dioxide.

c. Solid zinc is added to an aqueous solution containing dissolved hydrogen chloride to produce gaseous hydrogen that bubbles out of the solution and zinc chloride that remains dissolved in the water.

Richard Megna/Fundamental Photographs

Zinc metal reacts with hydrochloric acid to produce bubbles of hydrogen gas.

SOLUTION

a. In this case we have only one reactant, mercury(II) oxide. The name mercury(II) oxide means that the Hg^{2+} cation is present, so one O^{2-} ion is required for a zero net charge. Thus the formula is HgO, which is written HgO(s) in this case because it is given as a solid. The products are liquid mercury, written Hg(l), and gaseous oxygen, written $O_2(g)$. (Remember that oxygen exists as a diatomic molecule under normal conditions.) The unbalanced equation is

$$\underset{\text{Reactant}}{HgO(s)} \;\rightarrow\; \underset{\text{Products}}{Hg(l) \;+\; O_2(g)}$$

b. In this case, solid carbon, written C(s), reacts with oxygen gas, $O_2(g)$, to form gaseous carbon dioxide, which is written $CO_2(g)$. The equation (which happens to be balanced) is

$$\underset{\text{Reactants}}{C(s) \;+\; O_2(g)} \;\rightarrow\; \underset{\text{Product}}{CO_2(g)}$$

c. In this reaction solid zinc, Zn(s), is added to an aqueous solution of hydrogen chloride, which is written HCl(aq) and called hydrochloric acid. These are the reactants. The products of the reaction are gaseous hydrogen, $H_2(g)$, and aqueous zinc chloride. The name zinc chloride means that the Zn^{2+} ion is present, so two Cl^- ions are needed to achieve a zero net charge. Thus zinc chloride dissolved in water is written $ZnCl_2(aq)$. The unbalanced equation for the reaction is

Because Zn forms only the Zn^{2+} ion, a Roman numeral is usually not used. Thus $ZnCl_2$ is commonly called zinc chloride.

$$\underset{\text{Reactants}}{Zn(s) \;+\; HCl(aq)} \;\rightarrow\; \underset{\text{Products}}{H_2(g) \;+\; ZnCl_2(aq)}$$

Self-Check   EXERCISE 6.1   Identify the reactants and products and write the *unbalanced* equation (including symbols for states) for each of the following chemical reactions.

    a. Solid magnesium metal reacts with liquid water to form solid magnesium hydroxide and hydrogen gas.

    b. Solid ammonium dichromate (review Table 5.4 if this compound is unfamiliar) decomposes to solid chromium(III) oxide, gaseous nitrogen, and gaseous water.

    c. Gaseous ammonia reacts with gaseous oxygen to form gaseous nitrogen monoxide and gaseous water.

See Problems 6.13 through 6.34. ∎

6.3 Balancing Chemical Equations

OBJECTIVE: To learn how to write a balanced equation for a chemical reaction.

> Trial and error is often useful for solving problems. It's okay to make a few wrong turns before you get to the right answer.

As we saw in the previous section, an unbalanced chemical equation is not an accurate representation of the reaction that occurs. Whenever you see an equation for a reaction, you should ask yourself whether it is balanced. The principle that lies at the heart of the balancing process is that **atoms are conserved in a chemical reaction.** That is, atoms are neither created nor destroyed. They are just grouped differently. The same number of each type of atom is found among the reactants and among the products.

Chemists determine the identity of the reactants and products of a reaction by experimental observation. For example, when methane (natural gas) is burned in the presence of sufficient oxygen gas, the products are always carbon dioxide and water. **The identities (formulas) of the compounds must never be changed in balancing a chemical equation.** In other words, the subscripts in a formula cannot be changed, nor can atoms be added to or subtracted from a formula.

Most chemical equations can be balanced by trial and error—that is, by inspection. Keep trying until you find the numbers of reactants and products that give the same number of each type of atom on both sides of the arrow. For example, consider the reaction of hydrogen gas and oxygen gas to form liquid water. First, we write the unbalanced equation from the description of the reaction.

$$H_2(g) + O_2(g) \rightarrow H_2O(l)$$

We can see that this equation is unbalanced by counting the atoms on both sides of the arrow.

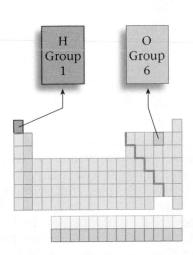

$$H_2(g) \quad + \quad O_2(g) \longrightarrow H_2O(l)$$

H H   O O   H / O / H

2 H   2 O   2 H, 1 O

Reactants	Products
2 H	2 H
2 O	1 O

We have one more oxygen atom in the reactants than in the product. Because we cannot create or destroy atoms and because we *cannot change the formulas* of the reactants or products, we must balance the equation by adding more molecules of reactants and/or products. In this case we need one more oxygen atom on the right, so we add another water molecule (which contains one O atom). Then we count all of the atoms again.

$$H_2(g) + O_2(g) \longrightarrow H_2O(l) + H_2O(l)$$

Reactants	Products
2 H	4 H
2 O	2 O

Totals: 2 H   2 O   4 H   2 O

We have balanced the oxygen atoms, but now the hydrogen atoms have become unbalanced. There are more hydrogen atoms on the right than on the left. We can solve this problem by adding another hydrogen molecule (H_2) to the reactant side.

$$H_2(g) + H_2(g) + O_2(g) \longrightarrow H_2O(l) + H_2O(l)$$

Reactants	Products
4 H	4 H
2 O	2 O

Totals: 4 H   2 O   4 H   2 O

The equation is now balanced. We have the same numbers of hydrogen and oxygen atoms represented on both sides of the arrow. Collecting like molecules, we write the balanced equation as

$$2H_2(g) + O_2(g) \rightarrow 2H_2O(l)$$

Consider next what happens if we multiply every part of this balanced equation by 2:

$$2 \times [2H_2(g) + O_2(g) \rightarrow 2H_2O(l)]$$

to give

$$4H_2(g) + 2O_2(g) \rightarrow 4H_2O(l)$$

This equation is balanced (count the atoms to verify this). In fact, we can multiply or divide *all parts* of the original balanced equation by any number to give a new balanced equation. Thus each chemical reaction has many possible balanced equations. Is one of the many possibilities preferred over the others? Yes.

The accepted convention is that the "best" balanced equation is the one with the *smallest integers (whole numbers)*. These integers are called the **coefficients** for the balanced equation. Therefore, for the reaction of hydrogen and oxygen to form water, the "correct" balanced equation is

$$2H_2(g) + O_2(g) \rightarrow 2H_2O(l)$$

The coefficients 2, 1 (never written), and 2, respectively, are the smallest *integers* that give a balanced equation for this reaction.

Next we will balance the equation for the reaction of liquid ethanol, C_2H_5OH, with oxygen gas to form gaseous carbon dioxide and water. This reaction, among many others, occurs in engines that burn a gasoline–ethanol mixture called gasohol.

The first step in obtaining the balanced equation for a reaction is always to identify the reactants and products from the description given for the reaction. In this case we are told that liquid ethanol, $C_2H_5OH(l)$, reacts

The Beetle That Shoots Straight

If someone said to you, "Name something that protects itself by spraying its enemies," your answer would almost certainly be "a skunk." Of course, you would be correct, but there is another correct answer—the bombardier beetle. When threatened, this beetle shoots a boiling stream of toxic chemicals at its enemy. How does this clever beetle accomplish this? Obviously, the boiling mixture cannot be stored inside the beetle's body all the time. Instead, when endangered, the beetle mixes chemicals that produce the hot spray. The chemicals involved are stored in two compartments. One compartment contains the chemicals hydrogen peroxide (H_2O_2) and methylhydroquinone ($C_7H_8O_2$). The key reaction is the decomposition of hydrogen peroxide to form oxygen gas and water:

$$2H_2O_2(aq) \rightarrow 2H_2O(l) + O_2(g)$$

Hydrogen peroxide also reacts with the hydroquinones to produce other compounds that become part of the toxic spray.

However, none of these reactions occurs very fast unless certain enzymes are present. (En-zymes are natural substances that speed up biologic reactions by means we will not discuss here.) When the beetle mixes the hydrogen peroxide and hydroquinones with the enzyme, the decomposition of H_2O_2 occurs rapidly, producing a hot mixture pressurized by the formation of oxygen gas. When the gas pressure becomes high enough, the hot spray is ejected in one long stream or in short bursts. The beetle has a highly accurate aim and can shoot several attackers with one batch of spray.

Thomas Eisner

A bombardier beetle defending itself.

with gaseous oxygen, $O_2(g)$, to produce gaseous carbon dioxide, $CO_2(g)$, and gaseous water, $H_2O(g)$. Therefore, the unbalanced equation is

$$C_2H_5OH(l) \quad + \quad O_2(g) \quad \rightarrow \quad CO_2(g) \quad + \quad H_2O(g)$$

| Liquid ethanol | Gaseous oxygen | Gaseous carbon dioxide | Gaseous water |

> In balancing equations, start by looking at the most complicated molecule.

When one molecule in an equation is more complicated (contains more elements) than the others, it is best to start with that molecule. The most complicated molecule here is C_2H_5OH, so we begin by considering the products that contain the atoms in C_2H_5OH. We start with carbon. The only product that contains carbon is CO_2. Because C_2H_5OH contains two carbon atoms, we place a 2 before the CO_2 to balance the carbon atoms.

C_2H_5OH

2 C, 6 H, 1 O

$$C_2H_5OH(l) + O_2(g) \longrightarrow 2CO_2(g) + H_2O(g)$$

2 C atoms   2 C atoms

Remember, we cannot change the formula of any reactant or product when we balance an equation. We can only place coefficients in front of the formulas.

Next we consider hydrogen. The only product containing hydrogen is H_2O. C_2H_5OH contains six hydrogen atoms, so we need six hydrogen atoms on the right. Because each H_2O contains two hydrogen atoms, we need three H_2O molecules to yield six hydrogen atoms. So we place a 3 before the H_2O.

$$C_2H_5OH(l) + O_2(g) \longrightarrow 2CO_2(g) + 3H_2O(g)$$
$$\underset{6\,H}{\underbrace{(5+1)\,H}} \qquad \underset{6\,H}{\underbrace{(3\times2)\,H}}$$

Finally, we count the oxygen atoms. On the left we have three oxygen atoms (one in C_2H_5OH and two in O_2), and on the right we have seven oxygen atoms (four in $2CO_2$ and three in $3H_2O$). We can correct this imbalance if we have three O_2 molecules on the left. That is, we place a coefficient of 3 before the O_2 to produce the balanced equation.

$$C_2H_5OH(l) + 3O_2(g) \longrightarrow 2CO_2(g) + 3H_2O(g)$$
$$\underset{7\,O}{\underbrace{1\,O \qquad (3\times2)\,O}} \qquad \underset{7\,O}{\underbrace{(2\times2)\,O \qquad 3\,O}}$$

At this point you may have a question: why did we choose O_2 on the left when we balanced the oxygen atoms? Why not use C_2H_5OH, which has an oxygen atom? The answer is that if we had changed the coefficient in front of C_2H_5OH, we would have unbalanced the hydrogen and carbon atoms. Now we count all of the atoms as a check to make sure the equation is balanced.

$$C_2H_5OH(l) + 3O_2(g) \longrightarrow 2CO_2(g) + 3H_2O(g)$$

Totals: 2 C   6 H   7 O | 2 C   7 O   6 H

The equation is now balanced. We have the same numbers of all types of atoms on both sides of the arrow. Notice that these coefficients are the smallest integers that give a balanced equation.

The process of writing and balancing the equation for a chemical reaction consists of several steps:

Reactants	Products
2 C	2 C
6 H	6 H
7 O	7 O

How to Write and Balance Equations

Step 1 Read the description of the chemical reaction. What are the reactants, the products, and their states? Write the appropriate formulas.

Step 2 Write the *unbalanced* equation that summarizes the information from step 1.

Step 3 Balance the equation by inspection, starting with the most complicated molecule. Proceed element by element to determine what coefficients are necessary so that the same number of each type of atom appears on both the reactant side and the product side. Do not change the identities (formulas) of any of the reactants or products.

(continued)

Step 4 Check to see that the coefficients used give the same number of each type of atom on both sides of the arrow. (Note that an "atom" may be present in an element, a compound, or an ion.) Also check to see that the coefficients used are the smallest integers that give the balanced equation. This can be done by determining whether all coefficients can be divided by the same integer to give a set of smaller *integer* coefficients.

EXAMPLE 6.2 | **Balancing Chemical Equations I**

For the following reaction, write the unbalanced equation and then balance the equation: solid potassium reacts with liquid water to form gaseous hydrogen and potassium hydroxide that dissolves in the water.

SOLUTION

Step 1 From the description given for the reaction, we know that the reactants are solid potassium, $K(s)$, and liquid water, $H_2O(l)$. The products are gaseous hydrogen, $H_2(g)$, and dissolved potassium hydroxide, $KOH(aq)$.

Step 2 The unbalanced equation for the reaction is

$$K(s) + H_2O(l) \rightarrow H_2(g) + KOH(aq)$$

Step 3 Although none of the reactants or products is very complicated, we will start with KOH because it contains the most elements (three). We will arbitrarily consider hydrogen first. Note that on the reactant side of the equation in step 2, there are two hydrogen atoms but on the product side there are three. If we place a coefficient of 2 in front of both H_2O and KOH, we now have four H atoms on each side.

$$K(s) + 2H_2O(l) \rightarrow H_2(g) + 2KOH(aq)$$

| 4 H | 2 H | 2 H |
| atoms | atoms | atoms |

Also note that the oxygen atoms balance.

$$K(s) + 2H_2O(l) \rightarrow H_2(g) + 2KOH(aq)$$

| 2 O | 2 O |
| atoms | atoms |

However, the K atoms do not balance; we have one on the left and two on the right. We can fix this easily by placing a coefficient of 2 in front of $K(s)$ to give the balanced equation:

$$2K(s) + 2H_2O(l) \rightarrow H_2(g) + 2KOH(aq)$$

Step 4

CHECK: There are 2 K, 4 H, and 2 O on both sides of the arrow, and the coefficients are the smallest integers that give a balanced equation. We know this because we cannot divide through by a given integer to give a set of smaller *integer* (whole-number) coefficients. For example, if we divide all of the coefficients by 2, we get

$$K(s) + H_2O(l) \rightarrow \tfrac{1}{2}H_2(g) + KOH(aq)$$

This is not acceptable because the coefficient for H_2 is not an integer. ■

Reactants	Products
2 K	2 K
4 H	4 H
2 O	2 O

EXAMPLE 6.3 | Balancing Chemical Equations II

Under appropriate conditions at 1000 °C, ammonia gas reacts with oxygen gas to produce gaseous nitrogen monoxide (common name, nitric oxide) and gaseous water. Write the unbalanced and balanced equations for this reaction.

SOLUTION

Step 1 The reactants are gaseous ammonia, $NH_3(g)$, and gaseous oxygen, $O_2(g)$. The products are gaseous nitrogen monoxide, $NO(g)$, and gaseous water, $H_2O(g)$.

Reactants	Products
1 N	1 N
3 H	2 H
2 O	2 O

Step 2 The unbalanced equation for the reaction is

$$NH_3(g) + O_2(g) \rightarrow NO(g) + H_2O(g)$$

Step 3 In this equation there is no molecule that is obviously the most complicated. Three molecules contain two elements, so we arbitrarily start with NH_3. We arbitrarily begin by looking at hydrogen. A coefficient of 2 for NH_3 and a coefficient of 3 for H_2O give six atoms of hydrogen on both sides.

$$2NH_3(g) + O_2(g) \rightarrow NO(g) + 3H_2O(g)$$
$$\underbrace{\qquad}_{6\ H} \qquad\qquad\qquad \underbrace{\qquad}_{6\ H}$$

We can balance the nitrogen by giving NO a coefficient of 2.

$$2NH_3(g) + O_2(g) \rightarrow 2NO(g) + 3H_2O(g)$$
$$\underbrace{\qquad}_{2\ N} \qquad\qquad \underbrace{\qquad}_{2\ N}$$

$\frac{5}{2} = 2\frac{1}{2}$
O—O
O—O $2\frac{1}{2}\ O_2$
O—O   contains
O┼O   5 O atoms

Finally, we note that there are two atoms of oxygen on the left and five on the right. The oxygen can be balanced with a coefficient of $\frac{5}{2}$ for O_2, because $\frac{5}{2} \times O_2$ gives five oxygen atoms.

$$2NH_3(g) + \tfrac{5}{2}O_2(g) \rightarrow 2NO(g) + 3H_2O(g)$$
$$\underbrace{\qquad}_{5\ O} \quad \underbrace{\qquad}_{2\ O} \quad \underbrace{\qquad}_{3\ O}$$

However, the convention is to have integer (whole-number) coefficients, so we multiply the entire equation by 2.

$$2 \times [2NH_3(g) + \tfrac{5}{2}O_2(g) \rightarrow 2NO(g) + 3H_2O(g)]$$

or

$$2 \times 2NH_3(g) + 2 \times \tfrac{5}{2}O_2(g) \rightarrow 2 \times 2NO(g) + 2 \times 3H_2O(g)$$
$$4NH_3(g) + 5O_2(g) \rightarrow 4NO(g) + 6H_2O(g)$$

Reactants	Products
4 N	4 N
12 H	12 H
10 O	10 O

Step 4

CHECK: There are 4 N, 12 H, and 10 O atoms on both sides, so the equation is balanced. These coefficients are the smallest integers that give a balanced equation. That is, we cannot divide all coefficients by the same integer and obtain a smaller set of *integers*.

Self-Check EXERCISE 6.2 Propane, C_3H_8, a liquid at 25 °C under high pressure, is often used for gas grills and as a fuel in rural areas where there is no natural gas pipeline. When liquid propane is released from its storage tank, it changes to propane gas that reacts with oxygen gas (it "burns") to give gaseous carbon dioxide and gaseous water. Write and balance the equation for this reaction.

HINT: This description of a chemical process contains many words, some of which are crucial to solving the problem and some of which are not. First sort out the important information and use symbols to represent it.

See Problems 6.37 through 6.44. ∎

EXAMPLE 6.4 | Balancing Chemical Equations III

Decorations on glass are produced by etching with hydrofluoric acid.

Reactants	Products
1 Si	1 Si
1 H	2 H
1 F	4 F
2 O	1 O

Reactants	Products
1 Si	1 Si
4 H	2 H
4 F	4 F
2 O	1 O

Reactants	Products
1 Si	1 Si
4 H	4 H
4 F	4 F
2 O	2 O

Glass is sometimes decorated by etching patterns on its surface. Etching occurs when hydrofluoric acid (an aqueous solution of HF) reacts with the silicon dioxide in the glass to form gaseous silicon tetrafluoride and liquid water. Write and balance the equation for this reaction.

SOLUTION

Step 1 From the description of the reaction we can identify the reactants:

hydrofluoric acid $\qquad$ HF(aq)

solid silicon dioxide $\qquad$ SiO$_2$(s)

and the products:

gaseous silicon tetrafluoride $\qquad$ SiF$_4$(g)

liquid water $\qquad$ H$_2$O(l)

Step 2 The unbalanced equation is

$$SiO_2(s) + HF(aq) \rightarrow SiF_4(g) + H_2O(l)$$

Step 3 There is no clear choice here for the most complicated molecule. We arbitrarily start with the elements in SiF$_4$. The silicon is balanced (one atom on each side), but the fluorine is not. To balance the fluorine, we need a coefficient of 4 before the HF.

$$SiO_2(s) + 4HF(aq) \rightarrow SiF_4(g) + H_2O(l)$$

Hydrogen and oxygen are not balanced. Because we have four hydrogen atoms on the left and two on the right, we place a 2 before the H$_2$O:

$$SiO_2(s) + 4HF(aq) \rightarrow SiF_4(g) + 2H_2O(l)$$

This balances the hydrogen *and* the oxygen (two atoms on each side).

Step 4

CHECK: $\quad$ SiO$_2$(s) + 4HF(aq) $\quad\rightarrow\quad$ SiF$_4$(g) + 2H$_2$O(l)

Totals: $\quad$ 1 Si, 2 O, 4 H, 4 F $\quad\rightarrow\quad$ 1 Si, 4 F, 4 H, 2 O

All atoms check, so the equation is balanced.

Self-Check $\quad$ **EXERCISE 6.3** $\quad$ Give the balanced equation for each of the following reactions.

a. When solid ammonium nitrite is heated, it produces nitrogen gas and water vapor.

b. Gaseous nitrogen monoxide (common name, nitric oxide) decomposes to produce dinitrogen monoxide gas (common name, nitrous oxide) and nitrogen dioxide gas.

c. Liquid nitric acid decomposes to reddish-brown nitrogen dioxide gas, liquid water, and oxygen gas. (This is why bottles of nitric acid become yellow upon standing.)

If you are having trouble writing formulas from names, review the appropriate sections of Chapter 5. It is very important that you are able to do this.

See Problems 6.37 through 6.44. ∎

CHAPTER 6 REVIEW

Key Terms

chemical reaction (6.2)
chemical equation (6.2)
reactant (6.2)
product (6.2)

balancing a chemical
   equation (6.2)
coefficient (6.3)

Summary

1. Chemical reactions usually give some kind of visual signal—a color changes, a solid forms, bubbles form, heat and/or flame is produced.

2. A chemical equation represents a chemical reaction. Reactants are shown on the left side of an arrow and products on the right. In a chemical reaction, atoms are neither created nor destroyed; they are merely rearranged. A balanced chemical equation gives the relative numbers of reactant and product molecules.

3. A chemical equation for a reaction can be balanced by using a systematic approach. First identify the reactants and products and write the formulas. Next write the unbalanced equation. Then balance by trial and error, starting with the most complicated molecule(s). Finally, check to be sure the equation is balanced.

Active Learning Questions

These questions are designed to be considered by groups of students in class. Often these questions work well for introducing a particular topic in class.

1. The following are actual student responses to the question: Why is it necessary to balance chemical equations?

   a. The chemicals will not react until you have added the correct ratios.
   b. The correct products will not form unless the right amounts of reactants have been added.
   c. A certain number of products cannot form without a certain number of reactants.
   d. The balanced equation tells you how much reactant you need, and allows you to predict how much product you will make.
   e. A ratio must be established for the reaction to occur as written.

   Justify the best choice, and, for choices you did not pick, explain what is wrong with them.

2. What information do we get from a formula? From an equation?

3. Given the equation for the reaction: $N_2 + H_2 \rightarrow NH_3$, draw a molecular diagram that represents the reaction (make sure it is balanced).

4. What do the subscripts in a chemical formula represent? What do the coefficients in a balanced chemical equation represent?

5. Can the subscripts in a chemical formula be fractions? Explain.

6. Can the coefficients in a balanced chemical equation be fractions? Explain.

7. Changing the subscripts of chemicals can mathematically balance the equations. Why is this unacceptable?

8. Table 6.1 lists some clues that a chemical reaction has occurred. However, these events do not necessarily prove the existence of a chemical change. Give an example for each of the clues that is not a chemical reaction but a physical change.

9. Use molecular-level drawings to show the difference between physical and chemical changes.

10. It is stated in Section 6.3 of the text that to balance equations by inspection you start "with the most complicated molecule." What does this mean? Why is it best to do this?

11. Which of the following statements concerning balanced chemical equations are true? There may be more than one true statement.

   a. Atoms are neither created nor destroyed.
   b. The coefficients indicate the mass ratios of the substances used.
   c. The sum of the coefficients on the reactant side always equals the sum of the coefficients on the product side.

12. Consider the generic chemical equation $aA + bB \rightarrow cC + dD$ (where a, b, c, and d represent coefficients for the chemicals A, B, C, and D, respectively).

   a. How many possible values are there for "c"? Explain your answer.
   b. How many possible values are there for "c/d"? Explain your answer.

13. How is the balancing of chemical equations related to the law of conservation of mass?

14. Which of the following correctly describes the balanced chemical equation given below? There may be

more than one true statement. If a statement is incorrect, explain why it is incorrect.

$$4Al + 3O_2 \rightarrow 2Al_2O_3$$

a. For every 4 atoms of aluminum that react with 6 atoms of oxygen, 2 molecules of aluminum oxide are produced.

b. For every 4 moles of aluminum that reacts with 3 moles of oxygen, 2 moles of aluminum (III) oxide is produced.

c. For every 4 grams of aluminum that reacts with 3 grams of oxygen, 2 grams of aluminum oxide is produced.

15. Which of the following correctly balances the chemical equation given below? There may be more than one correct balanced equation. If a balanced equation is incorrect, explain why it is incorrect.

$$CaO + C \rightarrow CaC_2 + CO_2$$

a. $CaO_2 + 3C \rightarrow CaC_2 + CO_2$

b. $2CaO + 5C \rightarrow 2CaC_2 + CO_2$

c. $CaO + 2\frac{1}{2}C \rightarrow CaC_2 + \frac{1}{2}CO_2$

d. $4CaO + 10C \rightarrow 4CaC_2 + 2CO_2$

16. The reaction of an element X ($\triangle$) with element Y ($\bigcirc$) is represented in the following diagram. Which of the elements best describes this reaction?

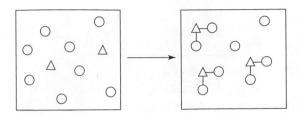

a. $3X + 8Y \rightarrow X_3Y_8$

b. $3X + 6Y \rightarrow X_3Y_6$

c. $X + 2Y \rightarrow XY_2$

d. $3X + 8Y \rightarrow 3XY_2 + 2Y$

OWL Questions and problems for this chapter are available in OWL. Use the OWL access card that came with this textbook to access assigned questions and problems for this chapter.

7 Reactions in Aqueous Solutions

● Chlorine in water reacting with potassium bromide. *(Richard Megna/Fundamental Photographs)*

A burning match involves several chemical reactions.

Royalty-Free Corbis

The chemical reactions that are most important to us occur in water—in aqueous solutions. Virtually all of the chemical reactions that keep each of us alive and well take place in the aqueous medium present in our bodies. For example, the oxygen you breathe dissolves in your blood, where it associates with the hemoglobin in the red blood cells. While attached to the hemoglobin it is transported to your cells, where it reacts with fuel (from the food you eat) to provide energy for living. However, the reaction between oxygen and fuel is not direct—the cells are not tiny furnaces. Instead, electrons are transferred from the fuel to a series of molecules that pass them along (this is called the respiratory chain) until they eventually reach oxygen. Many other reactions are also crucial to our health and well-being. You will see numerous examples of these as you continue your study of chemistry.

In this chapter we will study some common types of reactions that take place in water, and we will become familiar with some of the driving forces that make these reactions occur. We will also learn how to predict the products for these reactions and how to write various equations to describe them.

7.1 Predicting Whether a Reaction Will Occur

OBJECTIVE: To learn about some of the factors that cause reactions to occur.

In this text we have already seen many chemical reactions. Now let's consider an important question: Why does a chemical reaction occur? What causes reactants to "want" to form products? As chemists have studied reactions, they have recognized several "tendencies" in reactants that drive them to form products. That is, there are several "driving forces" that pull reactants toward products—changes that tend to make reactions go in the direction of the arrow. The most common of these driving forces are

1. Formation of a solid
2. Formation of water
3. Transfer of electrons
4. Formation of a gas

When two or more chemicals are brought together, if any of these things can occur, a chemical change (a reaction) is likely to take place. Accordingly, when we are confronted with a set of reactants and want to predict whether a reaction will occur and what products might form, we will consider these driving forces. They will help us organize our thoughts as we encounter new reactions.

7.2 Reactions in Which a Solid Forms

OBJECTIVE: To learn to identify the solid that forms in a precipitation reaction.

One driving force for a chemical reaction is the formation of a solid, a process called **precipitation.** The solid that forms is called a **precipitate,** and the reaction is known as a **precipitation reaction.** For example, when an aqueous (water) solution of potassium chromate, $K_2CrO_4(aq)$, which is yellow, is added to a colorless aqueous solution containing barium nitrate, $Ba(NO_3)_2(aq)$, a yellow solid forms (see Figure 7.1). The fact that a solid forms tells us that a reaction—a chemical change—has occurred. That is, we have a situation where

$$Reactants \rightarrow Products$$

What is the equation that describes this chemical change? To write the equation, we must decipher the identities of the reactants and products. The reactants have already been described: $K_2CrO_4(aq)$ and $Ba(NO_3)_2(aq)$. Is there some way in which we can predict the identities of the products? What is the yellow solid? The best way to predict the identity of this solid is to first *consider what products are possible*. To do this we need to know what chemical species are present in the solution that results when the reactant solutions are mixed. First, let's think about the nature of each reactant in an aqueous solution.

Figure 7.1

The precipitation reaction that occurs when yellow potassium chromate, $K_2CrO_4(aq)$, is mixed with a colorless barium nitrate solution, $Ba(NO_3)_2(aq)$.

▶ What Happens When an Ionic Compound Dissolves in Water?

The designation $Ba(NO_3)_2(aq)$ means that barium nitrate (a white solid) has been dissolved in water. Note from its formula that barium nitrate contains the Ba^{2+} and NO_3^- ions. *In virtually every case when a solid containing ions dissolves in water, the ions separate* and move around independently. That is, $Ba(NO_3)_2(aq)$ does not contain $Ba(NO_3)_2$ units. Rather, it contains separated Ba^{2+} and NO_3^- ions. In the solution there are two NO_3^- ions for every Ba^{2+} ion. Chemists know that separated ions are present in this solution because it is an excellent conductor of electricity (see Figure 7.2). Pure water does not conduct an electric current. Ions must be present in water for a current to flow.

When each unit of a substance that dissolves in water produces separated ions, the substance is called a **strong electrolyte.** Barium nitrate is a strong electrolyte in water, because each $Ba(NO_3)_2$ unit produces the separated ions (Ba^{2+}, NO_3^-, NO_3^-).

Similarly, aqueous K_2CrO_4 also behaves as a strong electrolyte. Potassium chromate contains the K^+ and CrO_4^{2-} ions, so an aqueous solution of

a

Pure water does not conduct an electric current. The lamp does not light.

b

When an ionic compound is dissolved in water, current flows and the lamp lights. The result of this experiment is strong evidence that ionic compounds dissolved in water exist in the form of separated ions.

Figure 7.2

Electrical conductivity of aqueous solutions.

potassium chromate (which is prepared by dissolving solid K_2CrO_4 in water) contains these separated ions. That is, $K_2CrO_4(aq)$ does not contain K_2CrO_4 units but instead contains K^+ cations and CrO_4^{2-} anions, which move around independently. (There are two K^+ ions for each CrO_4^{2-} ion.)

The idea introduced here is very important: when ionic compounds dissolve, the *resulting solution contains the separated ions.* Therefore, we can represent the mixing of $K_2CrO_4(aq)$ and $Ba(NO_3)_2(aq)$ in two ways. We usually write these reactants as

$$K_2CrO_4(aq) + Ba(NO_3)_2(aq) \rightarrow \text{Products}$$

However, a more accurate representation of the situation is

We can express this information in equation form as follows:

$$2K^+(aq) + CrO_4^{2-}(aq) + Ba^{2+}(aq) + 2NO_3^-(aq) \rightarrow \text{Products}$$

The ions in $K_2CrO_4\ (aq)$ The ions in $Ba(NO_3)_2(aq)$

Thus the *mixed solution* contains four types of ions: K^+, CrO_4^{2-}, Ba^{2+}, and NO_3^-. Now that we know what the reactants are, we can make some educated guesses about the possible products.

▶ How to Decide What Products Form

Which of these ions combine to form the yellow solid observed when the original solutions are mixed? This is not an easy question to answer. Even an experienced chemist is not sure what will happen in a new reaction. The chemist tries to think of the various possibilities, considers the likelihood of each possibility, and then makes a prediction (an educated guess). Only after identifying each product experimentally can the chemist be sure what reaction actually has taken place. However, an educated guess is very useful because it indicates what kinds of products are most likely. It gives us a place to start. So the best way to proceed is first to think of the various possibilities and then to decide which of them is most likely.

What are the possible products of the reaction between $K_2CrO_4(aq)$ and $Ba(NO_3)_2(aq)$ or, more accurately, what reaction can occur among the ions K^+, CrO_4^{2-}, Ba^{2+}, and NO_3^-? We already know some things that will help us decide. We know that a *solid compound must have a zero net charge*. This means that the product of our reaction must contain *both anions and cations* (negative and positive ions). For example, K^+ and Ba^{2+} could not combine to form the solid because such a solid would have a positive charge. Similarly, CrO_4^{2-} and NO_3^- could not combine to form a solid because that solid would have a negative charge.

Something else that will help us is an observation that chemists have made by examining many compounds: *most ionic materials contain only two types of ions*—one type of cation and one type of anion. This idea is illustrated by the following compounds (among many others):

Compound	Cation	Anion
NaCl	Na^+	Cl^-
KOH	K^+	OH^-
Na_2SO_4	Na^+	SO_4^{2-}
NH_4Cl	NH_4^+	Cl^-
Na_2CO_3	Na^+	CO_3^{2-}

All the possible combinations of a cation and an anion to form uncharged compounds from among the ions K^+, CrO_4^{2-}, Ba^{2+}, and NO_3^- are shown below:

	NO_3^-	CrO_4^{2-}
K^+	KNO_3	K_2CrO_4
Ba^{2+}	$Ba(NO_3)_2$	$BaCrO_4$

So the compounds that *might* make up the solid are

K_2CrO_4 $\qquad$ $BaCrO_4$

KNO_3 $\qquad$ $Ba(NO_3)_2$

Which of these possibilities is most likely to represent the yellow solid? We know it's not K_2CrO_4 or $Ba(NO_3)_2$; these are the reactants. They were present (dissolved) in the separate solutions that were mixed initially. The only real possibilities are KNO_3 and $BaCrO_4$. To decide which of these is more likely to represent the yellow solid, we need more facts. An experienced chemist, for example, knows that KNO_3 is a white solid. On the other hand, the CrO_4^{2-} ion is yellow. Therefore, the yellow solid most likely is $BaCrO_4$.

We have determined that one product of the reaction between $K_2CrO_4(aq)$ and $Ba(NO_3)_2(aq)$ is $BaCrO_4(s)$, but what happened to the K^+ and NO_3^- ions? The answer is that these ions are left dissolved in the solution. That is, KNO_3 does not form a solid when the K^+ and NO_3^- ions are present in water. In other words, if we took the white solid $KNO_3(s)$ and put it in water, it would totally dissolve (the white solid would "disappear," yielding a colorless solution). So when we mix $K_2CrO_4(aq)$ and $Ba(NO_3)_2(aq)$, $BaCrO_4(s)$ forms but KNO_3 is left behind in solution [we write it as $KNO_3(aq)$]. (If we poured the mixture through a filter to remove the solid $BaCrO_4$ and then evaporated all of the water, we would obtain the white solid KNO_3.)

After all this thinking, we can finally write the unbalanced equation for the precipitation reaction:

$$K_2CrO_4(aq) + Ba(NO_3)_2(aq) \rightarrow BaCrO_4(s) + KNO_3(aq)$$

We can represent this reaction in pictures as follows:

Note that the K^+ and NO_3^- ions are not involved in the chemical change. They remain dispersed in the water before and after the reaction.

▶ Using Solubility Rules

In the example considered above we were finally able to identify the products of the reaction by using two types of chemical knowledge:

1. Knowledge of facts

2. Knowledge of concepts

For example, knowing the colors of the various compounds proved very helpful. This represents factual knowledge. Awareness of the concept that solids always have a net charge of zero was also essential. These two kinds of knowledge allowed us to make a good guess about the identity of the solid that formed. As you continue to study chemistry, you will see that a balance of factual and conceptual knowledge is always required. You must both *memorize* important facts and *understand* crucial concepts to succeed.

In the present case we are dealing with a reaction in which an ionic solid forms—that is, a process in which ions that are dissolved in water combine to give a solid. We know that for a solid to form, both positive and negative ions must be present in relative numbers that give zero net charge. However, oppositely charged ions in water do not always react to form a solid, as we have seen for K^+ and NO_3^-. In addition, Na^+ and Cl^- can coexist in water in very large numbers with no formation of solid NaCl. In other words, when solid NaCl (common salt) is placed in water, it dissolves—the white solid "disappears" as the Na^+ and Cl^- ions are dispersed throughout the water. (You probably have observed this phenomenon in preparing salt water to cook food.) The following two statements, then, are really saying the same thing.

1. Solid NaCl is very soluble in water.

2. Solid NaCl does not form when one solution containing Na^+ is mixed with another solution containing Cl^-.

To predict whether a given pair of dissolved ions will form a solid when mixed, we must know some facts about the solubilities of various types of ionic compounds. In this text we will use the term **soluble solid** to mean a solid that readily dissolves in water; the solid "disappears" as the ions are dispersed in the water. The terms **insoluble solid** and **slightly soluble solid** are taken to mean the same thing: a solid where such a tiny amount dissolves in water that it is undetectable with the naked eye. The solubility information about common solids that is summarized in Table 7.1 is based on observations of the behavior of many compounds. This is factual knowledge that you will need to predict what will happen in chemical reactions where a solid might form. This information is summarized in Figure 7.3.

> Solids must contain both anions and cations in the relative numbers necessary to produce zero net charge.

Table 7.1   General Rules for Solubility of Ionic Compounds (Salts) in Water at 25 °C

1. Most nitrate (NO_3^-) salts are soluble.
2. Most salts of Na^+, K^+, and NH_4^+ are soluble.
3. Most chloride salts are soluble. Notable exceptions are AgCl, $PbCl_2$, and Hg_2Cl_2.
4. Most sulfate salts are soluble. Notable exceptions are $BaSO_4$, $PbSO_4$, and $CaSO_4$. *CrO_4^{2-} salts depend on cation*
5. Most hydroxide compounds are only slightly soluble.* The important exceptions are NaOH and KOH. $Ba(OH)_2$ and $Ca(OH)_2$ are only moderately soluble.
6. Most sulfide (S^{2-}), carbonate (CO_3^{2-}), and phosphate (PO_4^{3-}) salts are only slightly soluble.*

*The terms *insoluble* and *slightly soluble* really mean the same thing: such a tiny amount dissolves that it is not possible to detect it with the naked eye.

go Chemistry **Module 5: Predicting the Water Solubility of Common Ionic Compounds** covers concepts in this section.

Notice that in Table 7.1 and Figure 7.3 the term *salt* is used to mean *ionic compound*. Many chemists use the terms *salt* and *ionic compound* interchangeably. In Example 7.1, we will illustrate how to use the solubility rules to predict the products of reactions among ions.

EXAMPLE 7.1   Identifying Precipitates in Reactions Where a Solid Forms

$AgNO_3$ is usually called silver nitrate rather than silver(I) nitrate because silver forms only Ag^+.

When an aqueous solution of silver nitrate is added to an aqueous solution of potassium chloride, a white solid forms. Identify the white solid and write the balanced equation for the reaction that occurs.

SOLUTION

First let's use the description of the reaction to represent what we know:

$$AgNO_3(aq) + KCl(aq) \rightarrow \text{White solid}$$

Figure 7.3

Solubilities of common compounds.

Remember, try to determine the essential facts from the words and represent these facts by symbols or diagrams. To answer the main question (What is the white solid?), we must establish what ions are present in the mixed solution. That is, we must know what the reactants are really like. Remember that *when ionic substances dissolve in water, the ions separate.* So we can write the equation

$$Ag^+(aq) + NO_3^-(aq) + K^+(aq) + Cl^-(aq) \rightarrow \text{Products}$$

Ions in
AgNO$_3$(aq)

Ions in
KCl(aq)

or using pictures

AgNO$_3$(aq)   KCl(aq)   Products

to represent the ions present in the mixed solution before any reaction occurs. In summary:

AgNO$_3$(aq) + KCl(aq)   Solution contains

Now we will consider what solid *might* form from this collection of ions. Because the solid must contain both positive and negative ions, the possible compounds that can be assembled from this collection of ions are

	NO$_3^-$	Cl$^-$
Ag$^+$	AgNO$_3$	AgCl
K$^+$	KNO$_3$	KCl

AgNO$_3$	AgCl
KNO$_3$	KCl

AgNO$_3$ and KCl are the substances already dissolved in the reactant solutions, so we know that they do not represent the white solid product. We are left with two possibilities:

AgCl

KNO$_3$

Figure 7.4

Precipitation of silver chloride occurs when solutions of silver nitrate and potassium chloride are mixed. The K^+ and NO_3^- ions remain in solution.

Another way to obtain these two possibilities is by *ion interchange*. This means that in the reaction of $AgNO_3(aq)$ and $KCl(aq)$, we take the cation from one reactant and combine it with the anion of the other reactant.

$$Ag^+ + NO_3^- + K^+ + Cl^- \rightarrow Products$$

Possible solid products

Ion interchange also leads to the following possible solids:

$$AgCl \text{ or } KNO_3$$

To decide whether $AgCl$ or KNO_3 is the white solid, we need the solubility rules (Table 7.1). Rule 2 states that most salts containing K^+ are soluble in water. Rule 1 says that most nitrate salts (those containing NO_3^-) are soluble. So the salt KNO_3 is water-soluble. That is, when K^+ and NO_3^- are mixed in water, a solid (KNO_3) does *not* form.

On the other hand, Rule 3 states that although most chloride salts (salts that contain Cl^-) are soluble, $AgCl$ is an exception. That is, $AgCl(s)$ is insoluble in water. Thus the white solid must be $AgCl$. Now we can write

$$AgNO_3(aq) + KCl(aq) \rightarrow AgCl(s) + ?$$

What is the other product?

To form $AgCl(s)$, we have used the Ag^+ and Cl^- ions:

$$Ag^+(aq) + NO_3^-(aq) + K^+(aq) + Cl^-(aq) \rightarrow AgCl(s)$$

This leaves the K^+ and NO_3^- ions. What do they do? Nothing. Because KNO_3 is very soluble in water (Rules 1 and 2), the K^+ and NO_3^- ions remain separate in the water; the KNO_3 remains dissolved and we represent it as $KNO_3(aq)$. We can now write the full equation:

$$AgNO_3(aq) + KCl(aq) \rightarrow AgCl(s) + KNO_3(aq)$$

Figure 7.4 shows the precipitation of $AgCl(s)$ that occurs when this reaction takes place. In graphic form, the reaction is

The following strategy is useful for predicting what will occur when two solutions containing dissolved salts are mixed.

How to Predict Precipitates When Solutions of Two Ionic Compounds Are Mixed

Step 1 Write the reactants as they actually exist before any reaction occurs. Remember that when a salt dissolves, its ions separate.

Step 2 Consider the various solids that could form. To do this, simply *exchange the anions* of the added salts.

Step 3 Use the solubility rules (Table 7.1) to decide whether a solid forms and, if so, to predict the identity of the solid.

EXAMPLE 7.2 │ Using Solubility Rules to Predict the Products of Reactions

Using the solubility rules in Table 7.1, predict what will happen when the following solutions are mixed. Write the balanced equation for any reaction that occurs.

   a. $KNO_3(aq)$ and $BaCl_2(aq)$

   b. $Na_2SO_4(aq)$ and $Pb(NO_3)_2(aq)$

   c. $KOH(aq)$ and $Fe(NO_3)_3(aq)$

SOLUTION (a)

Step 1 $KNO_3(aq)$ represents an aqueous solution obtained by dissolving solid KNO_3 in water to give the ions $K^+(aq)$ and $NO_3^-(aq)$. Likewise, $BaCl_2(aq)$ is a solution formed by dissolving solid $BaCl_2$ in water to produce $Ba^{2+}(aq)$ and $Cl^-(aq)$. When these two solutions are mixed, the following ions will be present:

$$K^+, \quad NO_3^-, \quad Ba^{2+}, \quad Cl^-$$

From $KNO_3(aq)$ From $BaCl_2(aq)$

Step 2 To get the possible products, we exchange the anions.

$$K^+ \quad NO_3^- \quad Ba^{2+} \quad Cl^-$$

This yields the possibilities KCl and $Ba(NO_3)_2$. These are the solids that *might* form. Notice that two NO_3^- ions are needed to balance the 2+ charge on Ba^{2+}.

Step 3 The rules listed in Table 7.1 indicate that both KCl and $Ba(NO_3)_2$ are soluble in water. So no precipitate forms when $KNO_3(aq)$ and $BaCl_2(aq)$ are mixed. All of the ions remain dissolved in the solution. This means that no reaction takes place. That is, no chemical change occurs.

SOLUTION (b)

Step 1   The following ions are present in the mixed solution before any reaction occurs:

$$Na^+, \quad SO_4^{2-}, \quad Pb^{2+}, \quad NO_3^-$$

From Na$_2$SO$_4$(aq)   From Pb(NO$_3$)$_2$(aq)

Step 2   Exchanging anions

$$Na^+ \quad SO_4^{2-} \quad Pb^{2+} \quad NO_3^-$$

yields the *possible* solid products PbSO$_4$ and NaNO$_3$.

Step 3   Using Table 7.1, we see that NaNO$_3$ is soluble in water (Rules 1 and 2) but that PbSO$_4$ is only slightly soluble (Rule 4). Thus, when these solutions are mixed, solid PbSO$_4$ forms. The balanced reaction is

$$Na_2SO_4(aq) + Pb(NO_3)_2\,(aq) \rightarrow PbSO_4(s) + 2NaNO_3(aq)$$

Remains dissolved

which can be represented as

SOLUTION (c)

Step 1   The ions present in the mixed solution before any reaction occurs are

$$K^+, \quad OH^-, \quad Fe^{3+}, \quad NO_3^-$$

From KOH(aq)   From Fe(NO$_3$)$_3$(aq)

Step 2   Exchanging anions

$$K^+ \quad OH^- \quad Fe^{3+} \quad NO_3^-$$

yields the possible solid products KNO$_3$ and Fe(OH)$_3$.

Step 3   Rules 1 and 2 (Table 7.1) state that KNO$_3$ is soluble, whereas Fe(OH)$_3$ is only slightly soluble (Rule 5). Thus, when these solutions are mixed, solid Fe(OH)$_3$ forms. The balanced equation for the reaction is

$$3KOH(aq) + Fe(NO_3)_3(aq) \rightarrow Fe(OH)_3(s) + 3KNO_3(aq)$$

which can be represented as

Self-Check **EXERCISE 7.1** Predict whether a solid will form when the following pairs of solutions are mixed. If so, identify the solid and write the balanced equation for the reaction.

    a. $Ba(NO_3)_2(aq)$ and $NaCl(aq)$

    b. $Na_2S(aq)$ and $Cu(NO_3)_2(aq)$

    c. $NH_4Cl(aq)$ and $Pb(NO_3)_2(aq)$

See Problems 7.17 and 7.18. ■

7.3 Describing Reactions in Aqueous Solutions

OBJECTIVE: To learn to describe reactions in solutions by writing molecular, complete ionic, and net ionic equations.

Much important chemistry, including virtually all of the reactions that make life possible, occurs in aqueous solutions. We will now consider the types of equations used to represent reactions that occur in water. For example, as we saw earlier, when we mix aqueous potassium chromate with aqueous barium nitrate, a reaction occurs to form solid barium chromate and dissolved potassium nitrate. One way to represent this reaction is by the equation

$$K_2CrO_4(aq) + Ba(NO_3)_2(aq) \rightarrow BaCrO_4(s) + 2KNO_3(aq)$$

This is called the **molecular equation** for the reaction; it shows the complete formulas of all reactants and products. However, although this equation shows the reactants and products of the reaction, it does not give a very clear picture of what actually occurs in solution. As we have seen, aqueous solutions of potassium chromate, barium nitrate, and potassium nitrate contain the individual ions, not molecules as is implied by the molecular equation. Thus the **complete ionic equation,**

Ions from K_2CrO_4 Ions from $Ba(NO_3)_2$

$$2K^+(aq) + CrO_4^{2-}(aq) + Ba^{2+}(aq) + 2NO_3^-(aq) \rightarrow$$
$$BaCrO_4(s) + 2K^+(aq) + 2NO_3^-(aq)$$

better represents the actual forms of the reactants and products in solution. *In a complete ionic equation, all substances that are strong electrolytes are represented as ions.* Notice that $BaCrO_4$ is not written as the separate ions, because it is present as a solid; it is not dissolved.

A strong electrolyte is a substance that completely breaks apart into ions when dissolved in water. The resulting solution readily conducts an electric current.

go Chemistry **Module 6: Writing Net Ionic Equations** covers concepts in this section.

The complete ionic equation reveals that only some of the ions participate in the reaction. Notice that the K^+ and NO_3^- ions are present in solution both before and after the reaction. Ions such as these, which do not participate directly in a reaction in solution, are called **spectator ions.** The ions that participate in this reaction are the Ba^{2+} and CrO_4^{2-} ions, which combine to form solid $BaCrO_4$:

$$Ba^{2+}(aq) + CrO_4^{2-}(aq) \rightarrow BaCrO_4(s)$$

The net ionic equation includes only those components that undergo a change in the reaction.

This equation, called the **net ionic equation,** includes only those components that are directly involved in the reaction. Chemists usually write the net ionic equation for a reaction in solution, because it gives the actual forms of the reactants and products and includes only the species that undergo a change.

Types of Equations for Reactions in Aqueous Solutions

Three types of equations are used to describe reactions in solutions.

1. The *molecular equation* shows the overall reaction but not necessarily the actual forms of the reactants and products in solution.
2. The *complete ionic equation* represents all reactants and products that are strong electrolytes as ions. All reactants and products are included.
3. The *net ionic equation* includes only those components that undergo a change. Spectator ions are not included.

To make sure these ideas are clear, we will do another example. In Example 7.2 we considered the reaction between aqueous solutions of lead nitrate and sodium sulfate. The molecular equation for this reaction is

$$Pb(NO_3)_2(aq) + Na_2SO_4(aq) \rightarrow PbSO_4(s) + 2NaNO_3(aq)$$

Because any ionic compound that is dissolved in water is present as the separated ions, we can write the complete ionic equation as follows:

$$Pb^{2+}(aq) + 2NO_3^-(aq) + 2Na^+(aq) + SO_4^{2-}(aq) \rightarrow$$
$$PbSO_4(s) + 2Na^+(aq) + 2NO_3^-(aq)$$

The $PbSO_4$ is not written as separate ions because it is present as a solid. The ions that take part in the chemical change are the Pb^{2+} and the SO_4^{2-} ions, which combine to form solid $PbSO_4$. Thus the net ionic equation is

$$Pb^{2+}(aq) + SO_4^{2-}(aq) \rightarrow PbSO_4(s)$$

The Na^+ and NO_3^- ions do not undergo any chemical change; they are spectator ions.

EXAMPLE 7.3 | Writing Equations for Reactions

For each of the following reactions, write the molecular equation, the complete ionic equation, and the net ionic equation.

Because silver is present as Ag^+ in all of its common ionic compounds, we usually delete the (I) when naming silver compounds.

a. Aqueous sodium chloride is added to aqueous silver nitrate to form solid silver chloride plus aqueous sodium nitrate.

b. Aqueous potassium hydroxide is mixed with aqueous iron(III) nitrate to form solid iron(III) hydroxide and aqueous potassium nitrate.

SOLUTION

a. *Molecular equation:*

$$NaCl(aq) + AgNO_3(aq) \rightarrow AgCl(s) + NaNO_3(aq)$$

Complete ionic equation:

$$Na^+(aq) + Cl^-(aq) + Ag^+(aq) + NO_3^-(aq) \rightarrow$$
$$AgCl(s) + Na^+(aq) + NO_3^-(aq)$$

Net ionic equation:

$$Cl^-(aq) + Ag^+(aq) \rightarrow AgCl(s)$$

b. *Molecular equation:*

$$3KOH(aq) + Fe(NO_3)_3(aq) \rightarrow Fe(OH)_3(s) + 3KNO_3(aq)$$

Complete ionic equation:

$$3K^+(aq) + 3OH^-(aq) + Fe^{3+}(aq) + 3NO_3^-(aq) \rightarrow$$
$$Fe(OH)_3(s) + 3K^+(aq) + 3NO_3^-(aq)$$

Net ionic equation:

$$3OH^-(aq) + Fe^{3+}(aq) \rightarrow Fe(OH)_3(s)$$

Self-Check **EXERCISE 7.2**   For each of the following reactions, write the molecular equation, the complete ionic equation, and the net ionic equation.

a. Aqueous sodium sulfide is mixed with aqueous copper(II) nitrate to produce solid copper(II) sulfide and aqueous sodium nitrate.

b. Aqueous ammonium chloride and aqueous lead(II) nitrate react to form solid lead(II) chloride and aqueous ammonium nitrate.

See Problems 7.25 through 7.30. ■

7.4 Reactions That Form Water: Acids and Bases

OBJECTIVE: To learn the key characteristics of the reactions between strong acids and strong bases.

Don't taste chemicals!

In this section we encounter two very important classes of compounds: acids and bases. Acids were first associated with the sour taste of citrus fruits. In fact, the word *acid* comes from the Latin word *acidus,* which means "sour." Vinegar tastes sour because it is a dilute solution of acetic acid; citric acid is responsible for the sour taste of a lemon. Bases, sometimes called *alkalis,* are characterized by their bitter taste and slippery feel, like wet soap. Most commercial preparations for unclogging drains are highly basic.

Acids have been known for hundreds of years. For example, the *mineral acids* sulfuric acid, H_2SO_4, and nitric acid, HNO_3, so named because they were originally obtained by the treatment of minerals, were discovered around 1300. However, it was not until the late 1800s that the essential nature of acids was discovered by Svante Arrhenius, then a Swedish graduate student in physics.

Figure 7.5

When gaseous HCl is dissolved in water, each molecule dissociates to produce H⁺ and Cl⁻ ions. That is, HCl behaves as a strong electrolyte.

The Nobel Prize in chemistry was awarded to Arrhenius in 1903 for his studies of solution conductivity.

The Arrhenius definition of an acid: a substance that produces H⁺ ions in aqueous solution.

The marsh marigold is a beautiful but poisonous plant. Its toxicity results partly from the presence of erucic acid.

Arrhenius, who was trying to discover why only certain solutions could conduct an electric current, found that conductivity arose from the presence of ions. In his studies of solutions, Arrhenius observed that when the substances HCl, HNO_3, and H_2SO_4 were dissolved in water, they behaved as strong electrolytes. He suggested that this was the result of ionization reactions in water.

$$HCl \xrightarrow{H_2O} H^+(aq) + Cl^-(aq)$$

$$HNO_3 \xrightarrow{H_2O} H^+(aq) + NO_3^-(aq)$$

$$H_2SO_4 \xrightarrow{H_2O} H^+(aq) + HSO_4^-(aq)$$

Arrhenius proposed that an **acid** *is a substance that produces H⁺ ions (protons) when it is dissolved in water.*

Studies show that when HCl, HNO_3, and H_2SO_4 are placed in water, *virtually every molecule* dissociates to give ions. This means that when 100 molecules of HCl are dissolved in water, 100 H⁺ ions and 100 Cl⁻ ions are produced. Virtually no HCl molecules exist in aqueous solution (see Figure 7.5). Because these substances are strong electrolytes that produce H⁺ ions, they are called **strong acids.**

Arrhenius also found that *aqueous solutions that exhibit basic behavior always contain hydroxide ions.* He defined a **base** as a *substance that produces hydroxide ions (OH⁻) in water.* The base most commonly used in the chemical laboratory is sodium hydroxide, NaOH, which contains Na⁺ and OH⁻ ions and is very soluble in water. Sodium hydroxide, like all ionic substances, produces separated cations and anions when it is dissolved in water.

$$NaOH(s) \xrightarrow{H_2O} Na^+(aq) + OH^-(aq)$$

Although dissolved sodium hydroxide is usually represented as NaOH(*aq*), you should remember that the solution really contains separated Na⁺ and OH⁻ ions. In fact, for every 100 units of NaOH dissolved in water, 100 Na⁺ and 100 OH⁻ ions are produced.

Potassium hydroxide (KOH) has properties markedly similar to those of sodium hydroxide. It is very soluble in water and produces separated ions.

$$KOH(s) \xrightarrow{H_2O} K^+(aq) + OH^-(aq)$$

Because these hydroxide compounds are strong electrolytes that contain OH⁻ ions, they are called **strong bases.**

When strong acids and strong bases (hydroxides) are mixed, the fundamental chemical change that always occurs is that *H⁺ ions react with OH⁻ ions to form water.*

$$H^+(aq) + OH^-(aq) \rightarrow H_2O(l)$$

Water is a very stable compound, as evidenced by the abundance of it on the earth's surface. Therefore, when substances that can form water are mixed, there is a strong tendency for the reaction to occur. In particular, the hydroxide ion OH^- has a high affinity for H^+ ions, because water is produced in the reaction between these ions.

The tendency to form water is the second of the driving forces for reactions that we mentioned in Section 7.1. Any compound that produces OH^- ions in water reacts vigorously with any compound that can furnish H^+ ions to form H_2O. For example, the reaction between hydrochloric acid and aqueous sodium hydroxide is represented by the following molecular equation:

$$HCl(aq) + NaOH(aq) \rightarrow H_2O(l) + NaCl(aq)$$

> Hydrochloric acid is an aqueous solution that contains dissolved hydrogen chloride. It is a strong electrolyte.

Because HCl, NaOH, and NaCl exist as completely separated ions in water, the complete ionic equation for this reaction is

$$H^+(aq) + Cl^-(aq) + Na^+(aq) + OH^-(aq) \rightarrow H_2O(l) + Na^+(aq) + Cl^-(aq)$$

Notice that the Cl^- and Na^+ are spectator ions (they undergo no changes), so the net ionic equation is

$$H^+(aq) + OH^-(aq) \rightarrow H_2O(l)$$

Thus the only chemical change that occurs when these solutions are mixed is that water is formed from H^+ and OH^- ions.

EXAMPLE 7.4 | Writing Equations for Acid–Base Reactions

Nitric acid is a strong acid. Write the molecular, complete ionic, and net ionic equations for the reaction of aqueous nitric acid and aqueous potassium hydroxide.

SOLUTION

Molecular equation:

$$HNO_3(aq) + KOH(aq) \rightarrow H_2O(l) + KNO_3(aq)$$

Complete ionic equation:

$$H^+(aq) + NO_3^-(aq) + K^+(aq) + OH^-(aq) \rightarrow H_2O(l) + K^+(aq) + NO_3^-(aq)$$

Net ionic equation:

$$H^+(aq) + OH^-(aq) \rightarrow H_2O(l)$$

Note that K^+ and NO_3^- are spectator ions and that the formation of water is the driving force for this reaction. ∎

There are two important things to note as we examine the reaction of hydrochloric acid with aqueous sodium hydroxide and the reaction of nitric acid with aqueous potassium hydroxide.

1. The net ionic equation is the same in both cases; water is formed.

$$H^+(aq) + OH^-(aq) \rightarrow H_2O(l)$$

2. Besides water, which is *always a product* of the reaction of an acid with OH^-, the second product is an ionic compound, which might precipitate or remain dissolved, depending on its solubility.

$$HCl(aq) + NaOH(aq) \rightarrow H_2O(l) + NaCl(aq)$$
$$HNO_3(aq) + KOH(aq) \rightarrow H_2O(l) + KNO_3(aq)$$

> Dissolved ionic compounds

This ionic compound is called a **salt.** In the first case the salt is sodium chloride, and in the second case the salt is potassium nitrate. We can obtain these soluble salts in solid form (both are white solids) by evaporating the water.

Both strong acids and strong bases are strong electrolytes.

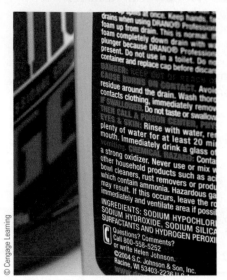

© Cengage Learning

Drano contains a strong base.

Summary of Strong Acids and Strong Bases

The following points about strong acids and strong bases are particularly important.

1. The common strong acids are aqueous solutions of HCl, HNO_3, and H_2SO_4.
2. A strong acid is a substance that completely dissociates (ionizes) in water. (Each molecule breaks up into an H^+ ion plus an anion.)
3. A strong base is a metal hydroxide compound that is very soluble in water. The most common strong bases are NaOH and KOH, which completely break up into separated ions (Na^+ and OH^- or K^+ and OH^-) when they are dissolved in water.
4. The net ionic equation for the reaction of a strong acid and a strong base (contains OH^-) is always the same: it shows the production of water.

$$H^+(aq) + OH^-(aq) \rightarrow H_2O(l)$$

5. In the reaction of a strong acid and a strong base, one product is always water and the other is always an ionic compound called a salt, which remains dissolved in the water. This salt can be obtained as a solid by evaporating the water.
6. The reaction of H^+ and OH^- is often called an acid–base reaction, where H^+ is the acidic ion and OH^- is the basic ion.

7.5 Reactions of Metals with Nonmetals (Oxidation–Reduction)

OBJECTIVES: To learn the general characteristics of a reaction between a metal and a nonmetal. • To understand electron transfer as a driving force for a chemical reaction.

In Chapter 4 we spent considerable time discussing ionic compounds—compounds formed in the reaction of a metal and a nonmetal. A typical example is sodium chloride, formed by the reaction of sodium metal and chlorine gas:

$$2Na(s) + Cl_2(g) \rightarrow 2NaCl(s)$$

Let's examine what happens in this reaction. Sodium metal is composed of sodium atoms, each of which has a net charge of zero. (The positive charges of the eleven protons in its nucleus are exactly balanced by the negative charges on the eleven electrons.) Similarly, the chlorine molecule consists of two uncharged chlorine atoms (each has seventeen protons and seventeen electrons). However, in the product (sodium chloride), the sodium is present as Na^+ and the chlorine as Cl^-. By what process do the neutral atoms become ions? The answer is that one electron is transferred from each sodium atom to each chlorine atom.

$$Na + Cl \rightarrow Na^+ + Cl^-$$

After the electron transfer, each sodium has ten electrons and eleven protons (a net charge of 1+), and each chlorine has eighteen electrons and seventeen protons (a net charge of 1−).

Figure 7.6

The thermite reaction gives off so much heat that the iron formed is molten.

This equation is read, "An aluminum atom yields an aluminum ion with a 3+ charge and three electrons."

Thus the reaction of a metal with a nonmetal to form an ionic compound involves the transfer of one or more electrons from the metal (which forms a cation) to the nonmetal (which forms an anion). This tendency to transfer electrons from metals to nonmetals is the third driving force for reactions that we listed in Section 7.1. A reaction that *involves a transfer of electrons* is called an **oxidation–reduction reaction.**

There are many examples of oxidation–reduction reactions in which a metal reacts with a nonmetal to form an ionic compound. Consider the reaction of magnesium metal with oxygen,

$$2Mg(s) + O_2(g) \rightarrow 2MgO(s)$$

which produces a bright, white light useful in camera flash units. Note that the reactants contain uncharged atoms, but the product contains ions:

MgO
Contains Mg^{2+}, O^{2-}

Therefore, in this reaction, each magnesium atom loses two electrons ($Mg \rightarrow Mg^{2+} + 2e^-$) and each oxygen atom gains two electrons ($O + 2e^- \rightarrow O^{2-}$). We might represent this reaction as follows:

Another example is

$$2Al(s) + Fe_2O_3(s) \rightarrow 2Fe(s) + Al_2O_3(s)$$

which is a reaction (called the thermite reaction) that produces so much energy (heat) that the iron is initially formed as a liquid (see Figure 7.6). In this case the aluminum is originally present as the elemental metal (which contains uncharged Al atoms) and ends up in Al_2O_3, where it is present as Al^{3+} cations (the $2Al^{3+}$ ions just balance the charge of the $3O^{2-}$ ions). Therefore, in the reaction each aluminum atom loses three electrons.

$$Al \rightarrow Al^{3+} + 3e^-$$

The opposite process occurs with the iron, which is initially present as Fe^{3+} ions in Fe_2O_3 and ends up as uncharged atoms in the elemental iron. Thus each iron cation gains three electrons to form an uncharged atom:

$$Fe^{3+} + 3e^- \rightarrow Fe$$

We can represent this reaction in schematic form as follows:

Figure 7.7

When powdered aluminum and iodine (shown in the foreground) are mixed (and a little water added), they react vigorously.

EXAMPLE 7.5 | **Identifying Electron Transfer in Oxidation–Reduction Reactions**

For each of the following reactions, show how electrons are gained and lost.

a. $2Al(s) + 3I_2(s) \rightarrow 2AlI_3(s)$ (This reaction is shown in Figure 7.7. Note the purple "smoke," which is excess I_2 being driven off by the heat.)

b. $2Cs(s) + F_2(g) \rightarrow 2CsF(s)$

SOLUTION

a. In AlI_3 the ions are Al^{3+} and I^- (aluminum always forms Al^{3+}, and iodine always forms I^-). In $Al(s)$ the aluminum is present as uncharged atoms. Thus aluminum goes from Al to Al^{3+} by losing three electrons ($Al \rightarrow Al^{3+} + 3e^-$). In I_2 each iodine atom is uncharged. Thus each iodine atom goes from I to I^- by gaining one electron ($I + e^- \rightarrow I^-$). A schematic for this reaction is

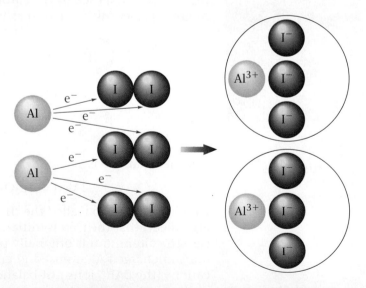

b. In CsF the ions present are Cs^+ and F^-. Cesium metal, $Cs(s)$, contains uncharged cesium atoms, and fluorine gas, $F_2(g)$, contains uncharged fluorine atoms. Thus in the reaction each cesium atom loses one electron ($Cs \rightarrow Cs^+ + e^-$) and each fluorine atom gains one electron ($F + e^- \rightarrow F^-$). The schematic for this reaction is

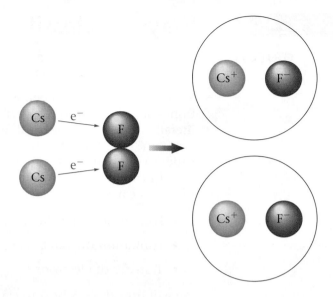

Self-Check   **EXERCISE 7.3**   For each reaction, show how electrons are gained and lost.

a. $2Na(s) + Br_2(l) \rightarrow 2NaBr(s)$

b. $2Ca(s) + O_2(g) \rightarrow 2CaO(s)$

See Problems 7.47 and 7.48. ∎

So far we have emphasized electron transfer (oxidation–reduction) reactions that involve a metal and a nonmetal. Electron transfer reactions can also take place between two nonmetals. We will not discuss these reactions in detail here. All we will say at this point is that one sure sign of an oxidation–reduction reaction between nonmetals is the presence of oxygen, $O_2(g)$, as a reactant or product. In fact, oxidation got its name from oxygen. Thus the reactions

$$CH_4(g) + 2O_2(g) \rightarrow CO_2(g) + 2H_2O(g)$$

and

$$2SO_2(g) + O_2(g) \rightarrow 2SO_3(g)$$

are electron transfer reactions, even though it is not obvious at this point.

We can summarize what we have learned about oxidation–reduction reactions as follows:

Characteristics of Oxidation–Reduction Reactions

1. When a metal reacts with a nonmetal, an ionic compound is formed. The ions are formed when the metal transfers one or more electrons to the nonmetal, the metal atom becoming a cation and the nonmetal atom becoming an anion. *Therefore, a metal–nonmetal reaction can always be assumed to be an oxidation–reduction reaction, which involves electron transfer.*

2. Two nonmetals can also undergo an oxidation–reduction reaction. At this point we can recognize these cases only by looking for O_2 as a reactant or product. When two nonmetals react, the compound formed is not ionic.

7.6 Ways to Classify Reactions

OBJECTIVE: To learn various classification schemes for reactions.

So far in our study of chemistry we have seen many, many chemical reactions—and this is just Chapter 7. In the world around us and in our bodies, literally millions of chemical reactions are taking place. Obviously, we need a system for putting reactions into meaningful classes that will make them easier to remember and easier to understand.

In Chapter 7 we have so far considered the following "driving forces" for chemical reactions:

- Formation of a solid

- Formation of water

- Transfer of electrons

We will now discuss how to classify reactions involving these processes. For example, in the reaction

$$K_2CrO_4(aq) + Ba(NO_3)_2(aq) \rightarrow BaCrO_4(s) + 2KNO_3(aq)$$

Solution   Solution   Solid   Solution
formed

solid $BaCrO_4$ (a precipitate) is formed. Because the *formation of a solid when two solutions are mixed* is called *precipitation,* we call this a **precipitation reaction.**

Notice in this reaction that two anions (NO_3^- and CrO_4^{2-}) are simply exchanged. Note that CrO_4^{2-} was originally associated with K^+ in K_2CrO_4 and that NO_3^- was associated with Ba^{2+} in $Ba(NO_3)_2$. In the products these associations are reversed. Because of this double exchange, we sometimes call this reaction a double-exchange reaction or **double-displacement reaction.** We might represent such a reaction as

$$AB + CD \rightarrow AD + CB$$

So we can classify a reaction such as this one as a precipitation reaction or as a double-displacement reaction. Either name is correct, but the former is more commonly used by chemists.

In this chapter we have also considered reactions in which water is formed when a strong acid is mixed with a strong base. All of these reactions had the same net ionic equation:

$$H^+(aq) + OH^-(aq) \rightarrow H_2O(l)$$

The H^+ ion comes from a strong acid, such as $HCl(aq)$ or $HNO_3(aq)$, and the origin of the OH^- ion is a strong base, such as $NaOH(aq)$ or $KOH(aq)$. An example is

$$HCl(aq) + KOH(aq) \rightarrow H_2O(l) + KCl(aq)$$

We classify these reactions as **acid–base reactions.** You can recognize this as an acid–base reaction because it *involves an H^+ ion that ends up in the product water.*

The third driving force is electron transfer. We see evidence of this driving force particularly in the "desire" of a metal to donate electrons to nonmetals. An example is

$$2Li(s) + F_2(g) \rightarrow 2LiF(s)$$

where each lithium atom loses one electron to form Li^+, and each fluorine atom gains one electron to form the F^- ion. The process of electron transfer is also called oxidation–reduction. Thus we classify the preceding reaction as an **oxidation–reduction reaction.**

An additional driving force for chemical reactions that we have not yet discussed is *formation of a gas.* A reaction in aqueous solution that forms a gas (which escapes as bubbles) is pulled toward the products by this event. An example is the reaction

$$2HCl(aq) + Na_2CO_3(aq) \rightarrow CO_2(g) + H_2O(l) + NaCl(aq)$$

for which the net ionic equation is

$$2H^+(aq) + CO_3^{2-}(aq) \rightarrow CO_2(g) + H_2O(l)$$

Note that this reaction forms carbon dioxide gas as well as water, so it illustrates two of the driving forces that we have considered. Because this reaction involves H^+ that ends up in the product water, we classify it as an acid–base reaction.

Consider another reaction that forms a gas:

$$Zn(s) + 2HCl(aq) \rightarrow H_2(g) + ZnCl_2(aq)$$

How might we classify this reaction? A careful look at the reactants and products shows the following:

$Zn(s)$	+	$2HCl(aq)$	$\rightarrow$	$H_2(g)$	+	$ZnCl_2(aq)$
Contains uncharged Zn atoms		Really $2H^+(aq) + 2Cl^-(aq)$		Contains uncharged H atoms		Really $Zn^{2+}(aq) + 2Cl^-(aq)$

Note that in the reactant zinc metal, Zn exists as uncharged atoms, whereas in the product it exists as Zn^{2+}. Thus each Zn atom loses two electrons. Where have these electrons gone? They have been transferred to two H^+ ions to form H_2. The schematic for this reaction is

| Zn metal | Solution of HCl | H_2 molecule | Solution of $ZnCl_2$ |

This is an electron transfer process, so the reaction can be classified as an oxidation–reduction reaction.

Another way this reaction is sometimes classified is based on the fact that a *single* type of anion (Cl^-) has been exchanged between H^+ and Zn^{2+}. That is, Cl^- is originally associated with H^+ in HCl and ends up associated with Zn^{2+} in the product $ZnCl_2$. We can call this a *single-replacement reaction* in contrast to double-displacement reactions, in which two types of anions are exchanged. We can represent a single replacement as

$$A + BC \rightarrow B + AC$$

Oxidation–Reduction Reactions Launch the Space Shuttle

Launching into space a vehicle that weighs millions of pounds requires unimaginable quantities of energy—all furnished by oxidation–reduction reactions.

Notice from Figure 7.8 that three cylindrical objects are attached to the shuttle orbiter. In the center is a tank about 28 feet in diameter and 154 feet long that contains liquid oxygen and liquid hydrogen (in separate compartments). These fuels are fed to the orbiter's rocket engines, where they react to form water and release a huge quantity of energy.

$$2H_2 + O_2 \rightarrow 2H_2O + energy$$

Note that we can recognize this reaction as an oxidation–reduction reaction because O_2 is a reactant.

Two solid-fuel rockets 12 feet in diameter and 150 feet long are also attached to the or-biter. Each rocket contains 1.1 million pounds of fuel: ammonium perchlorate (NH_4ClO_4) and powdered aluminum mixed with a binder ("glue"). Because the rockets are so large, they are built in segments and assembled at the launch site as shown in Figure 7.9. Each segment is filled with the syrupy propellant (Figure 7.10), which then solidifies to a consistency much like that of a hard rubber eraser.

The oxidation–reduction reaction between the ammonium perchlorate and the aluminum is represented as follows:

$$3NH_4ClO_4(s) + 3Al(s) \rightarrow Al_2O_3(s) + AlCl_3(s) + 3NO(g) + 6H_2O(g) + energy$$

It produces temperatures of about 5700 °F and 3.3 million pounds of thrust in each rocket.

Thus we can see that oxidation–reduction reactions furnish the energy to launch the space shuttle.

Figure 7.8

For launch, the space shuttle orbiter is attached to two solid-fuel rockets (left and right) and a fuel tank (center) that supplies hydrogen and oxygen to the orbiter's engines. *(Reprinted with permission from* Chemical and Engineering News, *September 19, 1988. Copyright © 1988 American Chemical Society.)*

Figure 7.9

The solid-fuel rockets are assembled from segments to make loading the fuel more convenient. *(Reprinted with permission from* Chemical and Engineering News, *September 19, 1988. Copyright © 1988 American Chemical Society.)*

Figure 7.10

A rocket segment being filled with the propellant mixture.

7.7 Other Ways to Classify Reactions

OBJECTIVE: To consider additional classes of chemical reactions.

So far in this chapter we have classified chemical reactions in several ways. The most commonly used of these classifications are

- Precipitation reactions

- Acid–base reactions

- Oxidation–reduction reactions

However, there are still other ways to classify reactions that you may encounter in your future studies of chemistry. We will consider several of these in this section.

▶ Combustion Reactions

Many chemical reactions that involve oxygen produce energy (heat) so rapidly that a flame results. Such reactions are called **combustion reactions.** We have considered some of these reactions previously. For example, the methane in natural gas reacts with oxygen according to the following balanced equation:

$$CH_4(g) + 2O_2(g) \rightarrow CO_2(g) + 2H_2O(g)$$

This reaction produces the flame of the common laboratory burner and is used to heat most homes in the United States. Recall that we originally classified this reaction as an oxidation–reduction reaction in Section 7.5. Thus we can say that the reaction of methane with oxygen is both an oxidation–reduction reaction and a combustion reaction. Combustion reactions, in fact, are a special class of oxidation–reduction reactions (see Figure 7.11).

There are many combustion reactions, most of which are used to provide heat or electricity for homes or businesses or energy for transportation. Some examples are:

- Combustion of propane (used to heat some rural homes)

$$C_3H_8(g) + 5O_2(g) \rightarrow 3CO_2(g) + 4H_2O(g)$$

Figure 7.11

Classes of reactions. Combustion reactions are a special type of oxidation–reduction reaction.

- Combustion of gasoline* (used to power cars and trucks)

$$2C_8H_{18}(l) + 25O_2(g) \rightarrow 16CO_2(g) + 18H_2O(g)$$

- Combustion of coal* (used to generate electricity)

$$C(s) + O_2(g) \rightarrow CO_2(g)$$

▶ Synthesis (Combination) Reactions

One of the most important activities in chemistry is the synthesis of new compounds. Each of our lives has been greatly affected by synthetic compounds such as plastic, polyester, and aspirin. When a given compound is formed from simpler materials, we call this a **synthesis** (or **combination**) **reaction.**

In many cases synthesis reactions start with elements, as shown by the following examples:

- Synthesis of water $2H_2(g) + O_2(g) \rightarrow 2H_2O(l)$

- Synthesis of carbon dioxide $C(s) + O_2(g) \rightarrow CO_2(g)$

- Synthesis of nitrogen monoxide $N_2(g) + O_2(g) \rightarrow 2NO(g)$

Notice that each of these reactions involves oxygen, so each can be classified as an oxidation–reduction reaction. The first two reactions are also commonly called combustion reactions because they produce flames. The reaction of hydrogen with oxygen to produce water, then, can be classified three ways: as an oxidation–reduction reaction, as a combustion reaction, and as a synthesis reaction.

There are also many synthesis reactions that do not involve oxygen:

- Synthesis of sodium chloride $2Na(s) + Cl_2(g) \rightarrow 2NaCl(s)$

- Synthesis of magnesium fluoride $Mg(s) + F_2(g) \rightarrow MgF_2(s)$

We have discussed the formation of sodium chloride before and have noted that it is an oxidation–reduction reaction; uncharged sodium atoms lose electrons to form Na^+ ions, and uncharged chlorine atoms gain electrons to form Cl^- ions. The synthesis of magnesium fluoride is also an oxidation–reduction reaction because Mg^{2+} and F^- ions are produced from the uncharged atoms.

We have seen that synthesis reactions in which the reactants are elements are oxidation–reduction reactions as well. In fact, we can think of these synthesis reactions as another subclass of the oxidation–reduction class of reactions.

▶ Decomposition Reactions

In many cases a compound can be broken down into simpler compounds or all the way to the component elements. This is usually accomplished by heating or by the application of an electric current. Such reactions are called **decomposition reactions.** We have discussed decomposition reactions before, including

- Decomposition of water

$$2H_2O(l) \xrightarrow{\text{Electric current}} 2H_2(g) + O_2(g)$$

Formation of the colorful plastics used in these zippers is an example of a synthetic reaction.

*This substance is really a complex mixture of compounds, but the reaction shown is representative of what takes place.

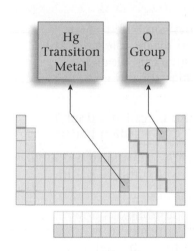

Figure 7.12

Summary of classes of reactions.

- Decomposition of mercury(II) oxide

$$2HgO(s) \xrightarrow{\text{Heat}} 2Hg(l) + O_2(g)$$

Because O_2 is involved in the first reaction, we recognize it as an oxidation–reduction reaction. In the second reaction, HgO, which contains Hg^{2+} and O^{2-} ions, is decomposed to the elements, which contain uncharged atoms. In this process each Hg^{2+} gains two electrons and each O^{2-} loses two electrons, so this is both a decomposition reaction and an oxidation–reduction reaction.

A decomposition reaction, in which a compound is broken down into its elements, is just the opposite of the synthesis (combination) reaction, in which elements combine to form the compound. For example, we have just discussed the synthesis of sodium chloride from its elements. Sodium chloride can be decomposed into its elements by melting it and passing an electric current through it:

$$2NaCl(l) \xrightarrow[\text{current}]{\text{Electric}} 2Na(l) + Cl_2(g)$$

There are other schemes for classifying reactions that we have not considered. However, we have covered many of the classifications that are commonly used by chemists as they pursue their science in laboratories and industrial plants.

It should be apparent that many important reactions can be classified as oxidation–reduction reactions. As shown in Figure 7.12, various types of reactions can be viewed as subclasses of the overall oxidation–reduction category.

EXAMPLE 7.6 | Classifying Reactions

Classify each of the following reactions in as many ways as possible.

a. $2K(s) + Cl_2(g) \rightarrow 2KCl(s)$

b. $Fe_2O_3(s) + 2Al(s) \rightarrow Al_2O_3(s) + 2Fe(s)$

   c. $2Mg(s) + O_2(g) \rightarrow 2MgO(s)$

   d. $HNO_3(aq) + NaOH(aq) \rightarrow H_2O(l) + NaNO_3(aq)$

   e. $KBr(aq) + AgNO_3(aq) \rightarrow AgBr(s) + KNO_3(aq)$

   f. $PbO_2(s) \rightarrow Pb(s) + O_2(g)$

SOLUTION

a. This is both a synthesis reaction (elements combine to form a compound) and an oxidation–reduction reaction (uncharged potassium and chlorine atoms are changed to K^+ and Cl^- ions in KCl).

b. This is an oxidation–reduction reaction. Iron is present in $Fe_2O_3(s)$ as Fe^{3+} ions and in elemental iron, Fe(s), as uncharged atoms. So each Fe^{3+} must gain three electrons to form Fe. The reverse happens to aluminum, which is present initially as uncharged aluminum atoms, each of which loses three electrons to give Al^{3+} ions in Al_2O_3. Note that this reaction might also be called a single-replacement reaction because O is switched from Fe to Al.

c. This is both a synthesis reaction (elements combine to form a compound) and an oxidation–reduction reaction (each magnesium atom loses two electrons to give Mg^{2+} ions in MgO, and each oxygen atom gains two electrons to give O^{2-} in MgO).

d. This is an acid–base reaction. It might also be called a double-displacement reaction because NO_3^- and OH^- "switch partners."

e. This is a precipitation reaction that might also be called a double-displacement reaction in which the anions Br^- and NO_3^- are exchanged.

f. This is a decomposition reaction (a compound breaks down into elements). It also is an oxidation–reduction reaction, because the ions in PbO_2 (Pb^{4+} and O^{2-}) are changed to uncharged atoms in the elements Pb(s) and $O_2(g)$. That is, electrons are transferred from O^{2-} to Pb^{4+} in the reaction.

(Self-Check) **EXERCISE 7.4**

Classify each of the following reactions in as many ways as possible.

   a. $4NH_3(g) + 5O_2(g) \rightarrow 4NO(g) + 6H_2O(g)$

   b. $S_8(s) + 8O_2(g) \rightarrow 8SO_2(g)$

   c. $2Al(s) + 3Cl_2(g) \rightarrow 2AlCl_3(s)$

   d. $2AlN(s) \rightarrow 2Al(s) + N_2(g)$

   e. $BaCl_2(aq) + Na_2SO_4(aq) \rightarrow BaSO_4(s) + 2NaCl(aq)$

   f. $2Cs(s) + Br_2(l) \rightarrow 2CsBr(s)$

   g. $KOH(aq) + HCl(aq) \rightarrow H_2O(l) + KCl(aq)$

   h. $2C_2H_2(g) + 5O_2(g) \rightarrow 4CO_2(g) + 2H_2O(l)$

See Problems 7.53 and 7.54. ∎

CHAPTER 7 REVIEW

Key Terms

precipitation (7.2)
precipitate (7.2)
precipitation
  reaction (7.2, 7.6)
strong electrolyte (7.2)
soluble solid (7.2)
insoluble (slightly
  soluble) solid (7.2)
molecular equation (7.3)
complete ionic
  equation (7.3)
spectator ions (7.3)
net ionic equation (7.3)
acid (7.4)
strong acid (7.4)
base (7.4)

strong base (7.4)
salt (7.4)
oxidation–reduction
  reaction (7.5, 7.6)
precipation reaction (7.6)
double-displacement
  reaction (7.6)
acid–base
  reaction (7.6)
combustion
  reaction (7.7)
synthesis (combination)
  reaction (7.7)
decomposition
  reaction (7.7)

Summary

1. Four driving forces that favor chemical change (chemical reaction) are formation of a solid, formation of water, transfer of electrons, and formation of a gas.

2. A reaction where a solid forms is called a precipitation reaction. General rules on solubility help predict whether a solid—and what solid—will form when two solutions are mixed.

3. Three types of equations are used to describe reactions in solution: (1) the molecular equation, which shows the complete formulas of all reactants and products; (2) the complete ionic equation, in which all reactants and products that are strong electrolytes are shown as ions; and (3) the net ionic equation, which includes only those components of the solution that undergo a change. Spectator ions (those ions that remain unchanged in a reaction) are not included in a net ionic equation.

4. A strong acid is a compound in which virtually every molecule dissociates in water to give an H^+ ion and an anion. Similarly, a strong base is a metal hydroxide compound that is soluble in water, giving OH^- ions and cations. The products of the reaction of a strong acid and a strong base are water and a salt.

5. Reactions of metals and nonmetals involve a transfer of electrons and are called oxidation–reduction reactions. A reaction between a nonmetal and oxygen is also an oxidation–reduction reaction. Combustion reactions involve oxygen and are a subgroup of oxidation–reduction reactions.

6. When a given compound is formed from simpler materials, such as elements, the reaction is called a synthesis or combination reaction. The reverse process, which occurs when a compound is broken down into its component elements, is called a decomposition reaction. These reactions are also subgroups of oxidation–reduction reactions.

Active Learning Questions

These questions are designed to be considered by groups of students in class. Often these questions work well for introducing a particular topic in class.

1. Consider the mixing of aqueous solutions of lead(II) nitrate and sodium iodide to form a solid.

   a. Name the possible products, and determine the formulas of these possible products.
   b. What is the precipitate? How do you know?
   c. Must the subscript for an ion in a reactant stay the same as the subscript of that ion in a product? Explain your answer.

2. Assume a highly magnified view of a solution of HCl that allows you to "see" the HCl. Draw this magnified view. If you dropped in a piece of magnesium, the magnesium would disappear and hydrogen gas would be released. Represent this change using symbols for the elements, and write the balanced equation.

3. Why is the formation of a solid evidence of a chemical reaction? Use a molecular-level drawing in your explanation.

4. Sketch molecular-level drawings to differentiate between two soluble compounds: one that is a strong electrolyte, and one that is not an electrolyte.

5. Mixing an aqueous solution of potassium nitrate with an aqueous solution of sodium chloride does not result in a chemical reaction. Why?

6. Why is the formation of water evidence of a chemical reaction? Use a molecular-level drawing in your explanation.

7. Use the Arrhenius definition of acids and bases to write the net ionic equation for the reaction of an acid with a base.

8. Why is the transfer of electrons evidence of a chemical reaction? Use a molecular-level drawing in your explanation.

9. Why is the formation of a gas evidence of a chemical reaction? Use a molecular-level drawing in your explanation.

10. Label each of the following statements as true or false. Explain your answers, and provide an example for each that supports your answer.

    a. All nonelectrolytes are insoluble.
    b. All insoluble substances are nonelectrolytes.
    c. All strong electrolytes are soluble.
    d. All soluble substances are strong electrolytes.

11. Look at Figure 7.2 in the text. It is possible for a weak electrolyte solution to cause the bulb to glow brighter than a strong electrolyte. Explain how this is possible.

12. What is the purpose of spectator ions? If they are not present as part of the reaction, why are they present at all?

13. Which of the following **must** be an oxidation–reduction reaction? Explain your answer, and include an example oxidation–reduction reaction for all that apply.

   a. A metal reacts with a nonmetal.
   b. A precipitation reaction.
   c. An acid–base reaction.

14. If an element is a reactant or product in a chemical reaction, the reaction must be an oxidation–reduction reaction. Why is this true?

15. Match each name below with the following microscopic pictures of that compound in aqueous solution.

a. barium nitrate   c. potassium carbonate
b. sodium chloride   d. magnesium sulfate

Which picture best represents $HNO_3(aq)$? Why aren't any of the pictures a good representation of $HC_2H_3O_2(aq)$?

16. On the basis of the general solubility rules given in Table 7.1, predict the identity of the precipitate that forms when aqueous solutions of the following substances are mixed. If no precipitate is likely, indicate which rules apply.

17. Write the balanced formula and net ionic equation for the reaction that occurs when the contents of the two beakers are added together. What colors represent the spectator ions in each reaction?

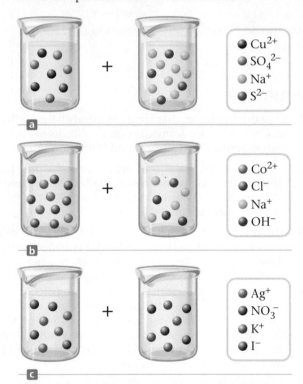

8

Chemical Composition

● These glass bottles contain silicon dioxide.
(James L. Amos/Peter Arnold)

The Enzo Ferrari has a body made of carbon fiber composite materials.

One very important chemical activity is the synthesis of new substances. Nylon, the artificial sweetener aspartame (Nutra-Sweet®), Kevlar used in bulletproof vests and the body parts of exotic cars, polyvinyl chloride (PVC) for plastic water pipes, Teflon, Nitinol (the alloy that remembers its shape even after being severely distorted), and so many other materials that make our lives easier—all originated in some chemist's laboratory. Some of the new materials have truly amazing properties such as the plastic that listens and talks, described in the "Chemistry in Focus" on page 206. When a chemist makes a new substance, the first order of business is to identify it. What is its composition? What is its chemical formula?

In this chapter we will learn to determine a compound's formula. Before we can do that, however, we need to think about counting atoms. How do we determine the number of each type of atom in a substance so that we can write its formula? Of course, atoms are too small to count individually. As we will see in this chapter, we typically count atoms by weighing them. So let us first consider the general principle of counting by weighing.

8.1   Counting by Weighing

OBJECTIVE: To understand the concept of average mass and explore how counting can be done by weighing.

Suppose you work in a candy store that sells gourmet jelly beans by the bean. People come in and ask for 50 beans, 100 beans, 1000 beans, and so on, and you have to count them out—a tedious process at best. As a good problem solver, you try to come up with a better system. It occurs to you that it might be far more efficient to buy a scale and count the jelly beans by weighing them. How can you count jelly beans by weighing them? What information about the individual beans do you need to know?

Assume that all of the jelly beans are identical and that each has a mass of 5 g. If a customer asks for 1000 jelly beans, what mass of jelly beans would be required? Each bean has a mass of 5 g, so you would need 1000 beans × 5 g/bean, or 5000 g (5 kg). It takes just a few seconds to weigh out 5 kg of jelly beans. It would take much longer to count out 1000 of them.

In reality, jelly beans are not identical. For example, let's assume that you weigh 10 beans individually and get the following results:

Bean	Mass
1	5.1 g
2	5.2 g
3	5.0 g
4	4.8 g
5	4.9 g
6	5.0 g
7	5.0 g
8	5.1 g
9	4.9 g
10	5.0 g

Plastic That Talks and Listens!

Imagine a plastic so "smart" that it can be used to sense a baby's breath, measure the force of a karate punch, sense the presence of a person 100 ft away, or make a balloon that sings. There is a plastic film capable of doing all these things. It's called **polyvinylidene difluoride (PVDF)**, which has the structure

When this polymer is processed in a particular way, it becomes piezoelectric and pyroelectric. A *piezoelectric* substance produces an electric current when it is physically deformed or, alternatively, undergoes a deformation when a current is applied. A *pyroelectric* material is one that develops an electrical potential in response to a change in its temperature.

Because PVDF is piezoelectric, it can be used to construct a paper-thin microphone; it responds to sound by producing a current proportional to the deformation caused by the sound waves. A ribbon of PVDF plastic one-quarter of an inch wide could be strung along a hallway and used to listen to all the conversations going on as people walk through. On the other hand, electric pulses can be applied to the PVDF film to produce a speaker. A strip of PVDF film glued to the inside of a balloon can play any song stored on a microchip attached to the film—hence a balloon that can sing happy birthday at a party. The PVDF film also can be used to construct a sleep apnea monitor, which, when placed beside the mouth of a sleeping infant, will set off an alarm if the breathing stops, thus helping to prevent sudden infant death syndrome (SIDS). The same type of film is used by the U.S. Olympic karate team to measure the force of kicks and punches as the team trains. Also, gluing two strips of film together gives a material that curls in response to a current, creating an artificial muscle. In addition, because the PVDF film is pyroelectric, it responds to the infrared (heat) radiation emitted by a human as far away as 100 ft, making it useful for burglar alarm systems. Making the PVDF polymer piezoelectric and pyroelectric requires some very special processing, which makes it costly ($10 per square foot), but this seems a small price to pay for its near-magical properties.

Can we count these nonidentical beans by weighing? Yes. The key piece of information we need is the *average mass* of the jelly beans. Let's compute the average mass for our 10-bean sample.

$$\text{Average mass} = \frac{\text{total mass of beans}}{\text{number of beans}}$$

$$= \frac{5.1\text{ g} + 5.2\text{ g} + 5.0\text{ g} + 4.8\text{ g} + 4.9\text{ g} + 5.0\text{ g} + 5.0\text{ g} + 5.1\text{ g} + 4.9\text{ g} + 5.0\text{ g}}{10}$$

$$= \frac{50.0}{10} = 5.0\text{ g}$$

The average mass of a jelly bean is 5.0 g. Thus, to count out 1000 beans, we need to weigh out 5000 g of beans. This sample of beans, in which the beans have an average mass of 5.0 g, can be treated exactly like a sample where all of the beans are identical. Objects do not need to have identical masses to be counted by weighing. We simply need to know the average mass of the objects. For purposes of counting, the objects *behave as though they were all identical*, as though they each actually had the average mass.

Suppose a customer comes into the store and says, "I want to buy a bag of candy for each of my kids. One of them likes jelly beans and the other one likes mints. Please put a scoopful of jelly beans in a bag and a scoopful of mints in another bag." Then the customer recognizes a problem. "Wait! My kids will fight unless I bring home exactly the same number of candies for each one. Both bags must have the same number of pieces because they'll definitely count them and compare. But I'm really in a hurry, so we don't have time to count them here. Is there a simple way you can be sure the bags will contain the same number of candies?"

You need to solve this problem quickly. Suppose you know the average masses of the two kinds of candy:

$$\text{Jelly beans:} \quad \text{average mass} = 5 \text{ g}$$
$$\text{Mints:} \quad \text{average mass} = 15 \text{ g}$$

You fill the scoop with jelly beans and dump them onto the scale, which reads 500 g. Now the key question: What mass of mints do you need to give the same number of mints as there are jelly beans in 500 g of jelly beans? Comparing the average masses of the jelly beans (5 g) and mints (15 g), you realize that each mint has three times the mass of each jelly bean:

$$\frac{15 \text{ g}}{5 \text{ g}} = 3$$

This means that you must weigh out an amount of mints that is three times the mass of the jelly beans:

$$3 \times 500 \text{ g} = 1500 \text{ g}$$

You weigh out 1500 g of mints and put them in a bag. The customer leaves with your assurance that both the bag containing 500 g of jelly beans and the bag containing 1500 g of mints contain the same number of candies.

In solving this problem, you have discovered a principle that is very important in chemistry: two samples containing different types of components, A and B, both *contain the same number of components if the ratio of the sample masses is the same as the ratio of the masses of the individual components* of A and B.

Let's illustrate this rather intimidating statement by using the example we just discussed. The individual components have the masses 5 g (jelly beans) and 15 g (mints). Consider several cases.

- Each sample contains 1 component:

$$\text{Mass of mint} = 15 \text{ g}$$
$$\text{Mass of jelly bean} = 5 \text{ g}$$

- Each sample contains 10 components:

$$10 \text{ mints} \times \frac{15 \text{ g}}{\text{mint}} = 150 \text{ g of mints}$$
$$10 \text{ jelly beans} \times \frac{5 \text{ g}}{\text{jelly bean}} = 50 \text{ g of jelly beans}$$

- Each sample contains 100 components:

$$100 \text{ mints} \times \frac{15 \text{ g}}{\text{mint}} = 1500 \text{ g of mints}$$
$$100 \text{ jelly beans} \times \frac{5 \text{ g}}{\text{jelly bean}} = 500 \text{ g of jelly beans}$$

Note in each case that the ratio of the masses is always 3 to 1:

$$\frac{1500}{500} = \frac{150}{50} = \frac{15}{5} = \frac{3}{1}$$

This is the ratio of the masses of the individual components:

$$\frac{\text{Mass of mint}}{\text{Mass of jelly bean}} = \frac{15}{5} = \frac{3}{1}$$

Any two samples, one of mints and one of jelly beans, that have a *mass ratio* of $15/5 = 3/1$ will contain the same number of components. And these same ideas apply also to atoms, as we will see in the next section.

8.2 Atomic Masses: Counting Atoms by Weighing

OBJECTIVE: To understand atomic mass and its experimental determination.

In Chapter 6 we considered the balanced equation for the reaction of solid carbon and gaseous oxygen to form gaseous carbon dioxide:

$$C(s) + O_2(g) \rightarrow CO_2(g)$$

Now suppose you have a small pile of solid carbon and want to know how many oxygen molecules are required to convert all of this carbon into carbon dioxide. The balanced equation tells us that one oxygen molecule is required for each carbon atom.

$$C(s) \quad + \quad O_2(g) \quad \rightarrow \quad CO_2(g)$$
1 atom reacts with 1 molecule to yield 1 molecule

To determine the number of oxygen molecules required, we must know how many carbon atoms are present in the pile of carbon. But individual atoms are far too small to see. We must learn to count atoms by weighing samples containing large numbers of them.

In the last section we saw that we can easily count things like jelly beans and mints by weighing. Exactly the same principles can be applied to counting atoms.

Because atoms are so tiny, the normal units of mass—the gram and the kilogram—are much too large to be convenient. For example, the mass of a single carbon atom is 1.99×10^{-23} g. To avoid using terms like 10^{-23} when describing the mass of an atom, scientists have defined a much smaller unit of mass called the **atomic mass unit**, which is abbreviated **amu.** In terms of grams,

$$1 \text{ amu} = 1.66 \times 10^{-24} \text{ g}$$

Now let's return to our problem of counting carbon atoms. To count carbon atoms by weighing, we need to know the mass of individual atoms, just as we needed to know the mass of the individual jelly beans. Recall from Chapter 4 that the atoms of a given element exist as isotopes. The isotopes of carbon are $^{12}_{6}C$, $^{13}_{6}C$, and $^{14}_{6}C$. Any sample of carbon contains a mixture of these isotopes, always in the same proportions. Each of these isotopes has a slightly different mass. Therefore, just as with the nonidentical jelly beans, we need to use an average mass for the carbon atoms. The **average atomic mass** for carbon atoms is 12.01 amu. This means that any sample of carbon from nature *can be treated as though it were composed of identical carbon atoms,* each with a mass of 12.01 amu. Now that we know the average mass of the carbon atom, we can count carbon atoms by weighing samples of natural carbon. For example, what mass of natural carbon must we take to have 1000 carbon atoms present? Because 12.01 amu is the average mass,

$$\text{Mass of 1000 natural carbon atoms} = (1000 \text{ atoms})\left(12.01 \frac{\text{amu}}{\text{atom}}\right)$$

$$= 12{,}010 \text{ amu} = 12.01 \times 10^3 \text{ amu}$$

Now let's assume that when we weigh the pile of natural carbon mentioned earlier, the result is 3.00×10^{20} amu. How many carbon atoms are present in this sample? We know that an average carbon atom has the mass 12.01 amu, so we can compute the number of carbon atoms by using the equivalence statement

$$1 \text{ carbon atom} = 12.01 \text{ amu}$$

to construct the appropriate conversion factor,

$$\frac{1 \text{ carbon atom}}{12.01 \text{ amu}}$$

The calculation is carried out as follows:

$$3.00 \times 10^{20} \text{ amu} \times \frac{1 \text{ carbon atom}}{12.01 \text{ amu}} = 2.50 \times 10^{19} \text{ carbon atoms}$$

The principles we have just discussed for carbon apply to all the other elements as well. All the elements as found in nature typically consist of a mixture of various isotopes. So to count the atoms in a sample of a given element by weighing, we must know the mass of the sample and the average mass for that element. Some average masses for common elements are listed in Table 8.1.

Table 8.1 Average Atomic Mass Values for Some Common Elements

Element	Average Atomic Mass (amu)
Hydrogen	1.008
Carbon	12.01
Nitrogen	14.01
Oxygen	16.00
Sodium	22.99
Aluminum	26.98

EXAMPLE 8.1 | Calculating Mass Using Atomic Mass Units (amu)

Calculate the mass, in amu, of a sample of aluminum that contains 75 atoms.

SOLUTION

To solve this problem we use the average mass for an aluminum atom: 26.98 amu. We set up the equivalence statement:

$$1 \text{ Al atom} = 26.98 \text{ amu}$$

It gives the conversion factor we need:

$$75 \text{ Al atoms} \times \frac{26.98 \text{ amu}}{1 \text{ Al atom}} = 2024 \text{ amu}$$

Self-Check **EXERCISE 8.1** Calculate the mass of a sample that contains 23 nitrogen atoms.

See Problems 8.5 and 8.8. ∎

The opposite calculation can also be carried out. That is, if we know the mass of a sample, we can determine the number of atoms present. This procedure is illustrated in Example 8.2.

EXAMPLE 8.2   Calculating the Number of Atoms from the Mass

Calculate the number of sodium atoms present in a sample that has a mass of 1172.49 amu.

SOLUTION

We can solve this problem by using the average atomic mass for sodium (see Table 8.1) of 22.99 amu. The appropriate equivalence statement is

$$1 \text{ Na atom} = 22.99 \text{ amu}$$

which gives the conversion factor we need:

$$1172.49 \text{ amu} \times \frac{1 \text{ Na atom}}{22.99 \text{ amu}} = 51.00 \text{ Na atoms}$$

Self-Check   EXERCISE 8.2   Calculate the number of oxygen atoms in a sample that has a mass of 288 amu.

See Problems 8.6 and 8.7. ■

To summarize, we have seen that we can count atoms by weighing if we know the average atomic mass for that type of atom. This is one of the fundamental operations in chemistry, as we will see in the next section.

The average atomic mass for each element is listed in tables found inside the front cover of this book. Chemists often call these values the *atomic weights* for the elements, although this terminology is passing out of use.

8.3   The Mole

OBJECTIVES: To understand the mole concept and Avogadro's number. • To learn to convert among moles, mass, and number of atoms in a given sample.

In the previous section we used atomic mass units for mass, but these are extremely small units. In the laboratory a much larger unit, the gram, is the convenient unit for mass. In this section we will learn to count atoms in samples with masses given in grams.

Let's assume we have a sample of aluminum that has a mass of 26.98 g. What mass of copper contains exactly the same number of atoms as this sample of aluminum?

To answer this question, we need to know the average atomic masses for aluminum (26.98 amu) and copper (63.55 amu). Which atom has the greater atomic mass, aluminum or copper? The answer is copper. If we have 26.98 g of aluminum, do we need more or less than 26.98 g of copper to have the same number of copper atoms as aluminum atoms? We need more than 26.98 g of copper because each copper atom has a greater mass than each aluminum atom. Therefore, a given number of copper atoms will weigh more than an equal number of aluminum atoms. How much copper do we need? Because the average masses of aluminum and copper atoms are 26.98 amu and 63.55 amu, respectively, 26.98 g of aluminum and 63.55 g of copper

Ken O'Donoghue

| Lead bar | Silver bars | Pile of copper |
| 207.2 g | 107.9 g | 63.55 g |

Figure 8.1

All these samples of pure elements contain the *same number* (a mole) of atoms: 6.022×10^{23} atoms.

Ken O'Donoghue

Figure 8.2

One-mole samples of iron (nails), iodine crystals, liquid mercury, and powdered sulfur.

> This definition of the mole is slightly different from the SI definition but is used because it is easier to understand at this point.

> Avogadro's number (to four significant figures) is 6.022×10^{23}. One mole of *anything* is 6.022×10^{23} units of that substance.

contain exactly the same number of atoms. So we need 63.55 g of copper. As we saw in the first section when we were discussing candy, *samples in which the ratio of the masses is the same as the ratio of the masses of the individual atoms always contain the same number of atoms.* In the case just considered, the ratios are

$$\frac{26.98 \text{ g}}{63.55 \text{ g}} = \frac{26.98 \text{ amu}}{63.55 \text{ amu}}$$

Ratio of   Ratio of
sample   atomic
masses   masses

Therefore, 26.98 g of aluminum contains the same number of aluminum atoms as 63.55 g of copper contains copper atoms.

Now compare carbon (average atomic mass, 12.01 amu) and helium (average atomic mass, 4.003 amu). A sample of 12.01 g of carbon contains the same number of atoms as 4.003 g of helium. In fact, if we weigh out samples of all the elements such that each sample has a mass equal to that element's average atomic mass in grams, these samples all contain the same number of atoms (Figure 8.1). This number (the number of atoms present in all of these samples) assumes special importance in chemistry. It is called the mole, the unit all chemists use in describing numbers of atoms. The **mole** (abbreviated mol) can be defined as *the number equal to the number of carbon atoms in 12.01 grams of carbon.* Techniques for counting atoms very precisely have been used to determine this number to be 6.022×10^{23}. This number is called **Avogadro's number.** *One mole of something consists of 6.022×10^{23} units of that substance.* Just as a dozen eggs is 12 eggs, a mole of eggs is 6.022×10^{23} eggs. And a mole of water contains 6.022×10^{23} H_2O molecules.

The magnitude of the number 6.022×10^{23} is very difficult to imagine. To give you some idea, 1 mole of seconds represents a span of time 4 million times as long as the earth has already existed! One mole of marbles is enough to cover the entire earth to a depth of 50 miles! However, because atoms are so tiny, a mole of atoms or molecules is a perfectly manageable quantity to use in a reaction (Figure 8.2).

How do we use the mole in chemical calculations? Recall that Avogadro's number is defined such that a 12.01-g sample of carbon contains 6.022×10^{23} atoms. By the same token, because the average atomic mass of hydrogen is 1.008 amu (Table 8.1), 1.008 g of hydrogen contains 6.022×10^{23} hydrogen atoms. Similarly, 26.98 g of aluminum contains 6.022×10^{23} aluminum atoms. The point is that a sample of *any* element that weighs a number of grams equal to the average atomic mass of that element contains 6.022×10^{23} atoms (1 mol) of that element.

Table 8.2 shows the masses of several elements that contain 1 mole of atoms.

Table 8.2	Comparison of 1-Mol Samples of Various Elements	
Element	Number of Atoms Present	Mass of Sample (g)
Aluminum	6.022×10^{23}	26.98
Gold	6.022×10^{23}	196.97
Iron	6.022×10^{23}	55.85
Sulfur	6.022×10^{23}	32.07
Boron	6.022×10^{23}	10.81
Xenon	6.022×10^{23}	131.3

The mass of 1 mole of an element is equal to its average atomic mass in grams.

In summary, *a sample of an element with a mass equal to that element's average atomic mass expressed in grams contains 1 mole of atoms.*

To do chemical calculations, you *must* understand what the mole means and how to determine the number of moles in a given mass of a substance. However, before we do any calculations, let's be sure that the process of counting by weighing is clear. Consider the following "bag" of H atoms (symbolized by dots), which contains 1 mole (6.022×10^{23}) of H atoms and has a mass of 1.008 g. Assume the bag itself has no mass.

Contains 1 mol H atoms (6.022×10^{23} atoms)

Sample A
Mass = 1.008 g

Now consider another "bag" of hydrogen atoms in which the number of hydrogen atoms is unknown.

Contains an unknown number of H atoms

Sample B

A 1-mol sample of graphite (a form of carbon) weighs 12.01 g

We want to find out how many H atoms are present in sample ("bag") B. How can we do that? We can do it by weighing the sample. We find the mass of sample B to be 0.500 g.

How does this measured mass help us determine the number of atoms in sample B? We know that 1 mole of H atoms has a mass of 1.008 g. Sample B has a mass of 0.500 g, which is approximately half the mass of a mole of H atoms.

Sample A Mass = 1.008 g		Sample B Mass = 0.500 g
Contains 1 mol H atoms	→ Because the mass of B is about half the mass of A	Must contain about 1/2 mol H atoms

We carry out the actual calculation by using the equivalence statement

$$1 \text{ mol H atoms} = 1.008 \text{ g H}$$

to construct the conversion factor we need:

$$0.500 \text{ g H} \times \frac{1 \text{ mol H}}{1.008 \text{ g H}} = 0.496 \text{ mol H in sample B}$$

MATH SKILL BUILDER

In demonstrating how to solve problems requiring more than one step, we will often break the problem into smaller steps and report the answer to each step in the correct number of significant figures. While it may not always affect the final answer, it is a better idea to wait until the final step to round your answer to the correct number of significant figures.

Let's summarize. We know the mass of 1 mole of H atoms, so we can determine the number of moles of H atoms in any other sample of pure hydrogen by weighing the sample and *comparing* its mass to 1.008 g (the mass of 1 mole of H atoms). We can follow this same process for any element, because we know the mass of 1 mol for each of the elements.

Also, because we know that 1 mol is 6.022×10^{23} units, once we know the *moles* of atoms present, we can easily determine the *number* of atoms present. In the case considered above, we have approximately 0.5 mole of H atoms in sample B. This means that about 1/2 of 6×10^{23}, or 3×10^{23}, H atoms is present. We carry out the actual calculation by using the equivalence statement

$$1 \text{ mol} = 6.022 \times 10^{23}$$

to determine the conversion factor we need:

$$0.496 \text{ mol H atoms} \times \frac{6.022 \times 10^{23} \text{ H atoms}}{1 \text{ mol H atoms}}$$
$$= 2.99 \times 10^{23} \text{ H atoms in sample B}$$

These procedures are illustrated in Example 8.3.

EXAMPLE 8.3 **Calculating Moles and Number of Atoms**

Aluminum (Al), a metal with a high strength-to-weight ratio and a high resistance to corrosion, is often used for structures such as high-quality bicycle frames. Compute both the number of moles of atoms and the number of atoms in a 10.0-g sample of aluminum.

SOLUTION

In this case we want to change from mass to moles of atoms:

The mass of 1 mole (6.022×10^{23} atoms) of aluminum is 26.98 g. The sample we are considering has a mass of 10.0 g. Its mass is less than 26.98 g, so this sample contains less than 1 mole of aluminum atoms. We calculate the number of moles of aluminum atoms in 10.0 g by using the equivalence statement

$$1 \text{ mol Al} = 26.98 \text{ g Al}$$

A bicycle with an aluminum frame.

to construct the appropriate conversion factor:

$$10.0 \text{ g Al} \times \frac{1 \text{ mol Al}}{26.98 \text{ g Al}} = 0.371 \text{ mol Al}$$

Next we convert from moles of atoms to the number of atoms, using the equivalence statement

$$6.022 \times 10^{23} \text{ Al atoms} = 1 \text{ mol Al atoms}$$

We have

$$0.371 \text{ mol Al} \times \frac{6.022 \times 10^{23} \text{ Al atoms}}{1 \text{ mol Al}} = 2.23 \times 10^{23} \text{ Al atoms}$$

We can summarize this calculation as follows:

$$\boxed{10.0 \text{ g Al}} \quad \times \frac{1 \text{ mol}}{26.98 \text{ g}} \quad \Rightarrow \quad \boxed{0.371 \text{ mol Al}}$$

$$\boxed{0.371 \text{ mol Al atoms}} \quad \times \frac{6.022 \times 10^{23} \text{ Al atoms}}{\text{mol}} \quad \Rightarrow \quad \boxed{2.23 \times 10^{23} \text{ Al atoms}}$$

■

EXAMPLE 8.4   **Calculating the Number of Atoms**

A silicon chip used in an integrated circuit of a microcomputer has a mass of 5.68 mg. How many silicon (Si) atoms are present in this chip? The average atomic mass for silicon is 28.09 amu.

SOLUTION

Our strategy for doing this problem is to convert from milligrams of silicon to grams of silicon, then to moles of silicon, and finally to atoms of silicon:

| Milligrams of Si atoms | ⇒ | Grams of Si atoms | ⇒ | Moles of Si atoms | ⇒ | Number of Si atoms |

A silicon chip of the type used in electronic equipment.

where each arrow in the schematic represents a conversion factor. Because 1 g = 1000 mg, we have

$$5.68 \text{ mg Si} \times \frac{1 \text{ g Si}}{1000 \text{ mg Si}} = 5.68 \times 10^{-3} \text{ g Si}$$

Next, because the average mass of silicon is 28.09 amu, we know that 1 mole of Si atoms weighs 28.09 g. This leads to the equivalence statement

$$1 \text{ mol Si atoms} = 28.09 \text{ g Si}$$

Thus,

$$5.68 \times 10^{-3} \text{ g Si} \times \frac{1 \text{ mol Si}}{28.09 \text{ g Si}} = 2.02 \times 10^{-4} \text{ mol Si}$$

Using the definition of a mole (1 mol = 6.022×10^{23}), we have

$$2.02 \times 10^{-4} \text{ mol Si} \times \frac{6.022 \times 10^{23} \text{ atoms}}{1 \text{ mol Si}} = 1.22 \times 10^{20} \text{ Si atoms}$$

PhotoDisc/Getty Images

We can summarize this calculation as follows:

$$5.68 \text{ mg Si} \quad \times \quad \frac{1 \text{ g}}{1000 \text{ mg}} \quad \Rightarrow \quad 5.68 \times 10^{-3} \text{ g Si}$$

$$5.68 \times 10^{-3} \text{ g Si} \quad \times \quad \frac{1 \text{ mol}}{28.09 \text{ g}} \quad \Rightarrow \quad 2.02 \times 10^{-4} \text{ mol Si}$$

$$2.02 \times 10^{-4} \text{ mol Si} \quad \times \quad \frac{6.022 \times 10^{23} \text{ Si atoms}}{\text{mol}} \quad \Rightarrow \quad 1.22 \times 10^{20} \text{ Si atoms}$$

PROBLEM SOLVING: DOES THE ANSWER MAKE SENSE?

When you finish a problem, always think about the "reasonableness" of your answers. In Example 8.4, 5.68 mg of silicon is clearly much less than 1 mole of silicon (which has a mass of 28.09 g), so the final answer of 1.22×10^{20} atoms (compared to 6.022×10^{23} atoms in a mole) at least lies in the right direction. That is, 1.22×10^{20} atoms is a smaller number than 6.022×10^{23}. Also, always include the units as you perform calculations and make sure the correct units are obtained at the end. Paying careful attention to units and making this type of general check can help you detect errors such as an inverted conversion factor or a number that was incorrectly entered into your calculator.

As you can see, the problems are getting more complicated to solve. In the next section we will discuss strategies that will help you become a better problem solver.

Self-Check) EXERCISE 8.3

The values for the average masses of the atoms of the elements are listed inside the front cover of this book.

Chromium (Cr) is a metal that is added to steel to improve its resistance to corrosion (for example, to make stainless steel). Calculate both the number of moles in a sample of chromium containing 5.00×10^{20} atoms and the mass of the sample.

See Problems 8.19 through 8.24. ■

8.4 Learning to Solve Problems

OBJECTIVE: To understand how to solve problems by asking and answering a series of questions.

Imagine today is the first day of your new job. The problem is that you don't know how to get there. However, as luck would have it, a friend does know the way and offers to drive you. What should you do while you sit in the passenger seat? If your goal is simply to get to work today, you might not pay attention to how to get there. However, you will need to get there on your own tomorrow, so you should pay attention to distances, signs, and turns. The difference between these two approaches is the difference between taking a passive role (going along for the ride) and an active role (learning how to do it yourself). In this section, we will emphasize that you should take an active role in reading the text, especially the solutions to the practice problems.

One of the great rewards of studying chemistry is that you become a good problem solver. Being able to solve complex problems is a talent that will serve you well in all walks of life. It is our purpose in this text to help you learn to solve problems in a flexible, creative way based on understanding the fundamental ideas of chemistry. We call this approach **conceptual problem**

solving. The ultimate goal is to be able to solve new problems (that is, problems you have not seen before) on your own. In this text, we will provide problems, but instead of giving solutions for you to memorize, we will explain how to think about the solutions to the problems. Although the answers to these problems are important, it is even more important that you understand the process—the thinking necessary to get to the answer. At first we will be solving the problem for you (we will be "driving"). However, it is important that you do not take a passive role. While studying the solution, it is crucial that you interact—think through the problem with us, that is, take an active role so that eventually you can "drive" by yourself. Do not skip the discussion and jump to the answer. Usually, the solution involves asking a series of questions. Make sure that you understand each step in the process.

Although actively studying our solutions to the problems is helpful, at some point you will need to know how to think about these problems on your own. If we help you too much as you solve the problems, you won't really learn effectively. If we always "drive," you won't interact as meaningfully with the material. Eventually you need to learn to drive by yourself. Because of this, we will provide more help on the earlier problems and less as we proceed in later chapters. The goal is for you to learn how to solve a problem because you understand the main concepts and ideas in the problem.

Consider, for example, that you now know how to get from home to work. Does this mean that you can drive from work to home? Not necessarily, as you probably know from experience. If you have only memorized the directions from home to work and do not understand fundamental principles such as "I traveled north to get to the workplace, so my house is south of the workplace," you may find yourself stranded. Part of conceptual problem solving is understanding these fundamental principles.

Of course, there are many more places to go than from home to work and back. In a more complicated example, suppose you know how to get from your house to work (and back) and from your house to the library (and back). Can you get from work to the library without having to go back home? Probably not, if you have only memorized directions and you do not have a "big picture" of where your house, your workplace, and the library are relative to one another. Getting this big picture—a real understanding of the situation—is the other part of conceptual problem solving.

In conceptual problem solving, we let the problem guide us as we solve it. We ask a series of questions as we proceed and use our knowledge of fundamental principles to answer these questions. Learning this approach requires some patience, but the reward is that you become an effective solver of any new problem that confronts you in daily life or in your work in any field.

To help us as we proceed to solve a problem, the following organizing principles will be useful to us.

1. First, we need to read the problem and decide on the final goal. Then we sort through the facts given, focusing on keywords and often drawing a diagram of the problem. In this part of the analysis, we need to state the problem as simply and as visually as possible. We can summarize this process as **"Where Are We Going?"**

2. We need to work backward from the final goal in order to decide where to start. For example, in a stoichiometry problem we always start with the chemical reaction. Then as we proceed, we ask a series of questions, such as "What are the reactants and products?," "What is the balanced equation?," and "What are the amounts of the reactants?" Our understanding of the fundamental principles of chemistry will enable us to answer each of these simple questions

and eventually will lead us to the final solution. We can summarize this process as *"How Do We Get There?"*

3. Once we get the solution of the problem, then we ask ourselves: "Does it make sense?" That is, does our answer seem reasonable? We call this the **Reality Check.** It always pays to check your answer.

Using a conceptual approach to problem solving will enable you to develop real confidence as a problem solver. You will no longer panic when you see a problem that is different in some ways from those you have solved in the past. Although you might be frustrated at times as you learn this method, we guarantee that it will pay dividends later and should make your experience with chemistry a positive one that will prepare you for any career you choose.

To summarize, a creative problem solver has an understanding of fundamental principles and a big picture of the situation. One of our major goals in this text is to help you become a creative problem solver. We will do this first by giving you lots of guidance on how to solve problems. We will "drive," but we hope you will be paying attention instead of just "going along for the ride." As we move forward, we will gradually shift more of the responsibility to you. As you gain confidence in letting the problem guide you, you will be amazed at how effective you can be at solving some really complex problems, just like the ones you will confront in real life.

▶ An Example of Conceptual Problem Solving

Let's look at how conceptual problem solving works in practice. Because we used a driving analogy before, let's consider a problem about driving.

> Estimate the amount of money you would spend on gasoline to drive from New York, New York, to Los Angeles, California.

Where Are We Going?

The first thing we need to do is state the problem in words or as a diagram so that we understand the problem.

In this case, we are trying to estimate how much money we will spend on gasoline. How are we going to do this? We need to understand what factors cause us to spend more or less money. This requires us to ask, **"What Information Do We Need?,"** and **"What Do We Know?"**

Consider two people traveling in separate cars. Why might one person spend more money on gasoline than does the other person? In other words, if you were told that the two people spent different amounts of money on gasoline for a trip, what are some reasons you could give? Consider this, and write down some ideas before you continue reading.

Three factors that are important in this case are

- The price of a gallon of gasoline

- The distance of the trip between New York and Los Angeles

- The average gas mileage of the car we are driving

What do we know, or what are we given in the problem? In this problem, we are not given any of these values but are asked to estimate the cost of gasoline. So we need to estimate the required information. For example, the distance between New York and Los Angeles is about 3000 miles. The cost of gasoline varies over time and location, but a reasonable estimate is $2.00 a gallon. Gas mileage also varies, but we will assume it is about 30 miles per gallon.

Now that we have the necessary information, we will solve the problem.

How Do We Get There?

To set up the solution, we need to understand how the information affects our answer. Let's consider the relationship between the three factors we identified and our final answer.

- Price of gasoline: directly related. The more a gallon of gasoline costs, the more we will spend in total.

- Distance: directly related. The farther we travel, the more we will spend on gasoline.

- Gas mileage: inversely related. The better our gas mileage (the higher the number), the less we will spend on gasoline.

It should make sense, then, that we multiply the distance and price (because they are directly related) and then divide by the gas mileage (because it is inversely related). We will use dimensional analysis as discussed in Chapter 2. First let's determine how much gasoline we will need for our trip.

$$3000 \text{ miles} \times \frac{1 \text{ gal}}{30 \text{ miles}} = 100 \text{ gallons of gasoline}$$

Notice how the distance is in the numerator and the gas mileage is in the denominator, just as we determined they each should be. So, we will need about 100 gallons of gasoline. How much will this much gasoline cost?

$$100 \text{ gallons} \times \frac{\$2.00}{1 \text{ gallon}} = \$200$$

Notice that the price of a gallon of gas is in the numerator, just as predicted. So, given our information, we estimate the total cost of gasoline to be $200. The final step is to consider if this answer is reasonable.

REALITY CHECK   Does our answer make sense? This is always a good question to consider, and our answer will depend on our familiarity with the situation. Sometimes we may not have a good feel for what the answer should be, especially when we are learning a new concept. Other times we may have only a rough idea and may be able to claim that the answer seems reasonable, although we cannot say it is exactly right. This will usually be the case, and it is the case here if you are familiar with how much you spend on gasoline. For example, the price to fill up the tank for an average car (at $2.00 per gallon) is around $20 to $40. So, if our answer is under $100, we should be suspicious. An answer in the thousands of dollars is way too high. So, an answer in the hundreds of dollars seems reasonable.

8.5 Molar Mass

OBJECTIVES: To understand the definition of molar mass. • To learn to convert between moles and mass of a given sample of a chemical compound.

Note that when we say 1 mole of methane, we mean 1 mole of methane *molecules.*

A chemical compound is, fundamentally, a collection of atoms. For example, methane (the major component of natural gas) consists of molecules each containing one carbon atom and four hydrogen atoms (CH_4). How can we calculate the mass of 1 mole of methane? That is, what is the mass of 6.022×10^{23} CH_4 molecules? Because each CH_4 molecule contains one carbon atom and four hydrogen atoms, 1 mole of CH_4 molecules consists of

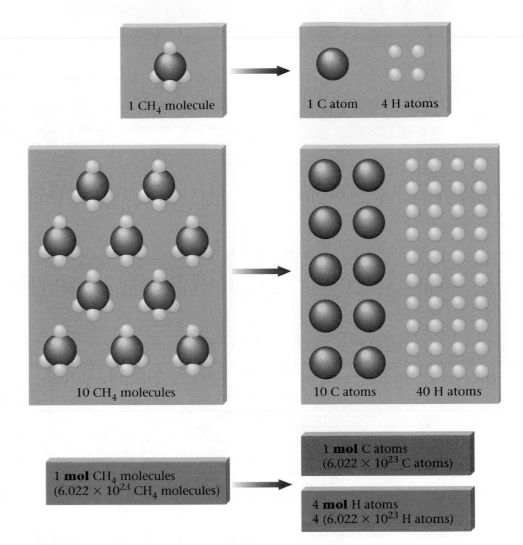

Figure 8.3

Various numbers of methane molecules showing their constituent atoms.

MATH SKILL BUILDER
Remember that the least number of decimal places limits the number of significant figures in addition.

A substance's molar mass (in grams) is the mass of 1 mole of that substance.

1 mole of carbon atoms and 4 moles of hydrogen atoms (Figure 8.3). The mass of 1 mole of methane can be found by summing the masses of carbon and hydrogen present:

$$\text{Mass of 1 mol C} = 1 \times 12.01 \text{ g} = 12.01 \text{ g}$$
$$\text{Mass of 4 mol H} = 4 \times 1.008 \text{ g} = \underline{4.032 \text{ g}}$$
$$\text{Mass of 1 mol CH}_4 \qquad\qquad = 16.04 \text{ g}$$

The quantity 16.04 g is called the molar mass for methane: the mass of 1 mole of CH_4 molecules. The **molar mass*** of any substance is the *mass (in grams) of 1 mole of the substance*. The molar mass is obtained by summing the masses of the component atoms.

EXAMPLE 8.5 | Calculating Molar Mass

Calculate the molar mass of sulfur dioxide, a gas produced when sulfur-containing fuels are burned. Unless "scrubbed" from the exhaust, sulfur dioxide can react with moisture in the atmosphere to produce acid rain.

*The term *molecular weight* was traditionally used instead of *molar mass*. The terms *molecular weight* and *molar mass* mean exactly the same thing. Because the term *molar mass* more accurately describes the concept, it will be used in this text.

SOLUTION

Where Are We Going?

We want to determine the molar mass of sulfur dioxide in units of g/mol.

What Do We Know?

- The formula for sulfur dioxide is SO_2, which means that 1 mole of SO_2 molecules contains 1 mole of sulfur atoms and 2 moles of oxygen atoms.

- We know the atomic masses of sulfur (32.07 g/mol) and oxygen (16.00 g/mol).

How Do We Get There?

We need to find the mass of 1 mole of SO_2 molecules, which is the molar mass of SO_2.

Mass of 1 mol S = 1 × 32.07   = 32.07 g
Mass of 2 mol O = 2 × 16.00   = 32.00 g
Mass of 1 mol SO_2 = 64.07 g = molar mass

The molar mass of SO_2 is 64.07 g. It represents the mass of 1 mole of SO_2 molecules.

REALITY CHECK   The answer is greater than the atomic masses of sulfur and oxygen. The units (g/mol) are correct, and the answer is reported to the correct number of significant figures (to two decimal places).

Self-Check **EXERCISE 8.4**   Polyvinyl chloride (called PVC), which is widely used for floor coverings ("vinyl") and for plastic pipes in plumbing systems, is made from a molecule with the formula C_2H_3Cl. Calculate the molar mass of this substance.

See Problems 8.27 through 8.30. ■

Some substances exist as a collection of ions rather than as separate molecules. For example, ordinary table salt, sodium chloride (NaCl), is composed of an array of Na^+ and Cl^- ions. There are no NaCl molecules present. In some books the term **formula weight** is used instead of molar mass for ionic compounds. However, in this book we will apply the term *molar mass* to both ionic and molecular substances.

To calculate the molar mass for sodium chloride, we must realize that 1 mole of NaCl contains 1 mole of Na^+ ions and 1 mole of Cl^- ions.

The mass of the electron is so small that Na^+ and Na have the same mass for our purposes, even though Na^+ has one electron fewer than Na. Also the mass of Cl virtually equals the mass of Cl^- even though it has one more electron than Cl.

Therefore, the molar mass (in grams) for sodium chloride represents the sum of the mass of 1 mole of sodium ions and the mass of 1 mole of chloride ions.

$$\begin{aligned}
\text{Mass of 1 mol } Na^+ &= 22.99 \text{ g}\\
\text{Mass of 1 mol } Cl^- &= 35.45 \text{ g}\\
\hline
\text{Mass of 1 mol } NaCl &= 58.44 \text{ g} = \text{molar mass}
\end{aligned}$$

The molar mass of $NaCl$ is 58.44 g. It represents the mass of 1 mole of sodium chloride.

EXAMPLE 8.6   Calculating Mass from Moles

Calcium carbonate, $CaCO_3$ (also called calcite), is the principal mineral found in limestone, marble, chalk, pearls, and the shells of marine animals such as clams.

   a. Calculate the molar mass of calcium carbonate.

   b. A certain sample of calcium carbonate contains 4.86 mol. What is the mass in grams of this sample?

SOLUTION

a. Where Are We Going?

We want to determine the molar mass of calcium carbonate in units of g/mol.

What Do We Know?

- The formula for calcium carbonate is $CaCO_3$. One mole of $CaCO_3$ contains 1 mole of Ca, 1 mole of C, and 3 moles of O.

- We know the atomic masses of calcium (40.08 g/mol), carbon (12.01 g/mol), and oxygen 16.00 g/mol).

How Do We Get There?

Calcium carbonate is an ionic compound composed of Ca^{2+} and CO_3^{2-} ions. One mole of calcium carbonate contains 1 mole of Ca^{2+} and 1 mole of CO_3^{2-} ions. We calculate the molar mass by summing the masses of the components.

$$\begin{aligned}
\text{Mass of 1 mol } Ca^{2+} &= 1 \times 40.08 \text{ g} = 40.08 \text{ g}\\
\text{Mass of 1 mol } CO_3^{2-} & \text{ (contains 1 mol C and 3 mol O):}\\
1 \text{ mol C} &= 1 \times 12.01 \text{ g} = 12.01 \text{ g}\\
3 \text{ mol O} &= 3 \times 16.00 \text{ g} = \underline{48.00 \text{ g}}\\
\text{Mass of 1 mol } CaCO_3 &= 100.09 \text{ g} = \text{molar mass}
\end{aligned}$$

REALITY CHECK   The answer is greater than the atomic masses of calcium, carbon, and oxygen. The units (g/mol) are correct, and the answer is reported to the correct number of significant figures (to two decimal places).

b. Where Are We Going?

We want to determine the mass of 4.86 moles of $CaCO_3$.

What Do We Know?

- From part a, we know that the molar mass of $CaCO_3$ is 100.09 g/mol.

- We have 4.86 mol $CaCO_3$.

How Do We Get There?

We determine the mass of 4.86 moles of $CaCO_3$ by using the molar mass.

$$4.86 \ \text{mol} \ \cancel{CaCO_3} \times \frac{100.09 \ \text{g} \ CaCO_3}{1 \ \text{mol} \ \cancel{CaCO_3}} = 486 \ \text{g} \ CaCO_3$$

This can be diagrammed as follows:

$$\boxed{4.86 \ \text{mol} \ CaCO_3} \ \times \ \frac{100.09 \ \text{g}}{\text{mol}} \ \blacktriangleright \ \boxed{486 \ \text{g} \ CaCO_3}$$

REALITY CHECK   We have a bit less than 5 moles of $CaCO_3$, which has a molar mass of about 100 g/mol. We should expect an answer a bit less than 500 g, so our answer makes sense. The number of significant figures in our answer (486 g) is three, as required by the initial number of moles (4.86 mol).

Self-Check   **EXERCISE 8.5**   Calculate the molar mass for sodium sulfate, Na_2SO_4. A sample of sodium sulfate with a mass of 300.0 g represents what number of moles of sodium sulfate?

> For average atomic masses, look inside the front cover of this book.

See Problems 8.35 through 8.38. ■

In summary, the molar mass of a substance can be obtained by summing the masses of the component atoms. The molar mass (in grams) represents the mass of 1 mole of the substance. Once we know the molar mass of a compound, we can compute the number of moles present in a sample of known mass. The reverse, of course, is also true as illustrated in Example 8.7.

EXAMPLE 8.7 | Calculating Moles from Mass

Juglone, a dye known for centuries, is produced from the husks of black walnuts. It is also a natural herbicide (weed killer) that kills off competitive plants around the black walnut tree but does not affect grass and other noncompetitive plants. The formula for juglone is $C_{10}H_6O_3$.

a. Calculate the molar mass of juglone.

b. A sample of 1.56 g of pure juglone was extracted from black walnut husks. How many moles of juglone does this sample represent?

SOLUTION

a. Where Are We Going?

We want to determine the molar mass of juglone in units of g/mol.

What Do We Know?

- The formula for juglone is $C_{10}H_6O_3$. One mole of juglone contains 10 moles of C, 6 moles of H, and 3 moles of O.

- We know the atomic masses of carbon (12.01 g/mol), hydrogen (1.008 g/mol), and oxygen (16.00 g/mol).

Black walnuts with and without their green hulls.

How Do We Get There?

The molar mass is obtained by summing the masses of the component atoms. In 1 mole of juglone there are 10 moles of carbon atoms, 6 moles of hydrogen atoms, and 3 moles of oxygen atoms.

$$\text{Mass of 10 mol C} = 10 \times 12.01 \text{ g} = 120.1 \text{ g}$$
$$\text{Mass of 6 mol H} = 6 \times 1.008 \text{ g} = 6.048 \text{ g}$$
$$\text{Mass of 3 mol O} = 3 \times 16.00 \text{ g} = \underline{48.00 \text{ g}}$$
$$\text{Mass of 1 mol C}_{10}\text{H}_6\text{O}_3 = 174.1 \text{ g} = \text{molar mass}$$

REALITY CHECK   Ten moles of carbon would have a mass of about 120 g, and our answer is higher than this. The units (g/mol) are correct, and the answer is reported to the correct number of significant figures (to two decimal places).

b. Where Are We Going?

We want to determine the number of moles of juglone in a sample with mass of 1.56 g.

What Do We Know?

- From part a, we know that the molar mass of juglone is 174.1 g/mol.
- We have 1.56 g of juglone.

How Do We Get There?

The mass of 1 mole of this compound is 174.1 g, so 1.56 g is much less than a mole. We can determine the exact fraction of a mole by using the equivalence statement

$$1 \text{ mol} = 174.1 \text{ g juglone}$$

to derive the appropriate conversion factor:

$$1.56 \text{ g juglone} \times \frac{1 \text{ mol juglone}}{174.1 \text{ g juglone}} = 0.00896 \text{ mol juglone}$$
$$= 8.96 \times 10^{-3} \text{ mol juglone}$$

$$\boxed{1.56 \text{ g juglone}} \times \frac{1 \text{ mol}}{174.1 \text{ g}} \Rightarrow \boxed{8.96 \times 10^{-3} \text{ mol juglone}}$$

REALITY CHECK   The mass of 1 mole of juglone is 174.1 g, so 1.56 g is much less than 1 mole. Our answer has units of mol, and the number of significant figures in our answer is three, as required by the initial mass of 1.56 g. ■

EXAMPLE 8.8   Calculating Number of Molecules

Isopentyl acetate, $C_7H_{14}O_2$, the compound responsible for the scent of bananas, can be produced commercially. Interestingly, bees release about 1 μg (1×10^{-6} g) of this compound when they sting. This attracts other bees, which then join the attack. How many moles and how many molecules of isopentyl acetate are released in a typical bee sting?

SOLUTION

Where Are We Going?

We want to determine the number of moles and the number of molecules of isopentyl acetate in a sample with mass of 1×10^{-6} g.

What Do We Know?

- The formula for isopentyl acetate is $C_7H_{14}O_2$.

- We know the atomic masses of carbon (12.01 g/mol), hydrogen (1.008 g/mol), and oxygen (16.00 g/mol).

- The mass of isopentyl acetate is 1×10^{-6} g.

- There are 6.022×10^{23} molecules in 1 mol.

How Do We Get There?

We are given a mass of isopentyl acetate and want the number of molecules, so we must first compute the molar mass.

$$7 \text{ mol C} \times 12.01 \frac{g}{mol} = 84.07 \text{ g C}$$

$$14 \text{ mol H} \times 1.008 \frac{g}{mol} = 14.11 \text{ g H}$$

$$2 \text{ mol O} \times 16.00 \frac{g}{mol} = \underline{32.00 \text{ g O}}$$

$$\text{Molar mass} = 130.18 \text{ g}$$

This means that 1 mole of isopentyl acetate (6.022×10^{23} molecules) has a mass of 130.18 g.

Next we determine the number of moles of isopentyl acetate in 1 μg, which is 1×10^{-6} g. To do this, we use the equivalence statement

$$\text{1 mol isopentyl acetate} = 130.18 \text{ g isopentyl acetate}$$

which yields the conversion factor we need:

$$1 \times 10^{-6} \text{ g } C_7H_{14}O_2 \times \frac{1 \text{ mol } C_7H_{14}O_2}{130.18 \text{ g } C_7H_{14}O_2} = 8 \times 10^{-9} \text{ mol } C_7H_{14}O_2$$

Using the equivalence statement 1 mol = 6.022×10^{23} units, we can determine the number of molecules:

$$8 \times 10^{-9} \text{ mol } C_7H_{14}O_2 \times \frac{6.022 \times 10^{23} \text{ molecules}}{1 \text{ mol } C_7H_{14}O_2} = 5 \times 10^{15} \text{ molecules}$$

This very large number of molecules is released in each bee sting.

REALITY CHECK   The mass of isopentyl acetate released in each sting (1×10^{-6} g) is much less than the mass of 1 mole of $C_7H_{14}O_2$, so the number of moles should be less than 1 mol, and it is (8×10^{-9} mol). The number of molecules should be much less than 6.022×10^{23}, and it is (5×10^{15} molecules).

Our answers have the proper units, and the number of significant figures in our answer is one, as required by the initial mass.

Self-Check   **EXERCISE 8.6**   The substance Teflon, the slippery coating on many frying pans, is made from the C_2F_4 molecule. Calculate the number of C_2F_4 units present in 135 g of Teflon.

See Problems 8.39 and 8.40. ■

8.6 Percent Composition of Compounds

OBJECTIVE: To learn to find the mass percent of an element in a given compound.

So far we have discussed the composition of compounds in terms of the numbers of constituent atoms. It is often useful to know a compound's composition in terms of the *masses* of its elements. We can obtain this information from the formula of the compound by comparing the mass of each element present in 1 mole of the compound to the total mass of 1 mole of the compound. The mass fraction for each element is calculated as follows:

MATH SKILL BUILDER

$$\text{Percent} = \frac{\text{Part}}{\text{Whole}} \times 100\%$$

$$\begin{array}{l}\text{Mass fraction} \\ \text{for a given} \\ \text{element}\end{array} = \frac{\text{mass of the element present in 1 mole of compound}}{\text{mass of 1 mole of compound}}$$

The mass fraction is converted to *mass percent* by multiplying by 100%.

We will illustrate this concept using the compound ethanol, an alcohol obtained by fermenting the sugar in grapes, corn, and other fruits and grains. Ethanol is often added to gasoline as an octane enhancer to form a fuel called gasohol. The added ethanol has the effect of increasing the octane of the gasoline and also lowering the carbon monoxide in automobile exhaust.

The formula for ethanol is written C_2H_5OH, although you might expect it to be written simply as C_2H_6O.

Note from its formula that each molecule of ethanol contains two carbon atoms, six hydrogen atoms, and one oxygen atom. This means that each mole of ethanol contains 2 moles of carbon atoms, 6 moles of hydrogen atoms, and 1 mole of oxygen atoms. We calculate the mass of each element present and the molar mass for ethanol as follows:

$$\text{Mass of C} = 2 \text{ mol} \times 12.01 \frac{\text{g}}{\text{mol}} = 24.02 \text{ g}$$

$$\text{Mass of H} = 6 \text{ mol} \times 1.008 \frac{\text{g}}{\text{mol}} = 6.048 \text{ g}$$

$$\text{Mass of O} = 1 \text{ mol} \times 16.00 \frac{\text{g}}{\text{mol}} = 16.00 \text{ g}$$

$$\text{Mass of 1 mol } C_2H_5OH = \overline{46.07 \text{ g}} = \text{molar mass}$$

The **mass percent** (sometimes called the weight percent) of carbon in ethanol can be computed by comparing the mass of carbon in 1 mole of ethanol with the total mass of 1 mole of ethanol and multiplying the result by 100%.

$$\text{Mass percent of C} = \frac{\text{mass of C in 1 mol } C_2H_5OH}{\text{mass of 1 mol } C_2H_5OH} \times 100\%$$

$$= \frac{24.02 \text{ g}}{46.07 \text{ g}} \times 100\% = 52.14\%$$

That is, ethanol contains 52.14% by mass of carbon. The mass percents of hydrogen and oxygen in ethanol are obtained in a similar manner.

$$\text{Mass percent of H} = \frac{\text{mass of H in 1 mol } C_2H_5OH}{\text{mass of 1 mol } C_2H_5OH} \times 100\%$$

$$= \frac{6.048 \text{ g}}{46.07 \text{ g}} \times 100\% = 13.13\%$$

MATH SKILL BUILDER

Sometimes, because of rounding-off effects, the sum of the mass percents in a compound is not exactly 100%.

$$\text{Mass percent of O} = \frac{\text{mass of O in 1 mol } C_2H_5OH}{\text{mass of 1 mol } C_2H_5OH} \times 100\%$$

$$= \frac{16.00 \text{ g}}{46.07 \text{ g}} \times 100\% = 34.73\%$$

The mass percents of all the elements in a compound add up to 100%, although rounding-off effects may produce a small deviation. Adding up the percentages is a good way to check the calculations. In this case, the sum of the mass percents is 52.14% + 13.13% + 34.73% = 100.00%.

EXAMPLE 8.9 | **Calculating Mass Percent**

Carvone is a substance that occurs in two forms, both of which have the same molecular formula ($C_{10}H_{14}O$) and molar mass. One type of carvone gives caraway seeds their characteristic smell; the other is responsible for the smell of spearmint oil. Compute the mass percent of each element in carvone.

SOLUTION

Where Are We Going?

We want to determine the mass percent of each element in carvone.

What Do We Know?

- The formula for carvone is $C_{10}H_{14}O$.
- We know the atomic masses of carbon (12.01 g/mol), hydrogen (1.008 g/mol), and oxygen (16.00 g/mol).
- The mass of isopentyl acetate is 1×10^{-6} g.
- There are 6.022×10^{23} molecules in 1 mole.

What Do We Need To Know?

- The mass of each element (we'll use 1 mol carvone)
- Molar mass of carvone

How Do We Get There?

Because the formula for carvone is $C_{10}H_{14}O$, the masses of the various elements in 1 mole of carvone are

$$\text{Mass of C in 1 mol} = 10 \text{ mol} \times 12.01 \frac{g}{mol} = 120.1 \text{ g}$$

$$\text{Mass of H in 1 mol} = 14 \text{ mol} \times 1.008 \frac{g}{mol} = 14.11 \text{ g}$$

$$\text{Mass of O in 1 mol} = 1 \text{ mol} \times 16.00 \frac{g}{mol} = 16.00 \text{ g}$$

$$\text{Mass of 1 mol } C_{10}H_{14}O = \overline{150.21 \text{ g}}$$
$$\text{Molar mass} = 150.2 \text{ g}$$

(rounding to the correct number of significant figures)

MATH SKILL BUILDER
The 120.1 limits the sum to one decimal place.

Next we find the fraction of the total mass contributed by each element and convert it to a percentage.

$$\text{Mass percent of C} = \frac{120.1 \text{ g C}}{150.2 \text{ g } C_{10}H_{14}O} \times 100\% = 79.96\%$$

$$\text{Mass percent of H} = \frac{14.11 \text{ g H}}{150.2 \text{ g } C_{10}H_{14}O} \times 100\% = 9.394\%$$

$$\text{Mass percent of O} = \frac{16.00 \text{ g O}}{150.2 \text{ g } C_{10}H_{14}O} \times 100\% = 10.65\%$$

REALITY CHECK   Add the individual mass percent values—they should total 100% within a small range due to rounding off. In this case, the percentages add up to 100.00%.

Self-Check **EXERCISE 8.7**   Penicillin, an important antibiotic (antibacterial agent), was discovered accidentally by the Scottish bacteriologist Alexander Fleming in 1928, although he was never able to isolate it as a pure compound. This and similar antibiotics have saved millions of lives that would otherwise have been lost to infections. Penicillin, like many of the molecules produced by living systems, is a large molecule containing many atoms. One type of penicillin, penicillin F, has the formula $C_{14}H_{20}N_2SO_4$. Compute the mass percent of each element in this compound.

See Problems 8.45 through 8.50. ∎

8.7   Formulas of Compounds

OBJECTIVE: To understand the meaning of empirical formulas of compounds.

Assume that you have mixed two solutions, and a solid product (a precipitate) forms. How can you find out what the solid is? What is its formula? There are several possible approaches you can take to answering these questions. For example, we saw in Chapter 7 that we can usually predict the identity of a precipitate formed when two solutions are mixed in a reaction of this type if we know some facts about the solubilities of ionic compounds.

However, although an experienced chemist can often predict the product expected in a chemical reaction, the only sure way to identify the product is to perform experiments. Usually we compare the physical properties of the product to the properties of known compounds.

Sometimes a chemical reaction gives a product that has never been obtained before. In such a case, a chemist determines what compound has been formed by determining which elements are present and how much of each. These data can be used to obtain the formula of the compound. In Section 8.6 we used the formula of the compound to determine the mass of each element present in a mole of the compound. To obtain the formula of an unknown compound, we do the opposite. That is, we use the measured masses of the elements present to determine the formula.

Recall that the formula of a compound represents the relative numbers of the various types of atoms present. For example, the molecular formula CO_2 tells us that for each carbon atom there are two oxygen atoms in each molecule of carbon dioxide. So to determine the formula of a substance we need to count the atoms. As we have seen in this chapter, we can do this by weighing. Suppose we know that a compound contains only the elements carbon, hydrogen, and oxygen, and we weigh out a 0.2015-g sample for analysis. Using methods we will not discuss here, we find that this 0.2015-g sample of compound contains 0.0806 g of carbon, 0.01353 g of hydrogen, and 0.1074 g of oxygen. We have just learned how to convert these masses to numbers of atoms by using the atomic mass of each element. We begin by converting to moles.

Carbon

$$(0.0806 \text{ g C}) \times \frac{1 \text{ mol C atoms}}{12.01 \text{ g C}} = 0.00671 \text{ mol C atoms}$$

Hydrogen

$$(0.01353 \text{ g H}) \times \frac{1 \text{ mol H atoms}}{1.008 \text{ g H}} = 0.01342 \text{ mol H atoms}$$

Oxygen

$$(0.1074 \text{ g O}) \times \frac{1 \text{ mol O atoms}}{16.00 \text{ g O}} = 0.006713 \text{ mol O atoms}$$

Let's review what we have established. We now know that 0.2015 g of the compound contains 0.00671 mole of C atoms, 0.01342 mole of H atoms, and 0.006713 mole of O atoms. Because 1 mol is 6.022×10^{23}, these quantities can be converted to actual numbers of atoms.

Carbon

$$(0.00671 \text{ mol C atoms}) \frac{(6.022 \times 10^{23} \text{ C atoms})}{1 \text{ mol C atoms}} = 4.04 \times 10^{21} \text{ C atoms}$$

Hydrogen

$$(0.01342 \text{ mol H atoms}) \frac{(6.022 \times 10^{23} \text{ H atoms})}{1 \text{ mol H atoms}} = 8.08 \times 10^{21} \text{ H atoms}$$

Oxygen

$$(0.006713 \text{ mol O atoms}) \frac{(6.022 \times 10^{23} \text{ O atoms})}{1 \text{ mol O atoms}} = 4.043 \times 10^{21} \text{ O atoms}$$

These are the numbers of the various types of atoms *in 0.2015 g of compound*. What do these numbers tell us about the formula of the compound? Note the following:

1. The compound contains the same number of C and O atoms.

2. There are twice as many H atoms as C atoms or O atoms.

We can represent this information by the formula CH_2O, which expresses the *relative* numbers of C, H, and O atoms present. Is this the true formula for the compound? In other words, is the compound made up of CH_2O molecules? It may be. However, it might also be made up of $C_2H_4O_2$ molecules, $C_3H_6O_3$ molecules, $C_4H_8O_4$ molecules, $C_5H_{10}O_5$ molecules, $C_6H_{12}O_6$ molecules, and so on. Note that each of these molecules has the required 1:2:1 ratio of carbon to hydrogen to oxygen atoms (the ratio shown by experiment to be present in the compound).

When we break a compound down into its separate elements and "count" the atoms present, we learn only the ratio of atoms—we get only the *relative* numbers of atoms. The formula of a compound that expresses the smallest whole-number ratio of the atoms present is called the **empirical formula** or *simplest formula*. A compound that contains the molecules $C_4H_8O_4$ has the same empirical formula as a compound that contains $C_6H_{12}O_6$ molecules. The empirical formula for both is CH_2O. The actual formula of a compound—the one that gives the composition of the molecules that are present—is called the **molecular formula**. The sugar called glucose is made of molecules with the molecular formula $C_6H_{12}O_6$ (Figure 8.4). Note from the molecular formula for glucose that the empirical formula is CH_2O. We can represent the molecular formula as a multiple (by 6) of the empirical formula:

$$C_6H_{12}O_6 = (CH_2O)_6$$

In the next section, we will explore in more detail how to calculate the empirical formula for a compound from the relative masses of the elements present. As we will see in Sections 8.8 and 8.9, we must know the molar mass of a compound to determine its molecular formula.

Figure 8.4

The glucose molecule. The molecular formula is $C_6H_{12}O_6$, as can be verified by counting the atoms. The empirical formula for glucose is CH_2O.

EXAMPLE 8.10 | Determining Empirical Formulas

In each case below, the molecular formula for a compound is given. Determine the empirical formula for each compound.

a. C_6H_6. This is the molecular formula for benzene, a liquid commonly used in industry as a starting material for many important products.

b. $C_{12}H_4Cl_4O_2$. This is the molecular formula for a substance commonly called dioxin, a powerful poison that sometimes occurs as a by-product in the production of other chemicals.

c. $C_6H_{16}N_2$. This is the molecular formula for one of the reactants used to produce nylon.

SOLUTION

a. $C_6H_6 = (CH)_6$; CH is the empirical formula. Each subscript in the empirical formula is multiplied by 6 to obtain the molecular formula.

b. $C_{12}H_4Cl_4O_2$; $C_{12}H_4Cl_4O_2 = (C_6H_2Cl_2O)_2$; $C_6H_2Cl_2O$ is the empirical formula. Each subscript in the empirical formula is multiplied by 2 to obtain the molecular formula.

c. $C_6H_{16}N_2 = (C_3H_8N)_2$; C_3H_8N is the empirical formula. Each subscript in the empirical formula is multiplied by 2 to obtain the molecular formula. ■

8.8 Calculation of Empirical Formulas

OBJECTIVE: To learn to calculate empirical formulas.

As we said in the previous section, one of the most important things we can learn about a new compound is its chemical formula. To calculate the empirical formula of a compound, we first determine the relative masses of the various elements that are present.

One way to do this is to measure the masses of elements that react to form the compound. For example, suppose we weigh out 0.2636 g of pure nickel metal into a crucible and heat this metal in the air so that the nickel can react with oxygen to form a nickel oxide compound. After the sample has cooled, we weigh it again and find its mass to be 0.3354 g. The gain in mass is due to the oxygen that reacts with the nickel to form the oxide. Therefore, the mass of oxygen present in the compound is the total mass of the product minus the mass of the nickel:

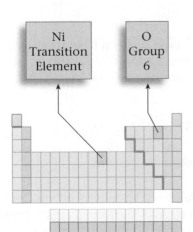

| Total mass of nickel oxide | − | Mass of nickel originally present | = | Mass of oxygen that reacted with the nickel |

or   0.3354 g   −   0.2636 g   =   0.0718 g

Note that the mass of nickel present in the compound is the nickel metal originally weighed out. So we know that the nickel oxide contains 0.2636 g of nickel and 0.0718 g of oxygen. What is the empirical formula of this compound?

To answer this question we must convert the masses to numbers of atoms, using atomic masses:

Four significant figures allowed.

Three significant figures allowed.

$$0.2636 \text{ g Ni} \times \frac{1 \text{ mol Ni atoms}}{58.69 \text{ g Ni}} = 0.004491 \text{ mol Ni atoms}$$

$$0.0718 \text{ g O} \times \frac{1 \text{ mol O atoms}}{16.00 \text{ g O}} = 0.00449 \text{ mol O atoms}$$

These mole quantities represent numbers of atoms (remember that a mole of atoms is 6.022×10^{23} atoms). It is clear from the moles of atoms that the compound contains an equal number of Ni and O atoms, so the formula is NiO. This is the *empirical formula*; it expresses the smallest whole-number (integer) ratio of atoms:

$$\frac{0.004491 \text{ mol Ni atoms}}{0.00449 \text{ mol O atoms}} = \frac{1 \text{ Ni}}{1 \text{ O}}$$

That is, this compound contains equal numbers of nickel atoms and oxygen atoms. We say the ratio of nickel atoms to oxygen atoms is 1:1 (1 to 1).

EXAMPLE 8.11   Calculating Empirical Formulas

An oxide of aluminum is formed by the reaction of 4.151 g of aluminum with 3.692 g of oxygen. Calculate the empirical formula for this compound.

SOLUTION

Where Are We Going?

We want to determine the empirical formula for the aluminum oxide, Al_xO_y. That is, we want to solve for x and y.

What Do We Know?

- The compound contains 4.151 g of aluminum and 3.692 g of oxygen.

- We know the atomic masses of aluminum (26.98 g/mol), and oxygen (16.00 g/mol).

What Do We Need To Know?

- x and y represent moles of atoms in 1 mole of the compound, so we need to determine the relative number of moles of Al and O.

How Do We Get There?

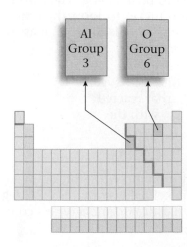

We need to know the relative numbers of each type of atom to write the formula, so we must convert these masses to moles of atoms to get the empirical formula. We carry out the conversion by using the atomic masses of the elements.

$$4.151 \text{ g Al} \times \frac{1 \text{ mol Al}}{26.98 \text{ g Al}} = 0.1539 \text{ mol Al atoms}$$

$$3.692 \text{ g O} \times \frac{1 \text{ mol O}}{16.00 \text{ g O}} = 0.2308 \text{ mol O atoms}$$

Because chemical formulas use only whole numbers, we next find the integer (whole-number) ratio of the atoms. To do this we start by dividing both numbers by the smallest of the two. This converts the smallest number to 1.

$$\frac{0.1539 \text{ mol Al}}{0.1539} = 1.000 \text{ mol Al atoms}$$

$$\frac{0.2308 \text{ mol O}}{0.1539} = 1.500 \text{ mol O atoms}$$

Note that dividing both numbers of moles of atoms by the *same* number does not change the *relative* numbers of oxygen and aluminum atoms. That is,

$$\frac{0.2308 \text{ mol O}}{0.1539 \text{ mol Al}} = \frac{1.500 \text{ mol O}}{1.000 \text{ mol Al}}$$

Thus we know that the compound contains 1.500 moles of O atoms for every 1.000 mole of Al atoms, or, in terms of individual atoms, we could say that the compound contains 1.500 O atoms for every 1.000 Al atom. However, because only *whole* atoms combine to form compounds, we must find a set of *whole numbers* to express the empirical formula. When we multiply both 1.000 and 1.500 by 2, we get the integers we need.

$$1.500 \text{ O} \times 2 = 3.000 = 3 \text{ O atoms}$$
$$1.000 \text{ Al} \times 2 = 2.000 = 2 \text{ Al atoms}$$

Therefore, this compound contains two Al atoms for every three O atoms, and the empirical formula is Al_2O_3. Note that the *ratio* of atoms in this compound is given by each of the following fractions:

We might express these data as:

$$Al_{1.000 \text{ mol}}O_{1.500 \text{ mol}}$$
or
$$Al_{2.000 \text{ mol}}O_{3.000 \text{ mol}}$$
or
$$Al_2O_3$$

$$\frac{0.2308 \text{ O}}{0.1539 \text{ Al}} = \frac{1.500 \text{ O}}{1.000 \text{ Al}} = \frac{\frac{3}{2}\text{O}}{1 \text{ Al}} = \frac{3 \text{ O}}{2 \text{ Al}}$$

The smallest whole-number ratio corresponds to the subscripts of the empirical formula, Al_2O_3.

REALITY CHECK The values for x and y are whole numbers. ∎

Sometimes the relative numbers of moles you get when you calculate an empirical formula will turn out to be nonintegers, as was the case in Example 8.11. When this happens, you must convert to the appropriate whole numbers. This is done by multiplying all the numbers by the same small integer, which can be found by trial and error. The multiplier needed is almost always between 1 and 6. We will now summarize what we have learned about calculating empirical formulas.

Steps for Determining the Empirical Formula of a Compound

Step 1 Obtain the mass of each element present (in grams).

Step 2 Determine the number of moles of each type of atom present.

Step 3 Divide the number of moles of each element by the smallest number of moles to convert the smallest number to 1. If all of the numbers so obtained are integers (whole numbers), these are the subscripts in the empirical formula. If one or more of these numbers are not integers, go on to step 4.

Step 4 Multiply the numbers you derived in step 3 by the smallest integer that will convert all of them to whole numbers. This set of whole numbers represents the subscripts in the empirical formula.

EXAMPLE 8.12 Calculating Empirical Formulas for Binary Compounds

When a 0.3546-g sample of vanadium metal is heated in air, it reacts with oxygen to achieve a final mass of 0.6330 g. Calculate the empirical formula of this vanadium oxide.

SOLUTION

Where Are We Going?

We want to determine the empirical formula for the vanadium oxide, V_xO_y. That is, we want to solve for x and y.

What Do We Know?

- The compound contains 0.3546 g of vanadium and has a total mass of 0.6330 g.

- We know the atomic masses of vanadium (50.94 g/mol) and oxygen (16.00 g/mol).

What Do We Need To Know?

- We need to know the mass of oxygen in the sample.

- x and y represent moles of atoms in 1 mole of the compound, so we need to determine the relative number of moles of V and O.

How Do We Get There?

Step 1 All the vanadium that was originally present will be found in the final compound, so we can calculate the mass of oxygen that reacted by taking the following difference:

Total mass of compound		Mass of vanadium in compound		Mass of oxygen in compound
	$-$		$=$	
0.6330 g	$-$	0.3546 g	$=$	0.2784 g

Step 2 Using the atomic masses (50.94 for V and 16.00 for O), we obtain

$$0.3546 \text{ g V} \times \frac{1 \text{ mol V atoms}}{50.94 \text{ g V}} = 0.006961 \text{ mol V atoms}$$

$$0.2784 \text{ g O} \times \frac{1 \text{ mol O atoms}}{16.00 \text{ g O}} = 0.01740 \text{ mol O atoms}$$

Step 3 Then we divide both numbers of moles by the smaller, 0.006961.

$$\frac{0.006961 \text{ mol V atoms}}{0.006961} = 1.000 \text{ mol V atoms}$$

$$\frac{0.01740 \text{ mol O atoms}}{0.006961} = 2.500 \text{ mol O atoms}$$

Because one of these numbers (2.500) is not an integer, we go on to step 4.

MATH SKILL BUILDER
$V_{1.000}O_{2.500}$ becomes V_2O_5.

Step 4 We note that $2 \times 2.500 = 5.000$ and $2 \times 1.000 = 2.000$, so we multiply both numbers by 2 to get integers.

$$2 \times 1.000 \text{ V} = 2.000 \text{ V} = 2 \text{ V}$$
$$2 \times 2.500 \text{ O} = 5.000 \text{ O} = 5 \text{ O}$$

This compound contains 2 V atoms for every 5 O atoms, and the empirical formula is V_2O_5.

REALITY CHECK   The values for x and y are whole numbers.

See Problems 8.61, 8.63, 8.65, and 8.66. ■

Self-Check **EXERCISE 8.8**   In a lab experiment it was observed that 0.6884 g of lead combines with 0.2356 g of chlorine to form a binary compound. Calculate the empirical formula of this compound.

The same procedures we have used for binary compounds also apply to compounds containing three or more elements, as Example 8.13 illustrates.

EXAMPLE 8.13   Calculating Empirical Formulas for Compounds Containing Three or More Elements

A sample of lead arsenate, an insecticide used against the potato beetle, contains 1.3813 g of lead, 0.00672 g of hydrogen, 0.4995 g of arsenic, and 0.4267 g of oxygen. Calculate the empirical formula for lead arsenate.

SOLUTION

Where Are We Going?

We want to determine the empirical formula for lead arsenate, $Pb_aH_bAs_cO_d$. That is, we want to solve for a, b, c, and d.

What Do We Know?

- The compound contains 1.3813 g of Pb, 0.00672 g of H, 0.4995 g of As, and 0.4267 g of O.

- We know the atomic masses of lead (207.2 g/mol), hydrogen (1.008 g/mol), arsenic (74.92 g/mol), and oxygen (16.00 g/mol).

What Do We Need To Know?

- a, b, c, and d represent moles of atoms in 1 mole of the compound, so we need to determine the relative number of moles of Pb, H, As, and O.

How Do We Get There?

Step 1   The compound contains 1.3813 g Pb, 0.00672 g H, 0.4995 g As, and 0.4267 g O.

Step 2   We use the atomic masses of the elements present to calculate the moles of each.

$$1.3813 \text{ g Pb} \times \frac{1 \text{ mol Pb}}{207.2 \text{ g Pb}} = 0.006667 \text{ mol Pb}$$

$$0.00672 \text{ g H} \times \frac{1 \text{ mol H}}{1.008 \text{ g H}} = 0.00667 \text{ mol H}$$

$$0.4995 \text{ g As} \times \frac{1 \text{ mol As}}{74.92 \text{ g As}} = 0.006667 \text{ mol As}$$

$$0.4267 \text{ g O} \times \frac{1 \text{ mol O}}{16.00 \text{ g O}} = 0.02667 \text{ mol O}$$

Step 3   Now we divide by the smallest number of moles.

$$\frac{0.006667 \text{ mol Pb}}{0.006667} = 1.000 \text{ mol Pb}$$

$$\frac{0.00667 \text{ mol H}}{0.006667} = 1.00 \text{ mol H}$$

$$\frac{0.006667 \text{ mol As}}{0.006667} = 1.000 \text{ mol As}$$

$$\frac{0.02667 \text{ mol O}}{0.006667} = 4.000 \text{ mol O}$$

The numbers of moles are all whole numbers, so the empirical formula is $PbHAsO_4$.

REALITY CHECK   The values for *a, b, c,* and *d* are whole numbers.

Self-Check   **EXERCISE 8.9**   Sevin, the commercial name for an insecticide used to protect crops such as cotton, vegetables, and fruit, is made from carbamic acid. A chemist analyzing a sample of carbamic acid finds 0.8007 g of carbon, 0.9333 g of nitrogen, 0.2016 g of hydrogen, and 2.133 g of oxygen. Determine the empirical formula for carbamic acid.

See Problems 8.57 and 8.59. ■

When a compound is analyzed to determine the relative amounts of the elements present, the results are usually given in terms of percentages by masses of the various elements. In Section 8.6 we learned to calculate the percent composition of a compound from its formula. Now we will do the opposite. Given the percent composition, we will calculate the empirical formula.

MATH SKILL BUILDER

Percent by mass for a given element means the grams of that element in 100 g of the compound.

To understand this procedure, you must understand the meaning of *percent*. Remember that percent means parts of a given component per 100 parts of the total mixture. For example, if a given compound is 15% carbon (by mass), the compound contains 15 g of carbon per 100 g of compound.

Calculation of the empirical formula of a compound when one is given its percent composition is illustrated in Example 8.14.

EXAMPLE 8.14 | Calculating Empirical Formulas from Percent Composition

Cisplatin, the common name for a platinum compound that is used to treat cancerous tumors, has the composition (mass percent) 65.02% platinum, 9.34% nitrogen, 2.02% hydrogen, and 23.63% chlorine. Calculate the empirical formula for cisplatin.

SOLUTION

Where Are We Going?

We want to determine the empirical formula for cisplatin, $Pt_aN_bH_cCl_d$. That is, we want to solve for *a, b, c,* and *d*.

What Do We Know?

- The compound has the composition (mass percent) 65.02% Pt, 9.34% N, 2.02% H, and 23.63% Cl.

- We know the atomic masses of platinum (195.1 g/mol), nitrogen (14.01 g/mol), hydrogen (1.008 g/mol), and chlorine (35.45 g/mol).

What Do We Need To Know?

- *a, b, c,* and *d* represent moles of atoms in 1 mole of the compound, so we need to determine the relative number of moles of Pt, N, H, and Cl.

- We have mass percent data, and to get to the number of moles we need to know the mass of each element (g) in the sample.

How Do We Get There?

Step 1   Determine how many grams of each element are present in 100 g of compound. Cisplatin is 65.02% platinum (by mass), which means there is 65.02 g of platinum (Pt) per 100.00 g of compound. Similarly, a 100.00-g sample of cisplatin contains 9.34 g of nitrogen (N), 2.02 g of hydrogen (H), and 26.63 g of chlorine (Cl).

   If we have a 100.00-g sample of cisplatin, we have 65.02 g Pt, 9.34 g N, 2.02 g H, and 23.63 g Cl.

Step 2   Determine the number of moles of each type of atom. We use the atomic masses to calculate moles.

$$65.02 \text{ g Pt} \times \frac{1 \text{ mol Pt}}{195.1 \text{ g Pt}} = 0.3333 \text{ mol Pt}$$

$$9.34 \text{ g N} \times \frac{1 \text{ mol N}}{14.01 \text{ g N}} = 0.667 \text{ mol N}$$

$$2.02 \text{ g H} \times \frac{1 \text{ mol H}}{1.008 \text{ g H}} = 2.00 \text{ mol H}$$

$$23.63 \text{ g Cl} \times \frac{1 \text{ mol Cl}}{35.45 \text{ g Cl}} = 0.6666 \text{ mol Cl}$$

Step 3   Divide through by the smallest number of moles.

$$\frac{0.3333 \text{ mol Pt}}{0.3333} = 1.000 \text{ mol Pt}$$

$$\frac{0.667 \text{ mol N}}{0.3333} = 2.00 \text{ mol N}$$

$$\frac{2.00 \text{ mol H}}{0.3333} = 6.01 \text{ mol H}$$

$$\frac{0.6666 \text{ mol Cl}}{0.3333} = 2.000 \text{ mol Cl}$$

The empirical formula for cisplatin is $PtN_2H_6Cl_2$. Note that the number for hydrogen is slightly greater than 6 because of rounding-off effects.

REALITY CHECK   The values for *a, b, c,* and *d* are whole numbers.

Self-Check EXERCISE 8.10   The most common form of nylon (Nylon-6) is 63.68% carbon, 12.38% nitrogen, 9.80% hydrogen, and 14.14% oxygen. Calculate the empirical formula for Nylon-6.

See Problems 8.67 through 8.74. ∎

   Note from Example 8.14 that once the percentages are converted to masses, this example is the same as earlier examples in which the masses were given directly.

8.9 Calculation of Molecular Formulas

OBJECTIVE: To learn to calculate the molecular formula of a compound, given its empirical formula and molar mass.

If we know the composition of a compound in terms of the masses (or mass percentages) of the elements present, we can calculate the empirical formula but not the molecular formula. For reasons that will become clear as we consider Example 8.15, to obtain the molecular formula we must know the molar mass. In this section we will consider compounds where both the percent composition and the molar mass are known.

EXAMPLE 8.15 | Calculating Molecular Formulas

A white powder is analyzed and found to have an empirical formula of P_2O_5. The compound has a molar mass of 283.88 g. What is the compound's molecular formula?

SOLUTION

Where Are We Going?

We want to determine the molecular formula (P_xO_y) for a compound. That is, we want to solve for x and y.

What Do We Know?

- The empirical formula of the compound is P_2O_5.

- The molar mass of the compound is 283.88 g/mol.

- We know the atomic masses of phosphorus (30.97 g/mol) and oxygen (16.00 g/mol).

- The molecular formula contains a whole number of empirical formula units. So, the molecular formula will be $P_{2x}O_{5y}$.

What Do We Need To Know?

- We need to know the empirical formula mass.

How Do We Get There?

To obtain the molecular formula, we must compare the empirical formula mass to the molar mass. The empirical formula mass for P_2O_5 is the mass of 1 mole of P_2O_5 units.

Atomic
mass of P

2 mol P: 2×30.97 g =   61.94 g
5 mol O: 5×16.00 g =   80.00 g
                          141.94 g   Mass of 1 mole of P_2O_5 units

Atomic
mass of O

Recall that the molecular formula contains a whole number of empirical formula units. That is,

Molecular formula = (empirical formula)$_n$

● = P

● = O

Figure 8.5

The structure of P_4O_{10} as a "ball-and-stick" model. This compound has a great affinity for water and is often used as a desiccant, or drying agent.

where n is a small whole number. Now, because

$$\text{Molecular formula} = n \times \text{empirical formula}$$

then

$$\text{Molar mass} = n \times \text{empirical formula mass}$$

Solving for n gives

$$n = \frac{\text{molar mass}}{\text{empirical formula mass}}$$

Thus, to determine the molecular formula, we first divide the molar mass by the empirical formula mass. This tells us how many empirical formula masses there are in one molar mass.

$$\frac{\text{Molar mass}}{\text{Empirical formula mass}} = \frac{283.88 \text{ g}}{141.94 \text{ g}} = 2$$

This result means that $n = 2$ for this compound, so the molecular formula consists of two empirical formula units, and the molecular formula is $(P_2O_5)_2$, or P_4O_{10}. The structure of this interesting compound is shown in Figure 8.5.

REALITY CHECK The values for x and y are whole numbers. Also, the ratio of P:O in the molecular formula (4:10) is 2:5.

Self-Check EXERCISE 8.11 A compound used as an additive for gasoline to help prevent engine knock shows the following percent composition:

71.65% Cl   24.27% C   4.07% H

The molar mass is known to be 98.96 g. Determine the empirical formula and the molecular formula for this compound.

See Problems 8.81 and 8.82. ■

It is important to realize that the molecular formula is always an integer multiple of the empirical formula. For example, the sugar glucose (see Figure 8.4) has the empirical formula CH_2O and the molecular formula $C_6H_{12}O_6$. In this case there are six empirical formula units in each glucose molecule:

$$(CH_2O)_6 = C_6H_{12}O_6$$

In general, we can represent the molecular formula in terms of the empirical formula as follows:

$$(\text{Empirical formula})_n = \text{molecular formula}$$

where n is an integer. If $n = 1$, the molecular formula is the same as the empirical formula. For example, for carbon dioxide the empirical formula (CO_2) and the molecular formula (CO_2) are the same, so $n = 1$. On the other hand, for tetraphosphorus decoxide the empirical formula is P_2O_5 and the molecular formula is $P_4O_{10} = (P_2O_5)_2$. In this case $n = 2$.

Key Terms

atomic mass unit
  (amu) (8.2)

average atomic mass (8.2)

mole (8.3)

Avogadro's number (8.3)

conceptual problem
  solving (8.4)

molar mass (8.5)

mass percent (8.6)

empirical formula (8.7)

molecular formula (8.7)

Summary

1. We can count individual units by weighing if we know the average mass of the units. Thus, when we know the average mass of the atoms of an element as that element occurs in nature, we can calculate the number of atoms in any given sample of that element by weighing the sample.

2. A mole is a unit of measure equal to 6.022×10^{23}, which is called Avogadro's number. One mole of any substance contains 6.022×10^{23} units.

3. One mole of an element has a mass equal to the element's atomic mass expressed in grams. The molar mass of any compound is the mass (in grams) of 1 mole of the compound and is the sum of the masses of the component atoms.

4. Percent composition consists of the mass percent of each element in a compound:

$$\text{Mass percent} = \frac{\text{mass of a given element in 1 mole of compound}}{\text{mass of 1 mole of compound}} \times 100\%$$

5. The empirical formula of a compound is the simplest whole-number ratio of the atoms present in the compound; it can be derived from the percent composition of the compound. The molecular formula is the exact formula of the molecules present; it is always an integer multiple of the empirical formula. The diagram in the left column summarizes these different ways of expressing the same information.

Active Learning Questions

These questions are designed to be considered by groups of students in class. Often these questions work well for introducing a particular topic in class.

1. In chemistry, what is meant by the term *mole?* What is the importance of the mole concept?

2. What is the difference between the empirical and molecular formulas of a compound? Can they ever be the same? Explain.

3. A substance A_2B is 60% A by mass. Calculate the percent B (by mass) for AB_2.

4. Give the formula for calcium phosphate and then answer the following questions:

  a. Calculate the percent composition of each of the elements in this compound.

  b. If you knew that there was 50.0 g of phosphorus in your sample, how many grams of calcium phosphate would you have? How many moles of calcium phosphate would this be? How many formula units of calcium phosphate?

5. How would you find the number of "chalk molecules" it takes to write your name on the board? Explain what you would need to do, and provide a sample calculation.

6. A 0.821-mol sample of a substance composed of diatomic molecules has a mass of 131.3 g. Identify this molecule.

7. How many molecules of water are there in a 10.0-g sample of water? How many hydrogen atoms are there in this sample?

8. What is the mass (in grams) of one molecule of ammonia?

9. Consider separate 100.0-g samples of each of the following: NH_3, N_2O, N_2H_4, HCN, HNO_3. Arrange these samples from largest mass of nitrogen to smallest mass of nitrogen and prove/explain your order.

10. A molecule has a mass of 4.65×10^{-23} g. Provide two possible chemical formulas for such a molecule.

11. Differentiate between the terms *atomic mass* and *molar mass*.

12. Consider Figure 4.19 in the text. Why is it that the formulas for ionic compound are always empirical formulas?

13. Why do we need to count atoms by weighing them?

14. The following claim is made in your text: 1 mole of marbles is enough to cover the entire earth to a depth of 50 miles.

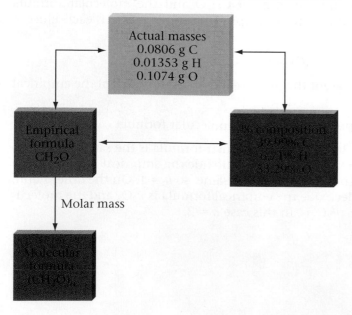

Actual masses
0.0806 g C
0.01353 g H
0.1074 g O

Empirical formula
CH_2O

% composition
39.99% C
6.71% H
53.29% O

Molar mass

Molecular formula
$(CH_2O)_n$

Provide mathematical support for this claim. Is it reasonably accurate?

15. Estimate the length of time it would take you to count to Avogadro's number. Provide mathematical support.

16. Suppose Avogadro's number was 1000 instead of 6.022×10^{23}. How, if at all, would this affect the relative masses on the periodic table? How, if at all, would this affect the absolute masses of the elements?

17. Estimate the number of atoms in your body and provide mathematical support. Because it is an estimate, it need not be exact, although you should choose your number wisely.

18. Consider separate equal mass samples of magnesium, zinc, and silver. Rank them from greatest to least number of atoms and support your answer.

19. You have a 20.0-g sample of silver metal. You are given 10.0 g of another metal and told that this sample contains twice the number of atoms as the sample of silver metal. Identify this metal.

20. How would you find the number of "ink molecules" it takes to write your name on a piece of paper with your pen? Explain what you would need to do, and provide a sample calculation.

21. True or false. The atom with the largest subscript in a formula is the atom with the largest percent by mass in the compound.

If true, explain why with an example. If false, explain why and provide a counterexample. In either case, provide mathematical support.

22. Which of the following compounds have the same empirical formulas?

H_2O and H_2O_2

a

N_2O_4 and NO_2

b

CO and CO_2

c

CH_4 and C_2H_6

d

23. The percent by mass of nitrogen is 46.7% for a species containing only nitrogen and oxygen. Which of the following could be this species?

N_2O_5 NO

a   **c**

NO_2 N_2O

b   **d**

24. Calculate the molar mass of the following substances.

○ H
● N

a

○ H
● N

b

25. Give the empirical formula for each of the compounds represented below.

a   **b**   **c**

○ H
● O
● N
● C
● P

d

⏺WL Questions and problems for this chapter are available in OWL. Use the OWL access card that came with this textbook to access assigned questions and problems for this chapter.

9

Chemical Quantities

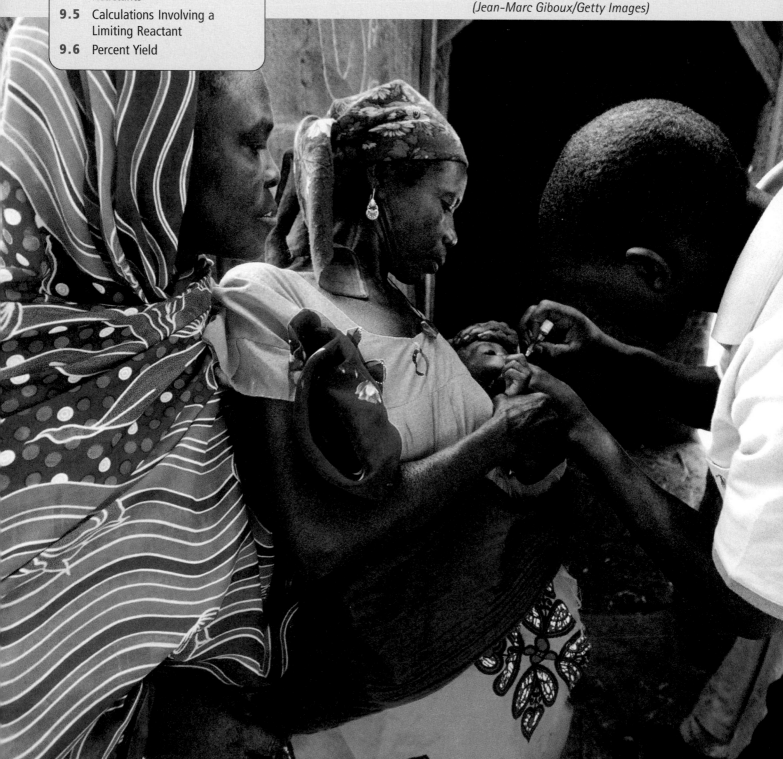

● Doctors administering the polio vaccine in Niger. *(Jean-Marc Giboux/Getty Images)*

More than 10 *billion* pounds of methanol is produced annually.

Suppose you work for a consumer advocate organization and you want to test a company's advertising claims about the effectiveness of its antacid. The company claims that its product neutralizes 10 times as much stomach acid per tablet as its nearest competitor. How would you test the validity of this claim?

Or suppose that after graduation you go to work for a chemical company that makes methanol (methyl alcohol), a substance used as a starting material for the manufacture of products such as antifreeze and aviation fuels. You are working with an experienced chemist who is trying to improve the company's process for making methanol from the reaction of gaseous hydrogen with carbon monoxide gas. The first day on the job, you are instructed to order enough hydrogen and carbon monoxide to produce 6.0 kg of methanol in a test run. How would you determine how much carbon monoxide and hydrogen you should order?

Methanol is a starting material for some jet fuels.

Royalty-Free Corbis

After you study this chapter, you will be able to answer these questions.

9.1 Information Given by Chemical Equations

OBJECTIVE: To understand the molecular and mass information given in a balanced equation.

Reactions are what chemistry is really all about. Recall from Chapter 6 that chemical changes are actually rearrangements of atom groupings that can be described by chemical equations. These chemical equations tell us the identities (formulas) of the reactants and products and also show how much of each reactant and product participates in the reaction. The numbers (coefficients) in the balanced chemical equation enable us to determine just how much product we can get from a given quantity of reactants. It is important to recognize that the coefficients in a balanced equation give us the *relative* numbers of molecules. That is, we are interested in the *ratio* of the coefficients, not individual coefficients.

To illustrate this idea, consider a nonchemical analogy. Suppose you are in charge of making deli sandwiches at a fast-food restaurant. A particular type of sandwich requires 2 slices of bread, 3 slices of meat, and 1 slice of cheese. We can represent making the sandwich with the following equation:

2 slices bread + 3 slices meat + 1 slice cheese → 1 sandwich

Your boss sends you to the store to get enough ingredients to make 50 sandwiches. How do you figure out how much of each ingredient to buy? Because you need enough to make 50 sandwiches, you multiply the preceding equation by 50.

50(2 slices bread) + 50(3 slices meat) + 50(1 slice cheese) → 50(1 sandwich)

That is

100 slices bread + 150 slices meat + 50 slices cheese → 50 sandwiches

Notice that the numbers 100:150:50 correspond to the ratio 2:3:1, which represents the coefficients in the "balanced equation" of making a sandwich. If you were asked to make any number of sandwiches, it would be easy to use the original sandwich equation to determine how much of each ingredient you need.

The equation for a chemical reaction gives you the same type of information. It indicates the relative numbers of reactant and product molecules involved in the reaction. Using the equation permits us to determine the amounts of reactants needed to give a certain amount of product or to predict how much product we can make from a given quantity of reactants.

goChemistry **Module 7: Simple Stoichiometry** covers concepts in this section.

To illustrate how this idea works with a chemistry example, consider the reaction between gaseous carbon monoxide and hydrogen to produce liquid methanol, $CH_3OH(l)$. The reactants and products are

$$\text{Unbalanced: } CO(g) + H_2(g) \rightarrow CH_3OH(l)$$
$$\quad\quad\quad\quad\quad\text{Reactants}\quad\quad\quad\text{Product}$$

Because atoms are just rearranged (not created or destroyed) in a chemical reaction, we must always balance a chemical equation. That is, we must choose coefficients that give the same number of each type of atom on both sides. Using the smallest set of integers that satisfies this condition gives the balanced equation

$$\text{Balanced: } CO(g) + 2H_2(g) \rightarrow CH_3OH(l)$$

CHECK: Reactants: 1 C, 1 O, 4 H; Products: 1 C, 1 O, 4 H

Again, the coefficients in a balanced equation give the *relative* numbers of molecules. That is, we could multiply this balanced equation by any number and still have a balanced equation. For example, we could multiply by 12:

$$12[CO(g) + 2H_2(g) \rightarrow CH_3OH(l)]$$

to obtain

$$12CO(g) + 24H_2(g) \rightarrow 12CH_3OH(l)$$

This is still a balanced equation (check to be sure). Because 12 represents a dozen, we could even describe the reaction in terms of dozens:

$$1 \text{ dozen } CO(g) + 2 \text{ dozen } H_2(g) \rightarrow 1 \text{ dozen } CH_3OH(l)$$

We could also multiply the original equation by a very large number, such as 6.022×10^{23}:

$$6.022 \times 10^{23}[CO(g) + 2H_2(g) \rightarrow CH_3OH(l)]$$

which leads to the equation

$$6.022 \times 10^{23} \, CO(g) + 2(6.022 \times 10^{23}) \, H_2(g) \rightarrow 6.022 \times 10^{23} \, CH_3OH(l)$$

One mole is 6.022×10^{23} units.

Just as 12 is called a dozen, chemists call 6.022×10^{23} a *mole* (abbreviated mol). Our equation, then, can be written in terms of moles:

$$1 \text{ mol } CO(g) + 2 \text{ mol } H_2(g) \rightarrow 1 \text{ mol } CH_3OH(l)$$

Table 9.1 Information Conveyed by the Balanced Equation for the Production of Methanol

$CO(g)$	$+$	$2H_2(g)$	$\rightarrow$	$CH_3OH(l)$
1 molecule CO	$+$	2 molecules H_2	$\rightarrow$	1 molecule CH_3OH
1 dozen CO molecules	$+$	2 dozen H_2 molecules	$\rightarrow$	1 dozen CH_3OH molecules
6.022×10^{23} CO molecules	$+$	$2(6.022 \times 10^{23})$ H_2 molecules	$\rightarrow$	6.022×10^{23} CH_3OH molecules
1 mol CO molecules	$+$	2 mol H_2 molecules	$\rightarrow$	1 mol CH_3OH molecules

Various ways of interpreting this balanced chemical equation are given in Table 9.1.

EXAMPLE 9.1 | **Relating Moles to Molecules in Chemical Equations**

Propane is often used as a fuel for outdoor grills.

Propane, C_3H_8, is a fuel commonly used for cooking on gas grills and for heating in rural areas where natural gas is unavailable. Propane reacts with oxygen gas to produce heat and the products carbon dioxide and water. This combustion reaction is represented by the unbalanced equation

$$C_3H_8(g) + O_2(g) \rightarrow CO_2(g) + H_2O(g)$$

Give the balanced equation for this reaction, and state the meaning of the equation in terms of numbers of molecules and moles of molecules.

SOLUTION

Using the techniques explained in Chapter 6, we can balance the equation.

$$C_3H_8(g) + 5O_2(g) \rightarrow 3CO_2(g) + 4H_2O(g)$$

CHECK: 3 C, 8 H, 10 O $\rightarrow$ 3 C, 8 H, 10 O

This equation can be interpreted in terms of molecules as follows:

  1 molecule of C_3H_8 reacts with 5 molecules of O_2 to give

  3 molecules of CO_2 plus 4 molecules of H_2O

or as follows in terms of moles (of molecules):

  1 mol C_3H_8 reacts with 5 mol O_2 to give 3 mol

  CO_2 plus 4 mol H_2O ∎

9.2 Mole–Mole Relationships

OBJECTIVE: To learn to use a balanced equation to determine relationships between moles of reactants and moles of products.

Now that we have discussed the meaning of a balanced chemical equation in terms of moles of reactants and products, we can use an equation to predict the moles of products that a given number of moles of reactants will yield. For example, consider the decomposition of water to give hydrogen and oxygen, which is represented by the following balanced equation:

$$2H_2O(l) \rightarrow 2H_2(g) + O_2(g)$$

This equation tells us that 2 mol H_2O yields 2 mol H_2 and 1 mol O_2.
  Now suppose that we have 4 mol water. If we decompose 4 mol water, how many moles of products do we get?

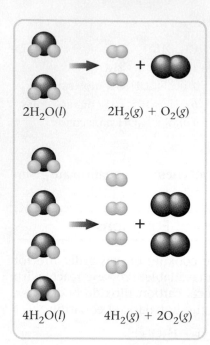

$2H_2O(l)$ $2H_2(g) + O_2(g)$

$4H_2O(l)$ $4H_2(g) + 2O_2(g)$

This equation with noninteger coefficients makes sense only if the equation means moles (of molecules) of the various reactants and products.

One way to answer this question is to multiply the entire equation by 2 (which will give us 4 mol H_2O).

$$2[2H_2O(l) \rightarrow 2H_2(g) + O_2(g)]$$
$$4H_2O(l) \rightarrow 4H_2(g) + 2O_2(g)$$

Now we can state that

4 mol H_2O yields 4 mol H_2 plus 2 mol O_2

which answers the question of how many moles of products we get with 4 mol H_2O.

Next, suppose we decompose 5.8 mol water. What numbers of moles of products are formed in this process? We could answer this question by rebalancing the chemical equation as follows: First, we divide *all coefficients* of the balanced equation

$$2H_2O(l) \rightarrow 2H_2(g) + O_2(g)$$

by 2, to give

$$H_2O(l) \rightarrow H_2(g) + \tfrac{1}{2}O_2(g)$$

Now, because we have 5.8 mol H_2O, we multiply this equation by 5.8.

$$5.8[H_2O(l) \rightarrow H_2(g) + \tfrac{1}{2}O_2(g)]$$

This gives

$$5.8H_2O(l) \rightarrow 5.8H_2(g) + 5.8(\tfrac{1}{2})O_2(g)$$
$$5.8H_2O(l) \rightarrow 5.8H_2(g) + 2.9O_2(g)$$

(Verify that this is a balanced equation.) Now we can state that

5.8 mol H_2O yields 5.8 mol H_2 plus 2.9 mol O_2

This procedure of rebalancing the equation to obtain the number of moles involved in a particular situation always works, but it can be cumbersome. In Example 9.2 we will develop a more convenient procedure, which uses conversion factors, or **mole ratios,** based on the balanced chemical equation.

EXAMPLE 9.2 | Determining Mole Ratios

What number of moles of O_2 will be produced by the decomposition of 5.8 moles of water?

SOLUTION

Where Are We Going?

We want to determine the number of moles of O_2 produced by the decomposition of 5.8 moles of H_2O.

What Do We Know?

- The balanced equation for the decomposition of water is

$$2H_2O \rightarrow 2H_2 + O_2$$

- We start with 5.8 mol H_2O.

How Do We Get There?

Our problem can be diagrammed as follows:

5.8 mol H_2O yields⟩ ? mol O_2

To answer this question, we need to know the relationship between moles of H_2O and moles of O_2 in the balanced equation (conventional form):

$$2H_2O(l) \rightarrow 2H_2(g) + O_2(g)$$

From this equation we can state that

> The statement 2 mol H_2O = 1 mol O_2 is obviously not true in a literal sense, but it correctly expresses the chemical equivalence between H_2O and O_2.

$$\boxed{2 \text{ mol } H_2O} \;\underrightarrow{\text{yields}}\; \boxed{1 \text{ mol } O_2}$$

which can be represented by the following equivalence statement:

$$2 \text{ mol } H_2O = 1 \text{ mol } O_2$$

We now want to use this equivalence statement to obtain the conversion factor (mole ratio) that we need. Because we want to go from moles of H_2O to moles of O_2, we need the mole ratio

$$\frac{1 \text{ mol } O_2}{2 \text{ mol } H_2O}$$

MATH SKILL BUILDER
For a review of equivalence statements and dimensional analysis, see Section 2.6.

so that mol H_2O will cancel in the conversion from moles of H_2O to moles of O_2.

$$5.8 \; \cancel{\text{mol } H_2O} \times \frac{1 \text{ mol } O_2}{2 \; \cancel{\text{mol } H_2O}} = 2.9 \text{ mol } O_2$$

So if we decompose 5.8 mol H_2O, we will get 2.9 mol O_2.

REALITY CHECK   Note that this is the same answer we obtained earlier when we rebalanced the equation to give

$$5.8 H_2O(l) \rightarrow 5.8 H_2(g) + 2.9 O_2(g) \; \blacksquare$$

We saw in Example 9.2 that to determine the moles of a product that can be formed from a specified number of moles of a reactant, we can use the balanced equation to obtain the appropriate mole ratio. We will now extend these ideas in Example 9.3.

EXAMPLE 9.3 | Using Mole Ratios in Calculations

Calculate the number of moles of oxygen required to react exactly with 4.30 moles of propane, C_3H_8, in the reaction described by the following balanced equation:

$$C_3H_8(g) + 5O_2(g) \rightarrow 3CO_2(g) + 4H_2O(g)$$

SOLUTION

Where Are We Going?

We want to determine the number of moles of O_2 required to react with 4.30 mol C_3H_8.

What Do We Know?

- The balanced equation for the reaction is

$$C_3H_8 + 5O_2 \rightarrow 3CO_2 + 4H_2O$$

- We start with 4.30 mol C_3H_8.

How Do We Get There?

In this case the problem can be stated as follows:

$$\boxed{4.30 \text{ mol } C_3H_8} \quad \text{requires} \quad \boxed{? \text{ mol } O_2}$$

To solve this problem, we need to consider the relationship between the reactants C_3H_8 and O_2. Using the balanced equation, we find that

$$1 \text{ mol } C_3H_8 \text{ requires } 5 \text{ mol } O_2$$

which can be represented by the equivalence statement

$$1 \text{ mol } C_3H_8 = 5 \text{ mol } O_2$$

This leads to the required mole ratio

$$\frac{5 \text{ mol } O_2}{1 \text{ mol } C_3H_8}$$

for converting from moles of C_3H_8 to moles of O_2. We construct the conversion ratio this way so that mol C_3H_8 cancels:

$$4.30 \text{ mol } C_3H_8 \times \frac{5 \text{ mol } O_2}{1 \text{ mol } C_3H_8} = 21.5 \text{ mol } O_2$$

We can now answer the original question:

$$4.30 \text{ mol } C_3H_8 \text{ requires } 21.5 \text{ mol } O_2$$

REALITY CHECK According to the balanced equation, more O_2 is required (by mol) than C_3H_8 by a factor of 5. With about 4 mol C_3H_8, we would expect about 20 mol of O_2, which is close to our answer.

Self-Check   **EXERCISE 9.1**   Calculate the moles of CO_2 formed when 4.30 moles of C_3H_8 reacts with the required 21.5 moles of O_2.

HINT: Use the moles of C_3H_8, and obtain the mole ratio between C_3H_8 and CO_2 from the balanced equation.

See Problems 9.15 and 9.16. ■

9.3 Mass Calculations

OBJECTIVE: To learn to relate masses of reactants and products in a chemical reaction.

In the last section we saw how to use the balanced equation for a reaction to calculate the numbers of moles of reactants and products for a particular case. However, moles represent numbers of molecules, and we cannot count molecules directly. In chemistry we count by weighing. Therefore, in this section we will review the procedures for converting between moles and masses and will see how these procedures are applied to chemical calculations.

To develop these procedures we will consider the reaction between powdered aluminum metal and finely ground iodine to produce aluminum iodide. The balanced equation for this vigorous chemical reaction is

$$2Al(s) + 3I_2(s) \rightarrow 2AlI_3(s)$$

Suppose we have 35.0 g of aluminum. What mass of I_2 should we weigh out to react exactly with this amount of aluminum?

To answer this question, let's use the problem-solving strategy discussed in Chapter 8.

Aluminum (*left*) and iodine (*right*), shown at the top, react vigorously to form aluminum iodide. The purple cloud results from excess iodine vaporized by the heat of the reaction.

Where Are We Going?

We want to find the mass of iodine (I_2) that will react with 35.0 g of aluminum (Al). We know from the balanced equation that

$$\text{2 mol Al requires 3 mol } I_2$$

This can be written as the mole ratio

$$\frac{3 \text{ mol } I_2}{2 \text{ mol Al}}$$

We can use this ratio to calculate moles of I_2 needed from the moles of Al present. However, this leads us to two questions:

1. How many moles of Al are present?
2. How do we convert moles of I_2 to mass of I_2 as required by the problem?

We need to be able to convert from grams to moles and from moles to grams.

How Do We Get There?

The problem states that we have 35.0 g of aluminum, so we must convert from grams to moles of aluminum. This is something we already know how to do. Using the table of average atomic masses inside the front cover of this book, we find the atomic mass of aluminum to be 26.98. This means that 1 mole of aluminum has a mass of 26.98 g. We can use the equivalence statement

$$\text{1 mol Al} = 26.98 \text{ g}$$

to find the moles of Al in 35.0 g.

$$35.0 \text{ g Al} \times \frac{1 \text{ mol Al}}{26.98 \text{ g Al}} = 1.30 \text{ mol Al}$$

Now that we have moles of Al, we can find the moles of I_2 required.

$$1.30 \text{ mol Al} \times \frac{3 \text{ mol } I_2}{2 \text{ mol Al}} = 1.95 \text{ mol } I_2$$

We now know the *moles* of I_2 required to react with the 1.30 moles of Al (35.0 g). The next step is to convert 1.95 moles of I_2 to grams so we will know how much to weigh out. We do this by using the molar mass of I_2. The atomic mass of iodine is 126.9 g (for 1 mole of I atoms), so the molar mass of I_2 is

$$2 \times 126.9 \text{ g/mol} = 253.8 \text{ g/mol} = \text{mass of 1 mol } I_2$$

Now we convert the 1.95 moles of I_2 to grams of I_2.

$$1.95 \text{ mol } I_2 \times \frac{253.8 \text{ g } I_2}{\text{mol } I_2} = 495 \text{ g } I_2$$

We have solved the problem. We need to weigh out 495 g of iodine (contains I_2 molecules) to react exactly with the 35.0 g of aluminum. We will further develop procedures for dealing with masses of reactants and products in Example 9.4.

REALITY CHECK We have determined that 495 g of I_2 is required to react with 35.0 g Al. Does this answer make sense? We know from the molar masses of Al and I_2 (26.98 g/mol and 253.8 g/mol) that the mass of 1 mol I_2 is almost 10 times as great as that of 1 mol Al. We also know that we need a greater number of moles of I_2 compared with Al (by a 3:2 ratio). So, we should expect to get a mass of I_2 that is well over 10 times as great as 35.0 g, and we did.

EXAMPLE 9.4 | **Using Mass–Mole Conversions with Mole Ratios**

Propane, C_3H_8, when used as a fuel, reacts with oxygen to produce carbon dioxide and water according to the following unbalanced equation:

$$C_3H_8(g) + O_2(g) \rightarrow CO_2(g) + H_2O(g)$$

What mass of oxygen will be required to react exactly with 96.1 g of propane?

SOLUTION

Where Are We Going?

We want to determine the mass of O_2 required to react exactly with 96.1 g C_3H_8.

What Do We Know?

- The unbalanced equation for the reaction is

$$C_3H_8 + O_2 \rightarrow CO_2 + H_2O.$$

- We start with 96.1 g mol C_3H_8.

- We know the atomic masses of carbon, hydrogen, and oxygen from the periodic table.

What Do We Need To Know?

- We need to know the balanced equation.

- We need the molar masses of O_2 and C_3H_8.

> Always balance the equation for the reaction first.

How Do We Get There?

To deal with the amounts of reactants and products, we first need the balanced equation for this reaction:

$$C_3H_8(g) + 5O_2(g) \rightarrow 3CO_2(g) + 4H_2O(g)$$

Our problem, in schematic form, is

96.1 g propane	requires	? grams O_2

> **MATH SKILL BUILDER**
> Remember that to show the correct significant figures in each step, we are rounding off after each calculation. In doing problems, you should carry extra numbers, rounding off only at the end.

Using the ideas we developed when we discussed the aluminum–iodine reaction, we will proceed as follows:

1. We are given the number of grams of propane, so we must convert to moles of propane (C_3H_8).

2. Then we can use the coefficients in the balanced equation to determine the moles of oxygen (O_2) required.

3. Finally, we will use the molar mass of O_2 to calculate grams of oxygen.

We can sketch this strategy as follows:

$$C_3H_8(g) \quad + \quad 5O_2(g) \quad \rightarrow \quad 3CO_2(g) \quad + \quad 4H_2O(g)$$

| 96.1 g C₃H₈ | | ? grams O₂ |

 Thus the first question we must answer is, *How many moles of propane are present in 96.1 g of propane?* The molar mass of propane is 44.09 g ($3 \times 12.01 + 8 \times 1.008$). The moles of propane present can be calculated as follows:

$$96.1 \text{ g } C_3H_8 \times \frac{1 \text{ mol } C_3H_8}{44.09 \text{ g } C_3H_8} = 2.18 \text{ mol } C_3H_8$$

 Next we recognize that each mole of propane reacts with 5 mol of oxygen. This gives us the equivalence statement

$$1 \text{ mol } C_3H_8 = 5 \text{ mol } O_2$$

from which we construct the mole ratio

$$\frac{5 \text{ mol } O_2}{1 \text{ mol } C_3H_8}$$

that we need to convert from moles of propane molecules to moles of oxygen molecules.

$$2.18 \text{ mol } C_3H_8 \times \frac{5 \text{ mol } O_2}{1 \text{ mol } C_3H_8} = 10.9 \text{ mol } O_2$$

Notice that the mole ratio is set up so that the moles of C_3H_8 cancel and the resulting units are moles of O_2.

 Because the original question asked for the *mass* of oxygen needed to react with 96.1 g of propane, we must convert the 10.9 mol O_2 to grams, using the molar mass of O_2 ($32.00 = 2 \times 16.00$).

$$10.9 \text{ mol } O_2 \times \frac{32.0 \text{ g } O_2}{1 \text{ mol } O_2} = 349 \text{ g } O_2$$

Therefore, 349 g of oxygen is required to burn 96.1 g of propane. We can summarize this problem by writing out a "conversion string" that shows how the problem was done.

$$96.1 \text{ g } C_3H_8 \times \frac{1 \text{ mol } C_3H_8}{44.09 \text{ g } C_3H_8} \times \frac{5 \text{ mol } O_2}{1 \text{ mol } C_3H_8} \times \frac{32.0 \text{ g } O_2}{1 \text{ mol } O_2} = 349 \text{ g } O_2$$

MATH SKILL BUILDER
Use units as a check to see that you have used the correct conversion factors (mole ratios).

This is a convenient way to make sure the final units are correct. The procedure we have followed is summarized below.

REALITY CHECK   According to the balanced equation, more O_2 is required (by moles) than C_3H_8 by a factor of 5. Because the molar mass of C_3H_8 is not much greater than that of O_2, we should expect that a greater mass of oxygen is required, and our answer confirms this.

Self-Check   EXERCISE 9.2   What mass of carbon dioxide is produced when 96.1 g of propane reacts with sufficient oxygen?

See Problems 9.23 through 9.26. ∎

Self-Check   EXERCISE 9.3   Calculate the mass of water formed by the complete reaction of 96.1 g of propane with oxygen.

See Problems 9.23 through 9.26. ∎

So far in this chapter, we have spent considerable time "thinking through" the procedures for calculating the masses of reactants and products in chemical reactions. We can summarize these procedures in the following steps:

Steps for Calculating the Masses of Reactants and Products in Chemical Reactions

Step 1 Balance the equation for the reaction.

Step 2 Convert the masses of reactants or products to moles.

Step 3 Use the balanced equation to set up the appropriate mole ratio(s).

Step 4 Use the mole ratio(s) to calculate the number of moles of the desired reactant or product.

Step 5 Convert from moles back to masses.

The process of using a chemical equation to calculate the relative masses of reactants and products involved in a reaction is called **stoichiometry** (pronounced stoi´ kē-ŏm´ i-trē). Chemists say that the balanced equation for a chemical reaction describes the stoichiometry of the reaction.

We will now consider a few more examples that involve chemical stoichiometry. Because real-world examples often involve very large or very small masses of chemicals that are most conveniently expressed by using scientific notation, we will deal with such a case in Example 9.5.

EXAMPLE 9.5 | Stoichiometric Calculations: Using Scientific Notation

For a review of writing formulas of ionic compounds, see Chapter 5.

Solid lithium hydroxide has been used in space vehicles to remove exhaled carbon dioxide from the living environment. The products are solid lithium carbonate and liquid water. What mass of gaseous carbon dioxide can 1.00×10^3 g of lithium hydroxide absorb?

SOLUTION

Where Are We Going?

We want to determine the mass of carbon dioxide absorbed by 1.00×10^3 g of lithium hydroxide.

What Do We Know?

- The names of the reactants and products.
- We start with 1.00×10^3 g of lithium hydroxide.
- We can obtain the atomic masses from the periodic table.

What Do We Need To Know?

- We need to know the balanced equation for the reaction, but we first have to write the formulas for the reactants and products.
- We need the molar masses of lithium hydroxide and carbon dioxide.

How Do We Get There?

Step 1 Using the description of the reaction, we can write the unbalanced equation

$$LiOH(s) + CO_2(g) \rightarrow Li_2CO_3(s) + H_2O(l)$$

The balanced equation is

$$2LiOH(s) + CO_2(g) \rightarrow Li_2CO_3(s) + H_2O(l)$$

Check this for yourself.

Step 2 We convert the given mass of LiOH to moles, using the molar mass of LiOH, which is 6.941 g + 16.00 g + 1.008 g = 23.95 g.

$$1.00 \times 10^3 \text{ g LiOH} \times \frac{1 \text{ mol LiOH}}{23.95 \text{ g LiOH}} = 41.8 \text{ mol LiOH}$$

Step 3 The appropriate mole ratio is

$$\frac{1 \text{ mol CO}_2}{2 \text{ mol LiOH}}$$

Astronaut Sidney M. Gutierrez changes the lithium hydroxide canisters on space shuttle *Columbia*.

Step 4   Using this mole ratio, we calculate the moles of CO_2 needed to react with the given mass of LiOH.

$$41.8 \text{ mol LiOH} \times \frac{1 \text{ mol } CO_2}{2 \text{ mol LiOH}} = 20.9 \text{ mol } CO_2$$

Step 5   We calculate the mass of CO_2 by using its molar mass (44.01 g).

$$20.9 \text{ mol } CO_2 \times \frac{44.01 \text{ g } CO_2}{1 \text{ mol } CO_2} = 920. \text{ g } CO_2 = 9.20 \times 10^2 \text{ g } CO_2$$

Thus 1.00×10^3 g of LiOH(s) can absorb 920. g of $CO_2(g)$.
We can summarize this problem as follows:

$$2\text{LiOH}(s) \quad + \quad CO_2(g) \quad \rightarrow \quad Li_2CO_3(s) \quad + \quad H_2O(l)$$

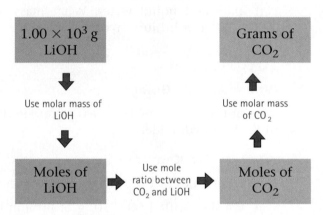

The conversion string is

$$1.00 \times 10^3 \text{ g LiOH} \times \frac{1 \text{ mol LiOH}}{23.95 \text{ g LiOH}} \times \frac{1 \text{ mol } CO_2}{2 \text{ mol LiOH}} \times \frac{44.01 \text{ g } CO_2}{1 \text{ mol } CO_2}$$
$$= 9.19 \times 10^2 \text{ g } CO_2$$

MATH SKILL BUILDER
Carrying extra significant figures and rounding off only at the end gives an answer of 919 g CO_2.

REALITY CHECK   According to the balanced equation, there is a 2:1 mole ratio of LiOH to CO_2. There is about a 1:2 molar mass ratio of LiOH:CO_2 (23.95:44.01). We should expect about the same mass of CO_2 as LiOH, and our answer confirms this (1000 g compared to 920 g).

Self-Check   EXERCISE 9.4   Hydrofluoric acid, an aqueous solution containing dissolved hydrogen fluoride, is used to etch glass by reacting with the silica, SiO_2, in the glass to produce gaseous silicon tetrafluoride and liquid water. The unbalanced equation is

$$HF(aq) + SiO_2(s) \rightarrow SiF_4(g) + H_2O(l)$$

a. Calculate the mass of hydrogen fluoride needed to react with 5.68 g of silica. *Hint:* Think carefully about this problem. What is the balanced equation for the reaction? What is given? What do you need to calculate? Sketch a map of the problem before you do the calculations.

b. Calculate the mass of water produced in the reaction described in part a.

See Problems 9.23 through 9.26. ∎

EXAMPLE 9.6 Stoichiometric Calculations: Comparing Two Reactions

Baking soda, $NaHCO_3$, is often used as an antacid. It neutralizes excess hydrochloric acid secreted by the stomach. The balanced equation for the reaction is

$$NaHCO_3(s) + HCl(aq) \rightarrow NaCl(aq) + H_2O(l) + CO_2(g)$$

Milk of magnesia, which is an aqueous suspension of magnesium hydroxide, $Mg(OH)_2$, is also used as an antacid. The balanced equation for the reaction is

$$Mg(OH)_2(s) + 2HCl(aq) \rightarrow 2H_2O(l) + MgCl_2(aq)$$

Which antacid can consume the most stomach acid, 1.00 g of $NaHCO_3$ or 1.00 g of $Mg(OH)_2$?

SOLUTION

Where Are We Going?

We want to compare the neutralizing power of two antacids, $NaHCO_3$ and $Mg(OH)_2$. In other words, how many moles of HCl will react with 1.00 g of each antacid?

What Do We Know?

- The balanced equations for the reactions.
- We start with 1.00 g each of $NaHCO_3$ and $Mg(OH)_2$.
- We can obtain atomic masses from the periodic table.

What Do We Need to Know?

- We need the molar masses of $NaHCO_3$ and $Mg(OH)_2$.

How Do We Get There?

The antacid that reacts with the larger number of moles of HCl is more effective because it will neutralize more moles of acid. A schematic for this procedure is

Notice that in this case we do not need to calculate how many grams of HCl react; we can answer the question with moles of HCl. We will now solve this problem for each antacid. Both of the equations are balanced, so we can proceed with the calculations.

Using the molar mass of $NaHCO_3$, which is 22.99 g + 1.008 g + 12.01 g + 3(16.00 g) = 84.01 g, we determine the moles of $NaHCO_3$ in 1.00 g of $NaHCO_3$.

Cars of the Future

There is a great deal of concern about how we are going to sustain our personal transportation system in the face of looming petroleum shortages (and the resultant high costs) and the challenges of global warming. The era of large gasoline-powered cars as the primary means of transportation in the United States seems to be drawing to a close. The fact that discoveries of petroleum are not keeping up with the rapidly increasing global demand for oil has caused skyrocketing prices. In addition, the combustion of gasoline produces carbon dioxide (about 1 lb of CO_2 per mile for many cars), which has been implicated in global warming.

So what will the car of the future in the United States be like? It seems that we are moving rapidly toward cars that have an electrical component as part of the power train. Hybrid cars, which use a small gasoline motor in conjunction with a powerful battery, have been quite successful. By supplementing the small gasoline engine, which would be inadequate by itself, with power from the battery, the typical hybrid gets 40 to 50 miles per gallon of gasoline. In this type of hybrid car, both the battery and the engine are used to power the wheels of the car as needed.

Another type of system that involves both a gasoline engine and a battery is the so-called "plug-in hybrid." In this car, the battery is the sole source of power to the car's wheels. The gasoline engine is only used to charge the battery as needed. One example of this type of car is the Chevrolet Volt, which is slated for production in 2010. The Volt is being designed to run about 40 miles on each battery charge. The car would be plugged into a normal household electric outlet overnight to recharge the battery. For trips longer than 40 miles, the gasoline engine would turn on to charge the battery.

Another type of "electrical car" being tested is one powered by a hydrogen–oxygen fuel cell. An example of such a car is the Honda FCX Clarity. The Clarity stores hydrogen in a tank that holds 4.1 kg of H_2 at a pressure of 5000 lb per square inch. The H_2 is sent to a fuel cell, where it reacts with oxygen from the air supplied by an air compressor. About 200 of these cars are going to be tested in Southern Califor-

$$1.00 \text{ g NaHCO}_3 \times \frac{1 \text{ mol NaHCO}_3}{84.01 \text{ g NaHCO}_3} = 0.0119 \text{ mol NaHCO}_3$$

$$= 1.19 \times 10^{-2} \text{ mol NaHCO}_3$$

Next we determine the moles of HCl, using the mole ratio $\dfrac{1 \text{ mol HCl}}{1 \text{ mol NaHCO}_3}$.

$$1.19 \times 10^{-2} \text{ mol NaHCO}_3 \times \frac{1 \text{ mol HCl}}{1 \text{ mol NaHCO}_3} = 1.19 \times 10^{-2} \text{ mol HCl}$$

Thus 1.00 g of $NaHCO_3$ neutralizes 1.19×10^{-2} mol HCl. We need to compare this to the number of moles of HCl that 1.00 g of $Mg(OH)_2$ neutralizes.

Using the molar mass of $Mg(OH)_2$, which is 24.31 g + 2(16.00 g) + 2(1.008 g) = 58.33 g, we determine the moles of $Mg(OH)_2$ in 1.00 g of $Mg(OH)_2$.

$$1.00 \text{ g Mg(OH)}_2 \times \frac{1 \text{ mol Mg(OH)}_2}{58.33 \text{ g Mg(OH)}_2} = 0.0171 \text{ mol Mg(OH)}_2$$

$$= 1.71 \times 10^{-2} \text{ mol Mg(OH)}_2$$

nia in the next 3 years, leased to people who live near one of the three 24-hour public hydrogen stations. The Clarity gets about 72 miles per kilogram of hydrogen. One obvious advantage of a car powered by an H_2/O_2 fuel cell is that the combustion product is only H_2O. However, there is a catch (it seems there is always a catch). Currently, 95% of hydrogen produced is obtained by natural gas (CH_4), and CO_2 is a by-product of this process. Intense research is now being conducted to find economically feasible ways to produce H_2 from water.

It appears that our cars of the future will have an electrical drive component. Whether it will involve a conventional battery or a fuel cell will depend on technological developments and costs.

Even model cars are becoming "green." The H-racer from Horizon Fuel Cell Technologies uses a hydrogen–oxygen fuel cell.

The Honda FCX Clarity is powered by a hydrogen–oxygen fuel cell.

To determine the moles of HCl that react with this amount of $Mg(OH)_2$, we use the mole ratio $\dfrac{2 \text{ mol HCl}}{1 \text{ mol Mg(OH)}_2}$.

$$1.71 \times 10^{-2} \text{ mol Mg(OH)}_2 \times \frac{2 \text{ mol HCl}}{1 \text{ mol Mg(OH)}_2} = 3.42 \times 10^{-2} \text{ mol HCl}$$

Therefore, 1.00 g of $Mg(OH)_2$ neutralizes 3.42×10^{-2} mol HCl. We have already calculated that 1.00 g of $NaHCO_3$ neutralizes only 1.19×10^{-2} mol HCl. Therefore, $Mg(OH)_2$ is a more effective antacid than $NaHCO_3$ on a mass basis.

Self-Check **EXERCISE 9.5** In Example 9.6 we answered one of the questions we posed in the introduction to this chapter. Now let's see if you can answer the other question posed there. Determine what mass of carbon monoxide and what mass of hydrogen are required to form 6.0 kg of methanol by the reaction

$$CO(g) + 2H_2(g) \rightarrow CH_3OH(l)$$

See Problem 9.39. ■

9.4   The Concept of Limiting Reactants

OBJECTIVE: To understand what is meant by the term "limiting reactant."

Earlier in this chapter, we discussed making sandwiches. Recall that the sandwich-making process could be described as follows:

2 pieces bread + 3 slices meat + 1 slice cheese → 1 sandwich

In our earlier discussion, we always purchased the ingredients in the correct ratios so that we used all the components, with nothing left over.

Now assume that you came to work one day and found the following quantities of ingredients:

20 slices of bread
24 slices of meat
12 slices of cheese

How many sandwiches can you make? What will be left over?

To solve this problem, let's see how many sandwiches we can make with each component.

Bread: $20 \text{ slices bread} \times \dfrac{1 \text{ sandwich}}{2 \text{ slices of bread}} = 10 \text{ sandwiches}$

Meat: $24 \text{ slices meat} \times \dfrac{1 \text{ sandwich}}{3 \text{ slices of meat}} = 8 \text{ sandwiches}$

Cheese: $12 \text{ slices cheese} \times \dfrac{1 \text{ sandwich}}{1 \text{ slice of cheese}} = 12 \text{ sandwiches}$

How many sandwiches can you make? The answer is 8. When you run out of meat, you must stop making sandwiches. The meat is the limiting ingredient.

What do you have left over? Making 8 sandwiches requires 16 pieces of bread. You started with 20 pieces, so you have 4 pieces of bread left. You also used 8 pieces of cheese for the 8 sandwiches, so you have $12 - 8 = 4$ pieces of cheese left.

In this example, the ingredient present in the largest number (the meat) was actually the component that limited the number of sandwiches you could make. This situation arose because each sandwich required 3 slices of meat—more than the quantity required of any other ingredient.

You probably have been dealing with limiting-reactant problems for most of your life. For example, suppose a lemonade recipe calls for 1 cup of sugar for every 6 lemons. You have 12 lemons and 3 cups of sugar. Which ingredient is limiting, the lemons or the sugar?*

▶ ## A Closer Look

goChemistry Module 8a: Stoichiometry and Limiting Reactants (Pt. 1) covers concepts in this section.

When molecules react with each other to form products, considerations very similar to those involved in making sandwiches arise. We can illustrate these ideas with the reaction of $N_2(g)$ and $H_2(g)$ to form $NH_3(g)$:

$$N_2(g) + 3H_2(g) \rightarrow 2NH_3(g)$$

---

*The ratio of lemons to sugar that the recipe calls for is 6 lemons to 1 cup of sugar. We can calculate the number of lemons required to "react with" the 3 cups of sugar as follows:

$$3 \text{ cups sugar} \times \frac{6 \text{ lemons}}{1 \text{ cup sugar}} = 18 \text{ lemons}$$

Thus 18 lemons would be required to use up 3 cups of sugar. However, we have only 12 lemons, so the lemons are limiting.

Consider the following container of $N_2(g)$ and $H_2(g)$:

What will this container look like if the reaction between N_2 and H_2 proceeds to completion? To answer this question, you need to remember that each N_2 requires 3 H_2 molecules to form 2 NH_3. To make things more clear, we will circle groups of reactants:

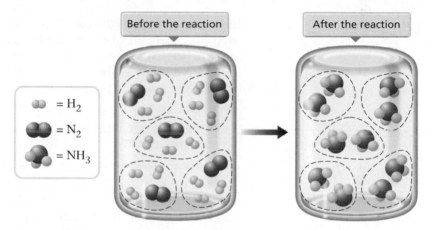

In this case, the mixture of N_2 and H_2 contained just the number of molecules needed to form NH_3 with nothing left over. That is, the ratio of the number of H_2 molecules to N_2 molecules was

$$\frac{15\ H_2}{5\ N_2} = \frac{3\ H_2}{1\ N_2}$$

This ratio exactly matches the numbers in the balanced equation

$$3H_2(g) + N_2(g) \rightarrow 2NH_3(g).$$

This type of mixture is called a *stoichiometric mixture*—one that contains the relative amounts of reactants that matches the numbers in the balanced equation. In this case, all reactants will be consumed to form products.

Now consider another container of $N_2(g)$ and $H_2(g)$:

What will the container look like if the reaction between $N_2(g)$ and $H_2(g)$ proceeds to completion? Remember that each N_2 requires 3 H_2. Circling groups of reactants, we have

In this case, the hydrogen (H_2) is limiting. That is, the H_2 molecules are used up before all of the N_2 molecules are consumed. In this situation, the amount of hydrogen limits the amount of product (ammonia) that can form—hydrogen is the limiting reactant. Some N_2 molecules are left over in this case because the reaction runs out of H_2 molecules first.

> To determine how much product can be formed from a given mixture of reactants, we have to look for the reactant that is limiting—the one that runs out first and thus limits the amount of product that can form.

In some cases, the mixture of reactants might be stoichiometric—that is, all reactants run out at the same time. In general, however, you cannot assume that a given mixture of reactants is a stoichiometric mixture, so you must determine whether one of the reactants is limiting.

> The reactant that runs out first and thus limits the amounts of products that can form is called the **limiting reactant (limiting reagent).**

To this point, we have considered examples where the numbers of reactant molecules could be counted. In "real life" you can't count the molecules directly—you can't see them, and, even if you could, there would be far too many to count. Instead, you must count by weighing. We must therefore explore how to find the limiting reactant, given the masses of the reactants.

9.5 Calculations Involving a Limiting Reactant

OBJECTIVES: To learn to recognize the limiting reactant in a reaction. • To learn to use the limiting reactant to do stoichiometric calculations.

Manufacturers of cars, bicycles, and appliances order parts in the same proportion as they are used in their products. For example, auto manufacturers order four times as many wheels as engines and bicycle manufacturers order twice as many pedals as seats. Likewise, when chemicals are mixed together so that they can undergo a reaction, they are often mixed in stoichiometric quantities—that is, in exactly the correct amounts so that all

Farmer Rodney Donala looks out over his corn fields in front of his 30,000-gallon tank (at right) of anhydrous ammonia, a liquid fertilizer.

goChemistry **Module 8b: Stoichiometry and Limiting Reactants (Pt. 2)** covers concepts in this section.

reactants "run out" (are used up) at the same time. To clarify this concept, we will consider the production of hydrogen for use in the manufacture of ammonia. Ammonia, a very important fertilizer itself and a starting material for other fertilizers, is made by combining nitrogen from the air with hydrogen. The hydrogen for this process is produced by the reaction of methane with water according to the balanced equation

$$CH_4(g) + H_2O(g) \rightarrow 3H_2(g) + CO(g)$$

Let's consider the question, *What mass of water is required to react exactly with 249 g of methane?* That is, how much water will just use up all of the 249 g of methane, leaving no methane or water remaining?

This problem requires the same strategies we developed in the previous section. Again, drawing a map of the problem is helpful.

We first convert the mass of CH_4 to moles, using the molar mass of CH_4 (16.04 g/mol).

$$249 \text{ g } CH_4 \times \frac{1 \text{ mol } CH_4}{16.04 \text{ g } CH_4} = 15.5 \text{ mol } CH_4$$

Because in the balanced equation 1 mol CH_4 reacts with 1 mol H_2O, we have

$$15.5 \text{ mol } CH_4 \times \frac{1 \text{ mol } H_2O}{1 \text{ mol } CH_4} = 15.5 \text{ mol } H_2O$$

Therefore, 15.5 mol H_2O will react exactly with the given mass of CH_4. Converting 15.5 mol H_2O to grams of H_2O (molar mass = 18.02 g/mol) gives

$$15.5 \text{ mol } H_2O \times \frac{18.02 \text{ g } H_2O}{1 \text{ mol } H_2O} = 279 \text{ g } H_2O$$

This result means that if 249 g of methane is mixed with 279 g of water, both reactants will "run out" at the same time. The reactants have been mixed in stoichiometric quantities.

If, on the other hand, 249 g of methane is mixed with 300 g of water, the methane will be consumed before the water runs out. The water will be in *excess*. In this case, the quantity of products formed will be determined by the quantity of methane present. Once the methane is consumed, no more products can be formed, even though some water still remains. In this situation, the amount of methane *limits* the amount of products that can be formed. Recall from Section 9.4 that we call such a reactant the limiting reactant or the limiting reagent. In any stoichiometry problem where reactants

The reactant that is consumed first limits the amounts of products that can form.

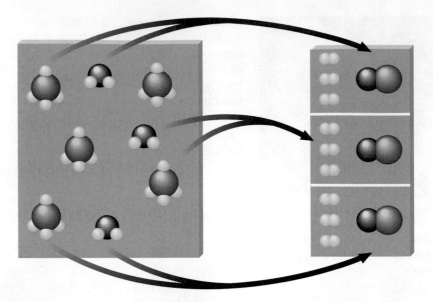

Figure 9.1

A mixture of 5CH$_4$ and 3H$_2$O molecules undergoes the reaction CH$_4$(g) + H$_2$O(g) → 3H$_2$(g) + CO(g). Note that the H$_2$O molecules are used up first, leaving two CH$_4$ molecules unreacted.

are not mixed in stoichiometric quantities, it is essential to determine which reactant is limiting to calculate correctly the amounts of products that will be formed. This concept is illustrated in Figure 9.1. Note from this figure that because there are fewer water molecules than CH$_4$ molecules, the water is consumed first. After the water molecules are gone, no more products can form. So in this case water is the limiting reactant.

EXAMPLE 9.7   **Stoichiometric Calculations: Identifying the Limiting Reactant**

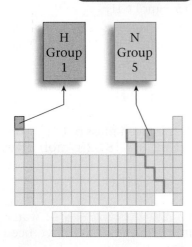

Suppose 25.0 kg (2.50 × 10^4 g) of nitrogen gas and 5.00 kg (5.00 × 10^3 g) of hydrogen gas are mixed and reacted to form ammonia. Calculate the mass of ammonia produced when this reaction is run to completion.

SOLUTION

Where Are We Going?

We want to determine the mass of ammonia produced given the masses of both reactants.

What Do We Know?

- The names of the reactants and products.
- We start with 2.50 × 10^4 g of nitrogen gas and 5.00 × 10^3 g of hydrogen gas.
- We can obtain the atomic masses from the periodic table.

What Do We Need To Know?

- We need to know the balanced equation for the reaction, but we first have to write the formulas for the reactants and products.
- We need the molar masses of nitrogen gas, hydrogen gas, and ammonia.
- We need to determine the limiting reactant.

How Do We Get There?

The unbalanced equation for this reaction is

$$N_2(g) + H_2(g) \rightarrow NH_3(g)$$

which leads to the balanced equation

$$N_2(g) + 3H_2(g) \rightarrow 2NH_3(g)$$

This problem is different from the others we have done so far in that we are mixing *specified amounts of two reactants* together. To know how much product forms, we must determine which reactant is consumed first. That is, we must determine which is the limiting reactant in this experiment. To do so we must add a step to our normal procedure. We can map this process as follows:

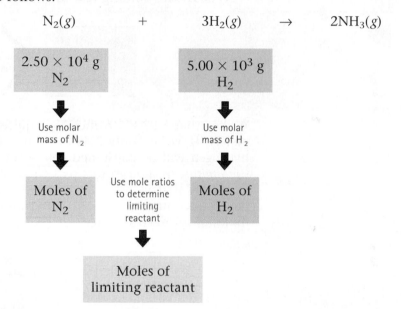

We will use the moles of the limiting reactant to calculate the moles and then the grams of the product.

$$N_2(g) \qquad + \qquad 3H_2(g) \qquad \rightarrow \qquad 2NH_3(g)$$

We first calculate the moles of the two reactants present:

$$2.50 \times 10^4 \text{ g } N_2 \times \frac{1 \text{ mol } N_2}{28.02 \text{ g } N_2} = 8.92 \times 10^2 \text{ mol } N_2$$

$$5.00 \times 10^3 \text{ g } H_2 \times \frac{1 \text{ mol } H_2}{2.016 \text{ g } H_2} = 2.48 \times 10^3 \text{ mol } H_2$$

Now we must determine which reactant is limiting (will be consumed first). We have 8.92×10^2 moles of N_2. Let's determine *how many moles of H_2 are required to react with this much N_2.* Because 1 mole of N_2 reacts with 3 moles of H_2, the number of moles of H_2 we need to react completely with 8.92×10^2 moles of N_2 is determined as follows:

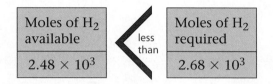

$$8.92 \times 10^2 \ \text{mol } N_2 \times \frac{3 \ \text{mol } H_2}{1 \ \text{mol } N_2} = 2.68 \times 10^3 \ \text{mol } H_2$$

Is N_2 or H_2 the limiting reactant? The answer comes from the comparison

Moles of H_2 available		Moles of H_2 required
2.48×10^3	less than	2.68×10^3

We see that 8.92×10^2 mol N_2 requires 2.68×10^3 mol H_2 to react completely. However, only 2.48×10^3 mol H_2 is present. This means that the hydrogen will be consumed before the nitrogen runs out, so hydrogen is the *limiting reactant* in this particular situation.

Note that in our effort to determine the limiting reactant, we could have started instead with the given amount of hydrogen and calculated the moles of nitrogen required.

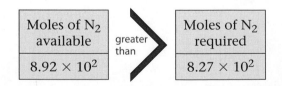

$$2.48 \times 10^3 \ \text{mol } H_2 \times \frac{1 \ \text{mol } N_2}{3 \ \text{mol } H_2} = 8.27 \times 10^2 \ \text{mol } N_2$$

Thus 2.48×10^3 mol H_2 requires 8.27×10^2 mol N_2. Because 8.92×10^2 mol N_2 is actually present, the nitrogen is in excess.

Moles of N_2 available		Moles of N_2 required
8.92×10^2	greater than	8.27×10^2

Always check to see which, if any, reactant is limiting when you are given the amounts of two or more reactants.

If nitrogen is in excess, hydrogen will "run out" first; again we find that hydrogen limits the amount of ammonia formed.

Because the moles of H_2 present are limiting, we must use this quantity to determine the moles of NH_3 that can form.

$$2.48 \times 10^3 \ \text{mol } H_2 \times \frac{2 \ \text{mol } NH_3}{3 \ \text{mol } H_2} = 1.65 \times 10^3 \ \text{mol } NH_3$$

Next we convert moles of NH_3 to mass of NH_3.

$$1.65 \times 10^3 \ \text{mol } NH_3 \times \frac{17.03 \ \text{g } NH_3}{1 \ \text{mol } NH_3} = 2.81 \times 10^4 \ \text{g } NH_3 = 28.1 \ \text{kg } NH_3$$

Therefore, 25.0 kg of N_2 and 5.00 kg of H_2 can form 28.1 kg of NH_3.

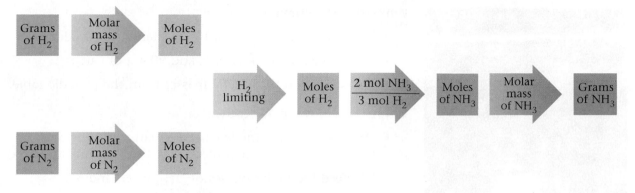

Figure 9.2

A map of the procedure used in Example 9.7.

REALITY CHECK   If neither reactant were limiting, we would expect an answer of 30.0 kg of NH_3 because mass is conserved (25.0 kg + 5.0 kg = 30.0 kg). Because one of the reactants (H_2 in this case) is limiting, the answer should be less than 30.0 kg, which it is. ■

The strategy used in Example 9.7 is summarized in Figure 9.2.

The following list summarizes the steps to take in solving stoichiometry problems in which the amounts of two (or more) reactants are given.

Steps for Solving Stoichiometry Problems Involving Limiting Reactants

Step 1 Write and balance the equation for the reaction.

Step 2 Convert known masses of reactants to moles.

Step 3 Using the numbers of moles of reactants and the appropriate mole ratios, determine which reactant is limiting.

Step 4 Using the amount of the limiting reactant and the appropriate mole ratios, compute the number of moles of the desired product.

Step 5 Convert from moles of product to grams of product, using the molar mass (if this is required by the problem).

EXAMPLE 9.8   ## Stoichiometric Calculations: Reactions Involving the Masses of Two Reactants

Nitrogen gas can be prepared by passing gaseous ammonia over solid copper(II) oxide at high temperatures. The other products of the reaction are solid copper and water vapor. How many grams of N_2 are formed when 18.1 g of NH_3 is reacted with 90.4 g of CuO?

SOLUTION

Where Are We Going?

We want to determine the mass of nitrogen produced given the masses of both reactants.

Copper(II) oxide reacting with ammonia in a heated tube.

What Do We Know?

- The names or formulas of the reactants and products.
- We start with 18.1 g of NH_3 and 90.4 g of CuO.
- We can obtain the atomic masses from the periodic table.

What Do We Need To Know?

- We need to know the balanced equation for the reaction, but we first have to write the formulas for the reactants and products.
- We need the molar masses of NH_3, CuO, and N_2.
- We need to determine the limiting reactant.

How Do We Get There?

Step 1   From the description of the problem, we obtain the following balanced equation:

$$2NH_3(g) + 3CuO(s) \rightarrow N_2(g) + 3Cu(s) + 3H_2O(g)$$

Step 2   Next, from the masses of reactants available we must compute the moles of NH_3 (molar mass = 17.03 g) and of CuO (molar mass = 79.55 g).

$$18.1 \text{ g NH}_3 \times \frac{1 \text{ mol NH}_3}{17.03 \text{ g NH}_3} = 1.06 \text{ mol NH}_3$$

$$90.4 \text{ g CuO} \times \frac{1 \text{ mol CuO}}{79.55 \text{ g CuO}} = 1.14 \text{ mol CuO}$$

Step 3   To determine which reactant is limiting, we use the mole ratio between CuO and NH_3.

$$1.06 \text{ mol NH}_3 \times \frac{3 \text{ mol CuO}}{2 \text{ mol NH}_3} = 1.59 \text{ mol CuO}$$

Then we compare how much CuO we have with how much of it we need.

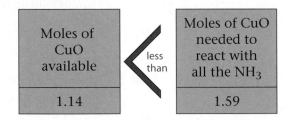

Therefore, 1.59 mol CuO is required to react with 1.06 mol NH_3, but only 1.14 mol CuO is actually present. So the amount of CuO is limiting; CuO will run out before NH_3 does.

Step 4   CuO is the limiting reactant, so we must use the amount of CuO in calculating the amount of N_2 formed. Using the mole ratio between CuO and N_2 from the balanced equation, we have

$$1.14 \text{ mol CuO} \times \frac{1 \text{ mol N}_2}{3 \text{ mol CuO}} = 0.380 \text{ mol N}_2$$

Step 5   Using the molar mass of N_2 (28.02), we can now calculate the mass of N_2 produced.

$$0.380 \text{ mol N}_2 \times \frac{28.02 \text{ g N}_2}{1 \text{ mol N}_2} = 10.6 \text{ g N}_2$$

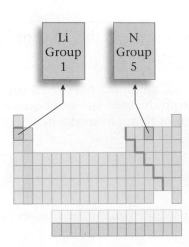

Self-Check **EXERCISE 9.6** Lithium nitride, an ionic compound containing the Li^+ and N^{3-} ions, is prepared by the reaction of lithium metal and nitrogen gas. Calculate the mass of lithium nitride formed from 56.0 g of nitrogen gas and 56.0 g of lithium in the unbalanced reaction

$$Li(s) + N_2(g) \rightarrow Li_3N(s)$$

See Problems 9.51 through 9.54. ■

9.6 Percent Yield

OBJECTIVE: To learn to calculate actual yield as a percentage of theoretical yield.

In the previous section we learned how to calculate the amount of products formed when specified amounts of reactants are mixed together. In doing these calculations, we used the fact that the amount of product is controlled by the limiting reactant. Products stop forming when one reactant runs out.

The amount of product calculated in this way is called the **theoretical yield** of that product. It is the amount of product predicted from the amounts of reactants used. For instance, in Example 9.8, 10.6 g of nitrogen represents the theoretical yield. This is the *maximum amount* of nitrogen that can be produced from the quantities of reactants used. Actually, however, the amount of product predicted (the theoretical yield) is seldom obtained. One reason for this is the presence of side reactions (other reactions that consume one or more of the reactants or products).

The *actual yield* of product, which is the amount of product *actually obtained,* is often compared to the theoretical yield. This comparison, usually expressed as a percentage, is called the **percent yield.**

> Percent yield is important as an indicator of the efficiency of a particular reaction.

$$\frac{\text{Actual yield}}{\text{Theoretical yield}} \times 100\% = \text{percent yield}$$

For example, *if* the reaction considered in Example 9.8 *actually* gave 6.63 g of nitrogen instead of the *predicted* 10.6 g, the percent yield of nitrogen would be

$$\frac{6.63 \text{ g } N_2}{10.6 \text{ g } N_2} \times 100\% = 62.5\%$$

EXAMPLE 9.9 | Stoichiometric Calculations: Determining Percent Yield

In Section 9.1, we saw that methanol can be produced by the reaction between carbon monoxide and hydrogen. Let's consider this process again. Suppose 68.5 kg (6.85×10^4 g) of $CO(g)$ is reacted with 8.60 kg (8.60×10^3 g) of $H_2(g)$.

   a. Calculate the theoretical yield of methanol.

   b. If 3.57×10^4 g of CH_3OH is actually produced, what is the percent yield of methanol?

SOLUTION (a)

Where Are We Going?

We want to determine the theoretical yield of methanol and the percent yield given an actual yield.

What Do We Know?

- From Section 9.1 we know the balanced equation is

$$2H_2 + CO \rightarrow CH_3OH$$

- We start with 6.85×10^4 g of CO and 8.60×10^3 g of H_2.
- We can obtain the atomic masses from the periodic table.

What Do We Need To Know?

- We need the molar masses of H_2, CO, and CH_3OH.
- We need to determine the limiting reactant.

How Do We Get There?

Step 1   The balanced equation is

$$2H_2(g) + CO(g) \rightarrow CH_3OH(l)$$

Step 2   Next we calculate the moles of reactants.

$$6.85 \times 10^4 \text{ g CO} \times \frac{1 \text{ mol CO}}{28.01 \text{ g CO}} = 2.45 \times 10^3 \text{ mol CO}$$

$$8.60 \times 10^3 \text{ g H}_2 \times \frac{1 \text{ mol H}_2}{2.016 \text{ g H}_2} = 4.27 \times 10^3 \text{ mol H}_2$$

Step 3   Now we determine which reactant is limiting. Using the mole ratio between CO and H_2 from the balanced equation, we have

$$2.45 \times 10^3 \text{ mol CO} \times \frac{2 \text{ mol H}_2}{1 \text{ mol CO}} = 4.90 \times 10^3 \text{ mol H}_2$$

Moles of H_2 present	less than	Moles of H_2 needed to react with all the CO
4.27×10^3		4.90×10^3

We see that 2.45×10^3 mol CO requires 4.90×10^3 mol H_2. Because only 4.27×10^3 mol H_2 is actually present, *H_2 is limiting*.

Step 4   We must therefore use the amount of H_2 and the mole ratio between H_2 and CH_3OH to determine the maximum amount of methanol that can be produced in the reaction.

$$4.27 \times 10^3 \text{ mol H}_2 \times \frac{1 \text{ mol CH}_3OH}{2 \text{ mol H}_2} = 2.14 \times 10^3 \text{ mol CH}_3OH$$

This represents the theoretical yield in moles.

Step 5   Using the molar mass of CH_3OH (32.04 g), we can calculate the theoretical yield in grams.

$$2.14 \times 10^3 \text{ mol CH}_3OH \times \frac{32.04 \text{ g CH}_3OH}{1 \text{ mol CH}_3OH} = 6.86 \times 10^4 \text{ g CH}_3OH$$

So, from the amounts of reactants given, the maximum amount of CH_3OH that can be formed is 6.86×10^4 g. This is the *theoretical yield*.

SOLUTION (b)

The percent yield is

$$\frac{\text{Actual yield (grams)}}{\text{Theoretical yield (grams)}} \times 100\% = \frac{3.57 \times 10^4 \text{ g CH}_3OH}{6.86 \times 10^4 \text{ g CH}_3OH} \times 100\% = 52.0\%$$

Self-Check **EXERCISE 9.7** Titanium(IV) oxide is a white compound used as a coloring pigment. In fact, the page you are now reading is white because of the presence of this compound in the paper. Solid titanium(IV) oxide can be prepared by reacting gaseous titanium(IV) chloride with oxygen gas. A second product of this reaction is chlorine gas.

$$TiCl_4(g) + O_2(g) \rightarrow TiO_2(s) + Cl_2(g)$$

a. Suppose 6.71×10^3 g of titanium(IV) chloride is reacted with 2.45×10^3 g of oxygen. Calculate the maximum mass of titanium(IV) oxide that can form.

b. If the percent yield of TiO_2 is 75%, what mass is actually formed?

See Problems 9.63 and 9.64. ■

CHAPTER 9 REVIEW

Key Terms

mole ratio (9.2)
stoichiometry (9.3)
limiting reactant
   (limiting reagent) (9.4)

theoretical yield (9.6)
percent yield (9.6)

Summary

1. A balanced equation relates the numbers of molecules of reactants and products. It can also be expressed in terms of the numbers of moles of reactants and products.

2. The process of using a chemical equation to calculate the relative amounts of reactants and products involved in the reaction is called doing stoichiometric calculations. To convert between moles of reactants and moles of products, we use mole ratios derived from the balanced equation.

3. Often reactants are not mixed in stoichiometric quantities (they do not "run out" at the same time). In that case, we must use the limiting reactant to calculate the amounts of products formed.

4. The actual yield of a reaction is usually less than its theoretical yield. The actual yield is often expressed as a percentage of the theoretical yield, which is called the percent yield.

Active Learning Questions

These questions are designed to be considered by groups of students in class. Often these questions work well for introducing a particular topic in class.

1. Relate Active Learning Question 2 from Chapter 2 to the concepts of chemical stoichiometry.

2. You are making cookies and are missing a key ingredient—eggs. You have plenty of the other ingredients, except that you have only 1.33 cups of butter and no eggs. You note that the recipe calls for 2 cups of butter and 3 eggs (plus the other ingredients) to make 6 dozen cookies. You telephone a friend and have him bring you some eggs.

a. How many eggs do you need?
b. If you use all the butter (and get enough eggs), how many cookies can you make?

Unfortunately, your friend hangs up before you tell him how many eggs you need. When he arrives, he has a surprise for you—to save time he has broken the eggs in a bowl for you. You ask him how many he brought, and he replies, "All of them, but I spilled some on the way over." You weigh the eggs and find that they weigh 62.1 g. Assuming that an average egg weighs 34.21 g:

c. How much butter is needed to react with all the eggs?
d. How many cookies can you make?
e. Which will you have left over, eggs or butter?
f. How much is left over?
g. Relate this question to the concepts of chemical stoichiometry.

3. Nitrogen (N_2) and hydrogen (H_2) react to form ammonia (NH_3). Consider the mixture of N_2 (●●) and H_2 (○○) in a closed container as illustrated below:

Assuming the reaction goes to completion, draw a representation of the product mixture. Explain how you arrived at this representation.

4. Which of the following equations best represents the reaction for Question 3?

a. $6N_2 + 6H_2 \rightarrow 4NH_3 + 4N_2$
b. $N_2 + H_2 \rightarrow NH_3$

c. $N + 3H \rightarrow NH_3$
d. $N_2 + 3H_2 \rightarrow 2NH_3$
e. $2N_2 + 6H_2 \rightarrow 4NH_3$

For choices you did not pick, explain what you feel is wrong with them, and justify the choice you did pick.

5. You know that chemical A reacts with chemical B. You react 10.0 g A with 10.0 g B. What information do you need to know to determine the amount of product that will be produced? Explain.

6. If 10.0 g of hydrogen gas is reacted with 10.0 g of oxygen gas according to the equation

$$2H_2 + O_2 \rightarrow 2H_2O$$

we should not expect to form 20.0 g of water. Why not? What mass of water can be produced with a complete reaction?

7. The limiting reactant in a reaction:

a. has the lowest coefficient in a balanced equation.
b. is the reactant for which you have the fewest number of moles.
c. has the lowest ratio: moles available/coefficient in the balanced equation.
d. has the lowest ratio: coefficient in the balanced equation/moles available.
d. None of the above.

For choices you did not pick, explain what you feel is wrong with them, and justify the choice you did pick.

8. Given the equation $3A + B \rightarrow C + D$, if 4 moles of A is reacted with 2 moles of B, which of the following is true?

a. The limiting reactant is the one with the higher molar mass.
b. A is the limiting reactant because you need 6 moles of A and have 4 moles.
c. B is the limiting reactant because you have fewer moles of B than moles of A.
d. B is the limiting reactant because three A molecules react with every one B molecule.
e. Neither reactant is limiting.

For choices you did not pick, explain what you feel is wrong with them, and justify the choice you did pick.

9. What happens to the weight of an iron bar when it rusts?

a. There is no change because mass is always conserved.
b. The weight increases.
c. The weight increases, but if the rust is scraped off, the bar has the original weight.
d. The weight decreases.

Justify your choice and, for choices you did not pick, explain what is wrong with them. Explain what it means for something to rust.

10. Consider the equation $2A + B \rightarrow A_2B$. If you mix 1.0 mole of A and 1.0 mole of B, how many moles of A_2B can be produced?

11. What is meant by the term *mole ratio*? Give an example of a mole ratio, and explain how it is used in solving a stoichiometry problem.

12. Which would produce a greater number of moles of product: a given amount of hydrogen gas reacting with an excess of oxygen gas to produce water, or the same amount of hydrogen gas reacting with an excess of nitrogen gas to make ammonia? Support your answer.

13. Consider a reaction represented by the following balanced equation

$$2A + 3B \rightarrow C + 4D$$

You find that it requires equal masses of A and B so that there are no reactants left over. Which of the following is true? Justify your choice.

a. The molar mass of A must be greater than the molar mass of B.
b. The molar mass of A must be less than the molar mass of B.
c. The molar mass of A must be the same as the molar mass of B.

14. Consider a chemical equation with two reactants forming one product. If you know the mass of each reactant, what else do you need to know to determine the mass of the product? Why isn't the mass necessarily the sum of the mass of the reactants? Provide a real example of such a reaction, and support your answer mathematically.

15. Consider the balanced chemical equation

$$A + 5B \rightarrow 3C + 4D$$

When equal masses of A and B are reacted, which is limiting, A or B? Justify your choice.

a. If the molar mass of A is greater than the molar mass of B, then A must be limiting.
b. If the molar mass of A is less than the molar mass of B, then A must be limiting.
c. If the molar mass of A is greater than the molar mass of B, then B must be limiting.
d. If the molar mass of A is less than the molar mass of B, then B must be limiting.

16. Which of the following reaction mixtures would produce the greatest amount of product, assuming all went to completion? Justify your choice.

Each involves the reaction symbolized by the equation

$$2H_2 + O_2 \rightarrow 2H_2O$$

a. 2 moles of H_2 and 2 moles of O_2.
b. 2 moles of H_2 and 3 moles of O_2.
c. 2 moles of H_2 and 1 mole of O_2.
d. 3 moles of H_2 and 1 mole of O_2.
e. Each would produce the same amount of product.

17. Baking powder is a mixture of cream of tartar ($KHC_4H_4O_6$) and baking soda ($NaHCO_3$). When it is placed in an oven at typical baking temperatures (as part of a cake, for example), it undergoes the following reaction (CO_2 makes the cake rise):

$$KHC_4H_4O_6(s) + NaHCO_3(s) \rightarrow$$
$$KNaC_4H_4O_6(s) + H_2O(g) + CO_2(g)$$

You decide to make a cake one day, and the recipe calls for baking powder. Unfortunately, you have no baking powder. You do have cream of tartar and baking soda, so you use stoichiometry to figure out how much of each to mix.

Of the following choices, which is the best way to make baking powder? The amounts given in the choices are in teaspoons (that is, you will use a teaspoon to measure the baking soda and cream of tartar). Justify your choice.

Assume a teaspoon of cream of tartar has the same mass as a teaspoon of baking soda.

a. Add equal amounts of baking soda and cream of tartar.
b. Add a bit more than twice as much cream of tartar as baking soda.
c. Add a bit more than twice as much baking soda as cream of tartar.
d. Add more cream of tartar than baking soda, but not quite twice as much.
e. Add more baking soda than cream of tartar, but not quite twice as much.

18. You have seven closed containers each with equal masses of chlorine gas (Cl_2). You add 10.0 g of sodium to the first sample, 20.0 g of sodium to the second sample, and so on (adding 70.0 g of sodium to the seventh sample). Sodium and chloride react to form sodium chloride according to the equation

$$2Na(s) + Cl_2(g) \rightarrow 2NaCl(s)$$

After each reaction is complete, you collect and measure the amount of sodium chloride formed. A graph of your results is shown below.

Answer the following questions:

a. Explain the shape of the graph.
b. Calculate the mass of NaCl formed when 20.0 g of sodium is used.
c. Calculate the mass of Cl_2 in each container.
d. Calculate the mass of NaCl formed when 50.0 g of sodium is used.
e. Identify the leftover reactant and determine its mass for parts b and d above.

19. You have a chemical in a sealed glass container filled with air. The setup is sitting on a balance as shown below. The chemical is ignited by means of a magnifying glass focusing sunlight on the reactant. After the chemical has completely burned, which of the following is true? Explain your answer.

a. The balance will read less than 250.0 g.
b. The balance will read 250.0 g.
c. The balance will read greater than 250.0 g.
d. Cannot be determined without knowing the identity of the chemical.

20. Consider an iron bar on a balance as shown.

As the iron bar rusts, which of the following is true? Explain your answer.

a. The balance will read less than 75.0 g.
b. The balance will read 75.0 g.
c. The balance will read greater than 75.0 g.
d. The balance will read greater than 75.0 g, but if the bar is removed, the rust scraped off, and the bar replaced, the balance will read 75.0 g.

21. Consider the reaction between NO(g) and O_2(g) represented below.

What is the balanced equation for this reaction, and what is the limiting reactant?

OWL Questions and problems for this chapter are available in OWL. Use the OWL access card that came with this textbook to access assigned questions and problems for this chapter.

10 Energy

● A hummingbird exerts a great deal of energy in order to hover. *(Image by © Raven Regan/ Design Pics/Corbis)*

Energy is at the center of our very existence as individuals and as a society. The food that we eat furnishes the energy to live, work, and play, just as the coal and oil consumed by manufacturing and transportation systems power our modern industrialized civilization.

Huge quantities of carbon-based fossil fuels have been available for the taking. This abundance of fuels has led to a world society with a huge appetite for energy, consuming millions of barrels of petroleum every day. We are now dangerously dependent on the dwindling supplies of oil, and this dependence is an important source of tension

Energy is a factor in all human activity.

among nations in today's world. In an incredibly short time we have moved from a period of ample and cheap supplies of petroleum to one of high prices and uncertain supplies. If our present standard of living is to be maintained, we must find alternatives to petroleum. To do this, we need to know the relationship between chemistry and energy, which we explore in this chapter.

10.1 The Nature of Energy

OBJECTIVE: To understand the general properties of energy.

Although energy is a familiar concept, it is difficult to define precisely. For our purposes we will define **energy** as *the ability to do work or produce heat.* We will define these terms below.

Energy can be classified as either potential or kinetic energy. **Potential energy** is energy due to position or composition. For example, water behind a dam has potential energy that can be converted to work when the water flows down through turbines, thereby creating electricity. Attractive and repulsive forces also lead to potential energy. The energy released when gasoline is burned results from differences in attractive forces between the nuclei and electrons in the reactants and products. The **kinetic energy** of an object is energy due to the motion of the object and depends on the mass of the object m and its velocity v: $\mathrm{KE} = \frac{1}{2}mv^2$.

One of the most important characteristics of energy is that it is conserved. The **law of conservation of energy** states *that energy can be converted from one form to another but can be neither created nor destroyed.* That is, the energy of the universe is constant.

Although the energy of the universe is constant, it can be readily converted from one form to another. Consider the two balls in Figure 10.1a. Ball A, because of its initially higher position, has more potential energy than ball B. When ball A is released, it moves down the hill and strikes ball B. Eventually, the arrangement shown in Figure 10.1b is achieved. What has happened in going from the initial to the final arrangement? The potential

Initial

a

In the initial positions, ball A has a higher potential energy than ball B.

Final

b

After A has rolled down the hill, the potential energy lost by A has been converted to random motions of the components of the hill (fractional heating) and to an increase in the potential energy of B.

Figure 10.1

energy of A has decreased because its position was lowered. However, this energy cannot disappear. Where is the energy lost by A?

Initially, the potential energy of A is changed to kinetic energy as the ball rolls down the hill. Part of this energy is transferred to B, causing it to be raised to a higher final position. Thus the potential energy of B has been increased, which means that **work** (force acting over a distance) has been performed on B. Because the final position of B is lower than the original position of A, however, some of the energy is still unaccounted for. Both balls in their final positions are at rest, so the missing energy cannot be attributed to their motions.

What has happened to the remaining energy? The answer lies in the interaction between the hill's surface and the ball. As ball A rolls down the hill, some of its kinetic energy is transferred to the surface of the hill as heat. This transfer of energy is called *frictional heating*. The temperature of the hill increases very slightly as the ball rolls down. Thus the energy stored in A in its original position (potential energy) is distributed to B through work and to the surface of the hill by heat.

Imagine that we perform this same experiment several times, varying the surface of the hill from very smooth to very rough. In rolling to the bottom of the hill (see Figure 10.1), A always loses the same amount of energy because its position always changes by exactly the same amount. The way that this energy transfer is divided between work and heat, however, depends on the specific conditions—the *pathway*. For example, the surface of the hill might be so rough that the energy of A is expended completely through frictional heating: A is moving so slowly when it hits B that it cannot move B to the next level. In this case, no work is done. Regardless of the condition of the hill's surface, the *total energy* transferred will be constant, although the amounts of heat and work will differ. Energy change is independent of the pathway, whereas work and heat are both dependent on the pathway.

This brings us to a very important idea, the state function. A **state function** is a property of the system that changes independently of its pathway. Let's consider a nonchemical example. Suppose you are traveling from Chicago to Denver. Which of the following are state functions?

- Distance traveled

- Change in elevation

Because the distance traveled depends on the route taken (that is, the *pathway* between Chicago and Denver), it is *not* a state function. On the other hand, the change in elevation depends only on the difference between Denver's elevation (5280 ft) and Chicago's elevation (580 ft). The change in elevation is always 5280 ft − 580 ft = 4700 ft; it does not depend on the route taken between the two cities.

We can also learn about state functions from the example illustrated in Figure 10.1. Because ball A always goes from its initial position on the hill to the bottom of the hill, its energy change is always the same, regardless of whether the hill is smooth or bumpy. This energy is a state function—a given change in energy is independent of the pathway of the process. In contrast, work and heat are *not* state functions. For a given change in the position of A, a smooth hill produces more work and less heat than a rough hill does. That is, for a given change in the position of A, the change in energy is always the same (state function) but the way the resulting energy is distributed as heat or work depends on the nature of the hill's surface (heat and work are not state functions).

10.2 Temperature and Heat

OBJECTIVE: To understand the concepts of temperature and heat.

Figure 10.2

Equal masses of hot water and cold water separated by a thin metal wall in an insulated box.

What does the temperature of a substance tell us about that substance? Put another way, how is warm water different from cold water? The answer lies in the motions of the water molecules. **Temperature** is a *measure of the random motions of the components of a substance.* That is, the H_2O molecules in warm water are moving around more rapidly than the H_2O molecules in cold water.

Consider an experiment in which we place 1.00 kg of hot water (90. °C) next to 1.00 kg of cold water (10. °C) in an insulated box. The water samples are separated from each other by a thin metal plate (see Figure 10.2). You already know what will happen: the hot water will cool down and the cold water will warm up.

Assuming that no energy is lost to the air, can we determine the final temperature of the two samples of water? Let's consider how to think about this problem.

First picture what is happening. Remember that the H_2O molecules in the hot water are moving faster than those in the cold water (see Figure 10.3). As a result, energy will be transferred through the metal wall from the hot water to the cold water. This energy transfer will cause the H_2O molecules in the hot water to slow down and the H_2O molecules in the cold water to speed up.

Thus we have a transfer of energy from the hot water to the cold water. This flow of energy is called heat. **Heat** can be defined as a *flow of energy due to a temperature difference.* What will eventually happen? The two water samples will reach the same temperature (see Figure 10.4). At this point, how does the energy lost by the hot water compare to the energy gained by the cold water? They must be the same (remember that energy is conserved).

We conclude that the final temperature is the average of the original temperatures:

$$T_{final} = \frac{T_{hot\,initial} + T_{cold\,initial}}{2} = \frac{90.\,°C + 10.\,°C}{2} = 50.\,°C$$

Hot water (90. °C) Cold water (10. °C)

Figure 10.3

The H_2O molecules in hot water have much greater random motions than the H_2O molecules in cold water.

Water (50. °C) Water (50. °C)

Figure 10.4

The water samples now have the same temperature (50. °C) and have the same random motions.

For the hot water, the temperature change is

Change in temperature (hot) = ΔT_{hot} = 90. °C − 50. °C = 40. °C

The temperature change for the cold water is

Change in temperature (cold) = ΔT_{cold} = 50. °C − 10. °C = 40. °C

In this example, the masses of hot water and cold water are equal. If they were unequal, this problem would be more complicated.

Let's summarize the ideas we have introduced in this section. Temperature is a measure of the random motions of the components of an object. Heat is a *flow* of energy due to a temperature difference. We say that the random motions of the components of an object constitute the *thermal energy* of that object. The flow of energy called heat is the way in which thermal energy is transferred from a hot object to a colder object.

10.3   Exothermic and Endothermic Processes

OBJECTIVE: To consider the direction of energy flow as heat.

In this section we will consider the energy changes that accompany chemical reactions. To explore this idea, let's consider the striking and burning of a match. Energy is clearly released through heat as the match burns. To discuss this reaction, we divide the universe into two parts: the system and the surroundings. The **system** is the part of the universe on which we wish to focus attention; the **surroundings** include everything else in the universe. In this case we define the system as the reactants and products of the reaction. The surroundings consist of the air in the room and anything else other than the reactants and products.

A burning match releases energy.

When a process results in the evolution of heat, it is said to be **exothermic** (*exo-* is a prefix meaning "out of"); that is, energy flows *out of the system.* For example, in the combustion of a match, energy flows out of the system as heat. Processes that absorb energy from the surroundings are said to be **endothermic.** When the heat flow moves *into a system,* the process is endothermic. Boiling water to form steam is a common endothermic process.

Where does the energy, released as heat, come from in an exothermic reaction? The answer lies in the difference in potential energies between the products and the reactants. Which has lower potential energy, the reactants or the products? We know that total energy is conserved and that energy flows from the system into the surroundings in an exothermic reaction. Thus *the energy gained by the surroundings must be equal to the energy lost by the system.* In the combustion of a match, the burned match has lost potential energy (in this case potential energy stored in the bonds of the reactants), which was transferred through heat to the surroundings (see Figure 10.5). The heat flow into the surroundings results from a lowering of the potential energy of the reaction system. *In any exothermic reaction, some of the potential energy stored in the chemical bonds is converted to thermal energy (random kinetic energy) via heat.*

Figure 10.5

The energy changes accompanying the burning of a match.

10.4 Thermodynamics

OBJECTIVE: To understand how energy flow affects internal energy.

The study of energy is called **thermodynamics.** The law of conservation of energy is often called the **first law of thermodynamics** and is stated as follows:

The energy of the universe is constant.

The **internal energy,** *E,* of a system can be defined most precisely as the sum of the kinetic and potential energies of all "particles" in the system. The internal energy of a system can be changed by a flow of work, heat, or both. That is,

$$\Delta E = q + w$$

where

Δ ("delta") means a change in the function that follows

q represents heat

w represents work

Thermodynamic quantities always consist of two parts: a *number,* giving the magnitude of the change, and a *sign,* indicating the direction of the flow. *The sign reflects the system's point of view.* For example, when a quantity of energy flows *into* the system via heat (an endothermic process), q is equal to $+x$, where the *positive* sign indicates that the *system's energy is increasing.* On the other hand, when energy flows *out of* the system via heat (an exothermic process), q is equal to $-x$, where the *negative* sign indicates that the *system's energy is decreasing.*

In this text the same conventions apply to the flow of work. If the system does work on the surroundings (energy flows out of the system), w is negative. If the surroundings do work on the system (energy flows into the system), w is positive. We define work from the system's point of view to be consistent for all thermodynamic quantities. That is, in this convention the signs of both q and w reflect what happens to the system; thus we use $\Delta E = q + w$.

$$\Delta E < 0 \qquad\qquad\qquad \Delta E > 0$$

10.5 Measuring Energy Changes

OBJECTIVE: To understand how heat is measured.

Diet drinks are now labeled as "low joule" instead of "low calorie" in European countries.

Earlier in this chapter we saw that when we heat a substance to a higher temperature, we increase the motions of the components of the substance—that is, we increase the thermal energy of the substance. Different materials respond differently to being heated. To explore this idea we need to introduce the common units of energy: the *calorie* and the *joule* (pronounced "jewel").

In the metric system the **calorie** is defined as the amount of energy (heat) required to raise the temperature of one gram of water by one Celsius degree. The "calorie" with which you are probably familiar is used to measure the energy content of food and is actually a kilocalorie (1000 calories), written with a capital C (Calorie) to distinguish it from the calorie used in chemistry. The **joule** (an SI unit) can be most conveniently defined in terms of the calorie:

$$1 \text{ calorie} = 4.184 \text{ joules}$$

or using the normal abbreviations

$$1 \text{ cal} = 4.184 \text{ J}$$

You need to be able to convert between calories and joules. We will consider that conversion process in Example 10.1.

EXAMPLE 10.1 | Converting Calories to Joules

Express 60.1 cal of energy in units of joules.

SOLUTION

By definition 1 cal = 4.184 J, so the conversion factor needed is $\dfrac{4.184 \text{ J}}{1 \text{ cal}}$, and the result is

$$60.1 \text{ cal} \times \frac{4.184 \text{ J}}{1 \text{ cal}} = 251 \text{ J}$$

Note that the 1 in the denominator is an exact number by definition and so does not limit the number of significant figures.

Self-Check EXERCISE 10.1   How many calories of energy correspond to 28.4 J?

See Problems 10.25 through 10.30. ∎

Coffee: Hot and Quick(lime)

Convenience and speed are the watchwords of our modern society. One new product that fits these requirements is a container of coffee that heats itself with no batteries needed. Consumers can now buy a 10-ounce container of Wolfgang Puck gourmet latte that heats itself to 145 °F in 6 minutes and stays hot for 30 minutes. What kind of chemical magic makes this happen? Pushing a button on the bottom of the container. This action allows water to mix with calcium oxide, or quicklime (see accompanying figure). The resulting reaction

$$CaO(s) + H_2O(l) \rightarrow Ca(OH)_2(s)$$

releases enough energy as heat to bring the coffee to a pleasant drinking temperature.

Other companies are experimenting with similar technology to heat liquids such as tea, hot chocolate, and soup.

A different reaction is now being used to heat MREs (meals ready-to-eat) for soldiers on the battlefield. In this case the energy to heat the meals is furnished by mixing magnesium iron oxide with water to produce an exothermic reaction.

Clearly, chemistry is "hot stuff."

Outer container holds beverage

Inner cone holds quicklime

Pu ck" holds water, fits inside the cone

Push button breaks the seal that combines water and quicklime, which generates heat

Now think about heating a substance from one temperature to another. How does the amount of substance heated affect the energy required? In 2 g of water there are twice as many molecules as in 1 g of water. It takes twice as much energy to change the temperature of 2 g of water by 1 °C, because we must change the motions of twice as many molecules in a 2-g sample as in a 1-g sample. Also, as we would expect, it takes twice as much energy to raise the temperature of a given sample of water by 2 degrees as it does to raise the temperature by 1 degree.

EXAMPLE 10.2 | Calculating Energy Requirements

Determine the amount of energy (heat) in joules required to raise the temperature of 7.40 g water from 29.0 °C to 46.0 °C.

SOLUTION

Where Are We Going?

We want to determine the amount of energy (heat in joules) needed to increase the temperature of 7.40 g water from 29.0 °C to 46.0 °C.

What Do We Know?

- The mass of water is 7.40 g, and the temperature is increased from 29.0 °C to 46.0 °C.

What Information Do We Need?

• The amount of heat needed to raise 1.00 g water by 1.00 °C. From the text we see that 4.184 J of energy is required.

How Do We Get There?

In solving any kind of problem, it is often useful to draw a diagram that represents the situation. In this case, we have 7.40 g of water that is to be heated from 29.0 °C to 46.0 °C.

7.40 g water $T = 29.0$ °C	⇨ ? energy	7.40 g water $T = 46.0$ °C

Our task is to determine how much energy is required to accomplish this task.

From the discussion in the text, we know that 4.184 J of energy is required to raise the temperature of *one* gram of water by *one* Celsius degree.

1.00 g water $T = 29.0$ °C	⇨ 4.184 J	1.00 g water $T = 30.0$ °C

Because in our case we have 7.40 g of water instead of 1.00 g, it will take 7.40 × 4.184 J to raise the temperature by one degree.

7.40 g water $T = 29.0$ °C	⇨ 7.40 × 4.184 J	7.40 g water $T = 30.0$ °C

However, we want to raise the temperature of our sample of water by more than 1 °C. In fact, the temperature change required is from 29.0 °C to 46.0 °C. This is a change of 17.0 °C (46.0 °C − 29.0 °C = 17.0 °C). Thus we will have to supply 17.0 times the energy necessary to raise the temperature of 7.40 g of water by 1 °C.

7.40 g water $T = 29.0$ °C	⇨ 17.0 × 7.40 × 4.184 J	7.40 g water $T = 46.0$ °C

This calculation is summarized as follows:

$$4.184 \frac{\text{J}}{\text{g} \, °\text{C}} \quad \times \quad 7.40 \text{ g} \quad \times \quad 17.0 \, °\text{C} \quad = \quad 526 \text{ J}$$

Energy per gram of water per degree of temperature	×	Actual grams of water	×	Actual temperature change	=	Energy required

MATH SKILL BUILDER
The result you will get on your calculator is 4.184 × 7.40 × 17.0 = 526.3472, which rounds off to 526.

We have shown that 526 J of energy (as heat) is required to raise the temperature of 7.40 g of water from 29.0 °C to 46.0 °C. Note that because 4.184 J of energy is required to heat 1 g of water by 1 °C, the units are J/g °C (joules per gram per Celsius degree).

REALITY CHECK   The units (J) are correct, and the answer is reported to the correct number of significant figures (three).

Self-Check **EXERCISE 10.2**   Calculate the joules of energy required to heat 454 g of water from 5.4 °C to 98.6 °C.

See Problems 10.31 through 10.36. ∎

Nature Has Hot Plants

The voodoo lily (Titan Arum) is a beautiful and seductive plant. The exotic-looking lily features an elaborate reproductive mechanism—a purple spike that can reach nearly 3 feet in length and is cloaked by a hoodlike leaf. But approach to the plant reveals bad news—it smells terrible!

Despite its antisocial odor, this putrid plant has fascinated biologists for many years because of its ability to generate heat. At the peak of its metabolic activity, the plant's blossom can be as much as 15 °C above its surrounding temperature. To generate this much heat, the metabolic rate of the plant must be close to that of a flying hummingbird!

What's the purpose of this intense heat production? For a plant

Titan Arum is reputedly the largest flower in the world.

faced with limited food supplies in the very competitive tropical climate where it grows, heat production seems like a great waste of energy. The answer to this mystery is that the voodoo lily is pollinated mainly by carrion-loving insects. Thus the lily prepares a malodorous mixture of chemicals characteristic of rotting meat, which it then "cooks" off into the surrounding air to attract flesh-feeding beetles and flies. Then, once the insects enter the pollination chamber, the high temperatures there (as high as 110 °F) cause the insects to remain very active to better carry out their pollination duties.

The voodoo lily is only one of many thermogenic (heat-producing) plants. These plants are of special interest to biologists because they provide opportunities to study metabolic reactions that are quite subtle in "normal" plants.

Table 10.1 The Specific Heat Capacities of Some Common Substances	
Substance	Specific Heat Capacity (J/g °C)
water (*l*)* (liquid)	4.184
water (*s*) (ice)	2.03
water (*g*) (steam)	2.0
aluminum (*s*)	0.89
iron (*s*)	0.45
mercury (*l*)	0.14
carbon (*s*)	0.71
silver (*s*)	0.24
gold (*s*)	0.13

*The symbols (*s*), (*l*), and (*g*) indicate the solid, liquid, and gaseous states, respectively.

So far we have seen that the energy (heat) required to change the temperature of a substance depends on

1. The amount of substance being heated (number of grams)

2. The temperature change (number of degrees)

There is, however, another important factor: the identity of the substance.

Different substances respond differently to being heated. We have seen that 4.184 J of energy raises the temperature of 1 g of water by 1 °C. In contrast, this same amount of energy applied to 1 g of gold raises its temperature by approximately 32 °C! The point is that some substances require relatively large amounts of energy to change their temperatures, whereas others require relatively little. Chemists describe this difference by saying that substances have different heat capacities. *The amount of energy required to change the temperature of one gram of a substance by one Celsius degree* is called its **specific heat capacity** or, more commonly, its *specific heat*. The specific heat capacities for several substances are listed in Table 10.1. You can see from the table that the specific heat capacity for water is very high compared to those of the other substances listed. This is why lakes and seas are much slower to respond to cooling or heating than are the surrounding land masses.

EXAMPLE 10.3   Calculations Involving Specific Heat Capacity

a. What quantity of energy (in joules) is required to heat a piece of iron weighing 1.3 g from 25 °C to 46 °C?

b. What is the answer in calories?

SOLUTION

Where Are We Going?

We want to determine the amount of energy (units of joules and calories) to increase the temperature of 1.3 g, of iron from 25° C to 46° C.

What Do We Know?

- The mass of iron is 1.3 g, and the temperature is increased from 25° C to 46° C.

What Information Do We Need?

- We need the specific heat capacity of iron and the conversion factor between joules and calories.

How Do We Get There?

a. It is helpful to draw the following diagram to represent the problem.

| 1.3 g iron $T = 25\,°C$ | ⇨ ? joules | 1.3 g iron $T = 46\,°C$ |

From Table 10.1 we see that the specific heat capacity of iron is 0.45 J/g °C. That is, it takes 0.45 J to raise the temperature of a 1-g piece of iron by 1 °C.

| 1.0 g iron $T = 25\,°C$ | ⇨ 0.45 J | 1.0 g iron $T = 26\,°C$ |

In this case our sample is 1.3 g, so 1.3 × 0.45 J is required for *each* degree of temperature increase.

| 1.3 g iron $T = 25\,°C$ | ⇨ 1.3 × 0.45 J | 1.3 g iron $T = 26\,°C$ |

Because the temperature increase is 21 °C (46 °C − 25 °C = 21 °C), the total amount of energy required is

$$0.45\,\frac{J}{g\,°C} \times 1.3\,g \times 21\,°C = 12\,J$$

| 1.3 g iron $T = 25\,°C$ | ⇨ 21 × 1.3 × 0.45 J | 1.3 g iron $T = 46\,°C$ |

Note that the final units are joules, as they should be.

b. To calculate this energy in calories, we can use the definition 1 cal = 4.184 J to construct the appropriate conversion factor. We want to change from joules to calories, so cal must be in the numerator and J in the denominator, where it cancels:

$$12\,J \times \frac{1\,cal}{4.184\,J} = 2.9\,cal$$

MATH SKILL BUILDER
The result you will get on your calculator is 0.45 × 1.3 × 21 = 12.285, which rounds off to 12.

Remember that 1 in this case is an exact number by definition and therefore does not limit the number of significant figures (the number 12 is limiting here).

REALITY CHECK   The units (joules and calories) are correct, and the answer is reported to the correct number of significant figures (two).

Self-Check **EXERCISE 10.3**   A 5.63-g sample of solid gold is heated from 21 °C to 32 °C. How much energy (in joules and calories) is required?

See Problems 10.31 through 10.36. ■

Note that in Example 10.3, to calculate the energy (heat) required, we took the product of the specific heat capacity, the sample size in grams, and the change in temperature in Celsius degrees.

$$\boxed{\text{Energy (heat) required } (Q)} = \boxed{\text{Specific heat capacity } (s)} \times \boxed{\begin{array}{c}\text{Mass } (m) \text{ in}\\ \text{grams of}\\ \text{sample}\end{array}} \times \boxed{\begin{array}{c}\text{Change in}\\ \text{temperature}\\ (\Delta T) \text{ in °C}\end{array}}$$

We can represent this by the following equation:

$$Q = s \times m \times \Delta T$$

MATH SKILL BUILDER
The symbol Δ (the Greek letter delta) is shorthand for "change in."

where

Q = energy (heat) required

s = specific heat capacity

m = mass of the sample in grams

ΔT = change in temperature in Celsius degrees

This equation always applies when a substance is being heated (or cooled) and no change of state occurs. Before you begin to use this equation, however, make sure you understand what it means.

EXAMPLE 10.4 | Specific Heat Capacity Calculations: Using the Equation

A 1.6-g sample of a metal that has the appearance of gold requires 5.8 J of energy to change its temperature from 23 °C to 41 °C. Is the metal pure gold?

SOLUTION

Where Are We Going?

We want to determine if a metal is gold.

What Do We Know?

- The mass of metal is 1.6 g, and 5.8 J of energy is required to increase the temperature from 23° C to 41° C.

What Information Do We Need?

- We need the specific heat capacity of gold.

Firewalking: Magic or Science?

For millennia people have been amazed at the ability of Eastern mystics to walk across beds of glowing coals without any apparent discomfort. Even in the United States, thousands of people have performed feats of firewalking as part of motivational seminars. How can this be possible? Do firewalkers have supernatural powers?

Actually, there are good scientific explanations of why firewalking is possible. First, human tissue is mainly composed of water, which has a relatively large specific heat capacity. This means that a large amount of energy must be transferred from the coals to change significantly the temperature of the feet. During the brief contact between feet and coals involved in firewalking, there is relatively little time for energy flow, so the feet do not reach a high enough temperature to cause damage.

Also, although the surface of the coals has a very high temperature, the red-hot layer is very thin. Therefore, the quantity of energy available to heat the feet is smaller than might be expected.

Thus, although firewalking is impressive, there are several scientific reasons why anyone with the proper training should be able to do it on a properly prepared bed of coals. (Don't try this on your own!)

AP Photo/Itsuo Inouye

A group of firewalkers in Japan.

How Do We Get There?

We can represent the data given in this problem by the following diagram:

| 1.6 g metal $T = 23\ °C$ | ⇨ 5.8 J | 1.6 g metal $T = 41\ °C$ |

$$\Delta T = 41\ °C - 23\ °C = 18\ °C$$

Using the data given, we can calculate the value of the specific heat capacity for the metal and compare this value to the one for gold given in Table 10.1. We know that

$$Q = s \times m \times \Delta T$$

or, pictorially,

| 1.6 g metal $T = 23\ °C$ | ⇨ 5.8 J = ? × 1.6 × 18 | 1.6 g metal $T = 41\ °C$ |

When we divide both sides of the equation

$$Q = s \times m \times \Delta T$$

by $m \times \Delta T$, we get

$$\frac{Q}{m \times \Delta T} = s$$

Thus, using the data given, we can calculate the value of s. In this case,

Q = energy (heat) required = 5.8 J

m = mass of the sample = 1.6 g

ΔT = change in temperature = 18 °C (41 °C − 23 °C = 18 °C)

Thus

$$s = \frac{Q}{m \times \Delta T} = \frac{5.8 \text{ J}}{(1.6 \text{ g})(18 \text{ °C})} = 0.20 \text{ J/g °C}$$

MATH SKILL BUILDER
The result you will get on your calculator is 5.8/(1.6 × 18) = 0.2013889, which rounds off to 0.20.

From Table 10.1, the specific heat capacity for gold is 0.13 J/g °C. Thus the metal must not be pure gold.

Self-Check EXERCISE 10.4   A 2.8-g sample of pure metal requires 10.1 J of energy to change its temperature from 21 °C to 36 °C. What is this metal? (Use Table 10.1.)

See Problems 10.31 through 10.36. ∎

10.6 Thermochemistry (Enthalpy)

OBJECTIVE: To consider the heat (enthalpy) of chemical reactions.

We have seen that some reactions are exothermic (produce heat energy) and other reactions are endothermic (absorb heat energy). Chemists also like to know exactly how much energy is produced or absorbed by a given reaction. To make that process more convenient, we have invented a special energy function called **enthalpy,** which is designated by H. For a reaction occurring under conditions of constant pressure, the change in enthalpy (ΔH) is equal to the energy that flows as heat. That is,

$$\Delta H_p = \text{heat}$$

where the subscript "p" indicates that the process has occurred under conditions of constant pressure and Δ means "a change in." Thus the enthalpy change for a reaction (that occurs at constant pressure) is the same as the heat for that reaction.

EXAMPLE 10.5 | Enthalpy

When 1 mole of methane (CH_4) is burned at constant pressure, 890 kJ of energy is released as heat. Calculate ΔH for a process in which a 5.8-g sample of methane is burned at constant pressure.

SOLUTION

Where Are We Going?

We want to determine ΔH for the reaction of 5.8 g of methane (CH_4) with oxygen at constant pressure.

What Do We Know?

- When 1 mol CH_4 is burned, 890 kJ of energy is released.

- We have 5.8 g of CH_4.

What Information Do We Need?

- Molar mass of methane, which we can get from the atomic masses of carbon (12.01 g/mol) and hydrogen (1.008 g/mol). The molar mass is 16.0 g/mol.

How Do We Get There?

At constant pressure, 890 kJ of energy per mole of CH_4 is produced as heat:

$$q_p = \Delta H = -890 \text{ kJ/mol } CH_4$$

Note that the minus sign indicates an exothermic process. In this case, a 5.8-g sample of CH_4 (molar mass = 16.0 g/mol) is burned. Since this amount is smaller than 1 mol, less than 890 kJ will be released as heat. The actual value can be calculated as follows:

$$5.8 \text{ g } CH_4 \times \frac{1 \text{ mol } CH_4}{16.0 \text{ g } CH_4} = 0.36 \text{ mol } CH_4$$

and

$$0.36 \text{ mol } CH_4 \times \frac{-890 \text{ kJ}}{\text{mol } CH_4} = -320 \text{ kJ}$$

Thus, when a 5.8-g sample of CH_4 is burned at constant pressure,

$$\Delta H = \text{heat flow} = -320 \text{ kJ}$$

REALITY CHECK   The mass of methane burned is less than 1 mol, so less than 890 kJ will be released as heat. The answer has two significant figures as required by the given quantities.

Self-Check EXERCISE 10.5   The reaction that occurs in the heat packs used to treat sports injuries is

$$4Fe(s) + 3O_2(g) \rightarrow 2Fe_2O_3(s) \qquad \Delta H = -1652 \text{ kJ}$$

How much heat is released when 1.00 g of Fe(s) is reacted with excess $O_2(g)$?

See Problems 10.40 and 10.41. ■

▶ Calorimetry

A **calorimeter** (see Figure 10.6) is a device used to determine the heat associated with a chemical reaction. The reaction is run in the calorimeter and the temperature change of the calorimeter is observed. Knowing the temperature change that occurs in the calorimeter and the heat capacity of the calorimeter enables us to calculate the heat energy released or absorbed by the reaction. Thus we can determine ΔH for the reaction.

Once we have measured the ΔH values for various reactions, we can use these data to *calculate* the ΔH values of other reactions. We will see how to carry out these calculations in the next section.

Thermometer

Styrofoam cover

Styrofoam cups

Stirrer

Figure 10.6

A coffee-cup calorimeter made of two Styrofoam cups.

Methane: An Important Energy Source

Methane is the main component of natural gas, a valuable fossil fuel. It is such a good fuel because the combustion of methane with oxygen

$$CH_4(g) + 2O_2(g) \rightarrow CO_2(g) + 2H_2O(g)$$

produces 55 kJ of energy per gram of methane. Natural gas, which is associated with petroleum deposits and contains as much as 97% methane, originated from the decomposition of plants in ancient forests that became buried in natural geological processes.

Although the methane in natural gas represents a tremendous source of energy for our civilization, an even more abundant source of methane lies in the depths of the ocean. The U.S. Geological Survey estimates that 320,000 trillion cubic feet of methane is trapped in the deep ocean near the United States. This amount is 200 times the amount of methane contained in the natural gas deposits in the United States. In the ocean, the methane is trapped in cavities formed by water molecules that are arranged very much like the water molecules in ice. These structures are called methane hydrates.

Although extraction of methane from the ocean floor offers tremendous potential benefits, it also carries risks. Methane is a "greenhouse gas"—its presence in the atmosphere helps to trap the heat from the sun. As a result, any accidental release of the methane from the ocean could produce serious warming of the earth's climate. As usual, environmental trade-offs accompany human activities.

John Pinkston and Laura Stern/USGS/Menlo Park

Flaming pieces of methane hydrate.

10.7 Hess's Law

OBJECTIVE: To understand Hess's law.

One of the most important characteristics of enthalpy is that it is a state function. That is, the change in enthalpy for a given process is independent of the pathway for the process. Consequently, *in going from a particular set of reactants to a particular set of products, the change in enthalpy is the same whether the reaction takes place in one step or in a series of steps.* This principle, which is known as **Hess's law,** can be illustrated by examining the oxidation of nitrogen to produce nitrogen dioxide. The overall reaction can be written in one step, where the enthalpy change is represented by ΔH_1.

$$N_2(g) + 2O_2(g) \rightarrow 2NO_2(g) \qquad \Delta H_1 = 68 \text{ kJ}$$

This reaction can also be carried out in two distinct steps, with the enthalpy changes being designated as ΔH_2 and ΔH_3:

$N_2(g) + O_2(g) \rightarrow 2NO(g)$	$\Delta H_2 = 180$ kJ
$2NO(g) + O_2(g) \rightarrow 2NO_2(g)$	$\Delta H_3 = -112$ kJ
Net reaction: $N_2(g) + 2O_2(g) \rightarrow 2NO_2(g)$	$\Delta H_2 + \Delta H_3 = 68$ kJ

Note that the sum of the two steps gives the net, or overall, reaction and that

$$\Delta H_1 = \Delta H_2 + \Delta H_3 = 68 \text{ kJ}$$

The importance of Hess's law is that it allows us to *calculate* heats of reaction that might be difficult or inconvenient to measure directly in a calorimeter.

▶ Characteristics of Enthalpy Changes

To use Hess's law to compute enthalpy changes for reactions, it is important to understand two characteristics of ΔH for a reaction:

1. If a reaction is reversed, the sign of ΔH is also reversed.

2. The magnitude of ΔH is directly proportional to the quantities of reactants and products in a reaction. If the coefficients in a balanced reaction are multiplied by an integer, the value of ΔH is multiplied by the same integer.

Both these rules follow in a straightforward way from the properties of enthalpy changes. The first rule can be explained by recalling that the *sign* of ΔH indicates the *direction* of the heat flow at constant pressure. If the direction of the reaction is reversed, the direction of the heat flow also will be reversed. To see this, consider the preparation of xenon tetrafluoride, which was the first binary compound made from a noble gas:

$$\text{Xe}(g) + 2\text{F}_2(g) \rightarrow \text{XeF}_4(s) \qquad \Delta H = -251 \text{ kJ}$$

This reaction is exothermic, and 251 kJ of energy flows into the surroundings as heat. On the other hand, if the colorless XeF_4 crystals are decomposed into the elements, according to the equation

$$\text{XeF}_4(s) \rightarrow \text{Xe}(g) + 2\text{F}_2(g)$$

the opposite energy flow occurs because 251 kJ of energy must be added to the system to produce this endothermic reaction. Thus, for this reaction, $\Delta H = +251 \text{ kJ}$.

The second rule comes from the fact that ΔH is an extensive property, depending on the amount of substances reacting. For example, since 251 kJ of energy is evolved for the reaction

$$\text{Xe}(g) + 2\text{F}_2(g) \rightarrow \text{XeF}_4(s)$$

then for a preparation involving twice the quantities of reactants and products, or

$$2\text{Xe}(g) + 4\text{F}_2(g) \rightarrow 2\text{XeF}_4(s)$$

twice as much heat would be evolved:

$$\Delta H = 2(-251 \text{ kJ}) = -502 \text{ kJ}$$

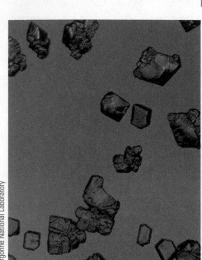

Crystals of xenon tetrafluoride, the first reported binary compound containing a noble gas element.

Argonne National Laboratory

EXAMPLE 10.6 | Hess's Law

Two forms of carbon are graphite, the soft, black, slippery material used in "lead" pencils and as a lubricant for locks, and diamond, the brilliant, hard gemstone. Using the enthalpies of combustion for graphite (−394 kJ/mol) and diamond (−396 kJ/mol), calculate ΔH for the conversion of graphite to diamond:

$$\text{C}_{\text{graphite}}(s) \rightarrow \text{C}_{\text{diamond}}(s)$$

SOLUTION

Where Are We Going?

We want to determine ΔH for the conversion of graphite to diamond.

What Do We Know?

The combustion reactions are

$$C_{graphite}(s) + O_2(g) \rightarrow CO_2(g) \qquad \Delta H = -394 \text{ kJ}$$
$$C_{diamond}(s) + O_2(g) \rightarrow CO_2(g) \qquad \Delta H = -396 \text{ kJ}$$

How Do We Get There?

Note that if we reverse the second reaction (which means we must change the sign of ΔH) and sum the two reactions, we obtain the desired reaction:

$$
\begin{aligned}
C_{graphite}(s) + O_2(g) &\rightarrow CO_2(g) & \Delta H &= -394 \text{ kJ} \\
CO_2(g) &\rightarrow C_{diamond}(s) + O_2(g) & \Delta H &= -(-396 \text{ kJ}) \\
\hline
C_{graphite}(s) &\rightarrow C_{diamond}(s) & \Delta H &= 2 \text{ kJ}
\end{aligned}
$$

Thus 2 kJ of energy is required to change 1 mole of graphite to diamond. This process is endothermic.

Self-Check EXERCISE 10.6   From the following information

$$S(s) + \tfrac{3}{2}O_2 \rightarrow SO_3(g) \qquad \Delta H = -395.2 \text{ kJ}$$
$$2SO_2(g) + O_2(g) \rightarrow 2SO_3(g) \qquad \Delta H = -198.2 \text{ kJ}$$

calculate ΔH for the reaction

$$S(s) + O_2(g) \rightarrow SO_2(g)$$

See Problems 10.45 through 10.48. ■

10.8   Quality Versus Quantity of Energy

OBJECTIVE: To see how the quality of energy changes as it is used.

One of the most important characteristics of energy is that it is conserved. Thus the total energy content of the universe will always be what it is now. If that is the case, why are we concerned about energy? For example, why should we worry about conserving our petroleum supply? Surprisingly, the "energy crisis" is not about the *quantity* of energy, but rather about the *quality* of energy. To understand this idea, consider an automobile trip from Chicago to Denver. Along the way you would put gasoline into the car to get to Denver. What happens to that energy? The energy stored in the bonds of the gasoline and of the oxygen that reacts with it is changed to thermal energy, which is spread along the highway to Denver. The total quantity of energy remains the same as before the trip but the energy concentrated in the gasoline becomes widely distributed in the environment:

$$\text{gasoline}(l) + O_2(g) \rightarrow CO_2(g) + H_2O(l) + \text{energy}$$

$\uparrow$ C_8H_{18} and other similar compounds

$\downarrow$ Spread along the highway, heating the road and the air

Which energy is easier to use to do work: the concentrated energy in the gasoline or the thermal energy spread from Chicago to Denver? Of course, the energy concentrated in the gasoline is more convenient to use.

This example illustrates a very important general principle: when we utilize energy to do work, we degrade its usefulness. In other words, when we use energy the *quality* of that energy (its ease of use) is lowered.

In summary,

Concentrated energy   Use the energy to do work   Spread energy

You may have heard someone mention the "heat death" of the universe. Eventually (many eons from now), all energy will be spread evenly throughout the universe and everything will be at the same temperature. At this point it will no longer be possible to do any work. The universe will be "dead."

We don't have to worry about the heat death of the universe anytime soon, of course, but we do need to think about conserving "quality" energy supplies. The energy stored in petroleum molecules got there over millions of years through plants and simple animals absorbing energy from the sun and using this energy to construct molecules. As these organisms died and became buried, natural processes changed them into the petroleum deposits we now access for our supplies of gasoline and natural gas.

Petroleum is highly valuable because it furnishes a convenient, concentrated source of energy. Unfortunately, we are using this fuel at a much faster rate than natural processes can replace it, so we are looking for new sources of energy. The most logical energy source is the sun. *Solar energy* refers to using the sun's energy directly to do productive work in our society. We will discuss energy supplies in the next section.

10.9 Energy and Our World

OBJECTIVE: To consider the energy resources of our world.

Woody plants, coal, petroleum, and natural gas provide a vast resource of energy that originally came from the sun. By the process of photosynthesis, plants store energy that can be claimed by burning the plants themselves or the decay products that have been converted over millions of years to **fossil fuels.** Although the United States currently depends heavily on petroleum for energy, this dependency is a relatively recent phenomenon, as shown in Figure 10.7. In this section we discuss some sources of energy and their effects on the environment.

▶ Petroleum and Natural Gas

Although how they were produced is not completely understood, petroleum and natural gas were most likely formed from the remains of marine organisms that lived approximately 500 million years ago. **Petroleum** is a thick, dark liquid composed mostly of compounds called *hydrocarbons* that contain carbon and hydrogen. (Carbon is unique among elements in the extent to which it can bond to itself to form chains of various lengths.) Table 10.2 gives the formulas and names for several common hydrocarbons. **Natural**

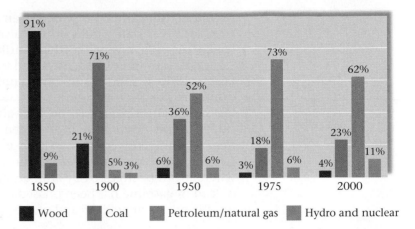

Figure 10.7

Energy sources used in the United States.

gas, usually associated with petroleum deposits, consists mostly of methane, but it also contains significant amounts of ethane, propane, and butane.

The composition of petroleum varies somewhat, but it includes mostly hydrocarbons having chains that contain from 5 to more than 25 carbons. To be used efficiently, the petroleum must be separated into fractions by boiling. The lighter molecules (having the lowest boiling points) can be boiled off, leaving the heavier ones behind. The commercial uses of various petroleum fractions are shown in Table 10.3.

The petroleum era began when the demand for lamp oil during the Industrial Revolution outstripped the traditional sources: animal fats and whale oil. In response to this increased demand, Edwin Drake drilled the first oil well in 1859 at Titusville, Pennsylvania. The petroleum from this well was refined to produce *kerosene* (fraction C_{10}–C_{18}), which served as an excellent lamp oil. *Gasoline* (fraction C_5–C_{10}) had limited use and was often discarded. This situation soon changed. The development of the electric light decreased the need for kerosene, and the advent of the "horseless carriage" with its gasoline-powered engine signaled the birth of the gasoline age.

As gasoline became more important, new ways were sought to increase the yield of gasoline obtained from each barrel of petroleum. William Burton invented a process at Standard Oil of Indiana called *pyrolytic (high-temperature) cracking.* In this process, the heavier molecules of the kerosene fraction are heated to about 700 °C, causing them to break (crack) into the smaller molecules of hydrocarbons in the gasoline fraction. As cars became larger, more efficient internal combustion engines were designed. Because of the uneven burning of the gasoline then available, these engines "knocked,"

Table 10.2 Names and Formulas for Some Common Hydrocarbons

Formula	Name
CH_4	Methane
C_2H_6	Ethane
C_3H_8	Propane
C_4H_{10}	Butane
C_5H_{12}	Pentane
C_6H_{14}	Hexane
C_7H_{16}	Heptane
C_8H_{18}	Octane

Table 10.3 Uses of the Various Petroleum Fractions

Petroleum Fraction in Terms of Numbers of Carbon Atoms	Major Uses
C_5–C_{10}	Gasoline
C_{10}–C_{18}	Kerosene
	Jet fuel
C_{15}–C_{25}	Diesel fuel
	Heating oil
	Lubricating oil
$>C_{25}$	Asphalt

producing unwanted noise and even engine damage. Intensive research to find additives that would promote smoother burning produced tetraethyl lead, $(C_2H_5)_4Pb$, a very effective "antiknock" agent.

The addition of tetraethyl lead to gasoline became a common practice, and by 1960, gasoline contained as much as 3g of lead per gallon. As we have discovered so often in recent years, technological advances can produce environmental problems. To prevent air pollution from automobile exhaust, catalytic converters have been added to car exhaust systems. The effectiveness of these converters, however, is destroyed by lead. The use of leaded gasoline also greatly increased the amount of lead in the environment, where it can be ingested by animals and humans. For these reasons, the use of lead in gasoline has been phased out, requiring extensive (and expensive) modifications of engines and of the gasoline refining process.

▶ Coal

Coal was formed from the remains of plants that were buried and subjected to high pressure and heat over long periods of time. Plant materials have a high content of cellulose, a complex molecule whose empirical formula is CH_2O but whose molar mass is approximately 500,000 g/mol. After the plants and trees that grew on the earth at various times and places died and were buried, chemical changes gradually lowered the oxygen and hydrogen content of the cellulose molecules. Coal "matures" through four stages: lignite, subbituminous, bituminous, and anthracite. Each stage has a higher carbon-to-oxygen and carbon-to-hydrogen ratio; that is, the relative carbon content gradually increases. Typical elemental compositions of the various coals are given in Table 10.4. The energy available from the combustion of a given mass of coal increases as the carbon content increases. Anthracite is the most valuable coal, and lignite is the least valuable.

Coal is an important and plentiful fuel in the United States, currently furnishing approximately 20% of our energy. As the supply of petroleum decreases, the share of the energy supply from coal could eventually increase to as high as 30%. However, coal is expensive and dangerous to mine underground, and the strip mining of fertile farmland in the Midwest or of scenic land in the West causes obvious problems. In addition, the burning of coal, especially high-sulfur coal, yields air pollutants such as sulfur dioxide, which, in turn, can lead to acid rain. However, even if coal were pure carbon, the carbon dioxide produced when it was burned would still have significant effects on the earth's climate.

Roland Weinrauch/dpa/Corbis

Table 10.4	Element Composition of Various Types of Coal				
	Mass Percent of Each Element				
Type of Coal	C	H	O	N	S
Lignite	71	4	23	1	1
Subbituminous	77	5	16	1	1
Bituminous	80	6	8	1	5
Anthracite	92	3	3	1	1

▶ **Effects of Carbon Dioxide on Climate**

The earth receives a tremendous quantity of radiant energy from the sun, about 30% of which is reflected back into space by the earth's atmosphere. The remaining energy passes through the atmosphere to the earth's surface. Some of this energy is absorbed by plants for photosynthesis and some by the oceans to evaporate water, but most of it is absorbed by soil, rocks, and water, increasing the temperature of the earth's surface. This energy is, in turn, radiated from the heated surface mainly as *infrared radiation,* often called *heat radiation.*

The atmosphere, like window glass, is transparent to visible light but does not allow all the infrared radiation to pass back into space. Molecules in the atmosphere, principally H_2O and CO_2, strongly absorb infrared radiation and radiate it back toward the earth, as shown in Figure 10.8. A net amount of thermal energy is retained by the earth's atmosphere, causing the earth to be much warmer than it would be without its atmosphere. In a way, the atmosphere acts like the glass of a greenhouse, which is transparent to visible light but absorbs infrared radiation, thus raising the temperature inside the building. This **greenhouse effect** is seen even more spectacularly on Venus, where the dense atmosphere is thought to be responsible for the high surface temperature of that planet.

Thus the temperature of the earth's surface is controlled to a significant extent by the carbon dioxide and water content of the atmosphere. The effect of atmospheric moisture (humidity) is readily apparent in the Midwest, for example. In summer, when the humidity is high, the heat of the sun is retained well into the night, giving very high nighttime temperatures. In winter, the coldest temperatures always occur on clear nights, when the low humidity allows efficient radiation of energy back into space.

The atmosphere's water content is controlled by the water cycle (evaporation and precipitation), and the average has remained constant over the years. However, as fossil fuels have been used more extensively, the carbon dioxide concentration has increased—up about 20% from 1880 to the present. Projections indicate that the carbon dioxide content of the atmosphere

Figure 10.8

The earth's atmosphere is transparent to visible light from the sun. This visible light strikes the earth, and part of it is changed to infrared radiation. The infrared radiation from the earth's surface is strongly absorbed by CO_2, H_2O, and other molecules present in smaller amounts (for example, CH_4 and N_2O) in the atmosphere. In effect, the atmosphere traps some of the energy, acting like the glass in a greenhouse and keeping the earth warmer than it would otherwise be.

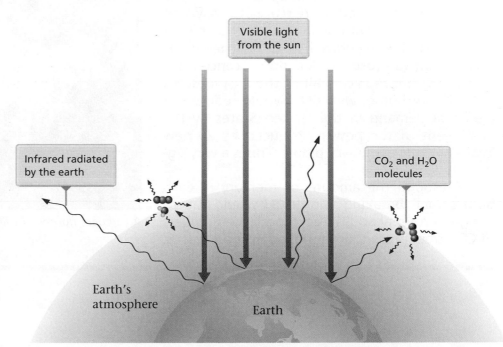

Visible light from the sun

Infrared radiated by the earth

CO_2 and H_2O molecules

Earth's atmosphere

Earth

Seeing the Light

We are about to have a revolution in lighting. The incandescent light bulb developed by Thomas Edison in the late nineteenth century still dominates our lighting systems. However, this is about to change because Edison's light bulb is so inefficient: about 95% of the energy goes to heat instead of light. In the United States, 22% of total electricity production goes for lighting, for a cost of about $58 million. Globally, illumination consumes about 19% of electricity, and demand for lighting is expected to grow by 60% in the next 25 years. Given energy prices and the problems associated with global warming, we must find more efficient lighting devices.

In the short term, the answer appears to be compact fluorescent lights (CFLs). These bulbs, which have a screw-type base, draw only about 20% as much energy as incandescent bulbs for a comparable amount of light production. Although they cost four times as much, CFLs last ten times as long as incandescent bulbs. CFLs produce light from a type of compound called a phosphor that coats the inner walls of the bulb. The phosphor is mixed with a small amount of mercury (about 5 mg per bulb). When the bulb is turned on, a beam of electrons is produced. The electrons are absorbed by mercury atoms, which are caused to emit ultraviolet (UV) light. This UV light is absorbed by phosphor, which then emits visible light (a process called fluorescence). It is estimated that replacing all of the incandescent bulbs in our homes with CFLs would reduce our electrical demand in the United States by the equivalent of the power produced by 20 new 1000-MW nuclear power plants. This is a very significant savings.

Although the amount of mercury in each bulb is small (breaking a single CFL would not endanger a normal adult), recycling large numbers of CFLs does present potential pollution hazards.

Research is now underway to find ways to alleviate this danger. For example, Professor Robert Hurt and his colleagues at Brown University have found that selenium prepared as tiny particles has a very high affinity for mercury and can be used in recycling operations to prevent dangerous occupational exposure to mercury.

Another type of lighting device that looks to be very valuable in the near future is the light-emitting diode (LED). An LED is a solid-state semiconductor designed to emit visible light when its electrons fall to lower energy levels. The tiny glowing light that indicates an audio system or television is on is an LED. In recent years, LEDs have been used in traffic lights, turn signals on cars, flashlights, and street lights. The use of LEDs for holiday lighting is rapidly increasing. It is estimated that LEDs eventually will reduce energy consumption for holiday lighting by 90%. The light production of LEDs per amount of energy consumed has increased dramatically in recent months, and the costs are decreasing steadily. Currently, LED lights are ten times more expensive than CFLs but last more than 15 years. Thus dramatic changes are occurring in the methods for lighting, and we all need to do our part to make our lives more energy efficient.

A compact fluorescent light bulb (CFL).

Figure 10.9

The atmospheric CO_2 concentration over the past 1000 years, based on ice core data and direct readings (since 1958). Note the dramatic increase in the past 100 years.

may be double in the twenty-first century what it was in 1880. This trend *could* increase the earth's average temperature by as much as 10 °C, causing dramatic changes in climate and greatly affecting the growth of food crops.

How well can we predict the long-term effects of carbon dioxide? Because weather has been studied for a period of time that is minuscule compared with the age of the earth, the factors that control the earth's climate in the long range are not clearly understood. For example, we do not understand what causes the earth's periodic ice ages. So it is difficult to estimate the effects of the increasing carbon dioxide levels.

In fact, the variation in the earth's average temperature over the past century is somewhat confusing. In the northern latitudes during the past century, the average temperature rose by 0.8 °C over a period of 60 years, then cooled by 0.5 °C during the next 25 years, and finally warmed by 0.2 °C in the succeeding 15 years. Such fluctuations do not match the steady increase in carbon dioxide. However, in southern latitudes and near the equator during the past century, the average temperature showed a steady rise totaling 0.4 °C. This figure is in reasonable agreement with the predicted effect of the increasing carbon dioxide concentration over that period. Another significant fact is that the last 10 years of the twentieth century have been the warmest decade on record.

Although the exact relationship between the carbon dioxide concentration in the atmosphere and the earth's temperature is not known at present, one thing is clear: The increase in the atmospheric concentration of carbon dioxide is quite dramatic (see Figure 10.9). We must consider the implications of this increase as we consider our future energy needs.

▶ **New Energy Sources**

As we search for the energy sources of the future, we need to consider economic, climatic, and supply factors. There are several potential energy sources: the sun (solar), nuclear processes (fission and fusion), biomass (plants), and synthetic fuels. Direct use of the sun's radiant energy to heat our homes and run our factories and transportation systems seems a sensible long-term goal. But what do we do now? Conservation of fossil fuels is one obvious step, but substitutes for fossil fuels also must be found. There is much research going on now to solve this problem.

10.10 Energy as a Driving Force

OBJECTIVE: To understand energy as a driving force for natural processes.

A major goal of science is to understand why things happen as they do. In particular, we are interested in the driving forces of nature. Why do things occur in a particular direction? For example, consider a log that has burned in a fireplace, producing ashes and heat energy. If you are sitting in front of the fireplace, you would be very surprised to see the ashes begin to absorb heat from the air and reconstruct themselves into the log. It just doesn't happen. That is, the process that always occurs is

$$log + O_2(g) \rightarrow CO_2(g) + H_2O(g) + ashes + energy$$

The reverse of this process

$$CO_2(g) + H_2O(g) + ashes + energy \rightarrow log + O_2(g)$$

never happens.

Consider another example. A gas is trapped in one end of a vessel as shown below.

When the valve is opened, what always happens? The gas spreads evenly throughout the entire container.

You would be very surprised to see the following process occur spontaneously:

So, why does this process

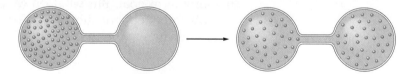

occur spontaneously but the reverse process

never occurs?

In many years of analyzing these and many other processes, scientists have discovered two very important driving forces:

- Energy spread
- Matter spread

Energy spread means that in a given process, concentrated energy is dispersed widely. This distribution happens every time an exothermic process occurs. For example, when a Bunsen burner burns, the energy stored in the fuel (natural gas—mostly methane) is dispersed into the surrounding air:

The energy that flows into the surroundings through heat increases the thermal motions of the molecules in the surroundings. In other words, this process increases the random motions of the molecules in the surroundings. *This always happens in every exothermic process.*

Matter spread means exactly what it says: the molecules of a substance are spread out and occupy a larger volume.

After looking at thousands of processes, scientists have concluded that these two factors are the important driving forces that cause events to occur. That is, processes are favored if they involve energy spread and matter spread.

Do these driving forces ever occur in opposition? Yes, they do—in many, many processes.

For example, consider ordinary table salt dissolving in water.

This process occurs spontaneously. You observe it every time you add salt to water to cook potatoes or pasta. Surprisingly, dissolving salt in water is *endothermic*. This process seems to go in the wrong direction—it involves energy concentration, not energy spread. Why does the salt dissolve? Because of matter spread. The Na^+ and Cl^- that are closely packed in the solid NaCl become spread around randomly in a much larger volume in the resulting solution. Salt dissolves in water because the favorable matter spread overcomes an unfavorable energy change.

▶ Entropy

Entropy is a function we have invented to keep track of the natural tendency for the components of the universe to become disordered—entropy (designated by the letter S) is a measure of disorder or randomness. As randomness increases, S increases. Which has lower entropy, solid water (ice) or gaseous water (steam)? Remember that ice contains closely packed, ordered H_2O molecules, and steam has widely dispersed, randomly moving H_2O molecules (see Figure 10.10). Thus ice has more order and a lower value of S.

What do you suppose happens to the disorder of the universe as energy spread and matter spread occur during a process?

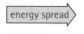 Faster random motions of the molecules in surroundings

 Components of matter are dispersed—they occupy a larger volume

It seems clear that both energy spread and matter spread lead to greater entropy (greater disorder) in the universe. This idea leads to a very important conclusion that is summarized in the **second law of thermodynamics:**

The entropy of the universe is always increasing.

A **spontaneous process** is one that occurs in nature without outside intervention—it happens "on its own." The second law of thermodynamics

Figure 10.10

Comparing the entropies of ice and steam.

Solid (Ice)   **Gas** (Steam)

helps us to understand why certain processes are spontaneous and others are not. It also helps us to understand the conditions necessary for a process to be spontaneous. For example, at 1 atm (1 atmosphere of pressure), ice will spontaneously melt above a temperature of 0 °C but not below this temperature. A process is spontaneous only if the entropy of the universe increases as a result of the process. That is, all processes that occur in the universe lead to a net increase in the disorder of the universe. As the universe "runs," it is always heading toward more disorder. We are plunging slowly but inevitably toward total randomness—the heat death of the universe. But don't despair; it will not happen soon.

CHAPTER 10 REVIEW

Key Terms

energy (10.1)	joule (10.5)
potential energy (10.1)	specific heat
kinetic energy (10.1)	capacity (10.5)
law of conservation of	enthalpy (10.6)
energy (10.1)	calorimeter (10.6)
work (10.1)	Hess's law (10.7)
state function (10.1)	fossil fuels (10.9)
temperature (10.2)	petroleum (10.9)
heat (10.2)	natural gas (10.9)
system (10.3)	coal (10.9)
surroundings (10.3)	greenhouse effect (10.9)
exothermic (10.3)	energy spread (10.10)
endothermic (10.3)	matter spread (10.10)
thermodynamics (10.4)	entropy (10.10)
first law of thermo-	second law of thermo-
dynamics (10.4)	dynamics (10.10)
internal energy (10.4)	spontaneous process
calorie (10.5)	(10.10)

Summary

1. One of the fundamental characteristics of energy is that it is conserved. Energy is changed in form but it is not produced or consumed in a process. Thermodynamics is the study of energy and its changes.

2. In a process some functions—called state functions—depend only on the beginning and final states of the system, not on the specific pathway followed. Energy is a state function. Other functions, such as heat and work, depend on the specific pathway followed and are not state functions.

3. The temperature of a substance indicates the vigor of the random motions of the components of that substance. The thermal energy of an object is the energy content of the object as produced by its random motions.

4. Heat is a flow of energy between two objects due to a temperature difference in the two objects. In an exothermic reaction, energy as heat flows out of the system into its surroundings. In an endothermic process, energy as heat flows from the surroundings into the system.

5. The internal energy of an object is the sum of the kinetic (due to motion) and potential (due to position) energies of the object. Internal energy can be changed by two types of energy flows, work (w) and heat (q): $\Delta E = q + w$.

6. A calorimeter is used to measure the heats of chemical reactions. The common units for heat are joules and calories.

7. The specific heat capacity of a substance (the energy required to change the temperature of one gram of the substance by one Celsius degree) is used to calculate temperature changes when a substance is heated.

8. The change in enthalpy for a process is equal to the heat for that process run at constant pressure.

9. Hess's law allows the calculation of the heat of a given reaction from known heats of related reactions.

10. Although energy is conserved in every process, the quality (usefulness) of the energy decreases with each use.

11. Our world has many sources of energy. The use of these sources affects the environment in various ways.

12. Natural processes occur in the direction that leads to an increase in the disorder (entropy) of the universe. The principal driving forces for processes are energy spread and matter spread.

Active Learning Questions

These questions are designed to be considered by groups of students in class. Often these questions work well for introducing a particular topic in class.

1. Look at Figure 10.1 in your text. Ball A has stopped moving. However, energy must be conserved. So what happened to the energy of ball A?

2. A friend of yours reads that the process of water freezing is exothermic. This friend tells you that this can't be true because exothermic implies "hot," and ice is cold. Is the process of water freezing exothermic? If so, explain this process so your friend can understand it. If not, explain why not.

3. You place hot metal into a beaker of cold water.

   a. Eventually what is true about the temperature of the metal compared to that of the water? Explain why this is true.
   b. Label this process as endothermic or exothermic if we consider the system to be

      i. the metal. Explain.
      ii. the water. Explain.

4. What does it mean when the heat for a process is reported with a negative sign?

5. You place 100.0 g of a hot metal in 100.0 g of cold water. Which substance (metal or water) undergoes a larger temperature change? Why is this?

6. Explain why aluminum cans make good storage containers for soft drinks. Styrofoam cups can be used to keep coffee hot and cola cold. How can this be?

7. In Section 10.7, two characteristics of enthalpy changes for reactions are listed. What are these characteristics? Explain why these characteristics are true.

8. What is the difference between *quality* and *quantity* of energy? Are both conserved? Is either conserved?

9. What is meant by the term *driving forces?* Why are *matter spread* and *energy spread* considered to be driving forces?

10. Give an example of a process in which *matter spread* is a driving force and an example of a process in which *energy spread* is a driving force, and explain each. These examples should be different from the ones given in the text.

11. Explain in your own words what is meant by the term *entropy.* Explain how both *matter spread* and *energy spread* are related to the concept of entropy.

12. Consider the processes

    $$H_2O(g) \rightarrow H_2O(l)$$
    $$H_2O(l) \rightarrow H_2O(g)$$

    a. Which process is favored by energy spread? Explain.
    b. Which process is favored by matter spread? Explain.
    c. How does temperature affect which process is favored? Explain.

13. What if energy was not conserved? How would this affect our lives?

14. The internal energy of a system is said to be the sum of the kinetic and potential energies of all the particles in the system. Section 10.1 discusses *potential energy* and *kinetic energy* in terms of a ball on a hill. Explain *potential energy* and *kinetic energy* for a chemical reaction.

15. Hydrogen gas and oxygen gas react violently to form water.

    a. Which is lower in energy: a mixture of hydrogen gases, or water? Explain.
    b. Sketch an energy-level diagram (like Figure 10.5) for this reaction and explain it.

16. Consider four 100.0-g samples of water, each in a separate beaker at 25.0 °C. Into each beaker you drop 10.0 g of a different metal that has been heated to 95.0 °C. Assuming no heat loss to the surroundings, which water sample will have the highest final temperature? Explain your answer.

    a. The water to which you have added aluminum (c = 0.89 J/g °C).
    b. The water to which you have added iron (c = 0.45 J/g °C).
    c. The water to which you have added copper (c = 0.20 J/g °C).
    d. The water to which you have added lead (c = 0.14 J/g °C).
    e. Because the masses of the metals are the same, the final temperatures would be the same.

17. For each of the following situations a–c, use the following choices i–iii to complete the statement "The final temperature of the water should be"

    i. Between 50 °C and 90 °C
    ii. 50 °C
    iii. Between 10 °C and 50 °C

    a. A 100.0-g sample of water at 90 °C is added to a 100.0-g sample of water at 10 °C.
    b. A 100.0-g sample of water at 90 °C is added to a 500.0-g sample of water at 10 °C.
    c. You have a Styrofoam cup with 50.0 g of water at 10 °C. You add a 50.0-g iron ball at 90 °C to the water.

18. How is Hess's law a restatement of the first law of thermodynamics?

19. Does the entropy of the system increase or decrease for each of the following? Explain.

    a. the evaporation of alcohol
    b. the freezing of water
    c. dissolving NaCl in water

20. Predict the sign of $\Delta S°$ for each of the following changes.

a.

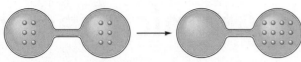

b. $AgCl(s) \rightarrow Ag^+(aq) + Cl^-(aq)$
c. $2H_2(g) + O_2 \rightarrow 2H_2O(l)$
d. $H_2O(l) \rightarrow H_2O(g)$

OWL Questions and problems for this chapter are available in OWL. Use the OWL access card that came with this textbook to access assigned questions and problems for this chapter.

11

Modern Atomic Theory

● The Aurora Australis from space. The colors are due to spectral emissions of nitrogen and oxygen. *(ISS-NASA/Science Faction)*

The concept of atoms is a very useful one. It explains many important observations, such as why compounds always have the same composition (a specific compound always contains the same types and numbers of atoms) and how chemical reactions occur (they involve a rearrangement of atoms).

Once chemists came to "believe" in atoms, a logical question followed: What are atoms like? What is the structure of an atom? In Chapter 4 we learned to picture the atom with a positively charged nucleus composed of protons and neutrons at its center and electrons moving around the nucleus in a space very large compared to the size of the nucleus.

In this chapter we will look at atomic structure in more detail. In particular, we will develop a picture of the electron arrangements in atoms—a picture that allows us to account for the chemistry of the various elements. Recall from our discussion of the periodic table in Chapter 4 that, although atoms exhibit a great variety of characteristics, certain elements can be grouped together because they behave similarly. For example, fluorine, chlorine, bromine, and iodine (the halogens) show great chemical similarities. Likewise, lithium, sodium, potassium, rubidium, and cesium (the alkali metals) exhibit many similar properties, and the noble gases (helium, neon, argon, krypton, xenon, and radon) are all very nonreactive. Although the members of each of these groups of elements show great similarity *within* the group, the differences in behavior *between* groups are striking. In this chapter we will see that it is the way the electrons are arranged in various atoms that accounts for these facts.

Alkali metals   Halogens   Noble gases

Owaki-Kulla/Corbis

A neon sign celebrating Route 66.

11.1 Rutherford's Atom

OBJECTIVE: To describe Rutherford's model of the atom.

Remember that in Chapter 4 we discussed the idea that an atom has a small positive core (called the nucleus) with negatively charged electrons moving around the nucleus in some way (Figure 11.1). This concept of a *nuclear atom* resulted from Ernest Rutherford's experiments in which he bombarded metal foil with α particles (see Section 4.5). Rutherford and his coworkers were able to show that the nucleus of the atom is composed of positively charged particles called *protons* and neutral particles called *neutrons*. Rutherford also found that the nucleus is apparently very small compared to the size of the entire atom. The electrons account for the rest of the atom.

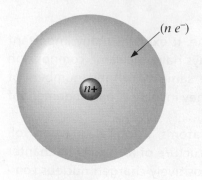

$(n\ e^-)$

$n+$

Figure 11.1

Rutherford's atom. The nuclear charge ($n+$) is balanced by the presence of n electrons moving in some way around the nucleus.

A major question left unanswered by Rutherford's work was, What are the electrons doing? That is, how are the electrons arranged and how do they move? Rutherford suggested that electrons might revolve around the nucleus like the planets revolve around the sun in our solar system. He couldn't explain, however, why the negative electrons aren't attracted into the positive nucleus, causing the atom to collapse.

At this point it became clear that more observations of the properties of atoms were needed to understand the structure of the atom more fully. To help us understand these observations, we need to discuss the nature of light and how it transmits energy.

11.2 Electromagnetic Radiation

OBJECTIVE: To explore the nature of electromagnetic radiation.

If you hold your hand a few inches from a brightly glowing light bulb, what do you feel? Your hand gets warm. The "light" from the bulb somehow transmits energy to your hand. The same thing happens if you move close to the glowing embers of wood in a fireplace—you receive energy that makes you feel warm. The energy you feel from the sun is a similar example.

In all three of these instances, energy is being transmitted from one place to another by light—more properly called **electromagnetic radiation.** Many kinds of electromagnetic radiation exist, including the X rays used to make images of bones, the "white" light from a light bulb, the microwaves used to cook hot dogs and other food, and the radio waves that transmit voices and music. How do these various types of electromagnetic radiation differ from one another? To answer this question we need to talk about waves. To explore the characteristics of waves, let's think about ocean waves. In Figure 11.2 a seagull is shown floating on the ocean and being raised and lowered by the motion of the water surface as waves pass by. Notice that the gull just moves up and down as the waves pass—it is not moved forward. A particular wave is characterized by three properties: *wavelength, frequency,* and *speed.*

Figure 11.2

A seagull floating on the ocean moves up and down as waves pass.

The **wavelength** (symbolized by the Greek letter lambda, λ) is the distance between two consecutive wave peaks (see Figure 11.3). The **frequency** of the wave (symbolized by the Greek letter nu, ν) indicates how many wave peaks pass a certain point per given time period. This idea can best be understood by thinking about how many times the seagull in Figure 11.2 goes up and down per minute. The *speed* of a wave indicates how fast a given peak travels through the water.

Although it is more difficult to picture than water waves, light (electromagnetic radiation) also travels as waves. The various types of electromagnetic radiation (X rays, microwaves, and so on) differ in their wavelengths. The classes of electromagnetic radiation are shown in Figure 11.4. Notice that X rays have very short wavelengths, whereas radiowaves have very long wavelengths.

λ

Figure 11.3

The wavelength of a wave is the distance between peaks.

Light as a Sex Attractant

Parrots, which are renowned for their vibrant colors, apparently have a secret weapon that enhances their colorful appearance—a phenomenon called fluorescence. Fluorescence occurs when a substance absorbs ultraviolet (UV) light, which is invisible to the human eye, and converts it to visible light. This phenomenon is widely used in interior lighting in which long tubes are coated with a fluorescent substance. The fluorescent coating absorbs UV light (produced in the interior of the tube) and emits intense white light, which consists of all wavelengths of visible light.

Interestingly, scientists have shown that parrots have fluorescent feathers that are used to attract the opposite sex. Note in the accompanying photos that a budgerigar parrot has certai

feathers that produce fluorescence. Kathryn E. Arnold of the University of Glasgow in Scotland examined the skins of 700 Australian parrots from museum collections and found that the feathers that showed fluorescence were always display feathers—ones that were fluffed or waggled during courtship. To test her theory that fluorescence is a significant aspect of parrot romance, Arnold studied the behavior of a parrot toward birds of the opposite sex. In some cases, the potential mate had a UV-blocking substance applied to its feathers, blocking its fluorescence. Arnold's study revealed that parrots always preferred partners that showed fluorescence over those in which the fluorescence was blocked. Perhaps on your next date you might consider wearing a shirt with some fluorescent decoration!

The back and front of a budgerigar parrot. In the photo at the right, the same parrot is seen under ultraviolet light.

Andrey K. Geim/High Field Magnet Laboratory/University of Nijmegen

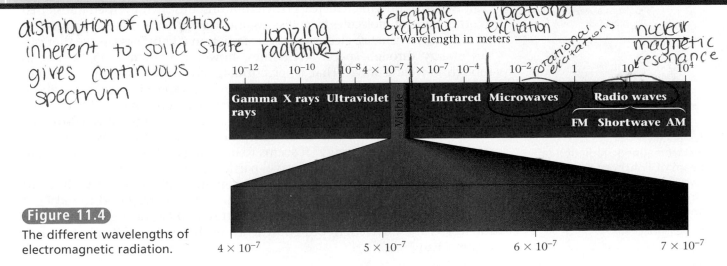

Figure 11.4

The different wavelengths of electromagnetic radiation.

Atmospheric Effects

The gaseous atmosphere of the earth is crucial to life in many different ways. One of the most important characteristics of the atmosphere is the way its molecules absorb radiation from the sun.

If it weren't for the protective nature of the atmosphere, the sun would "fry" us with its high-energy radiation. We are protected by the atmospheric ozone, a form of oxygen consisting of O_3 molecules, which absorbs high-energy radiation and thus prevents it from reaching the earth. This explains why we are so concerned that chemicals released into the atmosphere are destroying this high-altitude ozone.

The atmosphere also plays a central role in controlling the earth's temperature, a phenomenon called the *greenhouse effect*. The atmospheric gases CO_2, H_2O, CH_4, N_2O, and others do not absorb light in the visible region. Therefore, the visible light from the sun passes through the atmosphere to warm the earth. In turn, the earth radiates this energy back toward space as infrared radiation. (For example, think of the heat radiated from black asphalt on a hot summer day.) But the gases listed earlier are strong absorbers of *infrared* waves, and they reradiate some of this energy back toward the earth as shown in Figure 11.7. Thus these gases act as an insulating blanket keeping the earth much warmer than it would be without them. (If these gases were not present, all of the heat the earth radiates would be lost into space.)

However, there is a problem. When we burn fossil fuels (coal, petroleum, and natural gas), one of the products is CO_2. Because we use such huge quantities of fossil fuels, the CO_2 content in the atmosphere is increasing gradually but significantly. This should cause the earth to get warmer, eventually changing the weather patterns on the earth's surface and melting the polar ice caps, which would flood many low-lying areas.

Because the natural forces that control the earth's temperature are not very well understood at this point, it is difficult to decide whether the greenhouse warming has already started. But many scientists think it has. For example, the 1980s and 1990s were among the warmest years the earth has experienced since people started keeping records. Also, studies at the Scripps Institution of Oceanography indicate that the average temperatures of surface waters in the world's major oceans have risen since the 1960s in close

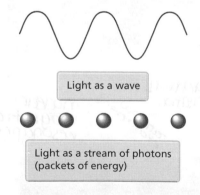

Figure 11.5

Electromagnetic radiation (a beam of light) can be pictured in two ways: as a wave and as a stream of individual packets of energy called photons.

Radiation provides an important means of energy transfer. For example, the energy from the sun reaches the earth mainly in the forms of visible and ultraviolet radiation. The glowing coals of a fireplace transmit heat energy by infrared radiation. In a microwave oven, the water molecules in food absorb microwave radiation, which increases their motions; this energy is then transferred to other types of molecules by collisions, increasing the food's temperature.

Thus we visualize electromagnetic radiation ("light") as a wave that carries energy through space. Sometimes, however, light doesn't behave as though it were a wave. That is, electromagnetic radiation can sometimes have properties that are characteristic of particles. (You will learn more about this idea in later courses.) Another way to think of a beam of light traveling through space, then, is as a stream of tiny packets of energy called **photons.**

What is the exact nature of light? Does it consist of waves or is it a stream of particles of energy? It seems to be both (see Figure 11.5). This situation is often referred to as the wave–particle nature of light.

Different wavelengths of electromagnetic radiation carry different amounts of energy. For example, the photons that correspond to red light carry less energy than the photons that correspond to blue light. In general, the longer the wavelength of light, the lower the energy of its photons (see Figure 11.6).

agreement with the predictions of models based on the increase in CO_2 concentrations. Studies also show that Arctic sea ice, the Greenland Ice Sheet, and various glaciers are melting much faster in recent years. These changes indicate that global warming is occurring.

The greenhouse effect is something we must watch closely. Controlling it may mean lowering our dependence on fossil fuels and increasing our reliance on nuclear, solar, or other power sources. In recent years, the trend has been in the opposite direction.

NASA

A composite satellite image of the earth's biomass constructed from the radiation given off by living matter over a multiyear period.

Figure 11.7

Certain gases in the earth's atmosphere absorb and re-emit some of the infrared (heat) radiation produced by the earth. This keeps the earth warmer than it would be otherwise.

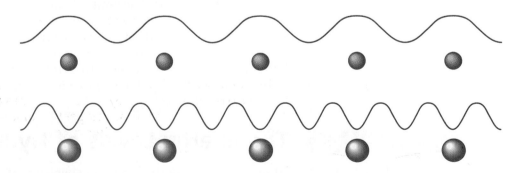

Figure 11.6

A photon of red light (relatively long wavelength) carries less energy than does a photon of blue light (relatively short wavelength).

11.3 Emission of Energy by Atoms

OBJECTIVE: To see how atoms emit light.

Consider the results of the experiment shown on page 328. This experiment is run by dissolving compounds containing the Li^+ ion, the Cu^{2+} ion, and the Na^+ ion in separate dishes containing methyl alcohol (with a little water added to help dissolve the compounds). The solutions are then set on fire.

Quantum concept applied

Figure 11.8

An excited lithium atom emitting a photon of red light to drop to a lower energy state.

When salts containing Li^+, Cu^{2+}, and Na^+ dissolved in methyl alcohol are set on fire, brilliant colors result: Li^+, red; Cu^{2+}, green; and Na^+, yellow.

Notice the brilliant colors that result. The solution containing Li^+ gives a beautiful, deep-red color, while the Cu^{2+} solution burns green. Notice that the Na^+ solution burns with a yellow–orange color, a color that should look familiar to you from the lights used in many parking lots. The color of these "sodium vapor lights" arises from the same source (the sodium atom) as the color of the burning solution containing Na^+ ions.

As we will see in more detail in the next section, the colors of these flames result from atoms in these solutions releasing energy by emitting visible light of specific wavelengths (that is, specific colors). The heat from the flame causes the atoms to absorb energy—we say that the atoms become *excited*. Some of this excess energy is then released in the form of light. The atom moves to a lower energy state as it emits a photon of light.

Lithium emits red light because its energy change corresponds to photons of red light (see Figure 11.8). Copper emits green light because it undergoes a different energy change than lithium; the energy change for copper corresponds to the energy of a photon of green light. Likewise, the energy change for sodium corresponds to a photon with a yellow–orange color.

To summarize, we have the following situation. When atoms receive energy from some source—they become excited—they can release this energy by emitting light. The emitted energy is carried away by a photon. Thus the energy of the photon corresponds exactly to the energy change experienced by the emitting atom. High-energy photons correspond to short-wavelength light and low-energy photons correspond to long-wavelength light. The photons of red light therefore carry less energy than the photons of blue light because red light has a longer wavelength than blue light does.

11.4   The Energy Levels of Hydrogen

OBJECTIVE: To understand how the emission spectrum of hydrogen demonstrates the quantized nature of energy.

An atom can lose energy by *emitting* a photon.

As we learned in the last section, an atom with excess energy is said to be in an *excited state*. An excited atom can release some or all of its excess energy by emitting a photon (a "particle" of electromagnetic radiation) and thus move to a lower energy state. The lowest possible energy state of an atom is called its *ground state*.

We can learn a great deal about the energy states of hydrogen atoms by observing the photons they emit. To understand the significance of this, you need to remember that the *different wavelengths of light carry different amounts*

Each photon of blue light carries a larger quantity of energy than a photon of red light.

A particular color (wavelength) of light carries a particular amount of energy per photon.

of energy per photon. Recall that a beam of red light has lower-energy photons than a beam of blue light.

When a hydrogen atom absorbs energy from some outside source, it uses this energy to enter an excited state. It can release this excess energy (go back to a lower state) by emitting a photon of light (Figure 11.9). We can picture this process in terms of the energy-level diagram shown in Figure 11.10. The important point here is that *the energy contained in the photon corresponds to the change in energy that the atom experiences* in going from the excited state to the lower state.

Consider the following experiment. Suppose we take a sample of H atoms and put a lot of energy into the system (as represented in Figure 11.9). When we study the photons of visible light emitted, we see only certain colors (Figure 11.11). That is, *only certain types of photons* are produced. We don't see all colors, which would add up to give "white light"; we see only selected colors. This is a very significant result. Let's discuss carefully what it means.

H atom

Excited-state H atom

Some H atoms absorb energy and become excited

Photon

Photon

Photon

The excited atoms emit photons of light and return to the ground state

a

A sample of H atoms receives energy from an external source, which causes some of the atoms to become excited (to possess excess energy).

b

The excited H atoms can release the excess energy by emitting photons. The energy of each emitted photon corresponds exactly to the energy lost by each excited atom.

Figure 11.9

Figure 11.10

When an excited H atom returns to a lower energy level, it emits a photon that contains the energy released by the atom. Thus the energy of the photon corresponds to the difference in energy between the two states.

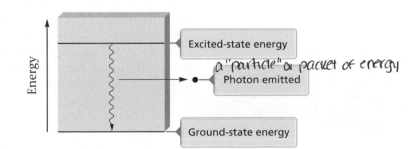

Energy

Excited-state energy

Photon emitted

a "particle" or packet of energy

Ground-state energy

Figure 11.11

When excited hydrogen atoms return to lower energy states, they emit photons of certain energies, and thus certain colors. Shown here are the colors and wavelengths (in nanometers) of the photons in the visible region that are emitted by excited hydrogen atoms.

410 nm   434 nm   486 nm   656 nm

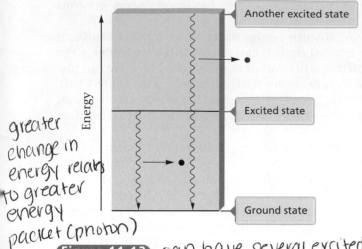

greater change in energy relates to greater energy packet (photon)

Figure 11.12 *can have several excited states*

Hydrogen atoms have several excited-state energy levels. The color of the photon emitted depends on the energy change that produces it. A larger energy change may correspond to a blue photon, whereas a smaller change may produce a red photon.

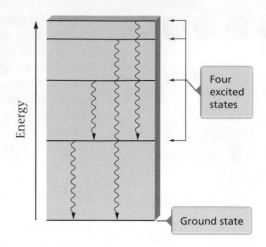

Figure 11.13

Each photon emitted by an excited hydrogen atom corresponds to a particular energy change in the hydrogen atom. In this diagram the horizontal lines represent discrete energy levels present in the hydrogen atom. A given H atom can exist in any of these energy states and can undergo energy changes to the ground state as well as to other excited states.

Figure 11.14

a Continuous energy levels. Any energy value is allowed.
b Discrete (quantized) energy levels. Only certain energy states are allowed.

Because only certain photons are emitted, we know that only certain energy changes are occurring (Figure 11.12). This means that the hydrogen atom must have *certain discrete energy levels* (Figure 11.13). Excited hydrogen atoms *always* emit photons with the same discrete colors (wavelengths)—those shown in Figure 11.11. They *never* emit photons with energies (colors) in between those shown. So we can conclude that all hydrogen atoms have the same set of discrete energy levels. We say the energy levels of hydrogen are **quantized.** That is, only *certain values are allowed.* Scientists have found that the energy levels of *all* atoms are quantized.

The quantized nature of the energy levels in atoms was a surprise when scientists discovered it. It had been assumed previously that an atom could exist at any energy level. That is, everyone had assumed that atoms could have a continuous set of energy levels rather than only certain discrete values (Figure 11.14). A useful analogy here is the contrast between the elevations allowed by a ramp, which vary continuously, and those allowed by a set of steps, which are discrete (Figure 11.15). The discovery of the quantized nature of energy has radically changed our view of the atom, as we will see in the next few sections.

Figure 11.15

The difference between continuous and quantized energy levels can be illustrated by comparing a flight of stairs with a ramp.

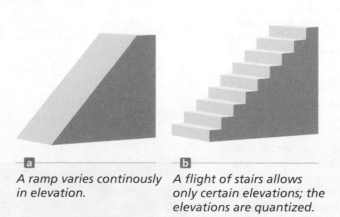

A ramp varies continuously in elevation.

A flight of stairs allows only certain elevations; the elevations are quantized.

quantum

11.5 The Bohr Model of the Atom

OBJECTIVE: To learn about Bohr's model of the hydrogen atom.

In 1911 at the age of twenty-five, Niels Bohr (Figure 11.16) received his Ph.D. in physics. He was convinced that the atom could be pictured as a small positive nucleus with electrons orbiting around it.

Over the next two years, Bohr constructed a model of the hydrogen atom with quantized energy levels that agreed with the hydrogen emission results we have just discussed. Bohr pictured the electron moving in circular orbits corresponding to the various allowed energy levels. He suggested that the electron could jump to a different orbit by absorbing or emitting a photon of light with exactly the correct energy content. Thus, in the Bohr atom, the energy levels in the hydrogen atom represented certain allowed circular orbits (Figure 11.17).

At first Bohr's model appeared very promising. It fit the hydrogen atom very well. However, when this model was applied to atoms other than hydrogen, it did not work. In fact, further experiments showed that the Bohr model is fundamentally incorrect. Although the Bohr model paved the way for later theories, it is important to realize that the current theory of atomic structure is not the same as the Bohr model. Electrons do *not* move around the nucleus in circular orbits like planets orbiting the sun. Surprisingly, as we shall see later in this chapter, we do not know exactly how the electrons move in an atom.

Figure 11.16

Niels Hendrik David Bohr (1885–1962) as a boy lived in the shadow of his younger brother Harald, who played on the 1908 Danish Olympic Soccer Team and later became a distinguished mathematician. In school, Bohr received his poorest marks in composition and struggled with writing during his entire life. In fact, he wrote so poorly that he was forced to dictate his Ph.D. thesis to his mother. He is one of the very few people who felt the need to write rough drafts of postcards. Nevertheless, Bohr was a brilliant physicist. After receiving his Ph.D. in Denmark, he constructed a quantum model for the hydrogen atom by the time he was 27. Even though his model later proved to be incorrect, Bohr remained a central figure in the drive to understand the atom. He was awarded the Nobel Prize in physics in 1922.

A historical connection of quantum theory to our atomic model (Max Planck)

only specific orbits allowed (specific energies)

Figure 11.17

The Bohr model of the hydrogen atom represented the electron as restricted to certain circular orbits around the nucleus.

11.6 The Wave Mechanical Model of the Atom

OBJECTIVE: To understand how the electron's position is represented in the wave mechanical model.

By the mid-1920s it had become apparent that the Bohr model was incorrect. Scientists needed to pursue a totally new approach. Two young physicists, Louis Victor de Broglie from France and Erwin Schrödinger from Austria, suggested that because light seems to have both wave and particle characteristics (it behaves simultaneously as a wave and as a stream of particles), the electron might also exhibit both of these characteristics. Although everyone

Louis Victor de Broglie

had assumed that the electron was a tiny particle, these scientists said it might be useful to find out whether it could be described as a wave.

When Schrödinger carried out a mathematical analysis based on this idea, he found that it led to a new model for the hydrogen atom that seemed to apply equally well to other atoms—something Bohr's model failed to do. We will now explore a general picture of this model, which is called the **wave mechanical model** of the atom.

In the Bohr model, the electron was assumed to move in circular orbits. In the wave mechanical model, on the other hand, the electron states are described by orbitals. *Orbitals are nothing like orbits.* To approximate the idea of an orbital, picture a single male firefly in a room in the center of which an open vial of female sex-attractant hormones is suspended. The room is extremely dark and there is a camera in one corner with its shutter open. Every time the firefly "flashes," the camera records a pinpoint of light and thus the firefly's position in the room at that moment. The firefly senses the sex attractant, and as you can imagine, it spends a lot of time at or close to it. However, now and then the insect flies randomly around the room.

When the film is taken out of the camera and developed, the picture will probably look like Figure 11.18. Because a picture is brightest where the film has been exposed to the most light, the color intensity at any given point tells us how often the firefly visited a given point in the room. Notice that, as we might expect, the firefly spent the most time near the room's center.

Now suppose you are watching the firefly in the dark room. You see it flash at a given point far from the center of the room. Where do you expect to see it next? There is really no way to be sure. The firefly's flight path is not precisely predictable. However, if you had seen the time-exposure picture of the firefly's activities (Figure 11.18), you would have some idea where to look next. Your best chance would be to look more toward the center of the room. Figure 11.18 suggests there is the highest probability (the highest odds, the greatest likelihood) of finding the firefly at any particular moment near the center of the room. You *can't be sure* the firefly will fly toward the center of the room, but it *probably* will. So the time-exposure picture is a kind of "probability map" of the firefly's flight pattern.

According to the wave mechanical model, the electron in the hydrogen atom can be pictured as being something like this firefly. Schrödinger found that he could not precisely describe the electron's path. His mathematics enabled him only to predict the probabilities of finding the electron at given points in space around the nucleus. In its ground state the hydrogen electron has a probability map like that shown in Figure 11.19. The more intense the color at a particular point, the more probable that the electron will be found at that point at a given instant. The model gives *no information about when* the electron occupies a certain point in space or *how it moves.* In fact, we have good reasons to believe that we can *never know* the details of electron motion, no matter how sophisticated our models may become. But one thing we feel confident about is that the electron *does not* orbit the nucleus in circles as Bohr suggested.

Figure 11.18

A representation of the photo of the firefly experiment. Remember that a picture is brightest where the film has been exposed to the most light. Thus the intensity of the color reflects how often the firefly visited a given point in the room. Notice that the brightest area is in the center of the room near the source of the sex attractant.

Figure 11.19

The probability map, or orbital, that describes the hydrogen electron in its lowest possible energy state. The more intense the color of a given dot, the more likely it is that the electron will be found at that point. We have no information about when the electron will be at a particular point or about how it moves. Note that the probability of the electron's presence is highest closest to the positive nucleus (located at the center of this diagram), as might be expected.

The Hydrogen Orbitals

OBJECTIVE: To learn about the shapes of orbitals designated by *s*, *p*, and *d*.

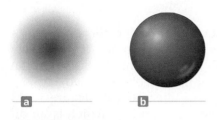

Figure 11.20

a The hydrogen 1*s* orbital.
b The size of the orbital is defined by a sphere that contains 90% of the total electron probability. That is, the electron can be found *inside* this sphere 90% of the time. The 1*s* orbital is often represented simply as a sphere. However, the most accurate picture of the orbital is the probability map represented in **a**.

The probability map for the hydrogen electron shown in Figure 11.19 is called an **orbital.** Although the probability of finding the electron decreases at greater distances from the nucleus, the probability of finding it even at great distances from the nucleus never becomes exactly zero. A useful analogy might be the lack of a sharp boundary between the earth's atmosphere and "outer space." The atmosphere fades away gradually, but there are always a few molecules present. Because the edge of an orbital is "fuzzy," an orbital does not have an exactly defined size. So chemists arbitrarily define its size as the sphere that contains 90% of the total electron probability (Figure 11.20b). This means that the electron spends 90% of the time inside this surface and 10% somewhere outside this surface. (Note that we are *not* saying the electron travels only on the *surface* of the sphere.) The orbital represented in Figure 11.20 is named the **1s orbital,** and it describes the hydrogen electron's lowest energy state (the ground state).

In Section 11.4 we saw that the hydrogen atom can absorb energy to transfer the electron to a higher energy state (an excited state). In terms of the obsolete Bohr model, this meant the electron was transferred to an orbit with a larger radius. In the wave mechanical model, these higher energy states correspond to different kinds of orbitals with different shapes.

At this point we need to stop and consider how the hydrogen atom is organized. Remember, we showed earlier that the hydrogen atom has discrete energy levels. We call these levels **principal energy levels** and label them with integers (Figure 11.21). Next we find that each of these levels is subdivided into **sublevels.** The following analogy should help you understand this. Picture an inverted triangle (Figure 11.22). We divide the principal levels into various numbers of sublevels. Principal level 1 consists of one sublevel, principal level 2 has two sublevels, principal level 3 has three sublevels, and principal level 4 has four sublevels.

Like our triangle, the principal energy levels in the hydrogen atom contain sublevels. As we will see presently, these sublevels contain spaces for the electron that we call orbitals. Principal energy level 1 consists of just one sublevel, or one type of orbital. The spherical shape of this orbital is shown in Figure 11.20. We label this orbital 1*s*. The number 1 is for the principal energy level, and *s* is a shorthand way to label a particular sublevel (type of orbital).

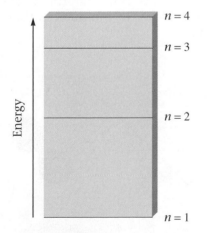

Figure 11.21

The first four principal energy levels in the hydrogen atom. Each level is assigned an integer, *n*.

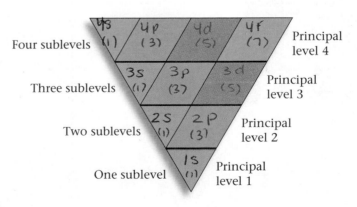

Figure 11.22

An illustration of how principal levels can be divided into sublevels.

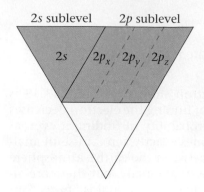

2s sublevel   2p sublevel

2s $2p_x$ $2p_y$ $2p_z$

The 1s orbital

Principal energy level 1

Shape

Figure 11.23

Principal level 2 shown divided into the 2s and 2p sublevels.

1s   2s

Figure 11.24

The relative sizes of the 1s and 2s orbitals of hydrogen.

Principal energy level 2 has two sublevels. (Note the correspondence between the principal energy level number and the number of sublevels.) These sublevels are labeled 2s and 2p. The 2s sublevel consists of one orbital (called the 2s), and the 2p sublevel consists of three orbitals (called $2p_x$, $2p_y$, and $2p_z$). Let's return to the inverted triangle to illustrate this. Figure 11.23 shows principal level 2 divided into the sublevels 2s and 2p (which is subdivided into $2p_x$, $2p_y$, and $2p_z$). The orbitals have the shapes shown in Figures 11.24 and 11.25. The 2s orbital is spherical like the 1s orbital but larger in size (see Figure 11.24). The three 2p orbitals are not spherical but have two "lobes." These orbitals are shown in Figure 11.25 both as electron probability maps and as surfaces that contain 90% of the total electron probability. Notice that the label x, y, or z on a given 2p orbital tells along which axis the lobes of that orbital are directed.

What we have learned so far about the hydrogen atom is summarized in Figure 11.26. Principal energy level 1 has one sublevel, which contains the 1s orbital. Principal energy level 2 contains two sublevels, one of which contains the 2s orbital and one of which contains the 2p orbitals (three of them). Note that each orbital is designated by a symbol or label. We summarize the information given by this label in the following box.

Orbital Labels

1. The number tells the principal energy level.
2. The letter tells the shape. The letter s means a spherical orbital; the letter p means a two-lobed orbital. The x, y, or z subscript on a p orbital label tells along which of the coordinate axes the two lobes lie.

One important characteristic of orbitals is that as the level number increases, the average distance of the electron in that orbital from the nucleus also increases. That is, when the hydrogen electron is in the 1s orbital (the ground state), it spends most of its time much closer to the nucleus than when it occupies the 2s orbital (an excited state).

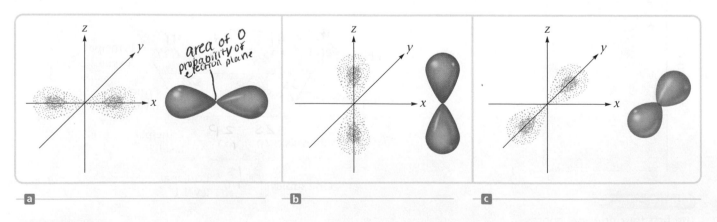

a   b   c

Figure 11.25

The three 2p orbitals: **a** $2p_x$, **b** $2p_z$, **c** $2p_y$. The x, y, or z label indicates along which axis the two lobes are directed. Each orbital is shown both as a probability map and as a surface that encloses 90% of the electron probability.

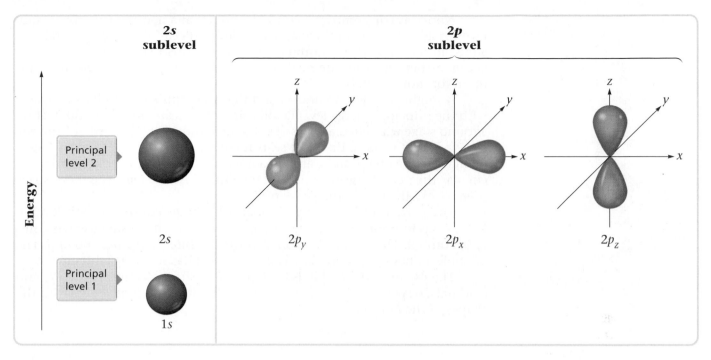

Figure 11.26
A diagram of principal energy levels 1 and 2 showing the shapes of orbitals that compose the sublevels.

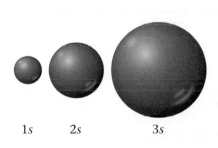

Figure 11.27

The relative sizes of the spherical 1s, 2s, and 3s orbitals of hydrogen.

You may be wondering at this point why hydrogen, which has only one electron, has more than one orbital. It is best to think of an orbital as a *potential space* for an electron. The hydrogen electron can occupy only a single orbital at a time, but the other orbitals are still available should the electron be transferred into one of them. For example, when a hydrogen atom is in its ground state (lowest possible energy state), the electron is in the 1s orbital. By adding the correct amount of energy (for example, a specific photon of light), we can excite the electron to the 2s orbital or to one of the 2p orbitals.

So far we have discussed only two of hydrogen's energy levels. There are many others. For example, level 3 has three sublevels (see Figure 11.22), which we label 3s, 3p, and 3d. The 3s sublevel contains a single 3s orbital, a spherical orbital larger than 1s and 2s (Figure 11.27). Sublevel 3p contains three orbitals: $3p_x$, $3p_y$, and $3p_z$, which are shaped like the 2p orbitals except that they are larger. The 3d sublevel contains five 3d orbitals with the shapes and labels shown in Figure 11.28. (You do not need to memorize the 3d orbital shapes and labels. They are shown for completeness.)

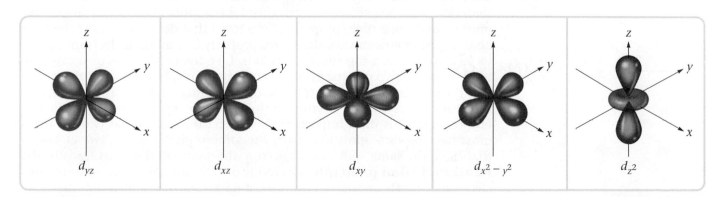

Figure 11.28
The shapes and labels of the five 3d orbitals.

Notice as you compare levels 1, 2, and 3 that a new type of orbital (sublevel) is added in each principal energy level. (Recall that the *p* orbitals are added in level 2 and the *d* orbitals in level 3.) This makes sense because in going farther out from the nucleus, there is more space available and thus room for more orbitals.

It might help you to understand that the number of orbitals increases with the principal energy level if you think of a theater in the round. Picture a round stage with circular rows of seats surrounding it. The farther from the stage a row of seats is, the more seats it contains because the circle is larger. Orbitals divide up the space around a nucleus somewhat like the seats in this circular theater. The greater the distance from the nucleus, the more space there is and the more orbitals we find.

The pattern of increasing numbers of orbitals continues with level 4. Level 4 has four sublevels labeled 4*s*, 4*p*, 4*d*, and 4*f*. The 4*s* sublevel has a single 4*s* orbital. The 4*p* sublevel contains three orbitals (4p_x, 4p_y, and 4p_z). The 4*d* sublevel has five 4*d* orbitals. The 4*f* sublevel has seven 4*f* orbitals.

The 4*s*, 4*p*, and 4*d* orbitals have the same shapes as the earlier *s*, *p*, and *d* orbitals, respectively, but are larger. We will not be concerned here with the shapes of the *f* orbitals.

11.8 The Wave Mechanical Model: Further Development

OBJECTIVES: To review the energy levels and orbitals of the wave mechanical model of the atom. • To learn about electron spin.

A model for the atom is of little use if it does not apply to all atoms. The Bohr model was discarded because it could be applied only to hydrogen. The wave mechanical model can be applied to all atoms in basically the same form as the one we have just used for hydrogen. In fact, the major triumph of this model is its ability to explain the periodic table of the elements. Recall that the elements on the periodic table are arranged in vertical groups, which contain elements that typically show similar chemical properties. The wave mechanical model of the atom allows us to explain, based on electron arrangements, why these similarities occur. We will see in due time how this is done.

Remember that an atom has as many electrons as it has protons to give it a zero overall charge. Therefore, all atoms beyond hydrogen have more than one electron. Before we can consider the atoms beyond hydrogen, we must describe one more property of electrons that determines how they can be arranged in an atom's orbitals. This property is spin. Each electron appears to be spinning as a top spins on its axis. Like the top, an electron can spin only in one of two directions. We often represent spin with an arrow: either ↑ or ↓. One arrow represents the electron spinning in the one direction, and the other represents the electron spinning in the opposite direction. For our purposes, what is most important about electron spin is that two electrons must have *opposite* spins to occupy the same orbital. That is, two electrons that have the same spin cannot occupy the same orbital. This leads to the **Pauli exclusion principle:** an atomic orbital can hold a maximum of two electrons, and those two electrons must have opposite spins.

Before we apply the wave mechanical model to atoms beyond hydrogen, we will summarize the model for convenient reference.

Principal Components of the Wave Mechanical Model of the Atom

1. Atoms have a series of energy levels called **principal energy levels,** which are designated by whole numbers symbolized by n; n can equal 1, 2, 3, 4, . . . Level 1 corresponds to $n = 1$, level 2 corresponds to $n = 2$, and so on.
2. The energy of the level increases as the value of n increases.
3. Each principal energy level contains one or more *types* of orbitals, called **sublevels.**
4. The number of sublevels present in a given principal energy level equals n. For example, level 1 contains one sublevel ($1s$); level 2 contains two sublevels (two types of orbitals), the $2s$ orbital and the three $2p$ orbitals; and so on. These are summarized in the following table. The number of each type of orbital is shown in parentheses.

n	Sublevels (Types of Orbitals) Present
1	$1s(1)$
2	$2s(1)$ $2p(3)$
3	$3s(1)$ $3p(3)$ $3d(5)$
4	$4s(1)$ $4p(3)$ $4d(5)$ $4f(7)$

5. The n value is always used to label the orbitals of a given principal level and is followed by a letter that indicates the type (shape) of the orbital. For example, the designation $3p$ means an orbital in level 3 that has two lobes (a p orbital always has two lobes).
6. An orbital can be empty or it can contain one or two electrons, but never more than two. If two electrons occupy the same orbital, they must have opposite spins.
7. The shape of an orbital does not indicate the details of electron movement. It indicates the probability distribution for an electron residing in that orbital.

EXAMPLE 11.1 | Understanding the Wave Mechanical Model of the Atom

Indicate whether each of the following statements about atomic structure is true or false.

   a. An s orbital is always spherical in shape.

   b. The $2s$ orbital is the same size as the $3s$ orbital.

   c. The number of lobes on a p orbital increases as n increases. That is, a $3p$ orbital has more lobes than a $2p$ orbital.

   d. Level 1 has one s orbital, level 2 has two s orbitals, level 3 has three s orbitals, and so on.

   e. The electron path is indicated by the surface of the orbital.

SOLUTION

   a. True. The size of the sphere increases as n increases, but the shape is always spherical.

b. False. The 3s orbital is larger (the electron is farther from the nucleus on average) than the 2s orbital.

c. False. A *p* orbital always has two lobes.

d. False. Each principal energy level has only one *s* orbital.

e. False. The electron is *somewhere inside* the orbital surface 90% of the time. The electron does not move around *on* this surface.

(Self-Check) **EXERCISE 11.1**   Define the following terms.

a. Bohr orbits

b. orbitals

c. orbital size

d. sublevel

See Problems 11.37 through 11.44. ∎

11.9 Electron Arrangements in the First Eighteen Atoms on the Periodic Table

OBJECTIVES: To understand how the principal energy levels fill with electrons in atoms beyond hydrogen. • To learn about valence electrons and core electrons.

We will now describe the electron arrangements in atoms with $Z = 1$ to $Z = 18$ by placing electrons in the various orbitals in the principal energy levels, starting with $n = 1$, and then continuing with $n = 2$, $n = 3$, and so on. For the first eighteen elements, the individual sublevels fill in the following order: 1s, then 2s, then 2p, then 3s, then 3p.

The most attractive orbital to an electron in an atom is always the 1s, because in this orbital the negatively charged electron is closer to the positively charged nucleus than in any other orbital. That is, the 1s orbital involves the space around the nucleus that is closest to the nucleus. As *n* increases, the orbital becomes larger—the electron, on average, occupies space farther from the nucleus.

So in its ground state hydrogen has its lone electron in the 1s orbital. This is commonly represented in two ways. First, we say that hydrogen has the electron arrangement, or **electron configuration,** $1s^1$. This just means there is one electron in the 1s orbital. We can also represent this configuration by using an **orbital diagram,** also called a **box diagram,** in which orbitals are represented by boxes grouped by sublevel with small arrows indicating the electrons. For *hydrogen,* the electron configuration and box diagram are

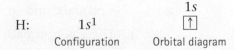

H: $1s^1$ ↑
                                    1s
        Configuration   Orbital diagram

The arrow represents an electron spinning in a particular direction. The next element is *helium,* $Z = 2$. It has two protons in its nucleus and so has two electrons. Because the 1s orbital is the most desirable, both electrons go there but with opposite spins. For helium, the electron configuration and box diagram are

He: $1s^2$ $\begin{array}{c} 1s \\ \boxed{\uparrow\downarrow} \end{array}$

The opposite electron spins are shown by the opposing arrows in the box.

Lithium ($Z = 3$) has three electrons, two of which go into the $1s$ orbital. That is, two electrons fill that orbital. The $1s$ orbital is the only orbital for $n = 1$, so the third electron must occupy an orbital with $n = 2$—in this case the $2s$ orbital. This gives a $1s^2 2s^1$ configuration. The electron configuration and box diagram are

Li: $1s^2 2s^1$ $\begin{array}{cc} 1s & 2s \\ \boxed{\uparrow\downarrow} & \boxed{\uparrow} \end{array}$

The next element, *beryllium,* has four electrons, which occupy the $1s$ and $2s$ orbitals with opposite spins.

Be: $1s^2 2s^2$ $\begin{array}{cc} 1s & 2s \\ \boxed{\uparrow\downarrow} & \boxed{\uparrow\downarrow} \end{array}$

Boron has five electrons, four of which occupy the $1s$ and $2s$ orbitals. The fifth electron goes into the second type of orbital with $n = 2$, one of the $2p$ orbitals.

B: $1s^2 2s^2 2p^1$ $\begin{array}{ccc} 1s & 2s & 2p \\ \boxed{\uparrow\downarrow} & \boxed{\uparrow\downarrow} & \boxed{\uparrow\,|\,|} \end{array}$

Because all the $2p$ orbitals have the same energy, it does not matter which $2p$ orbital the electron occupies.

Carbon, the next element, has six electrons: two electrons occupy the $1s$ orbital, two occupy the $2s$ orbital, and two occupy $2p$ orbitals. There are three $2p$ orbitals, so each of the mutually repulsive electrons occupies a different $2p$ orbital. For reasons we will not consider, in the separate $2p$ orbitals the electrons have the same spin.

The configuration for carbon could be written $1s^2 2s^2 2p^1 2p^1$ to indicate that the electrons occupy separate $2p$ orbitals. However, the configuration is usually given as $1s^2 2s^2 2p^2$, and it is understood that the electrons are in different $2p$ orbitals.

C: $1s^2 2s^2 2p^2$ $\begin{array}{ccc} 1s & 2s & 2p \\ \boxed{\uparrow\downarrow} & \boxed{\uparrow\downarrow} & \boxed{\uparrow\,\uparrow\,|} \end{array}$

Note the like spins for the unpaired electrons in the $2p$ orbitals.

The configuration for *nitrogen,* which has seven electrons, is $1s^2 2s^2 2p^3$. The three electrons in $2p$ orbitals occupy separate orbitals and have like spins.

N: $1s^2 2s^2 2p^3$ $\begin{array}{ccc} 1s & 2s & 2p \\ \boxed{\uparrow\downarrow} & \boxed{\uparrow\downarrow} & \boxed{\uparrow\,\uparrow\,\uparrow} \end{array}$

The configuration for *oxygen,* which has eight electrons, is $1s^2 2s^2 2p^4$. One of the $2p$ orbitals is now occupied by a pair of electrons with opposite spins, as required by the Pauli exclusion principle.

O: $1s^2 2s^2 2p^4$ $\begin{array}{ccc} 1s & 2s & 2p \\ \boxed{\uparrow\downarrow} & \boxed{\uparrow\downarrow} & \boxed{\uparrow\downarrow\,\uparrow\,\uparrow} \end{array}$

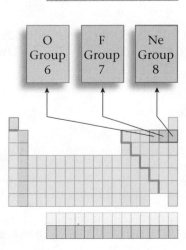

H $1s^1$										He $1s^2$
Li $2s^1$	Be $2s^2$				B $2p^1$	C $2p^2$	N $2p^3$	O $2p^4$	F $2p^5$	Ne $2p^6$
Na $3s^1$	Mg $3s^2$				Al $3p^1$	Si $3p^2$	P $3p^3$	S $3p^4$	Cl $3p^5$	Ar $3p^6$

Figure 11.29

The electron configurations in the sublevel last occupied for the first eighteen elements.

The electron configurations and orbital diagrams for *fluorine* (nine electrons) and *neon* (ten electrons) are

$$1s \quad 2s \quad\quad 2p$$

F: $1s^2 2s^2 2p^5$ ⇅ ⇅ ⇅⇅↑

Ne: $1s^2 2s^2 2p^6$ ⇅ ⇅ ⇅⇅⇅

With neon, the orbitals with $n = 1$ and $n = 2$ are completely filled.

For *sodium*, which has eleven electrons, the first ten electrons occupy the 1s, 2s, and 2p orbitals, and the eleventh electron must occupy the first orbital with $n = 3$, the 3s orbital. The electron configuration for sodium is $1s^2 2s^2 2p^6 3s^1$. To avoid writing the inner-level electrons, we often abbreviate the configuration $1s^2 2s^2 2p^6 3s^1$ as [Ne]$3s^1$, where [Ne] represents the electron configuration of neon, $1s^2 2s^2 2p^6$.

The orbital diagram for sodium is

$$1s \quad 2s \quad\quad 2p \quad\quad 3s$$

⇅ ⇅ ⇅⇅⇅ ↑

The next element, *magnesium*, $Z = 12$, has the electron configuration $1s^2 2s^2 2p^6 3s^2$, or [Ne]$3s^2$.

The next six elements, *aluminum* through *argon*, have electron configurations obtained by filling the 3p orbitals one electron at a time. Figure 11.29 summarizes the electron configurations of the first eighteen elements by giving the number of electrons in the type of orbital (sublevel) occupied last.

EXAMPLE 11.2 | **Writing Orbital Diagrams**

Write the orbital diagram for magnesium.

SOLUTION

Magnesium ($Z = 12$) has twelve electrons that are placed successively in the 1s, 2s, 2p, and 3s orbitals to give the electron configuration $1s^2 2s^2 2p^6 3s^2$. The orbital diagram is

$$1s \quad 2s \quad\quad 2p \quad\quad 3s$$

⇅ ⇅ ⇅⇅⇅ ⇅

Only occupied orbitals are shown here.

Self-Check EXERCISE 11.2 Write the complete electron configuration and the orbital diagram for each of the elements aluminum through argon.

See Problems 11.49 through 11.54. ∎

A Magnetic Moment

An anesthetized frog lies in the hollow core of an electromagnet. As the current in the coils of the magnet is increased, the frog magically rises and floats in midair (see photo). How can this happen? Is the electromagnet an antigravity machine? In fact, there is no magic going on here. This phenomenon demonstrates the magnetic properties of all matter. We know that iron magnets attract and repel each other depending on their relative orientations. Is a frog magnetic like a piece of iron? If a frog lands on a steel manhole cover, will it be trapped there by magnetic attractions? Of course not. The magnetism of the frog, as with most objects, shows up only in the presence of a strong inducing magnetic field. In other words, the powerful electromagnet surrounding the frog in the experiment described

A live frog levitated in a magnetic field.

Andrey K. Geim/High Field Magnet Laboratory/University of Nijmegen

above *induces* a magnetic field in the frog that opposes the inducing field. The opposing magnetic field in the frog repels the inducing field, and the frog lifts up until the magnetic force is balanced by the gravitational pull on its body. The frog then "floats" in air.

How can a frog be magnetic if it is not made of iron? It's the electrons. Frogs are composed of cells containing many kinds of molecules. Of course, these molecules are made of atoms—carbon atoms, nitrogen atoms, oxygen atoms, and other types. Each of these atoms contains electrons that are moving around the atomic nuclei. When these electrons sense a strong magnetic field, they respond by moving in a fashion that produces magnetic fields aligned to oppose the inducing field. This phenomenon is called *diamagnetism.*

All substances, animate and inanimate, because they are made of atoms, exhibit diamagnetism. Andre Geim and his colleagues at the University of Nijmegan, the Netherlands, have levitated frogs, grasshoppers, plants, and water droplets, among other objects. Geim says that, given a large enough electromagnet, even humans can be levitated. He notes, however, that constructing a magnet strong enough to float a human would be very expensive, and he sees no point in it. Geim does point out that inducing weightlessness with magnetic fields may be a good way to pretest experiments on weightlessness intended as research for future space flights—to see if the ideas fly as well as the objects.

At this point it is useful to introduce the concept of **valence electrons**—that is, *the electrons in the outermost (highest) principal energy level of an atom.* For example, nitrogen, which has the electron configuration $1s^2 2s^2 2p^3$, has electrons in principal levels 1 and 2. Therefore, level 2 (which has 2s and 2p sublevels) is the valence level of nitrogen, and the 2s and 2p electrons are the valence electrons. For the sodium atom (electron configuration $1s^2 2s^2 2p^6 3s^1$, or $[Ne]3s^1$), the valence electron is the electron in the 3s orbital, because in this case principal energy level 3 is the outermost level that contains an electron. The valence electrons are the most important electrons to chemists because, being the outermost electrons, they are the ones involved when atoms attach to each other (form bonds), as we will see in the next chapter. The inner electrons, which are known as **core electrons,** are not involved in bonding atoms to each other.

Note in Figure 11.29 that a very important pattern is developing: except for helium, *the atoms of elements in the same group (vertical column of the periodic table) have the same number of electrons in a given type of orbital* (sublevel), except that the orbitals are in different principal energy levels. Remember that the elements were originally organized into groups on the periodic table on the basis of similarities in chemical properties. Now we understand the reason behind these groupings. Elements with the same valence electron arrangement show very similar chemical behavior.

11.10 Electron Configurations and the Periodic Table

OBJECTIVE: To learn about the electron configurations of atoms with Z greater than 18.

In the previous section we saw that we can describe the atoms beyond hydrogen by simply filling the atomic orbitals starting with level $n = 1$ and working outward in order. This works fine until we reach the element *potassium* ($Z = 19$), which is the next element after argon. Because the $3p$ orbitals are fully occupied in argon, we might expect the next electron to go into a $3d$ orbital (recall that for $n = 3$ the sublevels are $3s$, $3p$, and $3d$). However, experiments show that the chemical properties of potassium are very similar to those of lithium and sodium. Because we have learned to associate similar chemical properties with similar valence-electron arrangements, we predict that the valence-electron configuration for potassium is $4s^1$, resembling sodium ($3s^1$) and lithium ($2s^1$). That is, we expect the last electron in potassium to occupy the $4s$ orbital instead of one of the $3d$ orbitals. This means that principal energy level 4 begins to fill before level 3 has been completed. This conclusion is confirmed by many types of experiments. So the electron configuration of potassium is

$$K: 1s^2 2s^2 2p^6 3s^2 3p^6 4s^1, \text{ or } [Ar]4s^1$$

The next element is *calcium*, with an additional electron that also occupies the $4s$ orbital.

$$Ca: 1s^2 2s^2 2p^6 3s^2 3p^6 4s^2, \text{ or } [Ar]4s^2$$

Figure 11.30

Partial electron configurations for the elements potassium through krypton. The transition metals shown in green (scandium through zinc) have the general configuration $[Ar]4s^2 3d^n$, except for chromium and copper.

The Chemistry of Bohrium

One of the best uses of the periodic table is to predict the properties of newly discovered elements. For example, the artificially synthesized element bohrium ($Z = 107$) is found in the same family as manganese, technetium, and rhenium and is expected to show chemistry similar to these elements. The problem, of course, is that only a few atoms of bohrium (Bh) can be made at a time and the atoms exist for only a very short time (about 17 seconds). It's a real challenge to study the chemistry of an element under these conditions. However, a team of nuclear chemists led by Heinz W. Gaggeler of the University of Bern in Switzerland isolated six atoms of ^{267}Bh and prepared the compound BhO_3Cl. Analysis of the decay products of this compound helped define the thermochemical properties of BhO_3Cl and showed that bohrium seems to behave as might be predicted from its position in the periodic table.

The 4s orbital is now full.

After calcium the next electrons go into the 3d orbitals to complete principal energy level 3. The elements that correspond to filling the 3d orbitals are called transition metals. Then the 4p orbitals fill. Figure 11.30 gives partial electron configurations for the elements potassium through krypton.

Note from Figure 11.30 that all of the transition metals have the general configuration $[Ar]4s^23d^n$ except chromium ($4s^13d^5$) and copper ($4s^13d^{10}$). The reasons for these exceptions are complex and will not be discussed here.

Instead of continuing to consider the elements individually, we will now look at the overall relationship between the periodic table and orbital filling. Figure 11.31 shows which type of orbital is filling in each area of the periodic table. Note the points in the box below.

Orbital Filling

1. In a principal energy level that has d orbitals, the s orbital from the *next* level fills before the d orbitals in the current level. That is, the $(n + 1)s$ orbitals always fill before the nd orbitals. For example, the 5s orbitals fill for rubidium and strontium before the 4d orbitals fill for the second row of transition metals (yttrium through cadmium).
2. After lanthanum, which has the electron configuration $[Xe]6s^25d^1$, a group of fourteen elements called the **lanthanide series,** or the lanthanides, occurs. This series of elements corresponds to the filling of the seven 4f orbitals.
3. After actinium, which has the configuration $[Rn]7s^26d^1$, a group of fourteen elements called the **actinide series,** or the actinides, occurs. This series corresponds to the filling of the seven 5f orbitals.
4. Except for helium, the group numbers indicate the sum of electrons in the ns and np orbitals in the highest principal energy level that contains electrons (where n is the number that indicates a particular principal energy level). These electrons are the valence electrons, the electrons in the outermost principal energy level of a given atom.

Electron configuration

Groups

Figure 11.31

The orbitals being filled for elements in various parts of the periodic table. Note that in going along a horizontal row (a period), the $(n + 1)s$ orbital fills before the nd orbital. The group label indicates the number of valence electrons (the number of s plus the number of p electrons in the highest occupied principal energy level) for the elements in each group.

*After the $6s$ orbital is full, one electron goes into a $5d$ orbital. This corresponds to the element lanthanum ([Xe]$6s^2 5d^1$). After lanthanum, the $4f$ orbitals fill with electrons.

**After the $7s$ orbital is full, one electron goes into $6d$. This is actinium ([Rn]$7s^2 6d^1$). The $5f$ orbitals then fill.

To help you further understand the connection between orbital filling and the periodic table, Figure 11.32 shows the orbitals in the order in which they fill.

A periodic table is almost always available to you. If you understand the relationship between the electron configuration of an element and its position on the periodic table, you can figure out the expected electron configuration of any atom.

EXAMPLE 11.3   Determining Electron Configurations

Using the periodic table inside the front cover of the text, give the electron configurations for sulfur (S), gallium (Ga), hafnium (Hf), and radium (Ra).

SOLUTION

Sulfur is element 16 and resides in Period 3, where the $3p$ orbitals are being filled (see Figure 11.33). Because sulfur is the fourth among the "$3p$ elements," it must have four $3p$ electrons. Sulfur's electron configuration is

$$S: 1s^2 2s^2 2p^6 3s^2 3p^4, \text{ or } [Ne]3s^2 3p^4$$

Gallium is element 31 in Period 4 just after the transition metals (see Figure 11.33). It is the first element in the "$4p$ series" and has a $4p^1$ arrangement. Gallium's electron configuration is

$$Ga: 1s^2 2s^2 2p^6 3s^2 3p^6 4s^2 3d^{10} 4p^1, \text{ or } [Ar]4s^2 3d^{10} 4p^1$$

Hafnium is element 72 and is found in Period 6, as shown in Figure 11.33. Note that it occurs just after the lanthanide series (see Figure 11.31).

Figure 11.32

A box diagram showing the order in which orbitals fill to produce the atoms in the periodic table. Each box can hold two electrons.

Figure 11.33

The positions of the elements considered in Example 11.3.

Thus the $4f$ orbitals are already filled. Hafnium is the second member of the $5d$ transition series and has two $5d$ electrons. Its electron configuration is

Hf: $1s^2 2s^2 2p^6 3s^2 3p^6 4s^2 3d^{10} 4p^6 5s^2 4d^{10} 5p^6 6s^2 4f^{14} 5d^2$, or $[Xe]6s^2 4f^{14} 5d^2$

Radium is element 88 and is in Period 7 (and Group 2), as shown in Figure 11.33. Thus radium has two electrons in the $7s$ orbital, and its electron configuration is

Ra: $1s^2 2s^2 2p^6 3s^2 3p^6 4s^2 3d^{10} 4p^6 5s^2 4d^{10} 5p^6 6s^2 4f^{14} 5d^{10} 6p^6 7s^2$, or $[Rn]7s^2$

Self-Check EXERCISE 11.3   Using the periodic table inside the front cover of the text, predict the electron configurations for fluorine, silicon, cesium, lead, and iodine. If you have trouble, use Figure 11.31.

See Problems 11.59 through 11.68. ■

▶ Summary of the Wave Mechanical Model and Valence-Electron Configurations

The concepts we have discussed in this chapter are very important. They allow us to make sense of a good deal of chemistry. When it was first observed that elements with similar properties occur periodically as the atomic number increases, chemists wondered why. Now we have an explanation. The wave mechanical model pictures the electrons in an atom as arranged in orbitals, with each orbital capable of holding two electrons. As we build up the atoms, the same types of orbitals recur in going from one principal energy level to another. This means that particular valence-electron configurations recur periodically. For reasons we will explore in the next chapter, elements with a particular type of valence configuration all show very similar chemical behavior. Thus groups of elements, such as the alkali metals, show similar chemistry because all the elements in that group have the same type of valence-electron arrangement. This concept, which explains so much chemistry, is the greatest contribution of the wave mechanical model to modern chemistry.

For reference, the valence-electron configurations for all the elements are shown on the periodic table in Figure 11.34. Note the following points:

1. The group labels for Groups 1, 2, 3, 4, 5, 6, 7, and 8 indicate the *total number* of valence electrons for the atoms in these groups. For

example, all the elements in Group 5 have the configuration ns^2np^3. (Any d electrons present are always in the next lower principal energy level than the valence electrons and so are not counted as valence electrons.)

2. The elements in Groups 1, 2, 3, 4, 5, 6, 7, and 8 are often called the **main-group elements,** or **representative elements.** Remember that every member of a given group (except for helium) has the same valence-electron configuration, except that the electrons are in different principal energy levels.

3. We will not be concerned in this text with the configurations for the f-transition elements (lanthanides and actinides), although they are included in Figure 11.34.

Representative Elements		d-Transition Elements										Representative Elements					Noble Gases

	1A ns^1	Group numbers																8A ns^2np^6
1	1 H $1s^1$	2A ns^2										3A ns^2np^1	4A ns^2np^2	5A ns^2np^3	6A ns^2np^4	7A ns^2np^5		2 He $1s^2$
2	3 Li $2s^1$	4 Be $2s^2$										5 B $2s^22p^1$	6 C $2s^22p^2$	7 N $2s^22p^3$	8 O $2s^22p^4$	9 F $2s^22p^5$	10 Ne $2s^22p^6$
3	11 Na $3s^1$	12 Mg $3s^2$										13 Al $3s^23p^1$	14 Si $3s^23p^2$	15 P $3s^23p^3$	16 S $3s^23p^4$	17 Cl $3s^23p^5$	18 Ar $3s^23p^6$
4	19 K $4s^1$	20 Ca $4s^2$	21 Sc $4s^23d^1$	22 Ti $4s^23d^2$	23 V $4s^23d^3$	24 Cr $4s^13d^5$	25 Mn $4s^23d^5$	26 Fe $4s^23d^6$	27 Co $4s^23d^7$	28 Ni $4s^23d^8$	29 Cu $4s^13d^{10}$	30 Zn $4s^23d^{10}$	31 Ga $4s^24p^1$	32 Ge $4s^24p^2$	33 As $4s^24p^3$	34 Se $4s^24p^4$	35 Br $4s^24p^5$	36 Kr $4s^24p^6$
5	37 Rb $5s^1$	38 Sr $5s^2$	39 Y $5s^24d^1$	40 Zr $5s^24d^2$	41 Nb $5s^14d^4$	42 Mo $5s^14d^5$	43 Tc $5s^14d^6$	44 Ru $5s^14d^7$	45 Rh $5s^14d^8$	46 Pd $4d^{10}$	47 Ag $5s^14d^{10}$	48 Cd $5s^24d^{10}$	49 In $5s^25p^1$	50 Sn $5s^25p^2$	51 Sb $5s^25p^3$	52 Te $5s^25p^4$	53 I $5s^25p^5$	54 Xe $5s^25p^6$
6	55 Cs $6s^1$	56 Ba $6s^2$	57 La* $6s^25d^1$	72 Hf $4f^{14}6s^25d^2$	73 Ta $6s^25d^3$	74 W $6s^25d^4$	75 Re $6s^25d^5$	76 Os $6s^25d^6$	77 Ir $6s^25d^7$	78 Pt $6s^15d^9$	79 Au $6s^15d^{10}$	80 Hg $6s^25d^{10}$	81 Tl $6s^26p^1$	82 Pb $6s^26p^2$	83 Bi $6s^26p^3$	84 Po $6s^26p^4$	85 At $6s^26p^5$	86 Rn $6s^26p^6$
7	87 Fr $7s^1$	88 Ra $7s^2$	89 Ac** $7s^26d^1$	104 Rf $7s^26d^2$	105 Db $7s^26d^3$	106 Sg $7s^26d^4$	107 Bh $7s^26d^5$	108 Hs $7s^26d^6$	109 Mt $7s^26d^7$	110 Ds $7s^26d^8$	111 Rg $7s^16d^{10}$	112 Uub $7s^26d^{10}$	113 Uut $7s^27p^1$	114 Uuq $7s^27p^2$	115 Uup $7s^27p^3$			118 Uuo $7s^27p^6$

Period number, highest occupied electron level

f-Transition Elements

*Lanthanides	58 Ce $6s^24f^{1}5d^1$	59 Pr $6s^24f^{3}5d^0$	60 Nd $6s^24f^{4}5d^0$	61 Pm $6s^24f^{5}5d^0$	62 Sm $6s^24f^{6}5d^0$	63 Eu $6s^24f^{7}5d^0$	64 Gd $6s^24f^{7}5d^1$	65 Tb $6s^24f^{9}5d^0$	66 Dy $6s^24f^{10}5d^0$	67 Ho $6s^24f^{11}5d^0$	68 Er $6s^24f^{12}5d^0$	69 Tm $6s^24f^{13}5d^0$	70 Yb $6s^24f^{14}5d^0$	71 Lu $6s^24f^{14}5d^1$
**Actinides	90 Th $7s^25f^{0}6d^2$	91 Pa $7s^25f^{2}6d^1$	92 U $7s^25f^{3}6d^1$	93 Np $7s^25f^{4}6d^1$	94 Pu $7s^25f^{6}6d^0$	95 Am $7s^25f^{7}6d^0$	96 Cm $7s^25f^{7}6d^1$	97 Bk $7s^25f^{9}6d^0$	98 Cf $7s^25f^{10}6d^0$	99 Es $7s^25f^{11}6d^0$	100 Fm $7s^25f^{12}6d^0$	101 Md $7s^25f^{13}6d^0$	102 No $7s^25f^{14}6d^0$	103 Lr $7s^25f^{14}6d^1$

Figure 11.34

The periodic table with atomic symbols, atomic numbers, and partial electron configurations.

11.11 Atomic Properties and the Periodic Table

OBJECTIVE: To understand the general trends in atomic properties in the periodic table.

With all of this talk about electron probability and orbitals, we must not lose sight of the fact that chemistry is still fundamentally a science based on the observed properties of substances. We know that wood burns, steel rusts, plants grow, sugar tastes sweet, and so on because we *observe* these phenomena. The atomic theory is an attempt to help us understand why these things occur. If we understand why, we can hope to better control the chemical events that are so crucial in our daily lives.

In the next chapter we will see how our ideas about atomic structure help us understand how and why atoms combine to form compounds. As we explore this, and as we use theories to explain other types of chemical behavior later in the text, it is important that we distinguish the observation (steel rusts) from the attempts to explain why the observed event occurs (theories). The observations remain the same over the decades, but the theories (our explanations) change as we gain a clearer understanding of how nature operates. A good example of this is the replacement of the Bohr model for atoms by the wave mechanical model.

Because the observed behavior of matter lies at the heart of chemistry, you need to understand thoroughly the characteristic properties of the various elements and the trends (systematic variations) that occur in those properties. To that end, we will now consider some especially important properties of atoms and see how they vary, horizontally and vertically, on the periodic table.

▶ Metals and Nonmetals

The most fundamental classification of the chemical elements is into metals and nonmetals. **Metals** typically have the following physical properties: a lustrous appearance, the ability to change shape without breaking (they can be pulled into a wire or pounded into a thin sheet), and excellent conductivity of heat and electricity. **Nonmetals** typically do not have these physical properties, although there are some exceptions. (For example, solid iodine is lustrous; the graphite form of carbon is an excellent conductor of electricity; and the diamond form of carbon is an excellent conductor of heat.) However, it is the *chemical* differences between metals and nonmetals that interest us the most: *metals tend to lose electrons to form positive ions, and nonmetals tend to gain electrons to form negative ions.* When a metal and a nonmetal react, a transfer of one or more electrons from the metal to the nonmetal often occurs.

Most of the elements are classified as metals, as is shown in Figure 11.35. Note that the metals are found on the left side and at the center of the periodic table. The relatively few nonmetals are in the upper-right corner of the table. A few elements exhibit both metallic and nonmetallic behavior; they are classified as **metalloids** or semimetals.

It is important to understand that simply being classified as a metal does not mean that an element behaves exactly like all other metals. For example, some metals can lose one or more electrons much more easily than others. In particular, cesium can give up its outermost electron (a 6*s* electron)

Gold leaf being applied to the dome of the courthouse in Huntington, West Virginia.

AP Photo/The Charleston Daily Mail/Chip Ellis

Figure 11.35

The classification of elements as metals, nonmetals, and metalloids.

more easily than can lithium (a 2s electron). In fact, for the alkali metals (Group 1) the ease of giving up an electron varies as follows:

$$Cs > Rb > K > Na > Li$$

Loses an
electron
most easily

Note that as we go down the group, the metals become more likely to lose an electron. This makes sense because as we go down the group, the electron being removed resides, on average, farther and farther from the nucleus. That is, the 6s electron lost from Cs is much farther from the attractive positive nucleus—and so is easier to remove—than the 2s electron that must be removed from a lithium atom.

The same trend is also seen in the Group 2 metals (alkaline earth metals): the farther down in the group the metal resides, the more likely it is to lose an electron.

Just as metals vary somewhat in their properties, so do nonmetals. In general, the elements that can most effectively pull electrons from metals occur in the upper-right corner of the periodic table.

As a general rule, we can say that the most chemically active metals appear in the lower-left region of the periodic table, whereas the most chemically active nonmetals appear in the upper-right region. The properties of the semimetals, or metalloids, lie between the metals and the nonmetals, as might be expected.

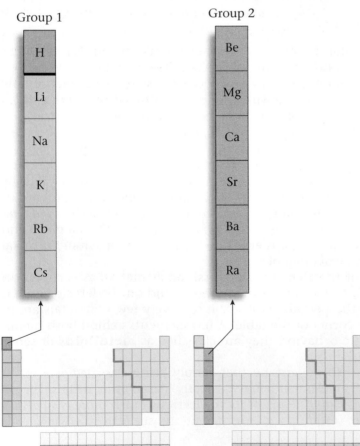

▶ **Ionization Energies**

The **ionization energy** of an atom is the energy required to remove an electron from an individual atom in the gas phase:

$$M(g) \quad \Rightarrow \quad M^+(g) + e^-$$

Ionization
energy

As we have noted, the most characteristic chemical property of a metal atom is losing electrons to nonmetals. Another way of saying this is to say that *metals have relatively low ionization*

Fireworks

The art of using mixtures of chemicals to produce explosives is an ancient one. Black powder—a mixture of potassium nitrate, charcoal, and sulfur—was being used in China well before A.D. 1000, and it has been used through the centuries in military explosives, in construction blasting, and for fireworks.

Before the nineteenth century, fireworks were confined mainly to rockets and loud bangs. Orange and yellow colors came from the presence of charcoal and iron filings. However, with the great advances in chemistry in the nineteenth century, new compounds found their way into fireworks. Salts of copper, strontium, and barium added brilliant colors. Magnesium and aluminum metals gave a dazzling white light.

How do fireworks produce their brilliant colors and loud bangs? Actually, only a handful of different chemicals are responsible for most of the spectacular effects. To produce the noise and flashes, an oxidizer (something with a strong affinity for electrons) is reacted with a metal such as magnesium or aluminum mixed with sulfur. The resulting reaction produces a brilliant flash, which is due to the aluminum or magnesium burning, and a loud report is produced by the rapidly expanding gases. For a color effect, an element with a colored flame is included.

Yellow colors in fireworks are due to sodium. Strontium salts give the red color familiar from

These brightly colored fireworks are the result of complex mixtures of chemicals.

highway safety flares. Barium salts give a green color.

Although you might think that the chemistry of fireworks is simple, achieving the vivid white flashes and the brilliant colors requires complex combinations of chemicals. For example, because the white flashes produce high flame temperatures, the colors tend to wash out. Another problem arises from the use of sodium salts. Because sodium produces an extremely bright yellow color, sodium salts cannot be used when other colors are desired. In short, the manufacture of fireworks that produce the desired effects and are also safe to handle requires very careful selection of chemicals.*

*The chemical mixtures in fireworks are very dangerous. *Do not* experiment with chemicals on your own.

---

energies—a relatively small amount of energy is needed to remove an electron from a typical metal.

Recall that metals at the bottom of a group lose electrons more easily than those at the top. In other words, ionization energies tend to decrease in going from the top to the bottom of a group.

In contrast to metals, nonmetals have relatively large ionization energies. Nonmetals tend to gain, not lose, electrons. Recall that metals appear on the left side of the periodic table and nonmetals appear on the right. Thus it is not surprising that ionization energies tend to increase from left to right across a given period on the periodic table.

In general, the elements that appear in the lower-left region of the periodic table have the lowest ionization energies (and are therefore the most chemically active metals). On the other hand, the elements with the highest ionization energies (the most chemically active nonmetals) occur in the upper-right region of the periodic table.

▶ Atomic Size

The sizes of atoms vary as shown in Figure 11.36. Notice that atoms get larger as we go down a group on the periodic table and that they get smaller as we go from left to right across a period.

We can understand the increase in size that we observe as we go down a group by remembering that as the principal energy level increases, the average distance of the electrons from the nucleus also increases. So atoms get bigger as electrons are added to larger principal energy levels.

Figure 11.36

Relative atomic sizes for selected atoms. Note that atomic size increases down a group and decreases across a period.

Explaining the decrease in **atomic size** across a period requires a little thought about the atoms in a given row (period) of the periodic table. Recall that the atoms in a particular period all have their outermost electrons in a given principal energy level. That is, the atoms in Period 1 have their outer electrons in the 1s orbital (principal energy level 1), the atoms in Period 2 have their outermost electrons in principal energy level 2 (2s and 2p orbitals), and so on (see Figure 11.31). Because all the orbitals in a given principal energy level are expected to be the same size, we might expect the atoms in a given period to be the same size. However, remember that the number of protons in the nucleus increases as we move from atom to atom in the period. The resulting increase in positive charge on the nucleus tends to pull the electrons closer to the nucleus. So instead of remaining the same size across a period as electrons are added in a given principal energy level, the atoms get smaller as the electron "cloud" is drawn in by the increasing nuclear charge.

C H A P T E R 11 REVIEW

Key Terms

electromagnetic radiation (11.2)

wavelength (11.2)

frequency (11.2)

photon (11.2)

quantized energy levels (11.4)

wave mechanical model (11.6)

orbital (11.7)

principal energy levels (11.7)

sublevels (11.7)

Pauli exclusion principle (11.8)

electron configuration (11.9)

orbital (box) diagram (11.9)

valence electrons (11.9)

core electrons (11.9)

lanthanide series (11.10)

actinide series (11.10)

main-group (representative) elements (11.10)

metals (11.11)

nonmetals (11.11)

metalloids (11.11)

ionization energy (11.11)

atomic size (11.11)

Summary

1. Energy travels through space by electromagnetic radiation ("light"), which can be characterized by the wavelength and frequency of the waves. Light can also be thought of as packets of energy called photons. Atoms can gain energy by absorbing a photon and can lose energy by emitting a photon.

2. The emissions of energy from hydrogen atoms produce only certain energies as hydrogen changes from a higher to a lower energy. This shows that the energy levels of hydrogen are quantized.

3. The Bohr model of the hydrogen atom postulated that the electron moved in circular orbits corresponding to the various allowed energy levels. Though it worked well for hydrogen, the Bohr model did not work for other atoms.

4. The wave mechanical model explains atoms by postulating that the electron has both wave and particle characteristics. Electron states are described by orbitals, which are probability maps indicating how likely it is to find the electron at a given point in space. The orbital size can be thought of as a surface containing 90% of the total electron probability.

5. According to the Pauli exclusion principle, an atomic orbital can hold a maximum of two electrons, and those electrons must have opposite spins.

6. Atoms have a series of energy levels, called principal energy levels (n), which contain one or more sublevels (types of orbitals). The number of sublevels increases with increasing n.

7. Valence electrons are the s and p electrons in the outermost principal energy level of an atom. Core electrons are the inner electrons of an atom.

8. Metals are found at the left and center of the periodic table. The most chemically active metals are found in the lower-left corner of the periodic table. The most chemically active nonmetals are located in the upper-right corner.

9. Ionization energy, the energy required to remove an electron from a gaseous atom, decreases going down a group and increases going from left to right across a period.

10. For the representative elements, atomic size increases going down a group but decreases going from left to right across a period.

Active Learning Questions

These questions are designed to be considered by groups of students in class. Often these questions work well for introducing a particular topic in class.

1. How does probability fit into the description of the atom?

2. What is meant by an *orbital?*

3. Account for the fact that the line that separates the metals from the nonmetals on the periodic table is diagonal downward to the right instead of horizontal or vertical.

4. Consider the following statements: "The ionization energy for the potassium atom is negative because when K loses an electron to become K^+, it achieves a noble gas electron configuration." Indicate everything that is correct in this statement. Indicate everything that is incorrect. Correct the mistaken information and explain the error.

5. In going across a row of the periodic table, protons and electrons are added and ionization energy generally increases. In going down a column of the periodic table, protons and electrons are also being added but ionization energy generally decreases. Explain.

6. Which is larger, the H $1s$ orbital or the Li $1s$ orbital? Why? Which has the larger radius, the H atom or the Li atom? Why?

7. True or false? The hydrogen atom has a $3s$ orbital. Explain.

8. Differentiate among the terms *energy level, sublevel,* and *orbital.*

9. Make sense of the fact that metals tend to lose electrons and nonmetals tend to gain electrons. Use the periodic table to support your answer.

10. Show how using the periodic table helps you find the expected electron configuration of any element.

For Questions 11–13, you will need to consider ionizations beyond the first ionization energy. For example, the second ionization energy is the energy to remove a second electron from an element.

11. Compare the first ionization energy of helium to its second ionization energy, remembering that both electrons come from the $1s$ orbital.

12. Which would you expect to have a larger second ionization energy, lithium or beryllium? Why?

13. The first four ionization energies for elements X and Y are shown below. The units are not kJ/mol.

	X	Y
first	170	200
second	350	400
third	1800	3500
fourth	2500	5000

Identify the elements X and Y. There may be more than one answer, so explain completely.

14. Explain what is meant by the term "excited state" as it applies to an electron. Is an electron in an excited state higher or lower in energy than an electron in the ground state? Is an electron in an excited state more or less stable than an electron in the ground state?

15. What does it mean when we say energy levels are *quantized?*

16. What evidence do we have that energy levels in an atom are quantized? State and explain the evidence.

17. Explain the hydrogen emission spectrum. Why is it significant that the color emitted is not white? How does the emission spectrum support the idea of quantized energy levels?

18. There are an infinite number of allowed transitions in the hydrogen atom. Why don't we see more lines in the emission spectrum for hydrogen?

19. You have learned that each orbital is allowed two electrons, and this pattern is evident on the periodic table. What if each orbital was allowed three electrons? How would this change the appearance of the periodic table? For example, what would be the atomic numbers of the noble gases?

20. Atom A has valence electrons that are lower in energy than the valence electrons of Atom B. Which atom has the higher ionization energy? Explain.

21. Consider the following waves representing electromagnetic radiation:

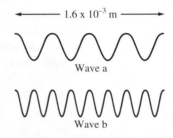

OWL Questions and problems for this chapter are available in OWL. Use the OWL access card that came with this textbook to access assigned questions and problems for this chapter.

12 Chemical Bonding

● The ionic structure of boron. *(Artem Oganov/Stony Brook University, New York)*

Diamond, composed of carbon atoms bonded together to produce one of the hardest materials known, makes a beautiful gemstone.

Tino Hammid

The world around us is composed almost entirely of compounds and mixtures of compounds. Rocks, coal, soil, petroleum, trees, and human beings are all complex mixtures of chemical compounds in which different kinds of atoms are bound together. Most of the pure elements found in the earth's crust also contain many atoms bound together. In a gold nugget each gold atom is bound to many other gold atoms, and in a diamond many carbon atoms are bonded very strongly to each other. Substances composed of unbound atoms do exist in nature, but they are very rare. (Examples include the argon atoms in the atmosphere and the helium atoms found in natural gas reserves.)

The manner in which atoms are bound together has a profound effect on the chemical and physical properties of substances. For example, both graphite and diamond are composed solely of carbon atoms. However, graphite is a soft, slippery material used as a lubricant in locks, and diamond is one of the hardest materials known, valuable both as a gemstone and in industrial cutting tools. Why do these materials, both composed solely of carbon atoms, have such different properties? The answer lies in the different ways in which the carbon atoms are bound to each other in these substances.

Molecular bonding and structure play the central role in determining the course of chemical reactions, many of which are vital to our survival. Most reactions in biological systems are very sensitive to the structures of the participating molecules; in fact, very subtle differences in shape sometimes serve to channel the chemical reaction one way rather than another. Molecules that act as drugs must have exactly the right structure to perform their functions correctly. Structure also plays a central role in our senses of smell and taste. Substances have a particular odor because they fit into the specially shaped receptors in our nasal passages. Taste is also dependent on molecular shape, as we discuss in the "Chemistry in Focus" on page 383.

To understand the behavior of natural materials, we must understand the nature of chemical bonding and the factors that control the structures of compounds. In this chapter, we will present various classes of compounds that illustrate the different types of bonds. We will then develop models to describe the structure and bonding that characterize the materials found in nature.

12.1   Types of Chemical Bonds

OBJECTIVES: To learn about ionic and covalent bonds and explain how they are formed.
• To learn about the polar covalent bond.

A water molecule.

What is a chemical bond? Although there are several possible ways to answer this question, we will define a **bond** as a force that holds groups of two or more atoms together and makes them function as a unit. For example, in water the fundamental unit is the H—O—H molecule, which we describe as being held together by the two O—H bonds. We can obtain information about the strength of a bond by measuring the energy required to break the bond, the **bond energy.**

Atoms can interact with one another in several ways to form aggregates. We will consider specific examples to illustrate the various types of chemical bonds.

In Chapter 7 we saw that when solid sodium chloride is dissolved in water, the resulting solution conducts electricity, a fact that convinces chemists that sodium chloride is composed of Na^+ and Cl^- ions. Thus, when sodium and chlorine react to form sodium chloride, electrons are transferred from the sodium atoms to the chlorine atoms to form Na^+ and Cl^- ions, which then aggregate to form solid sodium chloride. The resulting solid sodium chloride is a very sturdy material; it has a melting point of approximately 800 °C. The strong bonding forces present in sodium chloride result from the attractions among the closely packed, oppositely charged ions. This is an example of **ionic bonding.** Ionic substances are formed when an atom that loses electrons relatively easily reacts with an atom that has a high affinity for electrons. In other words, an **ionic compound** results when a metal reacts with a nonmetal.

We have seen that a bonding force develops when two very different types of atoms react to form oppositely charged ions. But how does a bonding force develop between two identical atoms? Let's explore this situation by considering what happens when two hydrogen atoms are brought close together, as shown in Figure 12.1. When hydrogen atoms are close together, the two electrons are simultaneously attracted to both nuclei. Note in Figure 12.1b how the electron probability increases between the two nuclei indicating that the electrons are shared by the two nuclei.

The type of bonding we encounter in the hydrogen molecule and in many other molecules where *electrons are shared by nuclei* is called **covalent bonding.** Note that in the H_2 molecule the electrons reside primarily in the space between the two nuclei, where they are attracted simultaneously by both protons. Although we will not go into detail about it here, the increased attractive forces in this area lead to the formation of the H_2 molecule from

Figure 12.1

The formation of a bond between two hydrogen atoms.

When two hydrogen atoms come close together, the two electrons are attracted simultaneously by both nuclei. This produces the bond. Note the relatively large electron probability between the nuclei indicating sharing of the electrons.

Two separate hydrogen atoms.

Hydrogen atoms sufficiently far apart to have no interaction

H atom

H atom

H_2 molecule

a

b

Figure 12.2

Probability representations of the electron sharing in HF.

Ionic and covalent bonds are the extreme bond types.

the two separated hydrogen atoms. When we say that a bond is formed between the hydrogen atoms, we mean that the H_2 molecule is more stable than two separated hydrogen atoms by a certain quantity of energy (the bond energy).

So far we have considered two extreme types of bonding. In ionic bonding, the participating atoms are so different that one or more electrons are transferred to form oppositely charged ions. The bonding results from the attractions among these ions. In covalent bonding, two identical atoms share electrons equally. The bonding results from the mutual attraction of the two nuclei for the shared electrons. Between these extremes are intermediate cases in which the atoms are not so different that electrons are completely transferred but are different enough so that unequal sharing of electrons results, forming what is called a **polar covalent bond.** The hydrogen fluoride (HF) molecule contains this type of bond, which produces the following charge distribution,

$$H\!-\!F$$
$$\delta^+ \quad \delta^-$$

where δ (delta) is used to indicate a partial or fractional charge.

The most logical explanation for the development of *bond polarity* (the partial positive and negative charges on the atoms in such molecules as HF) is that the electrons in the bonds are not shared equally. For example, we can account for the polarity of the HF molecule by assuming that the fluorine atom has a stronger attraction than the hydrogen atom for the shared electrons (Figure 12.2). Because bond polarity has important chemical implications, we find it useful to assign a number that indicates an atom's ability to attract shared electrons. In the next section we show how this is done.

12.2   Electronegativity

OBJECTIVE: To understand the nature of bonds and their relationship to electro-negativity.

We saw in the previous section that when a metal and a nonmetal react, one or more electrons are transferred from the metal to the nonmetal to give ionic bonding. On the other hand, two identical atoms react to form a covalent bond in which electrons are shared equally. When *different* nonmetals react, a bond forms in which electrons are shared *unequally*, giving a polar

Increasing electronegativity

Decreasing electronegativity

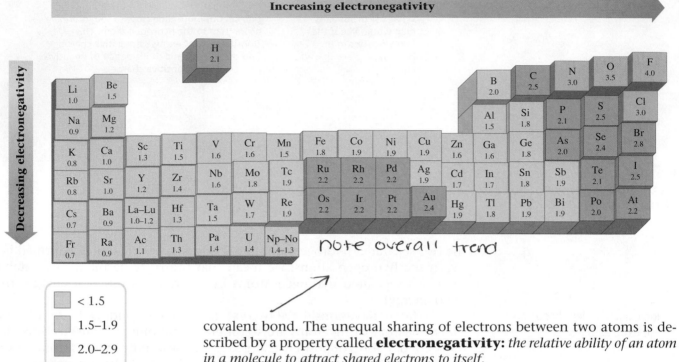

note overall trend

☐	< 1.5
☐	1.5–1.9
☐	2.0–2.9
☐	3.0–4.0

Figure 12.3

Electronegativity values for selected elements. Note that electronegativity generally increases across a period and decreases down a group. Note also that metals have relatively low electronegativity values and that nonmetals have relatively high values.

covalent bond. The unequal sharing of electrons between two atoms is described by a property called **electronegativity:** *the relative ability of an atom in a molecule to attract shared electrons to itself.*

Chemists determine electronegativity values for the elements (Figure 12.3) by measuring the polarities of the bonds between various atoms. Note that electronegativity generally increases going from left to right across a period and decreases going down a group for the representative elements. The range of electronegativity values is from 4.0 for fluorine to 0.7 for cesium and francium. Remember, the higher the atom's electronegativity value, the closer the shared electrons tend to be to that atom when it forms a bond.

The polarity of a bond depends on the *difference* between the electronegativity values of the atoms forming the bond. If the atoms have very similar electronegativities, the electrons are shared almost equally and the bond shows little polarity. If the atoms have very different electronegativity values, a very polar bond is formed. In extreme cases one or more electrons are actually transferred, forming ions and an ionic bond. For example, when an element from Group 1 (electronegativity values of about 0.8) reacts with an element from Group 7 (electronegativity values of about 3), ions are formed and an ionic substance results.

The relationship between electronegativity and bond type is shown in Table 12.1. The various types of bonds are summarized in Figure 12.4.

judge polarity of covalent bond

Table 12.1 The Relationship Between Electronegativity and Bond Type

Electronegativity Difference Between the Bonding Atoms	Bond Type	Covalent Character	Ionic Character
Zero	Covalent		
↓	↓		
Intermediate	Polar covalent	Decreases	Increases
↓	↓		
Large	Ionic		

a	b	c
A covalent bond formed between identical atoms.	A polar covalent bond, with both ionic and covalent components.	An ionic bond, with no electron sharing.

Figure 12.4

The three possible types of bonds.

EXAMPLE 12.1 **Using Electronegativity to Determine Bond Polarity**

Using the electronegativity values given in Figure 12.3, arrange the following bonds in order of increasing polarity: H—H, O—H, Cl—H, S—H, and F—H.

SOLUTION

The polarity of the bond increases as the difference in electronegativity increases. From the electronegativity values in Figure 12.3, the following variation in bond polarity is expected (the electronegativity value appears below each element).

Bond	Electronegativity Values	Difference in Electronegativity Values	Bond Type	Polarity
H—H	(2.1)(2.1)	2.1 − 2.1 = 0	Covalent	
S—H	(2.5)(2.1)	2.5 − 2.1 = 0.4	Polar covalent	
Cl—H	(3.0)(2.1)	3.0 − 2.1 = 0.9	Polar covalent	increasing
O—H	(3.5)(2.1)	3.5 − 2.1 = 1.4	Polar covalent	
F—H	(4.0)(2.1)	4.0 − 2.1 = 1.9	Polar covalent	

Therefore, in order of increasing polarity, we have

H—H   S—H   Cl—H   O—H   F—H

Least polar   Most polar

Self-Check EXERCISE 12.1 For each of the following pairs of bonds, choose the bond that will be more polar.

    a. H—P, H—C

    b. O—F, O—I

    c. N—O, S—O

    d. N—H, Si—H

See Problems 12.17 through 12.20. ■

Bond Polarity and Dipole Moments

OBJECTIVE: To understand bond polarity and how it is related to molecular polarity.

We saw in Section 12.1 that hydrogen fluoride has a positive end and a negative end. A molecule such as HF that has a center of positive charge and a center of negative charge is said to have a **dipole moment.** The dipolar character of a molecule is often represented by an arrow. This arrow points toward the negative charge center, and its tail indicates the positive center of charge:

δ^+ δ^-

Any diatomic (two-atom) molecule that has a polar bond has a dipole moment. Some polyatomic (more than two atoms) molecules also have dipole moments. For example, because the oxygen atom in the water molecule has a greater electronegativity than the hydrogen atoms, the electrons are not shared equally. This results in a charge distribution (Figure 12.5) that causes the molecule to behave as though it had two centers of charge—one positive and one negative. So the water molecule has a dipole moment.

The fact that the water molecule is polar (has a dipole moment) has a profound impact on its properties. In fact, it is not overly dramatic to state that the polarity of the water molecule is crucial to life as we know it on earth. Because water molecules are polar, they can surround and attract both positive and negative ions (Figure 12.6). These attractions allow ionic materials to dissolve in water. Also, the polarity of water molecules causes them to attract each other strongly (Figure 12.7). This means that much energy is required to change water from a liquid to a gas (the molecules must be separated from each other to undergo this change of state). Therefore, it is the polarity of the water molecule that causes water to remain a liquid at the temperatures on the earth's surface. If it were nonpolar, water would be a gas and the oceans would be empty.

δ^+ H

O $2\delta^-$

H

δ^+

a

The charge distribution in the water molecule. The oxygen has a charge of $2\delta^-$ because it pulls δ^- of charge from each hydrogen atom ($\delta^- + \delta^- = 2\delta^-$).

Center of positive charge

Center of negative charge

b

The water molecule behaves as if it had a positive end and a negative end, as indicated by the arrow.

Figure 12.5

we use this to understand solubility factors

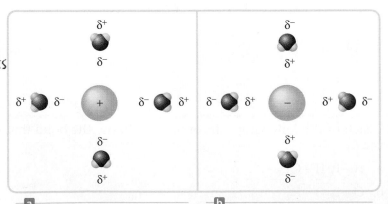

δ^+

δ^-

δ^+ δ^- + δ^- δ^+

δ^-

δ^+

δ^-

δ^+

δ^- δ^+ − δ^+ δ^-

δ^+

δ^-

a

Polar water molecules are strongly attracted to positive ions by their negative ends.

b

They are also strongly attracted to negative ions by their positive ends.

Figure 12.6

Figure 12.7

Polar water molecules are strongly attracted to each other.

these attractions are called hydrogen-bonding intermolecular forces of attraction

12.4 Stable Electron Configurations and Charges on Ions

OBJECTIVES: To learn about stable electron configurations. • To learn to predict the formulas of ionic compounds.

We have seen many times that when a metal and a nonmetal react to form an ionic compound, the metal atom loses one or more electrons to the non-metal. In Chapter 5, where binary ionic compounds were introduced, we saw that in these reactions, Group 1 metals always form 1+ cations, Group 2 metals always form 2+ cations, and aluminum in Group 3 always forms a 3+ cation. For the nonmetals, the Group 7 elements always form 1− anions, and the Group 6 elements always form 2− anions. This is further illustrated in Table 12.2.

Notice something very interesting about the ions in Table 12.2: they all have the electron configuration of neon, a noble gas. That is, sodium loses its one valence electron (the $3s$) to form Na^+, which has an [Ne] electron configuration. Likewise, Mg loses its two valence electrons to form Mg^{2+}, which also has an [Ne] electron configuration. On the other hand, the nonmetal atoms gain just the number of electrons needed for them to achieve the noble gas electron configuration. The O atom gains two electrons and the F atom gains one electron to give O^{2-} and F^-, respectively, both of which have the [Ne] electron configuration. We can summarize these observations as follows:

Electron Configurations of Ions

1. Representative (main-group) metals form ions by losing enough electrons to achieve the configuration of the previous noble gas (that is, the noble gas that occurs before the metal in question on the periodic table). For example, note from the periodic table inside the front cover of the text that neon is the noble gas previous to sodium and magnesium. Similarly, helium is the noble gas previous to lithium and beryllium.
2. Nonmetals form ions by gaining enough electrons to achieve the configuration of the next noble gas (that is, the noble gas that follows the element in question on the periodic table). For example, note that neon is the noble gas that follows oxygen and fluorine, and argon is the noble gas that follows sulfur and chlorine.

Table 12.2 The Formation of Ions by Metals and Nonmetals

		Electron Configuration	
Group	Ion Formation	Atom	Ion
1	$Na \rightarrow Na^+ + e^-$	$[Ne]3s^1 \xrightarrow{\;e^-\;lost\;}$	[Ne]
2	$Mg \rightarrow Mg^{2+} + 2e^-$	$[Ne]3s^2 \xrightarrow{\;2e^-\;lost\;}$	[Ne]
3	$Al \rightarrow Al^{3+} + 3e^-$	$[Ne]3s^23p^1 \xrightarrow{\;3e^-\;lost\;}$	[Ne]
6	$O + 2e^- \rightarrow O^{2-}$	$[He]2s^22p^4 + 2e^- \rightarrow [He]2s^22p^6 = $ [Ne]	
7	$F + e^- \rightarrow F^-$	$[He]2s^22p^5 + e^- \rightarrow [He]2s^22p^6 = $ [Ne]	

Atoms in stable compounds almost always have a noble gas electron configuration.

This brings us to an important general principle. In observing millions of stable compounds, chemists have learned that **in almost all stable chemical compounds of the representative elements, all of the atoms have achieved a noble gas electron configuration.** The importance of this observation cannot be overstated. It forms the basis for all of our fundamental ideas about why and how atoms bond to each other.

We have already seen this principle operating in the formation of ions (see Table 12.2). We can summarize this behavior as follows: when representative metals and nonmetals react, they transfer electrons in such a way that both the cation and the anion have noble gas electron configurations.

On the other hand, when nonmetals react with each other, they share electrons in ways that lead to a noble gas electron configuration for each atom in the resulting molecule. For example, oxygen ([He]$2s^2 2p^4$), which needs two more electrons to achieve an [Ne] configuration, can get these electrons by combining with two H atoms (each of which has one electron),

O: [He]   2s   2p

to form water, H_2O. This fills the valence orbitals of oxygen.

In addition, each H shares two electrons with the oxygen atom,

which fills the H 1s orbital, giving it a $1s^2$ or [He] electron configuration. We will have much more to say about covalent bonding in Section 12.6.

At this point let's summarize the ideas we have introduced so far.

Electron Configurations and Bonding

1. When a *nonmetal and a Group 1, 2, or 3 metal* react to form a binary ionic compound, the ions form in such a way that the valence-electron configuration of the *nonmetal* is *completed* to achieve the configuration of the *next* noble gas, and the valence orbitals of the *metal* are *emptied* to achieve the configuration of the *previous* noble gas. In this way both ions achieve noble gas electron configurations.
2. When *two nonmetals* react to form a covalent bond, they share electrons in a way that completes the valence-electron configurations of both atoms. That is, both nonmetals attain noble gas electron configurations by sharing electrons.

▶ Predicting Formulas of Ionic Compounds

Now that we know something about the electron configurations of atoms, we can explain *why* these various ions are formed.

To show how to predict what ions form when a metal reacts with a nonmetal, we will consider the formation of an ionic compound from calcium and oxygen. We can predict what compound will form by considering the valence electron configurations of the following two atoms:

Ca: [Ar]$4s^2$
O: [He]$2s^2 2p^4$

From Figure 12.3 we see that the electronegativity of oxygen (3.5) is much greater than that of calcium (1.0), giving a difference of 2.5. Because of this large difference, electrons are transferred from calcium to oxygen to form an oxygen anion and a calcium cation. How many electrons are transferred? We can base our prediction on the observation that noble gas configurations are the most stable. Note that oxygen needs two electrons to fill its valence orbitals ($2s$ and $2p$) and achieve the configuration of neon ($1s^2 2s^2 2p^6$), which is the next noble gas.

$$O + 2e^- \rightarrow O^{2-}$$
$$[He]2s^2 2p^4 + 2e^- \rightarrow [He]2s^2 2p^6, \text{ or } [Ne]$$

And by losing two electrons, calcium can achieve the configuration of argon (the previous noble gas).

$$Ca \rightarrow Ca^{2+} + 2e^-$$
$$[Ar]4s^2 \rightarrow [Ar] + 2e^-$$

Two electrons are therefore transferred as follows:

$$Ca + O \rightarrow Ca^{2+} + O^{2-}$$
$$2e^-$$

To predict the formula of the ionic compound, we use the fact that chemical compounds are always electrically neutral—they have the same total quantities of positive and negative charges. In this case we must have equal numbers of Ca^{2+} and O^{2-} ions, and the empirical formula of the compound is CaO.

The same principles can be applied to many other cases. For example, consider the compound formed from aluminum and oxygen. Aluminum has the electron configuration $[Ne]3s^2 3p^1$. To achieve the neon configuration, aluminum must lose three electrons, forming the Al^{3+} ion.

$$Al \rightarrow Al^{3+} + 3e^-$$
$$[Ne]3s^2 3p^1 \rightarrow [Ne] + 3e^-$$

3 × (2−) balances 2 × (3+).

Therefore, the ions will be Al^{3+} and O^{2-}. Because the compound must be electrically neutral, there will be three O^{2-} ions for every two Al^{3+} ions, and the compound has the empirical formula Al_2O_3.

Table 12.3 shows common elements that form ions with noble gas electron configurations in ionic compounds.

Notice that our discussion in this section refers to metals in Groups 1, 2, and 3 (the representative metals). The transition metals exhibit more complicated behavior (they form a variety of ions), which we will not be concerned with in this text.

Table 12.3 Common Ions with Noble Gas Configurations in Ionic Compounds

Group 1	Group 2	Group 3	Group 6	Group 7	Electron Configuration
Li^+	Be^{2+}				[He]
Na^+	Mg^{2+}	Al^{3+}	O^{2-}	F^-	[Ne]
K^+	Ca^{2+}		S^{2-}	Cl^-	[Ar]
Rb^+	Sr^{2+}		Se^{2-}	Br^-	[Kr]
Cs^+	Ba^{2+}		Te^{2-}	I^-	[Xe]

easier to lose electrons (e⁻) to become like

easier to gain electrons (e⁻) to become like

Ionic Bonding and Structures of Ionic Compounds

OBJECTIVES: To learn about ionic structures. • To understand factors governing ionic size.

When metals and nonmetals react, the resulting ionic compounds are very stable; large amounts of energy are required to "take them apart." For example, the melting point of sodium chloride is approximately 800 °C. The strong bonding in these ionic compounds results from the attractions among the oppositely charged cations and anions.

We write the formula of an ionic compound such as lithium fluoride simply as LiF, but this is really the empirical, or simplest, formula. The actual solid contains huge and equal numbers of Li^+ and F^- ions packed together in a way that maximizes the attractions of the oppositely charged ions. A representative part of the lithium fluoride structure is shown in Figure 12.8a. In this structure the larger F^- ions are packed together like hard spheres, and the much smaller Li^+ ions are interspersed regularly among the F^- ions. The structure shown in Figure 12.8b represents only a tiny part of the actual structure, which continues in all three dimensions with the same pattern as that shown.

> When spheres are packed together, they do not fill up all of the space. The spaces (holes) that are left can be occupied by smaller spheres.

The structures of virtually all binary ionic compounds can be explained by a model that involves packing the ions as though they were hard spheres. The larger spheres (usually the anions) are packed together, and the small ions occupy the interstices (spaces or holes) among them.

To understand the packing of ions it helps to realize that *a cation is always smaller than the parent atom, and an anion is always larger than the parent atom.* This makes sense because when a metal loses all of its valence electrons to form a cation, it gets much smaller. On the other hand, in forming an anion, a nonmetal gains enough electrons to achieve the next noble gas electron configuration and so becomes much larger. The relative sizes of the Group 1 and Group 7 atoms and their ions are shown in Figure 12.9.

a

This structure represents the ions as packed spheres.

b

This structure shows the positions (centers) of the ions. The spherical ions are packed in the way that maximizes the ionic attractions.

Figure 12.8

The structure of lithium fluoride.

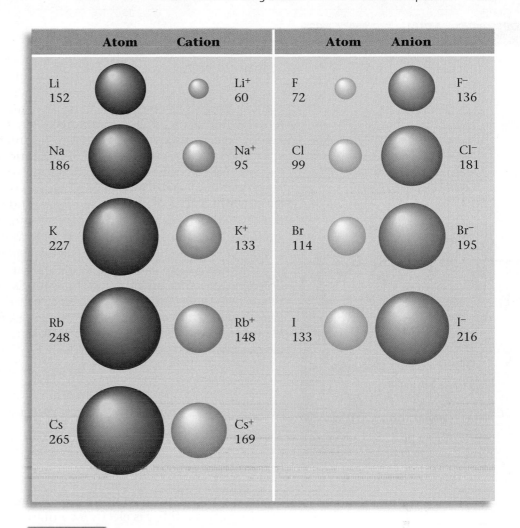

Atom	Cation	Atom	Anion
Li 152	Li$^+$ 60	F 72	F$^-$ 136
Na 186	Na$^+$ 95	Cl 99	Cl$^-$ 181
K 227	K$^+$ 133	Br 114	Br$^-$ 195
Rb 248	Rb$^+$ 148	I 133	I$^-$ 216
Cs 265	Cs$^+$ 169		

Figure 12.9

Relative sizes of some ions and their parent atoms. Note that cations are smaller and anions are larger than their parent atoms. The sizes (radii) are given in units of picometers (1 pm = 10^{-12} m).

▶ Ionic Compounds Containing Polyatomic Ions

So far in this chapter we have discussed only binary ionic compounds, which contain ions derived from single atoms. However, many compounds contain polyatomic ions: charged species composed of several atoms. For example, ammonium nitrate contains the NH_4^+ and NO_3^- ions. These ions with their opposite charges attract each other in the same way as do the simple ions in binary ionic compounds. However, the *individual* polyatomic ions are held together by covalent bonds, with all of the atoms behaving as a unit. For example, in the ammonium ion, NH_4^+, there are four N—H covalent bonds. Likewise the nitrate ion, NO_3^-, contains three covalent N—O bonds. Thus, although ammonium nitrate is an ionic compound because it contains the NH_4^+ and NO_3^- ions, it also contains covalent bonds in the individual polyatomic ions. When ammonium nitrate is dissolved in water, it behaves as a strong electrolyte like the binary ionic compounds sodium chloride and potassium bromide. As we saw in Chapter 7, this occurs because when an ionic solid dissolves, the ions are freed to move independently and can conduct an electric current.

The common polyatomic ions, which are listed in Table 5.4, are all held together by covalent bonds.

12.6 Lewis Structures

OBJECTIVE: To learn to write Lewis structures.

Remember that the electrons in the highest principal energy level of an atom are called the valence electrons.

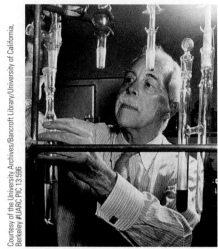

G. N. Lewis in his lab.

go Chemistry **Module 12: Drawing Lewis Electron Dot Structures** covers concepts in this section.

Bonding involves just the valence electrons of atoms. Valence electrons are transferred when a metal and a nonmetal react to form an ionic compound. Valence electrons are shared between nonmetals in covalent bonds.

The **Lewis structure** is a representation of a molecule that shows how the valence electrons are arranged among the atoms in the molecule. These representations are named after G. N. Lewis, who conceived the idea while lecturing to a class of general chemistry students in 1902. The rules for writing Lewis structures are based on observations of many molecules from which chemists have learned that the *most important requirement for the formation of a stable compound is that the atoms achieve noble gas electron configurations.*

We have already seen this rule operate in the reaction of metals and nonmetals to form binary ionic compounds. An example is the formation of KBr, where the K^+ ion has the [Ar] electron configuration and the Br^- ion has the [Kr] electron configuration. In writing Lewis structures, *we include only the valence electrons.* Using dots to represent valence electrons, we write the Lewis structure for KBr as follows:

$$K^+ \qquad\qquad [\,:\!\ddot{Br}\!:\,]^-$$

Noble gas configuration [Ar]   Noble gas configuration [Kr]

No dots are shown on the K^+ ion because it has lost its only valence electron (the $4s$ electron). The Br^- ion is shown with eight electrons because it has a filled valence shell.

Next we will consider Lewis structures for molecules with covalent bonds, involving nonmetals in the first and second periods. The principle of achieving a noble gas electron configuration applies to these elements as follows:

1. Hydrogen forms stable molecules where it shares two electrons. That is, it follows a **duet rule.** For example, when two hydrogen atoms, each with one electron, combine to form the H_2 molecule, we have

By sharing electrons, each hydrogen in H_2 has, in effect, two electrons; that is, each hydrogen has a filled valence shell.

2. Helium does not form bonds because its valence orbital is already filled; it is a noble gas. Helium has the electron configuration $1s^2$ and can be represented by the Lewis structure

$$He\!:$$

[He] configuration

3. The second-row nonmetals carbon through fluorine form stable molecules when they are surrounded by enough electrons to fill the valence orbitals—that is, the one *2s* and the three *2p* orbitals. Eight electrons are required to fill these orbitals, so these elements typically obey the **octet rule;** they are surrounded by eight electrons. An example is the F_2 molecule, which has the following Lewis structure:

> Carbon, nitrogen, oxygen, and fluorine almost always obey the octet rule in stable molecules.

F atom with seven valence electrons F_2 molecule   F atom with seven valence electrons

Note that each fluorine atom in F_2 is, in effect, surrounded by eight valence electrons, two of which are shared with the other atom. This is a **bonding pair** of electrons, as we discussed earlier. Each fluorine atom also has three pairs of electrons that are not involved in bonding. These are called **lone pairs** or **unshared pairs.**

4. Neon does not form bonds because it already has an octet of valence electrons (it is a noble gas). The Lewis structure is

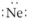

Note that only the valence electrons $(2s^2 2p^6)$ of the neon atom are represented by the Lewis structure. The $1s^2$ electrons are core electrons and are not shown.

> Lewis structures show only valence electrons.

Next we want to develop some general procedures for writing Lewis structures for molecules. Remember that Lewis structures involve only the valence electrons of atoms, so before we proceed, we will review the relationship of an element's position on the periodic table to the number of valence electrons it has. Recall that the group number gives the total number of valence electrons. For example, all Group 6 elements have six valence electrons (valence configuration $ns^2 np^4$).

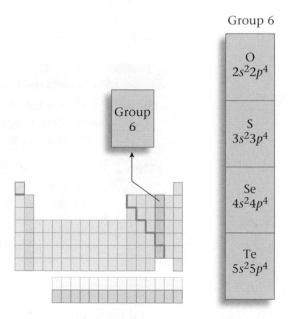

Similarly, all Group 7 elements have seven valence electrons (valence configuration ns^2np^5).

In writing the Lewis structure for a molecule, we need to keep the following things in mind:

1. We must include all the valence electrons from all atoms. The total number of electrons available is the sum of all the valence electrons from all the atoms in the molecule.

2. Atoms that are bonded to each other share one or more pairs of electrons.

3. The electrons are arranged so that each atom is surrounded by enough electrons to fill the valence orbitals of that atom. This means two electrons for hydrogen and eight electrons for second-row nonmetals.

The best way to make sure we arrive at the correct Lewis structure for a molecule is to use a systematic approach. We will use the approach summarized by the following rules.

Steps for Writing Lewis Structures

Step 1 Obtain the sum of the valence electrons from all of the atoms. Do not worry about keeping track of which electrons come from which atoms. It is the *total* number of valence electrons that is important.

Step 2 Use one pair of electrons to form a bond between each pair of bound atoms. For convenience, a line (instead of a pair of dots) is often used to indicate each pair of bonding electrons.

Step 3 Arrange the remaining electrons to satisfy the duet rule for hydrogen and the octet rule for each second-row element.

To see how these rules are applied, we will write the Lewis structures of several molecules.

To Bee or Not to Bee

One of the problems we face in modern society is how to detect illicit substances, such as drugs and explosives, in a convenient, accurate manner. Trained dogs are often used for this purpose because of their acute sense of smell. Now several researches are trying to determine whether insects, such as honeybees and wasps, can be even more effective chemical detectors. In fact, studies have shown that bees can be trained in just a few minutes to detect the smell of almost any chemical.

Scientists at Los Alamos National Laboratory in New Mexico are designing a portable device using bees that possibly could be used to sniff out drugs and bombs at airports, border crossings, and schools. They call their study the Stealthy Insect Sensor Project. The Los Alamos project is based on the idea that bees can be trained to associate the smell of a particular chemical with a sugary treat. Bees stick out their "tongues" when they detect a food source. By pairing a drop of sugar water with the scent of TNT (trinitrotoluene) or C-4 (composition 4) plastic explosive about six times, the bees can be trained to extend their proboscis at a whiff of the chemical alone. The bee bomb detector is about half the size of a shoe box and weighs 4 lb. Inside the box, bees are lined up in a row and strapped into straw-like tubes, then exposed to puffs of air as a camera monitors their reactions. The signals from the video camera are sent to a computer, which analyzes the bees' behavior and signals when the bees respond to the particular scent they have been trained to detect.

A project at the University of Georgia uses tiny parasitic wasps as a chemical detector. Wasps do not extend their tongues when they detect a scent. Instead, they communicate the discovery of a scent by body movements that the scientists call "dances." The device, called the Wasp Hound, contains a team of wasps in a hand-held ventilated cartridge that has a fan at one end to draw in air from outside. If the scent is one the wasps do not recognize, they continue flying randomly. However, if the scent is one the wasps have been conditioned to recognize, they cluster around the opening. A video camera paired with a computer analyzes their behavior and signals when a scent is detected.

The insect sensors are now undergoing field trials, which typically compare the effectiveness of insects to that of trained dogs. Initial results appear promising, but the effectiveness of these devices remains to be proved.

Los Alamos National Laboratory. Photo by Leroy Sanchez

A honeybee receives a fragrant reminder of its target scent each morning and responds by sticking out its proboscis.

EXAMPLE 12.2 | Writing Lewis Structures: Simple Molecules

Write the Lewis structure of the water molecule.

SOLUTION

We will follow the steps listed on page 372.

Step 1   Find the sum of the *valence* electrons for H_2O.

$$\underset{\substack{\uparrow \\ H \\ \text{(Group 1)}}}{1} + \underset{\substack{\uparrow \\ H \\ \text{(Group 1)}}}{1} + \underset{\substack{\uparrow \\ O \\ \text{(Group 6)}}}{6} = 8 \text{ valence electrons}$$

Step 2   Using a pair of electrons per bond, we draw in the two O—H bonds, using a line to indicate each pair of bonding electrons.

$$H—O—H$$

Note that

$$H—O—H \text{ represents } H : O : H$$

Step 3   We arrange the remaining electrons around the atoms to achieve a noble gas electron configuration for each atom. Four electrons have been used in forming the two bonds, so four electrons $(8 - 4)$ remain to be distributed. Each hydrogen is satisfied with two electrons (duet rule), but oxygen needs eight electrons to have a noble gas electron configuration. So the remaining four electrons are added to oxygen as two lone pairs. Dots are used to represent the lone pairs.

This is the correct Lewis structure for the water molecule. Each hydrogen shares two electrons, and the oxygen has four electrons and shares four to give a total of eight.

$$\underset{\substack{\uparrow \quad\quad \uparrow \quad\quad \uparrow \\ 2e^- \quad 8e^- \quad 2e^-}}{\text{H}\bigcirc\!\!\!\!\bigcirc\!\!\!\text{O}\!\!\!\bigcirc\!\!\!\!\bigcirc\text{H}}$$

Note that a line is used to represent a shared pair of electrons (bonding electrons) and dots are used to represent unshared pairs.

> $$H—\overset{..}{O}—H$$
>
> might also be drawn as
>
> $$H:\overset{..}{\underset{..}{O}}:H$$

Self-Check **EXERCISE 12.2**   Write the Lewis structure for HCl.

See Problems 12.59 through 12.62. ∎

12.7   Lewis Structures of Molecules with Multiple Bonds

OBJECTIVE: To learn how to write Lewis structures for molecules with multiple bonds.

Now let's write the Lewis structure for carbon dioxide.

Step 1   Summing the valence electrons gives

$$\underset{\substack{\uparrow \\ C \\ \text{(Group 4)}}}{4} + \underset{\substack{\uparrow \\ O \\ \text{(Group 6)}}}{6} + \underset{\substack{\uparrow \\ O \\ \text{(Group 6)}}}{6} = 16$$

CHEMISTRY *IN* FOCUS

Hiding Carbon Dioxide

As we discussed in Chapter 11 (see "Chemistry in Focus: Atmospheric Effects," page 326), global warming seems to be a reality. At the heart of this issue is the carbon dioxide produced by society's widespread use of fossil fuels. For example, in the United States, CO_2 makes up 81% of greenhouse gas emissions. Thirty percent of this CO_2 comes from coal-fired power plants used to produce electricity. One way to solve this problem would be to phase out coal-fired power plants. However, this outcome is not likely because the United States possesses so much coal (at least a 250-year supply) and coal is so cheap (about $0.03 per pound). Recognizing this fact, the U.S. government has instituted a research program to see if the CO_2 produced at power plants can be captured and sequestered (stored) underground in deep geological formations. The factors that need to be explored to determine whether sequestration is feasible are the capacities of underground storage sites and the chances that the sites will leak.

The injection of CO_2 into the earth's crust is already being undertaken by various oil companies. Since 1996, the Norwegian oil company Statoil has separated more than 1 million tons of CO_2 annually from natural gas and pumped it into a saltwater aquifer beneath the floor of the North Sea. In western Canada a group of oil companies has injected CO_2 from a North Dakota synthetic fuels plant into oil fields in an effort to increase oil recovery. The oil companies expect to store 22 million tons of CO_2 there and to produce 130 million barrels of oil over the next 20 years.

Sequestration of CO_2 has great potential as one method for decreasing the rate of global warming. Only time will tell whether it will work.

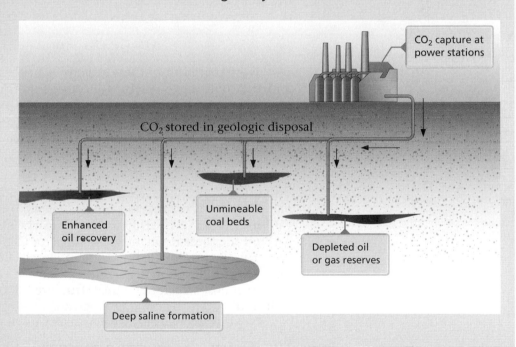

CO₂ capture at power stations

CO₂ stored in geologic disposal

Enhanced oil recovery

Unmineable coal beds

Depleted oil or gas reserves

Deep saline formation

O—C—O

represents

O:C:O

$:\ddot{O}—C—\ddot{O}:$

represents

$:\ddot{O}:C:\ddot{O}:$

Step 2 Form a bond between the carbon and each oxygen:

O—C—O

Step 3 Next, distribute the remaining electrons to achieve noble gas electron configurations on each atom. In this case twelve electrons (16 − 4) remain after the bonds are drawn. The distribution of these electrons is determined by a trial-and-error process. We have six pairs of electrons to distribute. Suppose we try three pairs on each oxygen to give

$:\ddot{O}—C—\ddot{O}:$

Is this correct? To answer this question we need to check two things:

1. The total number of electrons. There are sixteen valence electrons in this structure, which is the correct number.

2. The octet rule for each atom. Each oxygen has eight electrons around it, but the carbon has only four. This cannot be the correct Lewis structure.

How can we arrange the sixteen available electrons to achieve an octet for each atom? Suppose we place two shared pairs between the carbon and each oxygen:

Now each atom is surrounded by eight electrons, and the total number of electrons is sixteen, as required. This is the correct Lewis structure for carbon dioxide, which has two *double* bonds. A **single bond** involves two atoms sharing one electron pair. A **double bond** involves two atoms sharing two pairs of electrons.

$$\ddot{O}\!=\!C\!=\!\ddot{O}$$

represents

$$\ddot{O}::C::\ddot{O}$$

In considering the Lewis structure for CO_2, you may have come up with

$$:O\!\equiv\!C\!-\!\ddot{O}:\quad\text{or}\quad:\ddot{O}\!-\!C\!\equiv\!O:$$

Note that both of these structures have the required sixteen electrons and that both have octets of electrons around each atom (verify this for yourself). Both of these structures have a **triple bond** in which three electron pairs are shared. Are these valid Lewis structures for CO_2? Yes. So there really are three Lewis structures for CO_2:

$$:\ddot{O}\!-\!C\!\equiv\!O:\quad\ddot{O}\!=\!C\!=\!\ddot{O}\quad:O\!\equiv\!C\!-\!\ddot{O}:$$

$$:O\!\equiv\!C\!-\!\ddot{O}:$$

represents

$$:O:::C:\ddot{O}:$$

This brings us to a new term, **resonance.** A molecule shows resonance when *more than one Lewis structure can be drawn for the molecule.* In such a case we call the various Lewis structures **resonance structures.**

Of the three resonance structures for CO_2 shown above, the one in the center with two double bonds most closely fits our experimental information about the CO_2 molecule. In this text we will not be concerned about how to choose which resonance structure for a molecule gives the "best" description of that molecule's properties.

Next let's consider the Lewis structure of the CN^- (cyanide) ion.

Step 1   Summing the valence electrons, we have

$$CN^-$$
$$4 + 5 + 1 = 10$$

Note that the negative charge means an extra electron must be added.

Step 2   Draw a single bond (C—N).

Step 3   Next, we distribute the remaining electrons to achieve a noble gas configuration for each atom. Eight electrons remain to be distributed. We can try various possibilities, such as

$$\ddot{C}\!-\!\ddot{N}\quad\text{or}\quad:\ddot{C}\!-\!\ddot{N}:\quad\text{or}\quad:\ddot{C}\!-\!N:$$

These structures are incorrect. To show why none is a valid Lewis structure, count the electrons around the C and N atoms. In the left structure, neither

Broccoli—Miracle Food?

Eating the right foods is critical to our health. In particular, certain vegetables, although they do not enjoy a very jazzy image, seem especially important. A case in point is broccoli, a vegetable with a humble reputation that packs a powerful chemistry wallop.

Broccoli contains a chemical called sulforaphane, which has the following Lewis structure:

$$CH_3-S-(CH_2)_4-\overset{..}{N}=C=\overset{..}{\underset{..}{S}}:$$
$$\underset{:O:}{\overset{\|}{}}$$

Experiments indicate that sulforaphane furnishes protection against certain cancers by increasing the production of enzymes (called phase 2 enzymes) that "mop up" reactive molecules that can harm DNA. Sulforaphane also seems to combat bacteria. For example, among the most common harmful bacteria in humans is *Helicobacter pylori (H. pylori)*, which has been implicated in the development of several diseases of the stomach, including inflammation, cancer, and ulcers. Antibiotics are clearly the best treatment for *H. pylori* infections. However, especially in developing countries, where *H. pylori* is rampant, antibi-

otics are often too expensive to be available to the general population. In addition, the bacteria sometimes evade antibiotics by "hiding" in cells on the stomach walls and then reemerging after treatment ends.

Studies at Johns Hopkins in Baltimore and Vandoeuvre-les Nancy in France have shown that sulforaphane kills *H. pylori* (even when it has taken refuge in stomach-wall cells) at concentrations that are achievable by eating broccoli. The scientists at Johns Hopkins also found that sulforaphane seems to inhibit stomach cancer in mice. Although there are no guarantees that broccoli will keep you healthy, it might not hurt to add it to your diet.

$$:C\equiv N:$$

represents

$$:C:::N:$$

atom satisfies the octet rule. In the center structure, C has eight electrons but N has only four. In the right structure, the opposite is true. Remember that both atoms must simultaneously satisfy the octet rule. Therefore, the correct arrangement is

$$:C\equiv N:$$

(Satisfy yourself that both carbon and nitrogen have eight electrons.) In this case we have a triple bond between C and N, in which three electron pairs are shared. Because this is an anion, we indicate the charge outside of square brackets around the Lewis structure.

$$[:C\equiv N:]^-$$

In summary, sometimes we need double or triple bonds to satisfy the octet rule. Writing Lewis structures is a trial-and-error process. Start with single bonds between the bonded atoms and add multiple bonds as needed.

We will write the Lewis structure for NO_2^- in Example 12.3 to make sure the procedures for writing Lewis structures are clear.

EXAMPLE 12.3 **Writing Lewis Structures: Resonance Structures**

Write the Lewis structure for the NO_2^- anion.

SOLUTION

Step 1 Sum the valence electrons for NO_2^-.

$$\text{Valence electrons: } \underset{O}{6} + \underset{N}{5} + \underset{O}{6} + \underset{\substack{-1 \\ \text{charge}}}{1} = 18 \text{ electrons}$$

Step 2 Put in single bonds.

$$O\!-\!N\!-\!O$$

Step 3 Satisfy the octet rule. In placing the electrons, we find there are two Lewis structures that satisfy the octet rule:

$$[O\!=\!N\!-\!\ddot{O}\!:]^- \quad \text{and} \quad [:\!\ddot{O}\!-\!N\!=\!O\,]^-$$

Verify that each atom in these structures is surrounded by an octet of electrons. Try some other arrangements to see whether other structures exist in which the eighteen electrons can be used to satisfy the octet rule. It turns out that these are the only two that work. Note that this is another case where resonance occurs; there are two valid Lewis structures.

Self-Check EXERCISE 12.3 Ozone is a very important constituent of the atmosphere. At upper levels it protects us by absorbing high-energy radiation from the sun. Near the earth's surface it produces harmful air pollution. Write the Lewis structure for ozone, O_3.

See Problems 12.63 through 12.68. ■

Now let's consider a few more cases in Example 12.4.

EXAMPLE 12.4 **Writing Lewis Structures: Summary**

Give the Lewis structure for each of the following:

a. HF   e. CF_4

b. N_2 f. NO^+

c. NH_3 g. NO_3^-

d. CH_4

You may wonder how to decide which atom is the central atom in molecules of binary compounds. In cases where there is one atom of a given element and several atoms of a second element, the single atom is almost always the central atom of the molecule.

SOLUTION

In each case we apply the three steps for writing Lewis structures. Recall that lines are used to indicate shared electron pairs and that dots are used to indicate nonbonding pairs (lone pairs). The table on page 379 summarizes our results.

Self-Check EXERCISE 12.4 Write the Lewis structures for the following molecules:

a. NF_3 d. PH_3 g. NH_4^+

b. O_2 e. H_2S h. ClO_3^-

c. CO   f. SO_4^{2-} i. SO_2

See Problems 12.55 through 12.68. ■

Molecule or Ion	Total Valence Electrons	Draw Single Bonds	Calculate Number of Electrons Remaining	Use Remaining Electrons to Achieve Noble Gas Configurations	Check Atom	Check Electrons
a. HF	$1 + 7 = 8$	H—F	$8 - 2 = 6$	H—F̈:	H F	2 8
b. N_2	$5 + 5 = 10$	N—N	$10 - 2 = 8$	:N≡N:	N	8

c. NH_3	$5 + 3(1) = 8$	H—N—H
		    |
		    H
			$8 - 6 = 2$	H—N̈—H
				    |
				    H
					H
					N
						2
						8

d. CH_4	$4 + 4(1) = 8$	H
		    |
		H—C—H
		    |
		    H
			$8 - 8 = 0$	H
				    |
				H—C—H
				    |
				    H
					H
					C
						2
						8

e. CF_4	$4 + 4(7) = 32$	F
		    |
		F—C—F
		    |
		    F
			$32 - 8 = 24$	:F̈:
				    |
				:F̈—C—F̈:
				    |
				   :F̈:
					F
					C
						8
						8

f. NO^+	$5 + 6 - 1 = 10$	N—O	$10 - 2 = 8$	$[:N≡O:]^+$	N O	8 8
g. NO_3^-	$5 + 3(6) + 1 = 24$	$\begin{bmatrix} O \\ \| \\ N \\ O \quad O \end{bmatrix}$	$24 - 6 = 18$	$\begin{bmatrix} :\ddot{O}: \\ \| \\ N \\ .\ddot{O} \quad \ddot{O}. \end{bmatrix}^-$	N O	8 8
				NO₃⁻ shows resonance $\begin{bmatrix} :\ddot{O}: \\ \| \\ N \\ .\ddot{O} \quad \ddot{O}. \end{bmatrix}^-$	N O	8 8
				$\begin{bmatrix} :\ddot{O}: \\ \| \\ N \\ .\ddot{O} \quad \ddot{O}. \end{bmatrix}^-$	N O	8 8

Remember, when writing Lewis structures, you don't have to worry about which electrons come from which atoms in a molecule. It is best to think of a molecule as a new entity that uses all the available valence electrons from the various atoms to achieve the strongest possible bonds. Think of the valence electrons as belonging to the molecule, rather than to the individual atoms. Simply distribute all the valence electrons so that noble gas electron configurations are obtained for each atom, without regard to the origin of each particular electron.

▶ Some Exceptions to the Octet Rule

The idea that covalent bonding can be predicted by achieving noble gas electron configurations for all atoms is a simple and very successful idea. The rules we have used for Lewis structures describe correctly the bonding in

Donald Clegg

Figure 12.10

When liquid oxygen is poured between the poles of a magnet, it "sticks" until it boils away. This shows that the O_2 molecule has unpaired electrons (is paramagnetic).

most molecules. However, with such a simple model, some exceptions are inevitable. Boron, for example, tends to form compounds in which the boron atom has fewer than eight electrons around it—that is, it does not have a complete octet. Boron trifluoride, BF_3, a gas at normal temperatures and pressures, reacts very energetically with molecules such as water and ammonia that have unshared electron pairs (lone pairs).

The violent reactivity of BF_3 with electron-rich molecules arises because the boron atom is electron-deficient. The Lewis structure that seems most consistent with the properties of BF_3 (twenty-four valence electrons) is

Note that in this structure the boron atom has only six electrons around it. The octet rule for boron could be satisfied by drawing a structure with a double bond between the boron and one of the fluorines. However, experiments indicate that each B—F bond is a single bond in accordance with the above Lewis structure. This structure is also consistent with the reactivity of BF_3 with electron-rich molecules. For example, BF_3 reacts vigorously with NH_3 to form H_3NBF_3.

Note that in the product H_3NBF_3, which is very stable, boron has an octet of electrons.

It is also characteristic of beryllium to form molecules where the beryllium atom is electron-deficient.

The compounds containing the elements carbon, nitrogen, oxygen, and fluorine are accurately described by Lewis structures in the vast majority of cases. However, there are a few exceptions. One important example is the oxygen molecule, O_2. The following Lewis structure that satisfies the octet rule can be drawn for O_2 (see Self-Check Exercise 12.4).

However, this structure does not agree with the *observed behavior* of oxygen. For example, the photos in Figure 12.10 show that when liquid oxygen is poured between the poles of a strong magnet, it "sticks" there until it boils away. This provides clear evidence that oxygen is paramagnetic—that is, it contains unpaired electrons. However, the above Lewis structure shows only pairs of electrons. That is, no unpaired electrons are shown. There is no simple Lewis structure that satisfactorily explains the paramagnetism of the O_2 molecule.

Any molecule that contains an odd number of electrons does not conform to our rules for Lewis structures. For example, NO and NO_2 have eleven and seventeen valence electrons, respectively, and conventional Lewis structures cannot be drawn for these cases.

Even though there are exceptions, most molecules can be described by Lewis structures in which all the atoms have noble gas electron configurations, and this is a very useful model for chemists.

Paramagnetic substances have unpaired electrons and are drawn toward the space between a magnet's poles.

12.8 Molecular Structure

OBJECTIVE: To understand molecular structure and bond angles.

So far in this chapter we have considered the Lewis structures of molecules. These structures represent the arrangement of the *valence electrons* in a molecule. We use the word *structure* in another way when we talk about the **molecular structure** or **geometric structure** of a molecule. These terms refer to the three-dimensional arrangement of the *atoms* in a molecule. For example, the water molecule is known to have the molecular structure

$$\underset{H \quad\quad H}{O}$$

which is often called "bent" or "V-shaped." To describe the structure more precisely, we often specify the **bond angle.** For the H_2O molecule the bond angle is about 105°.

$$\underset{\underset{\sim 105°}{H \quad\quad H}}{O}$$

Computer graphic of a linear molecule containing three atoms

Computer graphic of a trigonal planar molecule

Computer graphic of a tetrahedral molecule

On the other hand, some molecules exhibit a **linear structure** (all atoms in a line). An example is the CO_2 molecule.

$$\underset{180°}{O{-}C{-}O}$$

Note that a linear molecule has a 180° bond angle.

A third type of molecular structure is illustrated by BF_3, which is planar or flat (all four atoms in the same plane) with 120° bond angles.

$$\underset{\underset{120°}{F \quad\quad\quad F}}{\overset{F}{\underset{120°\quad B \quad120°}{}}}$$

The name usually given to this structure is **trigonal planar structure,** although triangular might seem to make more sense.

Another type of molecular structure is illustrated by methane, CH_4. This molecule has the molecular structure shown in Figure 12.11, which is called a **tetrahedral structure** or a **tetrahedron.** The dashed lines shown connecting the H atoms define the four identical triangular faces of the tetrahedron.

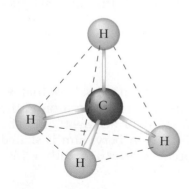

Figure 12.11

The tetrahedral molecular structure of methane. This representation is called a ball-and-stick model; the atoms are represented by balls and the bonds by sticks. The dashed lines show the outline of the tetrahedron.

In the next section we will discuss these various molecular structures in more detail. In that section we will learn how to predict the molecular structure of a molecule by looking at the molecule's Lewis structure.

12.9 Molecular Structure: The VSEPR Model

OBJECTIVE: To learn to predict molecular geometry from the number of electron pairs.

The structures of molecules play a very important role in determining their properties. For example, as we see in the "Chemistry in Focus" on page 383, taste is directly related to molecular structure. Structure is particularly important for biological molecules; a slight change in the structure of a large biomolecule can completely destroy its usefulness to a cell and may even change the cell from a normal one to a cancerous one.

Many experimental methods now exist for determining the molecular structure of a molecule—that is, the three-dimensional arrangement of the atoms. These methods must be used when accurate information about the structure is required. However, it is often useful to be able to predict the *approximate* molecular structure of a molecule. In this section we consider a simple model that allows us to do this. This model, called the **valence shell electron pair repulsion (VSEPR) model,** is useful for predicting the molecular structures of molecules formed from nonmetals. The main idea of this model is that *the structure around a given atom is determined by minimizing repulsions between electron pairs.* This means that the bonding and nonbonding electron pairs (lone pairs) around a given atom are positioned *as far apart as possible.* To see how this model works, we will first consider the molecule $BeCl_2$, which has the following Lewis structure (it is an exception to the octet rule):

$$:\!\ddot{Cl}\!-\!Be\!-\!\ddot{Cl}\!:$$

Note that there are two pairs of electrons around the beryllium atom. What arrangement of these electron pairs allows them to be as far apart as possible to minimize the repulsions? The best arrangement places the pairs on opposite sides of the beryllium atom at 180° from each other.

$$\overset{\displaystyle -Be-}{\underset{180°}{\smile}}$$

This is the maximum possible separation for two electron pairs. Now that we have determined the optimal arrangement of the electron pairs around the central atom, we can specify the molecular structure of $BeCl_2$—that is, the positions of the atoms. Because each electron pair on beryllium is shared with a chlorine atom, the molecule has a **linear structure** with a 180° bond angle.

$$:\!\ddot{Cl}\!-\!Be\!-\!\ddot{Cl}\!:$$
$$180°$$

Whenever two pairs of electrons are present around an atom, they should always be placed at an angle of 180° to each other to give a linear arrangement.

Next let's consider BF_3, which has the following Lewis structure (it is another exception to the octet rule):

$$:\!\ddot{F}\!:$$
$$\mid$$
$$:\!\ddot{F}\!-\!B\!-\!\ddot{F}\!:$$

Taste—It's the Structure That Counts

Why do certain substances taste sweet, sour, bitter, or salty? Of course, it has to do with the taste buds on our tongues. But how do these taste buds work? For example, why does sugar taste sweet to us? The answer to this question remains elusive, but it does seem clear that sweet taste depends on how certain molecules fit the "sweet receptors" in our taste buds.

One of the mysteries of the sweet taste sensation is the wide variety of molecules that taste sweet. For example, the many types of sugars include glucose and sucrose (table sugar). The first artificial sweetener was probably the Romans' sapa (see "Chemistry in Focus: Sugar of Lead" in Chapter 5), made by boiling wine in lead vessels to produce a syrup that contained lead acetate, $Pb(C_2H_3O_2)_2$, called sugar of lead because of its sweet taste. Other widely used modern artificial sweeteners include saccharin, aspartame, sucralose, and steviol, whose structures are shown in the accompanying figure. The structure of steviol is shown in simplified form. Each vertex represents a carbon atom, and not all of the hydrogen atoms are shown. Note the great disparity of structures for these sweet-tasting molecules. It's certainly not obvious which structural features trigger a sweet sensation when these molecules interact with the taste buds.

The pioneers in relating structure to sweet taste were two chemists, Robert S. Shallenberger and Terry E. Acree of Cornell University, who almost thirty years ago suggested that all sweet-tasting substances must contain a common feature they called a glycophore. They postulated that a glycophore always contains an atom or group of atoms that have available electrons located near a hydrogen atom attached to a relatively elec-tronegative atom. Murray Goodman, a chemist at the University of California at San Diego, expanded the definition of a glycophore to include a hydrophobic ("water-hating") region. Goodman finds that a "sweet molecule" tends to be L-shaped with positively and negatively charged regions on the upright of the L and a hydrophobic region on the base of the L. For a molecule to be sweet, the L must be planar. If the L is twisted in one direction, the molecule has a bitter taste. If the molecule is twisted in the other direction, the molecule is tasteless.

The latest model for the sweet-taste receptor, proposed by Piero Temussi of the University of Naples, postulates that there are four binding sites on the receptor that can be occupied independently. Small sweet-tasting molecules might bind to one of the sites, while a large molecule would bind to more than one site simultaneously.

So the search goes on for a better artificial sweetener. One thing's for sure; it all has to do with molecular structure.

Saccharin

Sucralose

Aspartame
(Nutra-Sweet™)

Steviol

Here the boron atom is surrounded by three pairs of electrons. What arrangement minimizes the repulsions among three pairs of electrons? Here the greatest distance between electron pairs is achieved by angles of 120°.

$$120° \quad 120°$$
$$\text{B}$$
$$120°$$

Because each of the electron pairs is shared with a fluorine atom, the mo-lec-ular structure is

This is a planar (flat) molecule with a triangular arrangement of F atoms, commonly described as a trigonal planar structure. *Whenever three pairs of electrons are present around an atom, they should always be placed at the corners of a triangle (in a plane at angles of 120° to each other).*

Next let's consider the methane molecule, which has the Lewis structure

$$H\text{—}\overset{\displaystyle H}{\underset{\displaystyle H}{C}}\text{—}H \quad \text{or} \quad H\!:\!\overset{\displaystyle H}{\underset{\displaystyle H}{\ddot{C}}}\!:\!H$$

There are four pairs of electrons around the central carbon atom. What arrangement of these electron pairs best minimizes the repulsions? First we try a square planar arrangement:

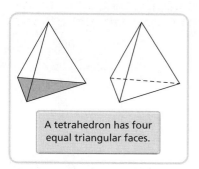

A tetrahedron has four equal triangular faces.

The carbon atom and the electron pairs are all in a plane represented by the surface of the paper, and the angles between the pairs are all 90°.

Is there another arrangement with angles greater than 90° that would put the electron pairs even farther away from each other? The answer is yes. We can get larger angles than 90° by using the following three-dimensional structure, which has angles of approximately 109.5°.

In this drawing the wedge indicates a position above the surface of the paper and the dashed lines indicate positions behind that surface. The solid line indicates a position on the surface of the page. The figure formed by connecting the lines is called a tetrahedron, so we call this arrangement of electron pairs the **tetrahedral arrangement.**

This is the maximum possible separation of four pairs around a given atom. *Whenever four pairs of electrons are present around an atom, they should always be placed at the corners of a tetrahedron (the tetrahedral arrangement).*

Now that we have the arrangement of electron pairs that gives the least repulsion, we can determine the positions of the atoms and thus the mo-lecular structure of CH_4. In methane each of the four electron pairs is shared between the carbon atom and a hydrogen atom. Thus the hydrogen atoms are placed as shown in Figure 12.12, and the molecule has a tetrahedral structure with the carbon atom at the center.

Recall that the main idea of the VSEPR model is to find the arrangement of electron pairs around the central atom that minimizes the repulsions. Then we can determine the *molecular structure* by knowing how the electron pairs are shared with the peripheral atoms. A systematic procedure for using the VSEPR model to predict the structure of a molecule is outlined below.

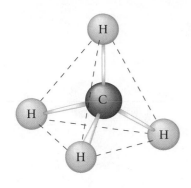

Figure 12.12

The molecular structure of methane. The tetrahedral arrangement of electron pairs produces a tetrahedral arrangement of hydrogen atoms.

Steps for Predicting Molecular Structure Using the VSEPR Model

Step 1 Draw the Lewis structure for the molecule.

Step 2 Count the electron pairs and arrange them in the way that minimizes repulsion (that is, put the pairs as far apart as possible).

Step 3 Determine the positions of the atoms from the way the electron pairs are shared.

Step 4 Determine the name of the molecular structure from the positions of the atoms.

EXAMPLE 12.5   Predicting Molecular Structure Using the VSEPR Model, I

Ammonia, NH_3, is used as a fertilizer (injected into the soil) and as a household cleaner (in aqueous solution). Predict the structure of ammonia using the VSEPR model.

SOLUTION

Step 1   Draw the Lewis structure.

$$H-\overset{\cdot\cdot}{N}-H$$
$$\vert$$
$$H$$

Step 2   Count the pairs of electrons and arrange them to minimize repulsions. The NH_3 molecule has four pairs of electrons around the N atom: three bonding pairs and one nonbonding pair. From the discussion of the methane molecule, we know that the best arrangement of four electron pairs is the tetrahedral structure shown in Figure 12.13a.

Step 3   Determine the positions of the atoms. The three H atoms share electron pairs as shown in Figure 12.13b.

Step 4   Name the molecular structure. It is very important to recognize that the name of the molecular structure is always based on the *positions of the atoms. The placement of the electron pairs determines the structure, but the name is based on the positions of the atoms.* Thus it is incorrect to say that the NH_3 molecule is tetrahedral. It has a tetrahedral arrangement of electron pairs but *not* a tetrahedral arrangement of atoms. The molecular structure of ammonia is a **trigonal pyramid** (one side is different from the other three) rather than a tetrahedron. ■

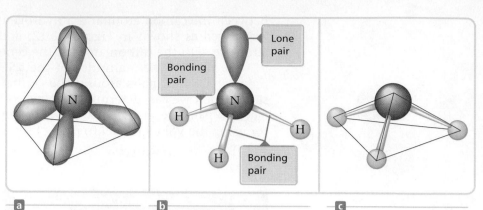

Figure 12.13

a The tetrahedral arrangement of electron pairs around the nitrogen atom in the ammonia molecule.

b Three of the electron pairs around nitrogen are shared with hydrogen atoms as shown, and one is a lone pair. Although the arrangement of electron pairs is tetrahedral, as in the methane molecule, the hydrogen atoms in the ammonia molecule occupy only three corners of the tetrahedron. A lone pair occupies the fourth corner.

c The NH_3 molecule has the trigonal pyramid structure (a pyramid with a triangle as a base).

EXAMPLE 12.6 | **Predicting Molecular Structure Using the VSEPR Model, II**

Describe the molecular structure of the water molecule.

SOLUTION

Step 1   The Lewis structure for water is

$$H-\ddot{O}-H$$

Step 2   There are four pairs of electrons: two bonding pairs and two nonbonding pairs. To minimize repulsions, these are best arranged in a tetrahedral structure as shown in Figure 12.14a.

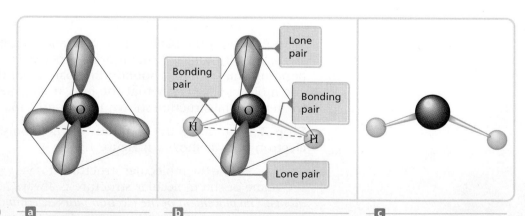

Figure 12.14

a The tetrahedral arrangement of the four electron pairs around oxygen in the water molecule.

b Two of the electron pairs are shared between oxygen and the hydrogen atoms, and two are lone pairs.

c The V-shaped molecular structure of the water molecule.

Step 3   Although H_2O has a tetrahedral arrangement of *electron pairs,* it is *not a tetrahedral molecule.* The *atoms* in the H_2O molecule form a V shape, as shown in Figure 12.14b and c.

Step 4   The molecular structure is called V-shaped or bent.

Self-Check EXERCISE 12.5   Predict the arrangement of electron pairs around the central atom. Then sketch and name the molecular structure for each of the following molecules or ions.

    a. NH_4^+ d. H_2S

    b. SO_4^{2-} e. ClO_3^-

    c. NF_3 f. BeF_2

See Problems 12.81 through 12.84. ■

The various molecules we have considered are summarized in Table 12.4 on the following page. Note the following general rules.

Rules for Predicting Molecular Structure Using the VSEPR Model

1. Two pairs of electrons on a central atom in a molecule are always placed 180° apart. This is a linear arrangement of pairs.
2. Three pairs of electrons on a central atom in a molecule are always placed 120° apart in the same plane as the central atom. This is a trigonal planar (triangular) arrangement of pairs.
3. Four pairs of electrons on a central atom in a molecule are always placed 109.5° apart. This is a tetrahedral arrangement of electron pairs.
4. When *every pair* of electrons on the central atom is *shared* with another atom, the molecular structure has the same name as the arrangement of electron pairs.

Number of Pairs	Name of Arrangement
2	linear
3	trigonal planar
4	tetrahedral

5. When one or more of the electron pairs around a central atom are unshared (lone pairs), the name for the molecular structure is *different* from that for the arrangement of electron pairs (see rows 4 and 5 in Table 12.4).

12.10 Molecular Structure: Molecules with Double Bonds

OBJECTIVE: To learn to apply the VSEPR model to molecules with double bonds.

Up to this point we have applied the VSEPR model only to molecules (and ions) that contain single bonds. In this section we will show that this model applies equally well to species with one or more double bonds. We will develop the procedures for dealing with molecules with double bonds by considering examples whose structures are known.

Table 12.4 Arrangements of Electron Pairs and the Resulting Molecular Structures for Two, Three, and Four Electron Pairs

Number of Electron Pairs	Bonds	Electron Pair Arrangement	Ball-and-Stick Model	Molecular Structure	Partial Lewis Structure	Ball-and-Stick Model
2	2	Linear	180°	Linear	A—B—A	Cl—Be—Cl
3	3	Trigonal planar (triangular)	120°	Trigonal planar (triangular)	A, B, A, A	F—B—F, F
4	4	Tetrahedral	109.5°	Tetrahedral	A, A—B—A, A	H, C, H, H, H
4	3	Tetrahedral	109.5°	Trigonal pyramid	A—B—A, A	H, N, H, H
4	2	Tetrahedral	109.5°	Bent or V-shaped	A—B—A	H, O, H

First we will examine the structure of carbon dioxide, a substance that may be contributing to the warming of the earth. The carbon dioxide molecule has the Lewis structure

$$\ddot{O}{=}C{=}\ddot{O}$$

as discussed in Section 12.7. Carbon dioxide is known by experiment to be a linear molecule. That is, it has a 180° bond angle.

Recall from Section 12.9 that two electron pairs around a central atom can minimize their mutual repulsions by taking positions on opposite sides of the atom (at 180° from each other). This causes a molecule like BeCl$_2$, which has the Lewis structure

$$:\ddot{\underset{..}{Cl}}{-}Be{-}\ddot{\underset{..}{Cl}}:$$

to have a linear structure. Now recall that CO$_2$ has two double bonds and is known to be linear, so the double bonds must be at 180° from each other. Therefore, we conclude that each double bond in this molecule acts *effectively* as one repulsive unit. This conclusion makes sense if we think of a bond in terms of an electron density "cloud" between two atoms. For example, we can picture the single bonds in BeCl$_2$ as follows:

Minimotor Molecule

Our modern society is characterized by a continual quest for miniaturization. Our computers, cell phones, portable music players, calculators, and many other devices have been greatly downsized over the last several years. The ultimate in miniaturization—machines made of single molecules. Although this idea sounds like an impossible dream, recent advances place us on the doorstep of such devices. For example, Hermann E. Gaub and his coworkers at the Center for Nanoscience at Ludwig-Maximilians University in Munich have just reported a single molecule that can do simple work.

Gaub and his associates constructed a polymer about 75 nm long by hooking together many light-sensitive molecules called azobenzenes:

Azobenzene is ideal for this application because its bonds are sensitive to specific wavelengths of light. When azobenzene absorbs light of 420 nm, it becomes extended; light at 365 nm causes the molecule to contract.

To make their tiny machine, the German scientists attached one end of the azobenzene polymer to a tiny, bendable lever similar to the tip of an atomic-force microscope. The other end of the polymer was attached to a glass surface. Flashes of 365-nm light caused the molecule to contract, bending the lever down and storing mechanical energy. Pulses of 420-nm radiation then extended the molecule, causing the lever to rise and releasing the stored energy. Eventually, one can imagine having the lever operate some part of a nanoscale machine. It seems we are getting close to the ultimate in miniature machines.

The minimum repulsion between these two electron density clouds occurs when they are on opposite sides of the Be atom (180° angle between them).

Each double bond in CO_2 involves the sharing of four electrons between the carbon atom and an oxygen atom. Thus we might expect the bonding cloud to be "fatter" than for a single bond:

However, the repulsive effects of these two clouds produce the same result as for single bonds; the bonding clouds have minimum repulsions when they are positioned on opposite sides of the carbon. The bond angle is 180°, and so the molecule is linear:

In summary, examination of CO_2 leads us to the conclusion that in using the VSEPR model for molecules with double bonds, each double bond should be treated the same as a single bond. In other words, although a double bond involves four electrons, these electrons are restricted to the space between a given pair of atoms. Therefore, these four electrons do not function as two independent pairs but are "tied together" to form one effective repulsive unit.

We reach this same conclusion by considering the known structures of other molecules that contain double bonds. For example, consider the ozone molecule, which has eighteen valence electrons and exhibits two resonance structures:

$$:\ddot{O}\!-\!\ddot{O}\!=\!\ddot{O}: \longleftrightarrow :\ddot{O}\!=\!\ddot{O}\!-\!\ddot{O}:$$

The ozone molecule is known to have a bond angle close to 120°. Recall that 120° angles represent the minimum repulsion for three pairs of electrons.

This indicates that the double bond in the ozone molecule is behaving as one effective repulsive unit:

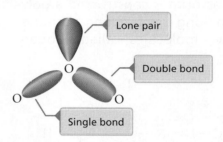

These and other examples lead us to the following rule: *When using the VSEPR model to predict the molecular geometry of a molecule, a double bond is counted the same as a single electron pair.*

Thus CO_2 has two "effective pairs" that lead to its linear structure, whereas O_3 has three "effective pairs" that lead to its bent structure with a 120° bond angle. Therefore, to use the VSEPR model for molecules (or ions) that have double bonds, we use the same steps as those given in Section 12.9, but we count any double bond the same as a single electron pair. Although we have not shown it here, triple bonds also count as one repulsive unit in applying the VSEPR model.

EXAMPLE 12.7 | **Predicting Molecular Structure Using the VSEPR Model, III**

Predict the structure of the nitrate ion.

SOLUTION

Step 1   The Lewis structures for NO_3^- are

$$\left[\begin{array}{c} :\ddot{O}: \\ \| \\ N \\ :\ddot{O}\!\diagup \quad \diagdown\!\ddot{O}: \end{array}\right]^{-} \leftrightarrow \left[\begin{array}{c} :\ddot{O}: \\ | \\ N \\ :\ddot{O}\!\diagup \quad \diagdown\!\ddot{O}: \end{array}\right]^{-} \leftrightarrow \left[\begin{array}{c} :\ddot{O}: \\ | \\ N \\ :\ddot{O}\!\diagup \quad \diagdown\!\ddot{O}: \end{array}\right]^{-}$$

Step 2   In each resonance structure there are effectively three pairs of electrons: the two single bonds and the double bond (which counts as one pair).

These three "effective pairs" will require a trigonal planar arrangement (120° angles).

Step 3   The atoms are all in a plane, with the nitrogen at the center and the three oxygens at the corners of a triangle (trigonal planar arrangement).

Step 4   The NO_3^- ion has a trigonal planar structure. ∎

CHAPTER 12 REVIEW

Key Terms

bond (12.1)
bond energy (12.1)
ionic bonding (12.1)
ionic compound (12.1)
covalent bonding (12.1)
polar covalent
  bond (12.1)
electronegativity (12.2)
dipole moment (12.3)
Lewis structure (12.6)
duet rule (12.6)
octet rule (12.6)
bonding pair (12.6)
lone (unshared)
  pair (12.6)
single bond (12.7)
double bond (12.7)

triple bond (12.7)
resonance (12.7)
resonance structure (12.7)
molecular (geometric)
  structure (12.8)
bond angle (12.8)
linear structure (12.8)
trigonal planar
  structure (12.8)
tetrahedral
  structure (12.8)
valence shell electron
  pair repulsion (VSEPR)
  model (12.9)
tetrahedral
  arrangement (12.9)
trigonal pyramid (12.9)

Summary

1. Chemical bonds hold groups of atoms together. They can be classified into several types. An ionic bond is formed when a transfer of electrons occurs to form ions; in a purely covalent bond, electrons are shared equally between identical atoms. Between these extremes lies the polar covalent bond, in which electrons are shared unequally between atoms with different electronegativities.

2. Electronegativity is defined as the relative ability of an atom in a molecule to attract the electrons shared in a bond. The difference in electronegativity values between the atoms involved in a bond determines the polarity of that bond.

3. In stable chemical compounds, the atoms tend to achieve a noble gas electron configuration. In the formation of a binary ionic compound involving representative elements, the valence-electron configuration of the nonmetal is completed: it achieves the configuration of the next noble gas. The valence orbitals of the metal are emptied to give the electron configuration of the previous noble gas. Two nonmetals share the valence electrons so that both atoms have completed valence-electron configurations (noble gas configurations).

4. Lewis structures are drawn to represent the arrangement of the valence electrons in a molecule. The rules for drawing Lewis structures are based on the observation that nonmetal atoms tend to achieve noble gas electron configurations by sharing electrons. This leads to a duet rule for hydrogen and to an octet rule for many other atoms.

5. Some molecules have more than one valid Lewis structure, a property called resonance. Although Lewis structures in which the atoms have noble gas electron configurations correctly describe most molecules, there are some notable exceptions, including O_2, NO, NO_2, and the molecules that contain Be and B.

6. The molecular structure of a molecule describes how the atoms are arranged in space.

7. The molecular structure of a molecule can be predicted by using the valence shell electron pair repulsion (VSEPR) model. This model bases its prediction on minimizing the repulsions among the electron pairs around an atom, which means arranging the electron pairs as far apart as possible.

Active Learning Questions

These questions are designed to be considered by groups of students in class. Often these questions work well for introducing a particular topic in class.

1. Using only the periodic table, predict the most stable ion for Na, Mg, Al, S, Cl, K, Ca, and Ga. Arrange these elements from largest to smallest radius and explain why the radius varies as it does.

2. Write the proper charges so that an alkali metal, a noble gas, and a halogen have the same electron configurations. What is the number of protons in each? The number of electrons in each? Arrange them from smallest to largest radii and explain your ordering rationale.

3. What is meant by a *chemical bond?*

4. Why do atoms form bonds with one another? What can make a molecule favored compared with the lone atoms?

5. How does a bond between Na and Cl differ from a bond between C and O? What about a bond between N and N?

6. In your own words, what is meant by the term *electronegativity?* What are the trends across and down the periodic table for electronegativity? Explain them, and describe how they are consistent with trends of ionization energy and atomic radii.

7. Explain the difference between ionic bonding and covalent bonding. How can we use the periodic table to help us determine the type of bonding between atoms?

8. True or false? In general, a larger atom has a smaller electronegativity. Explain.

9. Why is there an octet rule (and what does *octet* mean) in writing Lewis structures?

10. Does a Lewis structure tell which electrons came from which atoms? Explain.

11. If lithium and fluorine react, which has more attraction for an electron? Why?

12. In a bond between fluorine and iodine, which has more attraction for an electron? Why?

13. We use differences in electronegativity to account for certain properties of bond.
    What if all atoms had the same electronegativity values? How would bonding between atoms be affected? What are some differences we would notice?

14. Explain how you can use the periodic table to predict the formula of compounds.

15. Why do we only consider the valence electrons in drawing Lewis structures?

16. How do we determine the total number of valence electrons for an ion? Provide an example of an anion and a cation, and explain your answer.

17. What is the main idea in the valence shell electron pair repulsion (VSEPR) theory?

18. The molecules NH_3 and BF_3 have the same general formula (AB_3) but different shapes.

    a. Find the shape of each of the above molecules.
    b. Provide more examples of real molecules that have the same general formulas but different shapes.

19. How do we deal with multiple bonds in VSEPR theory?

20. In Section 12.10 of your text, the term "effective pairs" is used. What does this mean?

21. Consider the ions Sc^{3+}, Cl^-, K^+, Ca^{2+}, and S^{2-}. Match these ions to the following pictures that represent the relative sizes of the ions.

22. Write the name of each of the following shapes of molecules.

a —————— b —————— c ——————

13

Gases

A cluster balloonist at an Iowa fair. Cluster balloonists use a large number of relatively small helium balloons. *(AP Photo/The Daily Nonpareil/Ben DeVries)*

We live immersed in a gaseous solution. The earth's atmosphere is a mixture of gases that consists mainly of elemental nitrogen, N_2, and oxygen, O_2. The atmosphere both supports life and acts as a waste receptacle for the exhaust gases that accompany many industrial processes. The chemical reactions of these waste gases in the atmosphere lead to various types of pollution, including smog and acid rain. The two main sources of pollution are transportation and the production of electricity. The combustion of fuel in vehicles produces CO, CO_2, NO, and NO_2, along with unburned fragments of the petroleum used as fuel. The combustion of coal and petroleum in power plants produces NO_2 and SO_2 in the exhaust gases. These mixtures of chemicals can be activated by absorbing light to produce the photochemical smog that afflicts most large cities. The SO_2 in the air reacts with oxygen to produce SO_3 gas, which combines with water in the air to produce droplets of sulfuric acid (H_2SO_4), a major component of acid rain.

The gases in the atmosphere also shield us from harmful radiation from the sun and keep the earth warm by reflecting heat radiation back toward the earth. In fact, there is now great concern that an increase in atmospheric carbon dioxide, a product of the combustion of fossil fuels, is causing a dangerous warming of the earth. (See "Chemistry in Focus: Atmospheric Effects," in Chapter 11.)

In this chapter we will look carefully at the properties of gases. First, we will see how measurements of gas properties lead to various types of laws—statements that show how the properties are related to each other. Then we will construct a model to explain why gases behave as they do. This model will show how the behavior of the individual particles of a gas leads to the observed properties of the gas itself (a collection of many, many particles).

The study of gases provides an excellent example of the scientific method in action. It illustrates how observations lead to natural laws, which in turn can be accounted for by models.

Steve Fossett flies his balloon *Solo Spirit,* over the east coast of Australia during his attempt to make the first solo balloon flight around the world.

13.1 Pressure

OBJECTIVES: To learn about atmospheric pressure and how barometers work. • To learn the various units of pressure.

A gas uniformly fills any container, is easily compressed, and mixes completely with any other gas (see Section 3.1). One of the most obvious properties of a gas is that it exerts pressure on its surroundings. For example, when you blow up a balloon, the air inside pushes against the elastic sides of the balloon and keeps it firm.

Dry air (air from which the water vapor has been removed) is 78.1% N_2 molecules, 20.9% O_2 molecules, 0.9% Ar atoms, and 0.03% CO_2 molecules, along with smaller amounts of Ne, He, CH_4, Kr, and other trace components.

As a gas, water occupies 1200 times as much space as it does as a liquid at 25 °C and atmospheric pressure.

Soon after Torricelli died, a German physicist named Otto von Guericke invented an air pump. In a famous demonstration for the King of Prussia in 1683, Guericke placed two hemispheres together, pumped the air out of the resulting sphere through a valve, and showed that teams of horses could not pull the hemispheres apart. Then, after secretly opening the air valve, Guericke easily separated the hemispheres by hand. The King of Prussia was so impressed that he awarded Guericke a lifetime pension!

© Cengage Learning

a The pressure exerted by the gases in the atmosphere can be demonstrated by boiling water in a can and then turning off the heat and sealing the can.

b As the can cools, the water vapor condenses, lowering the gas pressure inside the can. This causes the can to crumple.

Figure 13.1

The gases most familiar to us form the earth's atmosphere. The pressure exerted by this gaseous mixture that we call air can be dramatically demonstrated by the experiment shown in Figure 13.1. A small volume of water is placed in a metal can and the water is boiled, which fills the can with steam. The can is then sealed and allowed to cool. Why does the can collapse as it cools? It is the atmospheric pressure that crumples the can. When the can is cooled after being sealed so that no air can flow in, the water vapor (steam) inside the can condenses to a very small volume of liquid water. As a gas, the water vapor filled the can, but when it is condensed to a liquid, the liquid does not come close to filling the can. The H_2O molecules formerly present as a gas are now collected in a much smaller volume of liquid, and there are very few molecules of gas left to exert pressure outward and counteract the air pressure. As a result, the pressure exerted by the gas molecules in the atmosphere smashes the can.

A device that measures atmospheric pressure, the **barometer,** was invented in 1643 by an Italian scientist named Evangelista Torricelli (1608–1647), who had been a student of the famous astronomer Galileo. Torricelli's barometer is constructed by filling a glass tube with liquid mercury and inverting it in a dish of mercury, as shown in Figure 13.2. Notice that a large quantity of mercury stays in the tube. In fact, at sea level the height of this column of mercury averages 760 mm. Why does this mercury stay in the tube, seemingly in defiance of gravity? Figure 13.2 illustrates how the pressure exerted by the atmospheric gases on the surface of mercury in the dish keeps the mercury in the tube.

Pressure = force / area

"Standard atmosphere" will support a 760 mm Hg

Empty space (a vacuum)

Hg

Weight of the mercury in the column

760 mm

Weight of the atmosphere (atmospheric pressure)

Figure 13.2

When a glass tube is filled with mercury and inverted in a dish of mercury at sea level, the mercury flows out of the tube until a column approximately 760 mm high remains (the height varies with atmospheric conditions). Note that the pressure of the atmosphere balances the weight of the column of mercury in the tube.

Atmospheric pressure results from the mass of the air being pulled toward the center of the earth by gravity—in other words, it results from the weight of the air. Changing weather conditions cause the atmospheric pressure to vary, so the height of the column of Hg supported by the atmosphere at sea level varies; it is not always 760 mm. The meteorologist who says a "low" is approaching means that the atmospheric pressure is going to decrease. This condition often occurs in conjunction with a storm.

Atmospheric pressure also varies with altitude. For example, when Torricelli's experiment is done in Breckenridge, Colorado (elevation 9600 feet), the atmosphere supports a column of mercury only about 520 mm high because the air is "thinner." That is, there is less air pushing down on the earth's surface at Breckenridge than at sea level.

▶ **Units of Pressure**

Mercury is used to measure pressure because of its high density. By way of comparison, the column of water required to measure a given pressure would be 13.6 times as high as a mercury column used for the same purpose.

Because instruments used for measuring pressure (see Figure 13.3) often contain mercury, the most commonly used units for pressure are based on the height of the mercury column (in millimeters) that the gas pressure can support. The unit **mm Hg** (millimeters of mercury) is often called the **torr** in honor of Torricelli. The terms *torr* and *mm Hg* are used interchangeably by chemists. A related unit for pressure is the **standard atmosphere** (abbreviated atm).

1 standard atmosphere = 1.000 atm = 760.0 mm Hg = 760.0 torr

The SI unit for pressure is the **pascal** (abbreviated Pa).

1 standard atmosphere = 101,325 Pa

Thus 1 atmosphere is about 100,000 or 10^5 pascals. Because the pascal is so small we will use it sparingly in this book. A unit of pressure that is employed

Figure 13.3

A device (called a manometer) for measuring the pressure of a gas in a container. The pressure of the gas is equal to *h* (the difference in mercury levels) in units of torr (equivalent to mm Hg).

Atmospheric pressure

Atmospheric pressure

h

h

Gas pressure less than atmospheric pressure

Gas pressure greater than atmospheric pressure

Hg

Hg

a

Gas pressure = atmospheric pressure – h.

b

Gas pressure = atmospheric pressure + h.

1.000 atm
760.0 mm Hg
760.0 torr
14.69 psi
101,325 Pa

in the engineering sciences and that we use for measuring tire pressure is pounds per square inch, abbreviated psi.

$$1.000 \text{ atm} = 14.69 \text{ psi}$$

Sometimes we need to convert from one unit of pressure to another. We do this by using conversion factors. The process is illustrated in Example 13.1.

EXAMPLE 13.1   Pressure Unit Conversions

Ken O'Donoghue

Checking the air pressure in a tire.

The pressure of the air in a tire is measured to be 28 psi. Represent this pressure in atmospheres, torr, and pascals.

SOLUTION

Where Are We Going?

We want to convert from units of pounds per square inch to units of atmospheres, torr, and pascals.

What Do We Know?

- 28 psi

What Information Do We Need?

- We need the equivalence statements for the units.

How Do We Get There?

To convert from pounds per square inch to atmospheres, we need the equivalence statement

$$1.000 \text{ atm} = 14.69 \text{ psi}$$

which leads to the conversion factor

$$\frac{1.000 \text{ atm}}{14.69 \text{ psi}}$$

$$28 \text{ psi} \times \frac{1.000 \text{ atm}}{14.69 \text{ psi}} = 1.9 \text{ atm}$$

To convert from atmospheres to torr, we use the equivalence statement

$$1.000 \text{ atm} = 760.0 \text{ torr}$$

which leads to the conversion factor

$$\frac{760.0 \text{ torr}}{1.000 \text{ atm}}$$

$$1.9 \text{ atm} \times \frac{760.0 \text{ torr}}{1.000 \text{ atm}} = 1.4 \times 10^3 \text{ torr}$$

MATH SKILL BUILDER
$1.9 \times 760.0 = 1444$
$1444 \quad \Rightarrow \quad 1400 = 1.4 \times 10^3$
Round off

To change from torr to pascals, we need the equivalence statement

$$1.000 \text{ atm} = 101,325 \text{ Pa}$$

which leads to the conversion factor

$$\frac{101,325 \text{ Pa}}{1.000 \text{ atm}}$$

MATH SKILL BUILDER
$1.9 \times 101,325 = 192,517.5$
$192,517.5 \quad \Rightarrow \quad 190,000 =$
1.9×10^5
Round off

$$1.9 \text{ atm} \times \frac{101,325 \text{ Pa}}{1.000 \text{ atm}} = 1.9 \times 10^5 \text{ Pa}$$

REALITY CHECK   The units on the answers are the units required.

Self-Check **EXERCISE 13.1** On a summer day in Breckenridge, Colorado, the atmospheric pressure is 525 mm Hg. What is this air pressure in atmospheres?

See Problems 13.7 through 13.12. ■

13.2 Pressure and Volume: Boyle's Law

OBJECTIVES: To understand the law that relates the pressure and volume of a gas. • To do calculations involving this law.

Gas   Gas   *h*

h

Hg

Figure 13.4

A J-tube similar to the one used by Boyle. The pressure on the trapped gas can be changed by adding or withdrawing mercury.

Mercury added

The first careful experiments on gases were performed by the Irish scientist Robert Boyle (1627–1691). Using a J-shaped tube closed at one end (Figure 13.4), which he reportedly set up in the multi-story entryway of his house, Boyle studied the relationship between the pressure of the trapped gas and its volume. Representative values from Boyle's experiments are given in Table 13.1. The units given for the volume (cubic inches) and pressure (inches of mercury) are the ones Boyle used. Keep in mind that the metric system was not in use at this time.

First let's examine Boyle's observations (Table 13.1) for general trends. Note that as the pressure increases, the volume of the trapped gas decreases. In fact, if you compare the data from experiments 1 and 4, you can see that as the pressure is doubled (from 29.1 to 58.2), the volume of the gas is halved (from 48.0 to 24.0). The same relationship can be seen in experiments 2 and 5 and in experiments 3 and 6 (approximately).

We can see the relationship between the volume of a gas and its pressure more clearly by looking at the product of the values of these two properties ($P \times V$) using Boyle's observations. This product is shown in the last column of Table 13.1. Note that for all the experiments,

$$P \times V = 1.4 \times 10^3 \text{ (in Hg)} \times \text{in.}^3$$

with only a slight variation due to experimental error. Other similar measurements on gases show the same behavior. This means that the relationship of the pressure and volume of a gas can be expressed in words as

pressure times volume equals a constant

For Boyle's law to hold, the amount of gas (moles) must not be changed. The temperature must also be constant.

The fact that the constant is sometimes 1.40×10^3 instead of 1.41×10^3 is due to experimental error (uncertainties in measuring the values of P and V).

Table 13.1 A Sample of Boyle's Observations (moles of gas and temperature both constant)

Experiment	Pressure (in Hg)	Volume (in.³)	Pressure × Volume (in Hg) × (in.³)	
			Actual	Rounded*
1	29.1	48.0	1396.8	1.40×10^3
2	35.3	40.0	1412.0	1.41×10^3
3	44.2	32.0	1414.4	1.41×10^3
4	58.2	24.0	1396.8	1.40×10^3
5	70.7	20.0	1414.0	1.41×10^3
6	87.2	16.0	1395.2	1.40×10^3
7	117.5	12.0	1410.0	1.41×10^3

*Three significant figures are allowed in the product because both of the numbers that are multiplied together have three significant figures.

Large pressure
Small volume

Small pressure
Large volume

Figure 13.5

A plot of P versus V from Boyle's data in Table 13.1.

or in terms of an equation as

$$PV = k$$

which is called **Boyle's law,** where k is a constant at a specific temperature for a given amount of gas. For the data we used from Boyle's experiment, $k = 1.41 \times 10^3$ (in Hg) $\times$ in.^{3}.

It is often easier to visualize the relationships between two properties if we make a graph. Figure 13.5 uses the data given in Table 13.1 to show how pressure is related to volume. This relationship, called a plot or a graph, shows that V decreases as P increases. When this type of relationship exists, we say that volume and pressure are inversely related or *inversely proportional;* when one increases, the other decreases. Boyle's law is illustrated by the gas samples in Figure 13.6.

Boyle's law means that if we know the volume of a gas at a given pressure, we can predict the new volume if the pressure is changed, *provided that neither the temperature nor the amount of gas is changed.* For example, if we represent the original pressure and volume as P_1 and V_1 and the final values as P_2 and V_2, using Boyle's law we can write

$$P_1V_1 = k$$

and

$$P_2V_2 = k$$

We can also say

$$P_1V_1 = k = P_2V_2$$

or simply

$$P_1V_1 = P_2V_2$$

This is really another way to write Boyle's law. We can solve for the final volume (V_2) by dividing both sides of the equation by P_2.

$$\frac{P_1V_1}{P_2} = \frac{P_2V_2}{P_2}$$

Canceling the P_2 terms on the right gives

$$\frac{P_1}{P_2} \times V_1 = V_2$$

Figure 13.6

Illustration of Boyle's law. These three containers contain the same number of molecules. At 298 K, $P \times V = 1$ L atm in all three containers.

or

$$V_2 = V_1 \times \frac{P_1}{P_2}$$

This equation tells us that we can calculate the new gas volume (V_2) by multiplying the original volume (V_1) by the ratio of the original pressure to the final pressure (P_1/P_2), as illustrated in Example 13.2.

EXAMPLE 13.2   **Calculating Volume Using Boyle's Law**

Freon-12 (the common name for the compound CCl_2F_2) was widely used in refrigeration systems, but has now been replaced by other compounds that do not lead to the breakdown of the protective ozone in the upper atmosphere. Consider a 1.5-L sample of gaseous CCl_2F_2 at a pressure of 56 torr. If pressure is changed to 150 torr at a constant temperature,

a. Will the volume of the gas increase or decrease?

b. What will be the new volume of the gas?

SOLUTION

Where Are We Going?

We want to determine if the volume will increase or decrease when the pressure is changed, and we want to calculate the new volume.

What Do We Know?

• We know the initial and final pressures and the initial volume.

• The amount of gas and temperature are held constant.

• Boyle's law: $P_1V_1 = P_2V_2$

How Do We Get There?

a. As the first step in a gas law problem, always write down the information given, in the form of a table showing the initial and final conditions.

Initial Conditions	Final Conditions
P_1 = 56 torr	P_2 = 150 torr
V_1 = 1.5 L	V_2 = ?

Drawing a picture also is often helpful. Notice that the pressure is increased from 56 torr to 150 torr, so the volume must decrease:

P_1V_1 ➡ P_2V_2

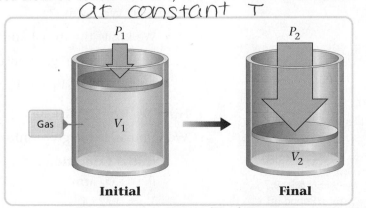

at constant T

increase pressure → decrease volume

We can verify this by using Boyle's law in the form

$$V_2 = V_1 \times \frac{P_1}{P_2}$$

Note that V_2 is obtained by "correcting" V_1 using the ratio P_1/P_2. Because P_1 is less than P_2, the ratio P_1/P_2 is a fraction that is less than 1. Thus V_2 must be a fraction of (smaller than) V_1; the volume decreases.

> The fact that the volume decreases in Example 13.2 makes sense because the pressure was increased. *To help catch errors, make it a habit to check whether an answer to a problem makes physical sense.*

b. We calculate V_2 as follows:

$$V_2 = V_1 \times \frac{P_1}{P_2} = 1.5 \text{ L} \times \frac{56 \text{ torr}}{150 \text{ torr}} = 0.56 \text{ L}$$

REALITY CHECK Because the pressure increases, we expect the volume to decrease. The pressure increased by almost a factor of three, and the volume decreased by about a factor of three.

Self-Check EXERCISE 13.2 A sample of neon to be used in a neon sign has a volume of 1.51 L at a pressure of 635 torr. Calculate the volume of the gas after it is pumped into the glass tubes of the sign, where it shows a pressure of 785 torr.

See Problems 13.21 and 13.22. ∎

EXAMPLE 13.3 Calculating Pressure Using Boyle's Law

In an automobile engine the gaseous fuel–air mixture enters the cylinder and is compressed by a moving piston before it is ignited. In a certain engine the initial cylinder volume is 0.725 L. After the piston moves up, the volume is 0.075 L. The fuel–air mixture initially has a pressure of 1.00 atm. Calculate the pressure of the compressed fuel–air mixture, assuming that both the temperature and the amount of gas remain constant.

SOLUTION

Where Are We Going?

We want to determine the new pressure of a fuel–air mixture that has undergone a volume change.

What Do We Know?

- We know the initial and final volumes and the initial pressure.
- The amount of gas and temperature are held constant.
- Boyle's law: $P_1V_1 = P_2V_2$

How Do We Get There?

We summarize the given information in the following table:

> **MATH SKILL BUILDER**
> $$P_1V_1 = P_2V_2$$
> $$\frac{P_1V_1}{V_2} = \frac{P_2 \cancel{V_2}}{\cancel{V_2}}$$
> $$P_1 \times \frac{V_1}{V_2} = P_2$$
> $$\frac{0.725}{0.075} = 9.666\ldots$$
> $$9.666 \Rightarrow 9.7$$
> Round off

Initial Conditions	Final Conditions
$P_1 = 1.00$ atm	$P_2 = ?$
$V_1 = 0.725$ L	$V_2 = 0.075$ L

Neon signs in Hong Kong.

Dave Jacobs/Stone/Getty Images

Then we solve Boyle's law in the form $P_1V_1 = P_2V_2$ for P_2 by dividing both sides by V_2 to give the equation

$$P_2 = P_1 \times \frac{V_1}{V_2} = 1.00 \text{ atm} \times \frac{0.725 \text{ L}}{0.075 \text{ L}} = 9.7 \text{ atm}$$

REALITY CHECK Because the volume decreases, we expect the pressure to increase. The volume decreased by about a factor of 10, and the pressure increased by about a factor of 10. ∎

13.3 Volume and Temperature: Charles's Law

OBJECTIVES: To learn about absolute zero. • To learn about the law relating the volume and temperature of a sample of gas at constant moles and pressure, and to do calculations involving that law.

In the century following Boyle's findings, scientists continued to study the properties of gases. The French physicist Jacques Charles (1746–1823), who was the first person to fill a balloon with hydrogen gas and who made the first solo balloon flight, showed that the volume of a given amount of gas (at constant pressure) increases with the temperature of the gas. That is, the volume increases when the temperature increases. A plot of the volume of a given sample of gas (at constant pressure) versus its temperature (in Celsius degrees) gives a straight line. This type of relationship is called *linear,* and this behavior is shown for several gases in Figure 13.7.

The solid lines in Figure 13.7 are based on actual measurements of temperature and volume for the gases listed. As we cool the gases they eventually liquefy, so we cannot determine any experimental points below this

John A. Rizzo/PhotoDisc/Getty Images

The air in a balloon expands when it is heated. This means that some of the air escapes from the balloon, lowering the air density inside and thus making the balloon buoyant.

note: each sample of gas contains different # of moles

Temperatures such as 0.00000002 K have been obtained in the laboratory, but 0 K has never been reached.

Figure 13.7

Plots of V (L) versus T (°C) for several gases. Note that each sample of gas contains a different number of moles to spread out the plots.

Figure 13.8

Plots of V versus T as in Figure 13.7, except that here the Kelvin scale is used for temperature.

temperature. However, when we extend each straight line (which is called *extrapolation* and is shown here by a dashed line), something very interesting happens. *All* of the lines extrapolate to zero volume at the same temperature: −273 °C. This suggests that −273 °C is the lowest possible temperature, because a negative volume is physically impossible. In fact, experiments have shown that matter cannot be cooled to temperatures lower than −273 °C. Therefore, this temperature is defined as **absolute zero** on the Kelvin scale.

When the volumes of the gases shown in Figure 13.7 are plotted against temperature on the Kelvin scale rather than the Celsius scale, the plots shown in Figure 13.8 result. These plots show that the volume of each gas is *directly proportional to the temperature* (in kelvins) and extrapolates to zero when the temperature is 0 K. Let's illustrate this statement with an example. Suppose we have 1 L of gas at 300 K. When we double the temperature of this gas to 600 K (without changing its pressure), the volume also doubles, to 2 L. Verify this type of behavior by looking carefully at the lines for various gases shown in Figure 13.8.

The direct proportionality between volume and temperature (in kelvins) is represented by the equation known as **Charles's law:**

$$V = bT$$

where T is in kelvins and b is the proportionality constant. Charles's law holds for a given sample of gas at constant pressure. It tells us that (for a given amount of gas at a given pressure) the volume of the gas is directly proportional to the temperature on the Kelvin scale:

$$V = bT \quad \text{or} \quad \frac{V}{T} = b = \text{constant}$$

Notice that in the second form, this equation states that the *ratio* of V to T (in kelvins) must be constant. (This is shown for helium in the margin.) Thus, when we triple the temperature (in kelvins) of a sample of gas, the volume of the gas triples as well.

$$\frac{V}{T} = \frac{3 \times V}{3 \times T} = b = \text{constant}$$

From Figure 13.8 for Helium		
V (L)	T (K)	b
0.7	100	0.01
1.7	200	0.01
2.7	300	0.01
3.7	400	0.01
5.7	600	0.01

We can also write Charles's law in terms of V_1 and T_1 (the initial conditions) and V_2 and T_2 (the final conditions).

$$\frac{V_1}{T_1} = b \quad \text{and} \quad \frac{V_2}{T_2} = b$$

> Charles's law in the form $V_1/T_1 = V_2/T_2$ applies only when both the amount of gas (moles) and the pressure are constant.

Thus

$$\frac{V_1}{T_1} = \frac{V_2}{T_2}$$

We will illustrate the use of this equation in Examples 13.4 and 13.5.

EXAMPLE 13.4 | **Calculating Volume Using Charles's Law, I**

A 2.0-L sample of air is collected at 298 K and then cooled to 278 K. The pressure is held constant at 1.0 atm.

   a. Does the volume increase or decrease?

   b. Calculate the volume of the air at 278 K.

SOLUTION

Where Are We Going?

We want to determine if the volume will increase or decrease when the temperature is changed, and we want to calculate the new volume.

What Do We Know?

- We know the initial and final temperatures and the initial volume.

- The amount of gas and pressure are held constant.

- Charles's law: $\dfrac{V_1}{T_1} = \dfrac{V_2}{T_2}$.

How Do We Get There?

   a. Because the gas is cooled, the volume of the gas must decrease:

$$\frac{V}{T} = \text{constant}$$

> *T* is decreased, so *V* must decrease to maintain a constant ratio.

   b. To calculate the new volume, V_2, we will use Charles's law in the form

$$\frac{V_1}{T_1} = \frac{V_2}{T_2}$$

> $\dfrac{V_1}{T_1}$ ➡ $\dfrac{V_2}{T_2}$
>
> Temperature smaller, volume smaller

We are given the following information:

Initial Conditions	Final Conditions
$T_1 = 298$ K	$T_2 = 278$ K
$V_1 = 2.0$ L	$V_2 = ?$

We want to solve the equation

$$\frac{V_1}{T_1} = \frac{V_2}{T_2}$$

for V_2. We can do this by multiplying both sides by T_2 and canceling.

$$T_2 \times \frac{V_1}{T_1} = \frac{V_2}{\cancel{T_2}} \times \cancel{T_2} = V_2$$

Thus

$$V_2 = T_2 \times \frac{V_1}{T_1} = 278 \cancel{K} \times \frac{2.0\ L}{298 \cancel{K}} = 1.9\ L$$

REALITY CHECK   Because the temperature decreases, we expect the volume to decrease. The temperature decreased slightly, so we would expect the volume to decrease slightly. ■

EXAMPLE 13.5 | **Calculating Volume Using Charles's Law, II**

A sample of gas at 15 °C (at 1 atm) has a volume of 2.58 L. The temperature is then raised to 38 °C (at 1 atm).

  a. Does the volume of the gas increase or decrease?

  b. Calculate the new volume.

SOLUTION

Where Are We Going?

We want to determine if the volume will increase or decrease when the temperature is changed, and we want to calculate the new volume.

What Do We Know?

  • We know the initial and final temperatures and the initial volume.

  • The amount of gas and pressure are held constant.

  • Charles's law: $\dfrac{V_1}{T_1} = \dfrac{V_2}{T_2}$.

How Do We Get There?

  a. In this case we have a given sample (constant amount) of gas that is heated from 15 °C to 38 °C *while the pressure is held constant*. We know from Charles's law that the volume of a given sample of gas is directly proportional to the temperature (at constant pressure). So the increase in temperature will *increase* the volume; the new volume will be greater than 2.58 L.

  b. To calculate the new volume, we use Charles's law in the form

$$\frac{V_1}{T_1} = \frac{V_2}{T_2}$$

We are given the following information:

Initial Conditions	Final Conditions
$T_1 = 15$ °C	$T_2 = 38$ °C
$V_1 = 2.58$ L	$V_2 = ?$

As is often the case, the temperatures are given in Celsius degrees. However, for us to use Charles's law, the temperature *must be in kelvins*. Thus we must convert by adding 273 to each temperature.

Researchers take samples from a steaming volcanic vent at Mount Baker in Washington.

Initial Conditions	Final Conditions
$T_1 = 15\ °C = 15 + 273$	$T_2 = 38\ °C = 38 + 273$
$\qquad = 288\ K$	$\qquad = 311\ K$
$V_1 = 2.58\ L$	$V_2 = ?$

Solving for V_2 gives

$$V_2 = V_1 \times \frac{T_2}{T_1} = 2.58\ L\left(\frac{311\ \cancel{K}}{288\ \cancel{K}}\right) = 2.79\ L$$

REALITY CHECK   Because the temperature increases, we expect the volume to increase.

Self-Check EXERCISE 13.3   A child blows a bubble that contains air at 28 °C and has a volume of 23 cm³ at 1 atm. As the bubble rises, it encounters a pocket of cold air (temperature 18 °C). If there is no change in pressure, will the bubble get larger or smaller as the air inside cools to 18 °C? Calculate the new volume of the bubble.

See Problems 13.29 and 13.30. ∎

Notice from Example 13.5 that we adjust the volume of a gas for a temperature change by multiplying the original volume by the ratio of the Kelvin temperatures—final (T_2) over initial (T_1). Remember to check whether your answer makes sense. When the temperature increases (at constant pressure), the volume must increase, and vice versa.

EXAMPLE 13.6 | Calculating Temperature Using Charles's Law

In former times, gas volume was used as a way to measure temperature using devices called gas thermometers. Consider a gas that has a volume of 0.675 L at 35 °C and 1 atm pressure. What is the temperature (in units of °C) of a room where this gas has a volume of 0.535 L at 1 atm pressure?

SOLUTION

Where Are We Going?

We want to determine the new temperature of a gas given that the volume has decreased at constant pressure.

What Do We Know?

- We know the initial and final volumes and the initial temperature.
- The amount of gas and pressure are held constant.
- Charles's law: $\dfrac{V_1}{T_1} = \dfrac{V_2}{T_2}$.

How Do We Get There?

The information given in the problem is

Initial Conditions	**Final Conditions**
$T_1 = 35\ °C = 35 + 273 = 308\ K$	$T_2 = ?$
$V_1 = 0.675\ L$	$V_2 = 0.535\ L$
$P_1 = 1\ atm$	$P_2 = 1\ atm$

The pressure remains constant, so we can use Charles's law in the form

$$\frac{V_1}{T_1} = \frac{V_2}{T_2}$$

and solve for T_2. First we multiply both sides by T_2.

$$T_2 \times \frac{V_1}{T_1} = \frac{V_2}{\cancel{T_2}} \times \cancel{T_2} = V_2$$

Next we multiply both sides by T_1.

$$\cancel{T_1} \times T_2 \times \frac{V_1}{\cancel{T_1}} = T_1 \times V_2$$

This gives

$$T_2 \times V_1 = T_1 \times V_2$$

Now we divide both sides by V_1 (multiply by $1/V_1$),

$$\frac{1}{\cancel{V_1}} \times T_2 \times \cancel{V_1} = \frac{1}{V_1} \times T_1 \times V_2$$

and obtain

$$T_2 = T_1 \times \frac{V_2}{V_1}$$

We have now isolated T_2 on one side of the equation, and we can do the calculation.

$$T_2 = T_1 \times \frac{V_2}{V_1} = (308\ K) \times \frac{0.535\ \cancel{L}}{0.675\ \cancel{L}} = 244\ K$$

To convert from units of K to units of °C, we subtract 273 from the Kelvin temperature.

$$T_{°C} = T_K - 273 = 244 - 273 = -29\ °C$$

The room is very cold; the new temperature is $-29\ °C$.

REALITY CHECK Because the volume is smaller, we expect the temperature to be lower. ■

13.4 Volume and Moles: Avogadro's Law

OBJECTIVE: To understand the law relating the volume and the number of moles of a sample of gas at constant temperature and pressure, and to do calculations involving this law.

What is the relationship between the volume of a gas and the number of molecules present in the gas sample? Experiments show that when the number of moles of gas is doubled (at constant temperature and pressure), the volume doubles. In other words, the volume of a gas is directly proportional to the number of moles if temperature and pressure remain constant. Figure 13.9 illustrates this relationship, which can also be represented by the equation

$$V = an \qquad \text{or} \qquad \frac{V}{n} = a$$

where V is the volume of the gas, n is the number of moles, and a is the proportionality constant. Note that this equation means that the ratio of V to n is constant as long as the temperature and pressure remain constant. Thus, when the number of moles of gas is increased by a factor of 5, the volume also increases by a factor of 5,

$$\frac{V}{n} = \frac{5 \times V}{5 \times n} = a = \text{constant}$$

and so on. In words, this equation means that *for a gas at constant temperature and pressure, the volume is directly proportional to the number of moles of gas.* This relationship is called **Avogadro's law** after the Italian scientist Amadeo Avogadro, who first postulated it in 1811.

For cases where the number of moles of gas is changed from an initial amount to another amount (at constant T and P), we can represent Avogadro's law as

$$\underset{\substack{\text{Initial} \\ \text{amount}}}{\frac{V_1}{n_1}} = a = \underset{\substack{\text{Final} \\ \text{amount}}}{\frac{V_2}{n_2}}$$

or

$$\frac{V_1}{n_1} = \frac{V_2}{n_2}$$

We will illustrate the use of this equation in Example 13.7.

[handwritten notes:] ♯. size of particle insignificant
- interaction between particles insignificant
- ideal behavior

Figure 13.9

The relationship between volume V and number of moles n. As the number of moles is increased from 1 to 2 (**a**) to (**b**), the volume doubles. When the number of moles is tripled (**c**), the volume is also tripled. The temperature and pressure remain the same in these cases.

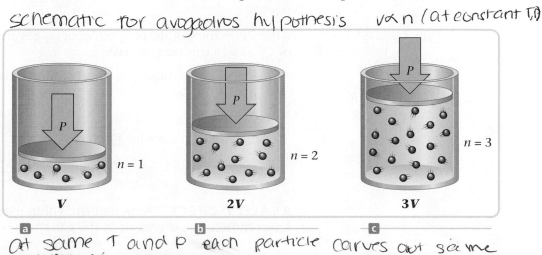

[handwritten:] schematic for avogadros hypothesis   v ∝ n (at constant T,P)

[handwritten below figure:] at same T and P each particle carves out same volume of space.

EXAMPLE 13.7 | Using Avogadro's Law in Calculations

Suppose we have a 12.2-L sample containing 0.50 mol oxygen gas, O_2, at a pressure of 1 atm and a temperature of 25 °C. If all of this O_2 is converted to ozone, O_3, at the same temperature and pressure, what will be the volume of the ozone formed?

SOLUTION

Where Are We Going?

We want to determine the volume of ozone (O_3) formed from 0.50 mol O_2 given the volume of oxygen.

What Do We Know?

- We know the initial number of moles of oxygen and the volume of oxygen.
- The temperature and pressure are held constant.
- Avogadro's law: $\dfrac{V_1}{T_1} = \dfrac{V_2}{T_2}$.

What Information Do We Need?

- We need the balanced equation for the reaction to determine the number of moles of ozone formed.

How Do We Get There?

To do this problem we need to compare the moles of gas originally present to the moles of gas present after the reaction. We know that 0.50 mol O_2 is present initially. To find out how many moles of O_3 will be present after the reaction, we need to use the balanced equation for the reaction.

$$3O_2(g) \rightarrow 2O_3(g)$$

We calculate the moles of O_3 produced by using the appropriate mole ratio from the balanced equation.

$$0.50 \ \cancel{\text{mol } O_2} \times \frac{2 \text{ mol } O_3}{3 \ \cancel{\text{mol } O_2}} = 0.33 \text{ mol } O_3$$

Avogadro's law states that

$$\frac{V_1}{n_1} = \frac{V_2}{n_2}$$

MATH SKILL BUILDER

$$\frac{V_1}{n_1} = \frac{V_2}{n_2}$$

$$n_2 \times \frac{V_1}{n_1} = \frac{V_2}{\cancel{n_2}} \times \cancel{n_2}$$

$$V_1 \times \frac{n_2}{n_1} = V_2$$

where V_1 is the volume of n_1 moles of O_2 gas and V_2 is the volume of n_2 moles of O_3 gas. In this case we have

Initial Conditions	Final Conditions
$n_1 = 0.50$ mol	$n_2 = 0.33$ mol
$V_1 = 12.2$ L	$V_2 = ?$

Solving Avogadro's law for V_2 gives

$$V_2 = V_1 \times \frac{n_2}{n_1} = 12.2 \text{ L} \left(\frac{0.33 \ \cancel{\text{mol}}}{0.50 \ \cancel{\text{mol}}} \right) = 8.1 \text{ L}$$

REALITY CHECK Note that the volume decreases, as it should, because fewer molecules are present in the gas after O_2 is converted to O_3.

Self-Check **EXERCISE 13.4** Consider two samples of nitrogen gas (composed of N_2 molecules). Sample 1 contains 1.5 mol N_2 and has a volume of 36.7 L at 25 °C and 1 atm. Sample 2 has a volume of 16.5 L at 25 °C and 1 atm. Calculate the number of moles of N_2 in Sample 2.

See Problems 13.41 through 13.44. ■

13.5 The Ideal Gas Law

OBJECTIVE: To understand the ideal gas law and use it in calculations.

We have considered three laws that describe the behavior of gases as it is revealed by experimental observations.

> Constant n means a constant number of moles of gas.

$$\text{Boyle's law:}\quad PV = k \quad\text{or}\quad V = \frac{k}{P}\text{ (at constant } T \text{ and } n)$$

$$\text{Charles's law:}\qquad\qquad V = bT \text{ (at constant } P \text{ and } n)$$

$$\text{Avogadro's law:}\qquad\quad V = an \text{ (at constant } T \text{ and } P)$$

These relationships, which show how the volume of a gas depends on pressure, temperature, and number of moles of gas present, can be combined as follows:

$$V = R\left(\frac{Tn}{P}\right)$$

where R is the combined proportionality constant and is called the **universal gas constant.** When the pressure is expressed in atmospheres and the volume is in liters, R always has the value 0.08206 L atm/K mol. We can rearrange the above equation by multiplying both sides by P,

> $$R = 0.08206 \frac{\text{L atm}}{\text{K mol}}$$

$$P \times V = \cancel{P} \times R\left(\frac{Tn}{\cancel{P}}\right)$$

to obtain the **ideal gas law** written in its usual form,

$$\boxed{PV = nRT}$$

The ideal gas law involves all the important characteristics of a gas: its pressure (P), volume (V), number of moles (n), and temperature (T). Knowledge of any three of these properties is enough to define completely the condition of the gas, because the fourth property can be determined from the ideal gas law.

It is important to recognize that the ideal gas law is based on experimental measurements of the properties of gases. A gas that obeys this equation is said to behave *ideally.* That is, this equation defines the behavior of an **ideal gas.** Most gases obey this equation closely at pressures of approximately 1 atm or lower, when the temperature is approximately 0 °C or higher. You should assume ideal gas behavior when working problems involving gases in this text.

The ideal gas law can be used to solve a variety of problems. Example 13.8 demonstrates one type, where you are asked to find one property characterizing the condition of a gas given the other three properties.

EXAMPLE 13.8 | Using the Ideal Gas Law in Calculations

A sample of hydrogen gas, H_2, has a volume of 8.56 L at a temperature of 0 °C and a pressure of 1.5 atm. Calculate the number of moles of H_2 present in this gas sample. (Assume that the gas behaves ideally.)

SOLUTION

Where Are We Going?

We want to determine the number of moles of hydrogen gas (H_2) present given conditions of temperature, pressure, and volume.

What Do We Know?

- We know the temperature, pressure, and volume of hydrogen gas.
- Ideal gas law: $PV = nRT$.

What Information Do We Need?

- $R = 0.08206$ L atm/mol K.

How Do We Get There?

In this problem we are given the pressure, volume, and temperature of the gas: $P = 1.5$ atm, $V = 8.56$ L, and $T = 0$ °C. Remember that the temperature must be changed to the Kelvin scale.

$$T = 0\,°C = 0 + 273 = 273\ K$$

We can calculate the number of moles of gas present by using the ideal gas law, $PV = nRT$. We solve for n by dividing both sides by RT:

$$\frac{PV}{RT} = n\,\frac{\cancel{RT}}{\cancel{RT}}$$

to give

$$\frac{PV}{RT} = n$$

Thus

$$n = \frac{PV}{RT} = \frac{(1.5\ \cancel{atm})(8.56\ \cancel{L})}{\left(0.08206\ \dfrac{\cancel{L}\ \cancel{atm}}{\cancel{K}\ mol}\right)(273\ \cancel{K})} = 0.57\ mol$$

Self-Check EXERCISE 13.5 A weather balloon contains 1.10×10^5 mol He and has a volume of 2.70×10^6 L at 1.00 atm pressure. Calculate the temperature of the helium in the balloon in kelvins and in Celsius degrees.

See Problems 13.53 through 13.60. ∎

EXAMPLE 13.9 | Ideal Gas Law Calculations Involving Conversion of Units

What volume is occupied by 0.250 mol carbon dioxide gas at 25 °C and 371 torr?

SOLUTION

Where Are We Going?

We want to determine the volume of carbon dioxide gas (CO_2) given the number of moles, pressure, and temperature.

What Do We Know?

- We know the number of moles, pressure, and temperature of the carbon dioxide.
- Ideal gas law: $PV = nRT$.

What Information Do We Need?

- $R = 0.08206$ L atm/mol K.

How Do We Get There?

We can use the ideal gas law to calculate the volume, but we must first convert pressure to atmospheres and temperature to the Kelvin scale.

$$P = 371 \text{ torr} = 371 \text{ torr} \times \frac{1.000 \text{ atm}}{760.0 \text{ torr}} = 0.488 \text{ atm}$$

$$T = 25 \,°C = 25 + 273 = 298 \text{ K}$$

We solve for V by dividing both sides of the ideal gas law ($PV = nRT$) by P.

$$V = \frac{nRT}{P} = \frac{(0.250 \text{ mol})\left(0.08206 \dfrac{\text{L atm}}{\text{K mol}}\right)(298 \text{ K})}{0.488 \text{ atm}} = 12.5 \text{ L}$$

The volume of the sample of CO_2 is 12.5 L.

> **MATH SKILL BUILDER**
>
> $$PV = nRT$$
> $$\frac{PV}{P} = \frac{nRT}{P}$$
> $$V = \frac{nRT}{P}$$

Self-Check EXERCISE 13.6   Radon, a radioactive gas formed naturally in the soil, can cause lung cancer. It can pose a hazard to humans by seeping into houses, and there is concern about this problem in many areas. A 1.5-mol sample of radon gas has a volume of 21.0 L at 33 °C. What is the pressure of the gas?

See Problems 13.53 through 13.60. ■

Note that R has units of L atm/K mol. Accordingly, whenever we use the ideal gas law, we must express the volume in units of liters, the temperature in kelvins, and the pressure in atmospheres. When we are given data in other units, we must first convert to the appropriate units.

The ideal gas law can also be used to calculate the changes that will occur when the conditions of the gas are changed as illustrated in Example 13.10.

EXAMPLE 13.10 | Using the Ideal Gas Law Under Changing Conditions

Suppose we have a 0.240-mol sample of ammonia gas at 25 °C with a volume of 3.5 L at a pressure of 1.68 atm. The gas is compressed to a volume of 1.35 L at 25 °C. Use the ideal gas law to calculate the final pressure.

SOLUTION

Where Are We Going?

We want to use the ideal gas law equation to determine the pressure of ammonia gas given a change in volume.

What Do We Know?

- We know the initial number of moles, pressure, volume, and temperature of the ammonia.
- We know the new volume.
- Ideal gas law: $PV = nRT$.

How Do We Get There?

In this case we have a sample of ammonia gas in which the conditions are changed. We are given the following information:

Initial Conditions	Final Conditions
$V_1 = 3.5$ L	$V_2 = 1.35$ L
$P_1 = 1.68$ atm	$P_2 = ?$
$T_1 = 25\ °C = 25 + 273 = 298$ K	$T_2 = 25\ °C = 25 + 273 = 298$ K
$n_1 = 0.240$ mol	$n_2 = 0.240$ mol

Note that both n and T remain constant—only P and V change. Thus we could simply use Boyle's law ($P_1V_1 = P_2V_2$) to solve for P_2. However, we will use the ideal gas law to solve this problem in order to introduce the idea that one equation—the ideal gas equation—can be used to do almost any gas problem. The key idea here is that in using the ideal gas law to describe a change in conditions for a gas, we always *solve the ideal gas equation in such a way that the variables that change are on one side of the equals sign and the constant terms are on the other side.* That is, we start with the ideal gas equation in the conventional form ($PV = nRT$) and rearrange it so that all the terms that change are moved to one side and all the terms that do not change are moved to the other side. In this case the pressure and volume change, and the temperature and number of moles remain constant (as does R, by definition). So we write the ideal gas law as

$$\underset{\text{Change}}{PV} \quad = \quad \underset{\text{Remain constant}}{nRT}$$

Because n, R, and T remain the same in this case, we can write $P_1V_1 = nRT$ and $P_2V_2 = nRT$. Combining these gives

$$P_1V_1 = nRT = P_2V_2 \quad \text{or} \quad P_1V_1 = P_2V_2$$

and

$$P_2 = P_1 \times \frac{V_1}{V_2} = (1.68 \text{ atm})\left(\frac{3.5 \cancel{L}}{1.35 \cancel{L}}\right) = 4.4 \text{ atm}$$

REALITY CHECK Does this answer make sense? The volume was decreased (at constant temperature and constant number of moles), which means that the pressure should increase, as the calculation indicates.

Self-Check EXERCISE 13.7 A sample of methane gas that has a volume of 3.8 L at 5 °C is heated to 86 °C at constant pressure. Calculate its new volume.

See Problems 13.61 and 13.62. ∎

Note that in solving Example 13.10, we actually obtained Boyle's law ($P_1V_1 = P_2V_2$) from the ideal gas equation. You might well ask, "Why go to all this trouble?" The idea is to learn to use the ideal gas equation to solve all types of gas law problems. This way you will never have to ask yourself, "Is this a Boyle's law problem or a Charles's law problem?"

We continue to practice using the ideal gas law in Example 13.11. Remember, the key idea is to rearrange the equation so that the quantities that change are moved to one side of the equation and those that remain constant are moved to the other.

EXAMPLE 13.11 | Calculating Volume Changes Using the Ideal Gas Law

A sample of diborane gas, B_2H_6, a substance that bursts into flames when exposed to air, has a pressure of 0.454 atm at a temperature of $-15\,°C$ and a volume of 3.48 L. If conditions are changed so that the temperature is $36\,°C$ and the pressure is 0.616 atm, what will be the new volume of the sample?

SOLUTION

Where Are We Going?

We want to use the ideal gas law equation to determine the volume of diborane gas.

What Do We Know?

- We know the initial pressure, volume, and temperature of the diborane gas.

- We know the new temperature and pressure.

- Ideal gas law: $PV = nRT$.

How Do We Get There?

We are given the following information:

Initial Conditions	Final Conditions
$P_1 = 0.454$ atm	$P_2 = 0.616$ atm
$V_1 = 3.48$ L	$V_2 = ?$
$T_1 = -15\,°C = 273 - 15 = 258$ K	$T_2 = 36\,°C = 273 + 36 = 309$ K

MATH SKILL BUILDER

$$PV = nRT$$

$$\frac{PV}{T} = \frac{nR\cancel{T}}{\cancel{T}}$$

$$\frac{PV}{T} = nR$$

Note that the value of n is not given. However, we know that n is constant (that is, $n_1 = n_2$) because no diborane gas is added or taken away. Thus, in this experiment, n is constant and P, V, and T change. Therefore, we rearrange the ideal gas equation ($PV = nRT$) by dividing both sides by T,

$$\underset{\text{Change}}{\frac{PV}{T}} = \underset{\text{Constant}}{nR}$$

which leads to the equation

$$\frac{P_1V_1}{T_1} = nR = \frac{P_2V_2}{T_2}$$

or

$$\frac{P_1V_1}{T_1} = \frac{P_2V_2}{T_2}$$

We can now solve for V_2 by dividing both sides by P_2 and multiplying both sides by T_2.

$$\frac{1}{P_2} \times \frac{P_1V_1}{T_1} = \frac{\cancel{P_2}V_2}{T_2} \times \frac{1}{\cancel{P_2}} = \frac{V_2}{T_2}$$

$$T_2 \times \frac{P_1V_1}{P_2T_1} = \frac{V_2}{\cancel{T_2}} \times \cancel{T_2} = V_2$$

That is,

$$\frac{T_2P_1V_1}{P_2T_1} = V_2$$

Snacks Need Chemistry, Too!

Have you ever wondered what makes popcorn pop? The popping is linked with the properties of gases. What happens when a gas is heated? Charles's law tells us that if the pressure is held constant, the volume of the gas must increase as the temperature is increased. But what happens if the gas being heated is trapped at a constant volume? We can see what happens by rearranging the ideal gas law ($PV = nRT$) as follows:

$$P = \left(\frac{nR}{V}\right)T$$

When n, R, and V are held constant, the pressure of a gas is directly proportional to the temperature. Thus, as the temperature of the trapped gas increases, its pressure also increases. This is exactly what happens inside a kernel of popcorn as it is heated. The moisture inside the kernel vaporized by the heat produces increasing pressure. The pressure finally becomes so great that the kernel breaks open, allowing the starch inside to expand to about 40 times its original size.

What's special about popcorn? Why does it pop while "regular" corn doesn't? William da Silva, a biologist at the University of Campinas in Brazil, has traced the "popability" of popcorn to its outer casing, called the pericarp. The molecules in the pericarp of popcorn, which are packed in a much more orderly way than in regular corn, transfer heat unusually quickly, producing a very fast pressure jump that pops the kernel. In addition, because the pericarp of popcorn is much thicker and stronger than that of regular corn, it can withstand more pressure, leading to a more explosive pop when the moment finally comes.

Popcorn popping.

© Cengage Learning

It is sometimes convenient to think in terms of the ratios of the initial temperature and pressure and the final temperature and pressure. That is,

$$V_2 = \frac{T_2 P_1 V_1}{T_1 P_2} = V_1 \times \frac{T_2}{T_1} \times \frac{P_1}{P_2}$$

Substituting the information given yields

$$V_2 = \frac{309 \text{ K}}{258 \text{ K}} \times \frac{0.454 \text{ atm}}{0.616 \text{ atm}} \times 3.48 \text{ L} = 3.07 \text{ L}$$

> *Always* convert the temperature to the Kelvin scale and the pressure to atmospheres when applying the ideal gas law.

Self-Check EXERCISE 13.8 A sample of argon gas with a volume of 11.0 L at a temperature of 13 °C and a pressure of 0.747 atm is heated to 56 °C and a pressure of 1.18 atm. Calculate the final volume.

See Problems 13.61 and 13.62. ■

The equation obtained in Example 13.11,

$$\frac{P_1 V_1}{T_1} = \frac{P_2 V_2}{T_2}$$

is often called the **combined gas law** equation. It holds when the amount of gas (moles) is held constant. While it may be convenient to remember this equation, it is not necessary because you can always use the ideal gas equation.

13.6 Dalton's Law of Partial Pressures

OBJECTIVE: To understand the relationship between the partial and total pressures of a gas mixture, and to use this relationship in calculations.

Many important gases contain a mixture of components. One notable example is air. Scuba divers who are going deeper than 150 feet use another important mixture, helium and oxygen. Normal air is not used because the nitrogen present dissolves in the blood in large quantities as a result of the high pressures experienced by the diver under several hundred feet of water. When the diver returns too quickly to the surface, the nitrogen bubbles out of the blood just as soda fizzes when it's opened, and the diver gets "the bends"—a very painful and potentially fatal condition. Because helium gas is only sparingly soluble in blood, it does not cause this problem.

Studies of gaseous mixtures show that each component behaves independently of the others. In other words, a given amount of oxygen exerts the same pressure in a 1.0-L vessel whether it is alone or in the presence of nitrogen (as in the air) or helium.

Among the first scientists to study mixtures of gases was John Dalton. In 1803, Dalton summarized his observations in this statement: *For a mixture of gases in a container, the total pressure exerted is the sum of the partial pressures of the gases present.* The **partial pressure** of a gas is the pressure that the gas would exert if it were alone in the container. This statement, known as **Dalton's law of partial pressures,** can be expressed as follows for a mixture containing three gases:

$$P_{total} = P_1 + P_2 + P_3$$

where the subscripts refer to the individual gases (gas 1, gas 2, and gas 3). The pressures P_1, P_2, and P_3 are the partial pressures; that is, each gas is responsible for only part of the total pressure (Figure 13.10).

Assuming that each gas behaves ideally, we can calculate the partial pressure of each gas from the ideal gas law:

$$P_1 = \frac{n_1RT}{V}, \quad P_2 = \frac{n_2RT}{V}, \quad P_3 = \frac{n_3RT}{V}$$

MATH SKILL BUILDER

$$PV = nRT$$

$$\frac{P\cancel{V}}{\cancel{V}} = \frac{nRT}{V}$$

$$P = \frac{nRT}{V}$$

Dalton's law of partial pressure $P_i = \dfrac{n_1 RT}{V} + \dfrac{n_2 RT}{V} \cdots$ $P_T = P_1 + P_2 + \cdots$

$P_1 + P_2$ 8.4 atm

6.0 atm

2.4 atm

| 0.50 mol H_2 | 5.0 L at 20 °C $P_{H_2} = 2.4$ atm |

| 1.25 mol He | 5.0 L at 20 °C $P_{He} = 6.0$ atm |

| 1.25 mol He + 0.50 mol H_2 / 1.75 mol gas | 5.0 L at 20 °C $P_{total} = P_{H_2} + P_{He}$ = 2.4 atm + 6.0 atm = 8.4 atm |

Figure 13.10

When two gases are present, the total pressure is the sum of the partial pressures of the gases.

assumes ideal behavior
-Kinetic molecular theory

8.4 atm   8.4 atm   8.4 atm

1.75 mol He

5.0 L at 20 °C
P_{total} = 8.4 atm

0.75 mol H_2
0.75 mol He
0.25 mol Ne
_____
1.75 mol

5.0 L at 20 °C
P_{total} = 8.4 atm

1.00 mol N_2
0.50 mol O_2
0.25 mol Ar
_____
1.75 mol

5.0 L at 20 °C
P_{total} = 8.4 atm

Figure 13.11

The total pressure of a mixture of gases depends on the number of moles of gas particles (atoms or molecules) present, not on the identities of the particles. Note that these three samples show the same total pressure because each contains 1.75 mol gas. The detailed nature of the mixture is unimportant.

The total pressure of the mixture, P_{total}, can be represented as

$$P_{total} = P_1 + P_2 + P_3 = \frac{n_1 RT}{V} + \frac{n_2 RT}{V} + \frac{n_3 RT}{V}$$

$$= n_1\left(\frac{RT}{V}\right) + n_2\left(\frac{RT}{V}\right) + n_3\left(\frac{RT}{V}\right)$$

$$= (n_1 + n_2 + n_3)\left(\frac{RT}{V}\right)$$

$$= n_{total}\left(\frac{RT}{V}\right)$$

where n_{total} is the sum of the numbers of moles of the gases in the mixture. Thus, for a mixture of ideal gases, it is the *total number of moles of particles* that is important, not the *identity* of the individual gas particles. This idea is illustrated in Figure 13.11.

The fact that the pressure exerted by an ideal gas is affected by the *number* of gas particles and is independent of the *nature* of the gas particles tells us two important things about ideal gases:

1. The volume of the individual gas particle (atom or molecule) must not be very important.

2. The forces among the particles must not be very important.

If these factors were important, the pressure of the gas would depend on the nature of the individual particles. For example, an argon atom is much larger than a helium atom. Yet 1.75 mol argon gas in a 5.0-L container at 20 °C exerts the same pressure as 1.75 mol helium gas in a 5.0-L container at 20 °C.

The same idea applies to the forces among the particles. Although the forces among gas particles depend on the nature of the particles, this seems to have little influence on the behavior of an ideal gas. We will see that these observations strongly influence the model that we will construct to explain ideal gas behavior.

EXAMPLE 13.12 | **Using Dalton's Law of Partial Pressures, I**

Mixtures of helium and oxygen are used in the "air" tanks of underwater divers for deep dives. For a particular dive, 12 L of O_2 at 25 °C and 1.0 atm and 46 L of He at 25 °C and 1.0 atm were both pumped into a 5.0-L tank. Calculate the partial pressure of each gas and the total pressure in the tank at 25 °C.

SOLUTION

Where Are We Going?

We want to determine the partial pressure of helium and oxygen and the total pressure in the tank.

What Do We Know?

- We know the initial volume, pressure, and temperature of both gases.
- We know the final volume of the tank.
- The temperature remains constant.
- Ideal gas law: $PV = nRT$.
- Dalton's law of partial pressures: $P_{total} = P_1 + P_2 + \ldots$

What Information Do We Need?

- $R = 0.08206$ L atm/mol K.

How Do We Get There?

Because the partial pressure of each gas depends on the moles of that gas present, we must first calculate the number of moles of each gas by using the ideal gas law in the form

$$n = \frac{PV}{RT}$$

From the above description we know that $P = 1.0$ atm, $V = 12$ L for O_2 and 46 L for He, and $T = 25 + 273 = 298$ K. Also, $R = 0.08206$ L atm/K mol (as always).

$$\text{Moles of } O_2 = n_{O_2} = \frac{(1.0 \text{ atm})(12 \text{ L})}{(0.08206 \text{ L atm/K mol})(298 \text{ K})} = 0.49 \text{ mol}$$

$$\text{Moles of He} = n_{He} = \frac{(1.0 \text{ atm})(46 \text{ L})}{(0.08206 \text{ L atm/K mol})(298 \text{ K})} = 1.9 \text{ mol}$$

The tank containing the mixture has a volume of 5.0 L, and the temperature is 25 °C (298 K). We can use these data and the ideal gas law to calculate the partial pressure of each gas.

$$P = \frac{nRT}{V}$$

$$P_{O_2} = \frac{(0.49 \text{ mol})(0.08206 \text{ L atm/K mol})(298 \text{ K})}{5.0 \text{ L}} = 2.4 \text{ atm}$$

$$P_{He} = \frac{(1.9 \text{ mol})(0.08206 \text{ L atm/K mol})(298 \text{ K})}{5.0 \text{ L}} = 9.3 \text{ atm}$$

The total pressure is the sum of the partial pressures.

$$P_{total} = P_{O_2} + P_{He} = 2.4 \text{ atm} + 9.3 \text{ atm} = 11.7 \text{ atm}$$

REALITY CHECK   The volume of each gas decreased, and the pressure of each gas increased. The partial pressure of helium is greater than that of oxygen, which makes sense because the initial temperatures and pressures of helium and oxygen were the same, but the initial volume of helium was much greater than that of oxygen.

MATH SKILL BUILDER

$$PV = nRT$$

$$\frac{PV}{RT} = \frac{n\cancel{R}\cancel{T}}{\cancel{R}\cancel{T}}$$

$$\frac{PV}{RT} = n$$

Divers use a mixture of oxygen and helium in their breathing tanks when diving to depths greater than 150 feet.

Kurt Amsler

Figure 13.12

The production of oxygen by thermal decomposition of $KClO_3$.

Self-Check EXERCISE 13.9

A 2.0-L flask contains a mixture of nitrogen gas and oxygen gas at 25 °C. The total pressure of the gaseous mixture is 0.91 atm, and the mixture is known to contain 0.050 mol N_2. Calculate the partial pressure of oxygen and the moles of oxygen present.

See Problems 13.67 through 13.70. ■

Table 13.2 The Vapor Pressure of Water as a Function of Temperature	
T (°C)	P (torr)
0.0	4.579
10.0	9.209
20.0	17.535
25.0	23.756
30.0	31.824
40.0	55.324
60.0	149.4
70.0	233.7
90.0	525.8

A mixture of gases occurs whenever a gas is collected by displacement of water. For example, Figure 13.12 shows the collection of the oxygen gas that is produced by the decomposition of solid potassium chlorate. The gas is collected by bubbling it into a bottle that is initially filled with water. Thus the gas in the bottle is really a mixture of water vapor and oxygen. (Water vapor is present because molecules of water escape from the surface of the liquid and collect as a gas in the space above the liquid.) Therefore, the total pressure exerted by this mixture is the sum of the partial pressure of the gas being collected and the partial pressure of the water vapor. The partial pressure of the water vapor is called the vapor pressure of water. Because water molecules are more likely to escape from hot water than from cold water, the *vapor pressure* of water increases with temperature. This is shown by the values of vapor pressure at various temperatures in Table 13.2.

EXAMPLE 13.13   Using Dalton's Law of Partial Pressures, II

A sample of solid potassium chlorate, $KClO_3$, was heated in a test tube (see Figure 13.12) and decomposed according to the reaction

$$2KClO_3(s) \rightarrow 2KCl(s) + 3O_2(g)$$

The oxygen produced was collected by displacement of water at 22 °C. The resulting mixture of O_2 and H_2O vapor had a total pressure of 754 torr and a volume of 0.650 L. Calculate the partial pressure of O_2 in the gas collected and the number of moles of O_2 present. The vapor pressure of water at 22 °C is 21 torr.

SOLUTION

Where Are We Going?

We want to determine the partial pressure of oxygen collected by water displacement and the number of moles of O_2 present.

What Do We Know?

• We know the temperature, total pressure, and volume of gas collected by water displacement.

- We know the vapor pressure of water at this temperature.
- Ideal gas law: $PV = nRT$.
- Dalton's law of partial pressures: $P_{total} = P_1 + P_2 + \ldots$

What Information Do We Need?

- $R = 0.08206$ L atm/mol K.

How Do We Get There?

We know the total pressure (754 torr) and the partial pressure of water (vapor pressure = 21 torr). We can find the partial pressure of O_2 from Dalton's law of partial pressures:

$$P_{total} = P_{O_2} + P_{H_2O} = P_{O_2} + 21 \text{ torr} = 754 \text{ torr}$$

or

$$P_{O_2} + 21 \text{ torr} = 754 \text{ torr}$$

We can solve for P_{O_2} by subtracting 21 torr from both sides of the equation.

$$P_{O_2} = 754 \text{ torr} - 21 \text{ torr} = 733 \text{ torr}$$

Next we solve the ideal gas law for the number of moles of O_2.

$$n_{O_2} = \frac{P_{O_2}V}{RT}$$

In this case, $P_{O_2} = 733$ torr. We change the pressure to atmospheres as follows:

$$\frac{733 \text{ torr}}{760 \text{ torr/atm}} = 0.964 \text{ atm}$$

Then,

$$V = 0.650 \text{ L}$$
$$T = 22 \,°C = 22 + 273 = 295 \text{ K}$$
$$R = 0.08206 \text{ L atm/K mol}$$

so

$$n_{O_2} = \frac{(0.964 \text{ atm})(0.650 \text{ L})}{(0.08206 \text{ L atm/K mol})(295 \text{ K})} = 2.59 \times 10^{-2} \text{ mol}$$

MATH SKILL BUILDER

$$PV = nRT$$
$$\frac{PV}{RT} = \frac{nRT}{RT}$$
$$\frac{PV}{RT} = n$$

Self-Check **EXERCISE 13.10** Consider a sample of hydrogen gas collected over water at 25 °C where the vapor pressure of water is 24 torr. The volume occupied by the gaseous mixture is 0.500 L, and the total pressure is 0.950 atm. Calculate the partial pressure of H_2 and the number of moles of H_2 present.

See Problems 13.71 through 13.74. ■

13.7 Laws and Models: A Review

OBJECTIVE: To understand the relationship between laws and models (theories).

In this chapter we have considered several properties of gases and have seen how the relationships among these properties can be expressed by various laws written in the form of mathematical equations. The most useful of these is the ideal gas equation, which relates all the important gas properties. However, under certain conditions gases do not obey the ideal gas equation. For example, at high pressures and/or low temperatures, the properties of gases deviate significantly from the predictions of the ideal gas equation. On the other hand, as the pressure is lowered and/or the temperature is increased, almost all gases show close agreement with the ideal gas equation. This means that an ideal gas is really a hypothetical substance. At low pressures and/or high temperatures, real gases *approach* the behavior expected for an ideal gas.

At this point we want to build a model (a theory) to explain *why* a gas behaves as it does. We want to answer the question, *What are the characteristics of the individual gas particles that cause a gas to behave as it does?* However, before we do this let's briefly review the scientific method. Recall that a law is a generalization about behavior that has been observed in many experiments. Laws are very useful; they allow us to predict the behavior of similar systems. For example, a chemist who prepares a new gaseous compound can assume that that substance will obey the ideal gas equation (at least at low *P* and/or high *T*).

However, laws do not tell us *why* nature behaves the way it does. Scientists try to answer this question by constructing theories (building models). The models in chemistry are speculations about how individual atoms or molecules (microscopic particles) cause the behavior of macroscopic systems (collections of atoms and molecules in large enough numbers so that we can observe them).

A model is considered successful if it explains known behavior and predicts correctly the results of future experiments. But a model can never be proved absolutely true. In fact, by its very nature *any model is an approximation* and is destined to be modified, at least in part. Models range from the simple (to predict approximate behavior) to the extraordinarily complex (to account precisely for observed behavior). In this text, we use relatively simple models that fit most experimental results.

13.8 The Kinetic Molecular Theory of Gases

OBJECTIVE: To understand the basic postulates of the kinetic molecular theory.

go Chemistry Module 16: Gas Law and the Kinetic Molecular Theory covers concepts in this section.

A relatively simple model that attempts to explain the behavior of an ideal gas is the **kinetic molecular theory.** This model is based on speculations about the behavior of the individual particles (atoms or molecules) in a gas. The assumptions (postulates) of the kinetic molecular theory can be stated as follows:

Postulates of the Kinetic Molecular Theory of Gases

1. Gases consist of tiny particles (atoms or molecules).
2. These particles are so small, compared with the distances between them, that the volume (size) of the individual particles can be assumed to be negligible (zero).
3. The particles are in constant random motion, colliding with the walls of the container. These collisions with the walls cause the pressure exerted by the gas.
4. The particles are assumed not to attract or to repel each other.
5. The average kinetic energy of the gas particles is directly proportional to the Kelvin temperature of the gas.

"ideal behavior" is more closely acheived with higher T and lower P

The kinetic energy referred to in postulate 5 is the energy associated with the motion of a particle. Kinetic energy (KE) is given by the equation $KE = \frac{1}{2}mv^2$, where m is the mass of the particle and v is the velocity (speed) of the particle. The greater the mass or velocity of a particle, the greater its kinetic energy. Postulate 5 means that if a gas is heated to higher temperatures, the average speed of the particles increases; therefore, their kinetic energy increases.

Although real gases do not conform exactly to the five assumptions listed here, we will see in the next section that these postulates do indeed explain *ideal* gas behavior—behavior shown by real gases at high temperatures and/or low pressures.

13.9 The Implications of the Kinetic Molecular Theory

OBJECTIVES: To understand the term *temperature*. • To learn how the kinetic molecular theory explains the gas laws.

In this section we will discuss the *qualitative* relationships between the kinetic molecular (KM) theory and the properties of gases. That is, without going into the mathematical details, we will show how the kinetic molecular theory explains some of the observed properties of gases.

▶ The Meaning of Temperature

In Chapter 2 we introduced temperature very practically as something we measure with a thermometer. We know that as the temperature of an object increases, the object feels "hotter" to the touch. But what does temperature really mean? How does matter change when it gets "hotter"? In Chapter 10 we introduced the idea that temperature is an index of molecular motion. The kinetic molecular theory allows us to further develop this concept. As postulate 5 of the KM theory states, the temperature of a gas reflects how rapidly, on average, its individual gas particles are moving. At high temperatures the particles move very fast and hit the walls of the container frequently, whereas at low temperatures the particles' motions are more sluggish and they collide with the walls of the container much less often. Therefore, temperature really is a measure of the motions of the gas particles. In fact, the Kelvin temperature of a gas is directly proportional to the average kinetic energy of the gas particles.

▶ The Relationship Between Pressure and Temperature

To see how the meaning of temperature given above helps to explain gas behavior, picture a gas in a rigid container. As the gas is heated to a higher temperature, the particles move faster, hitting the walls more often. And, of course, the impacts become more forceful as the particles move faster. If the pressure is due to collisions with the walls, the gas pressure should increase as temperature is increased.

Is this what we observe when we measure the pressure of a gas as it is heated? Yes. A given sample of gas in a rigid container (if the volume is not changed) shows an increase in pressure as its temperature is increased.

a A gas confined in a cylinder with a movable piston. The gas pressure P_{gas} is just balanced by the external pressure P_{ext}. That is, $P_{gas} = P_{ext}$.

b The temperature of the gas is increased at constant pressure P_{ext}. The increased particle motions at the higher temperature push back the piston, increasing the volume of the gas.

Figure 13.13

▶ The Relationship Between Volume and Temperature

Now picture the gas in a container with a movable piston. As shown in Figure 13.13a, the gas pressure P_{gas} is just balanced by an external pressure P_{ext}. What happens when we heat the gas to a higher temperature? As the temperature increases, the particles move faster, causing the gas pressure to increase. As soon as the gas pressure P_{gas} becomes greater than P_{ext} (the pressure holding the piston), the piston moves up until $P_{gas} = P_{ext}$. Therefore, the KM model predicts that the volume of the gas will increase as we raise its temperature at a constant pressure (Figure 13.13b). This agrees with experimental observations (as summarized by Charles's law).

EXAMPLE 13.14 | Using the Kinetic Molecular Theory to Explain Gas Law Observations

Use the KM theory to predict what will happen to the pressure of a gas when its volume is decreased (n and T constant). Does this prediction agree with the experimental observations?

SOLUTION

When we decrease the gas's volume (make the container smaller), the particles hit the walls more often because they do not have to travel so far between the walls. This would suggest an increase in pressure. This prediction on the basis of the model is in agreement with experimental observations of gas behavior (as summarized by Boyle's law). ■

In this section we have seen that the predictions of the kinetic molecular theory generally fit the behavior observed for gases. This makes it a useful and successful model.

13.10   Gas Stoichiometry

OBJECTIVES: To understand the molar volume of an ideal gas. • To learn the definition of STP. • To use these concepts and the ideal gas equation.

We have seen repeatedly in this chapter just how useful the ideal gas equation is. For example, if we know the pressure, volume, and temperature for a given sample of gas, we can calculate the number of moles present:

$n = PV/RT$. This fact makes it possible to do stoichiometric calculations for reactions involving gases. We will illustrate this process in Example 13.15.

EXAMPLE 13.15 | Gas Stoichiometry: Calculating Volume

Calculate the volume of oxygen gas produced at 1.00 atm and 25 °C by the complete decomposition of 10.5 g of potassium chlorate. The balanced equation for the reaction is

$$2KClO_3(s) \rightarrow 2KCl(s) + 3O_2(g)$$

SOLUTION

Where Are We Going?

We want to determine the volume of oxygen gas collected by the decomposition of $KClO_3$.

What Do We Know?

- We know the temperature and pressure of oxygen gas.
- We know the mass of $KClO_3$.
- The balanced equation: $2KClO_3(s) \rightarrow 2KCl(s) = 3O_2(g)$.
- Ideal gas law: $PV = nRT$.

What Information Do We Need?

- $R = 0.08206$ L atm/mol K.
- We need the number of moles of oxygen gas.
- Molar mass of $KClO_3$.

How Do We Get There?

This is a stoichiometry problem very much like the type we considered in Chapter 9. The only difference is that in this case, we want to calculate the volume of a gaseous product rather than the number of grams. To do so, we can use the relationship between moles and volume given by the ideal gas law.

We'll summarize the steps required to do this problem in the following schematic:

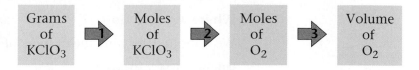

MATH SKILL BUILDER

$\frac{10.5}{122.6} = 0.085644$

$0.085644 \Rightarrow 0.0856$
Round
off

$0.0856 = 8.56 \times 10^{-2}$

Step 1 To find the moles of $KClO_3$ in 10.5 g, we use the molar mass of $KClO_3$ (122.6 g).

$$10.5 \text{ g } KClO_3 \times \frac{1 \text{ mol } KClO_3}{122.6 \text{ g } KClO_3} = 8.56 \times 10^{-2} \text{ mol } KClO_3$$

Step 2 To find the moles of O_2 produced, we use the mole ratio of O_2 to $KClO_3$ derived from the balanced equation.

$$8.56 \times 10^{-2} \text{ mol } KClO_3 \times \frac{3 \text{ mol } O_2}{2 \text{ mol } KClO_3} = 1.28 \times 10^{-1} \text{ mol } O_2$$

Step 3 To find the volume of oxygen produced, we use the ideal gas law $PV = nRT$, where

$$P = 1.00 \text{ atm}$$
$$V = ?$$
$$n = 1.28 \times 10^{-1} \text{ mol, the moles of } O_2 \text{ we calculated}$$
$$R = 0.08206 \text{ L atm/K mol}$$
$$T = 25\,°C = 25 + 273 = 298 \text{ K}$$

Solving the ideal gas law for V gives

$$V = \frac{nRT}{P} = \frac{(1.28 \times 10^{-1} \text{ mol})\left(0.08206 \frac{\text{L atm}}{\text{K mol}}\right)(298 \text{ K})}{1.00 \text{ atm}} = 3.13 \text{ L}$$

Thus 3.13 L of O_2 will be produced.

Self-Check EXERCISE 13.11 Calculate the volume of hydrogen produced at 1.50 atm and 19 °C by the reaction of 26.5 g of zinc with excess hydrochloric acid according to the balanced equation

$$Zn(s) + 2HCl(aq) \rightarrow ZnCl_2(aq) + H_2(g)$$

See Problems 13.85 through 13.92. ■

In dealing with the stoichiometry of reactions involving gases, it is useful to define the volume occupied by 1 mole of a gas under certain specified conditions. For 1 mole of an ideal gas at 0 °C (273 K) and 1 atm, the volume of the gas given by the ideal gas law is

$$V = \frac{nRT}{P} = \frac{(1.00 \text{ mol})(0.08206 \text{ L atm/K mol})(273 \text{ K})}{1.00 \text{ atm}} = 22.4 \text{ L}$$

This volume of 22.4 L is called the **molar volume** of an ideal gas.

The conditions 0 °C and 1 atm are called **standard temperature and pressure** (abbreviated **STP**). Properties of gases are often given under these conditions. Remember, the molar volume of an ideal gas is 22.4 L *at STP*. That is, 22.4 L contains 1 mole of an ideal gas at STP.

| STP: 0 °C and 1 atm |

EXAMPLE 13.16 | Gas Stoichiometry: Calculations Involving Gases at STP

A sample of nitrogen gas has a volume of 1.75 L at STP. How many moles of N_2 are present?

SOLUTION

Where Are We Going?

We want to determine the number of moles of nitrogen gas.

What Do We Know?

- The nitrogen gas has a volume of 1.75 L at STP.

What Information Do We Need?

- STP = 1.00 atm, 0 °C.
- At STP, 1 mole of an ideal gas occupies a volume of 22.4 L.

How Do We Get There?

We could solve this problem by using the ideal gas equation, but we can take a shortcut by using the molar volume of an ideal gas at STP. Because 1 mole of an ideal gas at STP has a volume of 22.4 L, a 1.75-L sample of N_2 at STP contains considerably less than 1 mol. We can find how many moles by using the equivalence statement

$$1.000 \text{ mol} = 22.4 \text{ L (STP)}$$

which leads to the conversion factor we need:

$$1.75 \text{ L } N_2 \times \frac{1.000 \text{ mol } N_2}{22.4 \text{ L } N_2} = 7.81 \times 10^{-2} \text{ mol } N_2$$

Self-Check EXERCISE 13.12 Ammonia is commonly used as a fertilizer to provide a source of nitrogen for plants. A sample of $NH_3(g)$ occupies a volume of 5.00 L at 25 °C and 15.0 atm. What volume will this sample occupy at STP?

See Problems 13.95 through 13.98. ∎

Standard conditions (STP) and molar volume are also useful in carrying out stoichiometric calculations on reactions involving gases, as shown in Example 13.17.

EXAMPLE 13.17 | Gas Stoichiometry: Reactions Involving Gases at STP

Quicklime, CaO, is produced by heating calcium carbonate, $CaCO_3$. Calculate the volume of CO_2 produced at STP from the decomposition of 152 g of $CaCO_3$ according to the reaction

$$CaCO_3(s) \rightarrow CaO(s) + CO_2(g)$$

SOLUTION

Where Are We Going?

We want to determine the volume of carbon dioxide produced from 152 g of $CaCO_3$.

What Do We Know?

- We know the temperature and pressure of carbon dioxide gas (STP).
- We know the mass of $CaCO_3$.
- The balanced equation: $CaCO_3(s) \rightarrow CaO(s) = CO_2(g)$.

What Information Do We Need?

- STP = 1.00 atm, 0 °C.
- At STP, 1 mole of an ideal gas occupies a volume of 22.4 L.
- We need the number of moles of carbon dioxide gas.
- Molar mass of $CaCO_3$.

How Do We Get There?

The strategy for solving this problem is summarized by the following schematic:

Step 1   Using the molar mass of $CaCO_3$ (100.1 g), we calculate the number of moles of $CaCO_3$.

$$152 \text{ g CaCO}_3 \times \frac{1 \text{ mol CaCO}_3}{100.1 \text{ g CaCO}_3} = 1.52 \text{ mol CaCO}_3$$

Step 2   Each mole of $CaCO_3$ produces 1 mole of CO_2, so 1.52 moles of CO_2 will be formed.

Step 3   We can convert the moles of CO_2 to volume by using the molar volume of an ideal gas, because the conditions are STP.

$$1.52 \text{ mol CO}_2 \times \frac{22.4 \text{ L CO}_2}{1 \text{ mol CO}_2} = 34.1 \text{ L CO}_2$$

Thus the decomposition of 152 g of $CaCO_3$ produces 34.1 L of CO_2 at STP. ■

> Remember that the molar volume of an ideal gas is 22.4 L at STP.

Note that the final step in Example 13.17 involves calculating the volume of gas from the number of moles. Because the conditions were specified as STP, we were able to use the molar volume of a gas at STP. If the conditions of a problem are different from STP, we must use the ideal gas law to compute the volume, as we did in Section 13.5.

CHAPTER 13 REVIEW

Key Terms

Summary

1. Atmospheric pressure is measured with a barometer. The most commonly used units of pressure are mm Hg (torr), atmospheres, and pascals (the SI unit).

2. Boyle's law states that the volume of a given amount of gas is inversely proportional to its pressure (at constant temperature): $PV = k$ or $P = k/V$. That is, as pressure increases, volume decreases.

3. Charles's law states that, for a given amount of gas at constant pressure, the volume is directly proportional to the temperature (in kelvins): $V = bT$. At -273 °C (0 K), the volume of a gas extrapolates to zero, and this temperature is called absolute zero.

4. Avogadro's law states that for a gas at constant temperature and pressure, the volume is directly proportional to the number of moles of gas: $V = an$.

5. These three laws can be combined into the ideal gas law, $PV = nRT$, where R is called the universal gas constant. This equation makes it possible to calculate any one of the properties—volume, pressure, temperature, or moles of gas present—given the other three. A gas that obeys this equation is said to behave ideally.

6. From the ideal gas equation we can derive the combined gas law,

$$\frac{P_1V_1}{T_1} = \frac{P_2V_2}{T_2}$$

which holds when the amount of gas (moles) remains constant.

7. The pressure of a gas mixture is described by Dalton's law of partial pressures, which states that the total pressure of the mixture of gases in a container is the sum of the partial pressures of the gases that make up the mixture.

8. The kinetic molecular theory of gases is a model that accounts for ideal gas behavior. This model assumes that a gas consists of tiny particles with negligible volumes, that there are no interactions among particles, and that the particles are in constant motion, colliding with the container walls to produce pressure.

Active Learning Questions

These questions are designed to be considered by groups of students in class. Often these questions work well for introducing a particular topic in class.

1. As you increase the temperature of a gas in a sealed, rigid container, what happens to the density of the gas? Would the results be the same if you did the same experiment in a container with a movable piston at a constant external pressure? Explain.

2. A diagram in a chemistry book shows a magnified view of a flask of air.

   What do you suppose is between the dots (which represent air molecules)?

   a. air   d. oxygen
   b. dust   e. nothing
   c. pollutants

3. If you put a drinking straw in water, place your finger over the opening, and lift the straw out of the water, some water stays in the straw. Explain.

4. A chemistry student relates the following story: I noticed my tires were a bit low and went to the gas station. As I was filling the tires I thought about the kinetic molecular (KM) theory. I noticed the tires because the volume was low, and I realized that I was increasing both the pressure and volume of the tires. "Hmmm," I thought, "that goes against what I learned in chemistry, where I was told pressure and volume are inversely proportional." What is the fault of the logic of the chemistry student in this situation? Explain under what conditions pressure and volume are inversely related (draw pictures and use the KM theory).

5. Chemicals X and Y (both gases) react to form the gas XY, but it takes some time for the reaction to occur. Both X and Y are placed in a container with a piston (free to move), and you note the volume. As the reaction occurs, what happens to the volume of the container? Explain your answer.

6. Which statement best explains why a hot-air balloon rises when the air in the balloon is heated?

   a. According to Charles's law, the temperature of a gas is directly related to its volume. Thus the volume of the balloon increases, decreasing the density.
   b. Hot air rises inside the balloon, which lifts the balloon.

   c. The temperature of a gas is directly related to its pressure. The pressure therefore increases, which lifts the balloon.
   d. Some of the gas escapes from the bottom of the balloon, thus decreasing the mass of gas in the balloon. This decreases the density of the gas in the balloon, which lifts the balloon.
   e. Temperature is related to the velocity of the gas molecules. Thus the molecules are moving faster, hitting the balloon more, and lifting the balloon.

   For choices you did not pick, explain what you feel is wrong with them, and justify the choice you did pick.

7. If you release a helium balloon, it soars upward and eventually pops. Explain this behavior.

8. If you have any two gases in different containers that are the same size at the same pressure and same temperature, what is true about the moles of each gas? Why is this true?

9. Using postulates of the kinetic molecular theory, give a molecular interpretation of Boyle's law, Charles's law, and Dalton's law of partial pressures.

10. Rationalize the following observations.

    a. Aerosol cans will explode if heated.
    b. You can drink through a soda straw.
    c. A thin-walled can will collapse when the air inside is removed by a vacuum pump.
    d. Manufacturers produce different types of tennis balls for high and low altitudes.

11. Show how Boyle's law and Charles's law are special cases of the ideal gas law.

12. Look at the demonstration discussed in Figure 13.1. How would this demonstration change if water was not added to the can? Explain.

13. How does Dalton's law of partial pressures help us with our model of ideal gases? That is, which postulates of the kinetic molecular theory does it support?

14. Draw molecular-level views that show the differences among solids, liquids, and gases.

15. Explain how increasing the number of moles of gas affects the pressure (assuming constant volume and temperature).

16. Explain how increasing the number of moles of gas affects the volume (assuming constant pressure and temperature).

17. Gases are said to exert pressure. Provide a molecular-level explanation for this.

18. Why is it incorrect to say that a sample of helium at 50 °C is twice as hot as a sample of helium at 25 °C?

19. We can use different units for pressure or volume, but we must use units of Kelvin for temperature. Why must we use the Kelvin temperature scale?

20. Estimate the mass of air at normal conditions that takes up the volume of your head. Provide support for your answer.

21. You are holding two balloons of the same volume. One balloon contains 1.0 g of helium. The other balloon contains neon. Calculate the mass of neon in the balloon.

22. You have helium gas in a two-bulbed container connected by a valve as shown below. Initially the valve is closed.

2.00 atm   3.00 atm

a. When the valve is opened, will the total pressure in the apparatus be less than 5.00 atm, equal to 5.00 atm, or greater than 5.00 atm? Explain your answer.
b. The left bulb has a volume of 9.00 L, and the right bulb has a volume of 3.00 L. Calculate the final pressure after the valve is opened.

23. Use the graphs below to answer the following questions.

a. Which of the above graphs best represents the relationship between the pressure and temperature (measured in kelvins) of 1 mole of an ideal gas?
b. Which of the above graphs best represents the relationship between the pressure and volume of 1 mole of an ideal gas?
c. Which of the above graphs best represents the relationship between the volume and temperature (measured in kelvins) of 1 mole of an ideal gas.

14

Liquids and Solids

● Ice, the solid form of water, provides recreation for this ice climber. *(Vandystadt/Tips Images)*

You have only to think about water to appreciate how different the three states of matter are. Flying, swimming, and ice skating are all done in contact with water in its various states. We swim in liquid water and skate on water in its solid form (ice). Airplanes fly in an atmosphere containing water in the gaseous state (water vapor). To allow these various activities, the arrangements of the water molecules must be significantly different in their gas, liquid, and solid forms.

In Chapter 13 we saw that the particles of a gas are far apart, are in rapid random motion, and have little effect on each other. Solids are obviously very different from gases. Gases have low densities, have high compressibilities, and completely fill a container. Solids have much greater densities than gases, are compressible only to a very slight extent, and are rigid; a solid maintains its shape regardless of its container. These properties indicate that the components of a solid are close together and exert large attractive forces on each other.

The properties of liquids lie somewhere between those of solids and of gases—but not midway between, as can be seen from some of the properties of the three states of water. For example, it takes about seven times more energy to change liquid water to steam (a gas) at 100 °C than to melt ice to form liquid water at 0 °C.

$$H_2O(s) \rightarrow H_2O(l) \quad \text{energy required} \cong 6 \text{ kJ/mol}$$
$$H_2O(l) \rightarrow H_2O(g) \quad \text{energy required} \cong 41 \text{ kJ/mol}$$

These values indicate that going from the liquid to the gaseous state involves a much greater change than going from the solid to the liquid. Therefore, we can conclude that the solid and liquid states are more similar than the liquid and gaseous states. This is also demonstrated by the densities of the three states of water (Table 14.1). Note that water in its gaseous state is about 2000 times less dense than in the solid and liquid states and that the latter two states have very similar densities.

We find in general that the liquid and solid states show many similarities and are strikingly different from the gaseous state (see Figure 14.1). The best way to picture the solid state is in terms of closely packed, highly ordered particles in contrast to the widely spaced, randomly arranged particles of a gas. The liquid state lies in between, but its properties indicate that it much more closely resembles the solid than the gaseous state. It is useful to picture a liquid in terms of particles that are generally quite close together, but with a more disordered arrangement than for the solid state and with some empty spaces. For most substances, the solid state has a higher density than the liquid, as Figure 14.1 suggests. However, water is an exception to this rule.

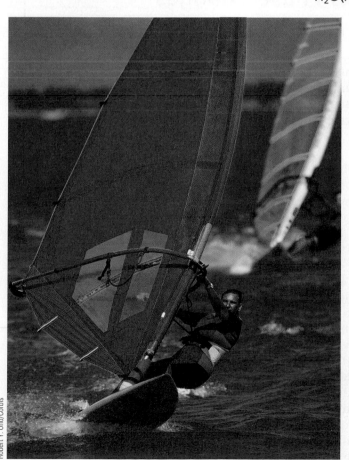
Wind surfers use liquid water for recreation.

Table 14.1	Densities of the Three States of Water
State	Density (g/cm³)
solid (0 °C, 1 atm)	0.9168
liquid (25 °C, 1 atm)	0.9971
gas (100 °C, 1 atm)	5.88×10^{-4}

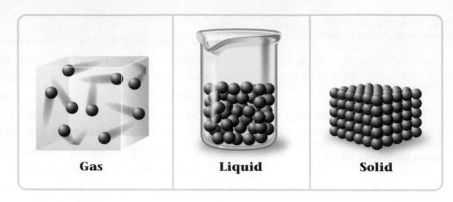

Figure 14.1

Representations of the gas, liquid, and solid states.

Ice has an unusual amount of empty space and so is less dense than liquid water, as indicated in Table 14.1.

In this chapter we will explore the important properties of liquids and solids. We will illustrate many of these properties by considering one of the earth's most important substances: water.

14.1   Water and Its Phase Changes

OBJECTIVE: To learn some of the important features of water.

In the world around us we see many solids (soil, rocks, trees, concrete, and so on), and we are immersed in the gases of the atmosphere. But the liquid we most commonly see is water; it is virtually everywhere, covering about 70% of the earth's surface. Approximately 97% of the earth's water is found in the oceans, which are actually mixtures of water and huge quantities of dissolved salts.

Water is one of the most important substances on earth. It is crucial for sustaining the reactions within our bodies that keep us alive, but it also affects our lives in many indirect ways. The oceans help moderate the earth's temperature. Water cools automobile engines and nuclear power plants. Water provides a means of transportation on the earth's surface and acts as a medium for the growth of the myriad creatures we use as food, and much more.

Pure water is a colorless, tasteless substance that at 1 atm pressure freezes to form a solid at 0 °C and vaporizes completely to form a gas at 100 °C. This means that (at 1 atm pressure) the liquid range of water occurs between the temperatures 0 °C and 100 °C.

What happens when we heat liquid water? First the temperature of the water rises. Just as with gas molecules, the motions of the water molecules increase as it is heated. Eventually the temperature of the water reaches 100 °C; now bubbles develop in the interior of the liquid, float to the surface, and burst—the boiling point has been reached. An interesting thing happens at the boiling point: even though heating continues, the temperature stays at 100 °C until all the water has changed to vapor. Only when all of the water has changed to the gaseous state does the temperature begin to rise

The water we drink often has a taste because of the substances dissolved in it. It is not pure water.

Figure 14.2

The heating/cooling curve for water heated or cooled at a constant rate. The plateau at the boiling point is longer than the plateau at the melting point, because it takes almost seven times as much energy (and thus seven times the heating time) to vaporize liquid water as to melt ice. Note that to make the diagram clear, the blue line is not drawn to scale. It actually takes more energy to melt ice and boil water than to heat water from 0 °C to 100 °C.

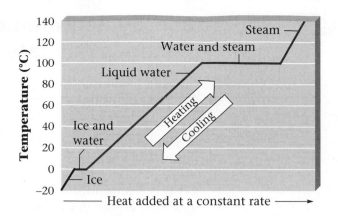

again. (We are now heating the vapor.) At 1 atm pressure, liquid water always changes to gaseous water at 100 °C, the **normal boiling point** for water.

The experiment just described is represented in Figure 14.2, which is called the **heating/cooling curve** for water. Going from left to right on this graph means energy is being added (heating). Going from right to left on the graph means that energy is being removed (cooling).

When liquid water is cooled, the temperature decreases until it reaches 0 °C, where the liquid begins to freeze (see Figure 14.2). The temperature remains at 0 °C until all the liquid water has changed to ice and then begins to drop again as cooling continues. At 1 atm pressure, water freezes (or, in the opposite process, ice melts) at 0 °C. This is called the **normal freezing point** of water. Liquid and solid water can coexist indefinitely if the temperature is held at 0 °C. However, at temperatures below 0 °C liquid water freezes, while at temperatures above 0 °C ice melts.

Interestingly, water expands when it freezes. That is, one gram of ice at 0 °C has a greater volume than one gram of liquid water at 0 °C. This has very important practical implications. For instance, water in a confined space can break its container when it freezes and expands. This accounts for the bursting of water pipes and engine blocks that are left unprotected in freezing weather.

The expansion of water when it freezes also explains why ice cubes float. Recall that density is defined as mass/volume. When one gram of liquid water freezes, its volume becomes greater (it expands). Therefore, the *density* of one gram of ice is less than the density of one gram of water, because in the case of ice we divide by a slightly larger volume. For example, at 0 °C the density of liquid water is

$$\frac{1.00 \text{ g}}{1.00 \text{ mL}} = 1.00 \text{ g/mL}$$

and the density of ice is

$$\frac{1.00 \text{ g}}{1.09 \text{ mL}} = 0.917 \text{ g/mL}$$

The lower density of ice also means that ice floats on the surface of lakes as they freeze, providing a layer of insulation that helps to prevent lakes and rivers from freezing solid in the winter. This means that aquatic life continues to have liquid water available through the winter.

14.2   Energy Requirements for the Changes of State

OBJECTIVES: To learn about interactions among water molecules. • To understand and use heat of fusion and heat of vaporization.

It is important to recognize that changes of state from solid to liquid and from liquid to gas are *physical* changes. No *chemical* bonds are broken in these processes. Ice, water, and steam all contain H_2O molecules. When water is boiled to form steam, water molecules are separated from each other (see Figure 14.3) but the individual molecules remain intact.

The bonding forces that hold the atoms of a molecule together are called **intramolecular** (within the molecule) **forces.** The forces that occur among molecules that cause them to aggregate to form a solid or a liquid are called **intermolecular** (between the molecules) **forces.** These two types of forces are illustrated in Figure 14.4.

> Remember that temperature is a measure of the random motions (average kinetic energy) of the particles in a substance.

It takes energy to melt ice and to vaporize water, because intermolecular forces between water molecules must be overcome. In ice the molecules are virtually locked in place, although they can vibrate about their positions. When energy is added, the vibrational motions increase, and the molecules eventually achieve the greater movement and disorder characteristic of liquid water. The ice has melted. As still more energy is added, the gaseous state is eventually reached, in which the individual molecules are far apart and interact relatively little. However, the gas still consists of water molecules. It would take *much* more energy to overcome the covalent bonds and decompose the water molecules into their component atoms.

The energy required to melt 1 mole of a substance is called the **molar heat of fusion.** For ice, the molar heat of fusion is 6.02 kJ/mol. The energy required to change 1 mole of liquid to its vapor is called the **molar heat of vaporization.** For water, the molar heat of vaporization is 40.6 kJ/mol at 100 °C. Notice in Figure 14.2 that the plateau that corresponds to the vaporization of water is much longer than that for the melting of ice. This occurs because it takes much more energy (almost seven times as much) to vaporize a mole of water than to melt a mole of ice. This is consistent with our models of solids, liquids, and gases (see Figure 14.1). In liquids, the particles (molecules) are relatively close together, so most of the intermolecular forces are still present. However, when the molecules go from the liquid to the gaseous state, they must be moved far apart. To separate the molecules enough to form a gas, virtually all of the intermolecular forces must be overcome, and this requires large quantities of energy.

Figure 14.3

Both liquid water and gaseous water contain H_2O molecules. In liquid water the H_2O molecules are close together, whereas in the gaseous state the molecules are widely separated. The bubbles contain gaseous water.

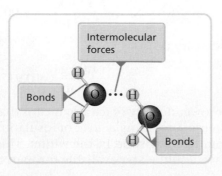

Figure 14.4

Intramolecular (bonding) forces exist between the atoms in a molecule and hold the molecule together. Intermolecular forces exist between molecules. These are the forces that cause water to condense to a liquid or form a solid at low enough temperatures. Intermolecular forces are typically much weaker than intramolecular forces.

CHEMISTRY *IN* FOCUS

Whales Need Changes of State

Sperm whales are prodigious divers. They commonly dive a mile or more into the ocean, hovering at that depth in search of schools of squid or fish. To remain motionless at a given depth, the whale must have the same density as the surrounding water. Because the density of seawater increases with depth, the sperm whale has a system that automatically increases its density as it dives. This system involves the spermaceti organ found in the whale's head. Spermaceti is a waxy substance with the formula

$$CH_3-(CH_2)_{15}-O-\underset{\underset{O}{\parallel}}{C}-(CH_2)_{14}-CH_3$$

which is a liquid above 30 °C. At the ocean surface the spermaceti in the whale's head is a liquid, warmed by the flow of blood through the spermaceti organ. When the whale dives, this blood flow decreases and the colder water causes the spermaceti to begin freezing. Because solid spermaceti is more dense than the liquid state, the sperm whale's density increases as it dives, matching the increase in the water's density.* When the whale wants to resurface, blood flow through the spermaceti organ increases, remelting the spermaceti and making the whale more buoyant. So the sperm whale's sophisticated density-regulating mechanism is based on a simple change of state.

*For most substances, the solid state is more dense than the liquid state. Water is an important exception.

A sperm whale.

EXAMPLE 14.1 | Calculating Energy Changes: Solid to Liquid

Calculate the energy required to melt 8.5 g of ice at 0 °C. The molar heat of fusion for ice is 6.02 kJ/mol.

SOLUTION

Where Are We Going?

We want to determine the energy (in kJ) required to melt 8.5 g of ice at 0 °C.

What Do We Know?

- We have 8.5 g of ice (H_2O) at 0 °C.
- The molar heat of fusion of ice is 6.02 kJ/mol.

What Information Do We Need?

- We need to know the number of moles of ice in 8.5 g.

How Do We Get There?

The molar heat of fusion is the energy required to melt *1 mole* of ice. In this problem we have 8.5 g of solid water. We must find out how many moles of ice this mass represents. Because the molar mass of water is $16 + 2(1) = 18$, we know that 1 mole of water has a mass of 18 g, so we can convert 8.5 g of H_2O to moles of H_2O.

$$8.5 \text{ g } H_2O \times \frac{1 \text{ mol } H_2O}{18 \text{ g } H_2O} = 0.47 \text{ mol } H_2O$$

Because 6.02 kJ of energy is required to melt a mole of solid water, our sample will take about half this amount (we have approximately half a mole of ice). To calculate the exact amount of energy required, we will use the equivalence statement

6.02 kJ required for 1 mol H_2O

which leads to the conversion factor we need:

$$0.47 \text{ mol } H_2O \times \frac{6.02 \text{ kJ}}{\text{mol } H_2O} = 2.8 \text{ kJ}$$

This can be represented symbolically as

REALITY CHECK Because we have just under half of 1 mole of ice, our answer should be about half the molar heat of fusion of ice. The answer of 2.8 kJ is just under half of 6.02 kJ, so this answer makes sense. ■

EXAMPLE 14.2 | Calculating Energy Changes: Liquid to Gas

Specific heat capacity was discussed in Section 10.5.

Calculate the energy (in kJ) required to heat 25 g of liquid water from 25 °C to 100. °C and change it to steam at 100. °C. The specific heat capacity of liquid water is 4.18 J/g °C, and the molar heat of vaporization of water is 40.6 kJ/mol.

SOLUTION

Where Are We Going?

We want to determine the energy (in kJ) required to heat and vaporize a given quantity of water.

What Do We Know?

- We have 25 g of H_2O at 25 °C. The water will be heated to 100. °C and then vaporized at 100. °C
- The specific heat capacity of liquid water is 4.18 J/g °C.
- The molar vaporization of water is 40.6 kJ/mol.
- $Q = s \times m \times \Delta T$.

What Information Do We Need?

- We need to know the number of moles of water in 25 g.

How Do We Get There?

This problem can be split into two parts: (1) heating the water to its boiling point and (2) converting the liquid water to vapor at the boiling point.

Step 1: Heating to Boiling   We must first supply energy to heat the liquid water from 25 °C to 100. °C. Because 4.18 J is required to heat *one* gram of water by *one* Celsius degree, we must multiply by both the mass of water (25 g) and the temperature change (100. °C − 25 °C = 75 °C),

Energy required (Q)	=	Specific heat capacity (s)	×	Mass of water (m)	×	Temperature change (ΔT)

which we can represent by the equation

$$Q = s \times m \times \mathrm{D}T$$

Thus

$$Q = 4.18 \frac{J}{g\,°C} \times 25\,g \times 75\,°C = 7.8 \times 10^3\,J$$

Energy required   Specific   Mass   Temperature
to heat 25 g of   heat   of   change
water from 25 °C   capacity   water
to 100. °C

$$= 7.8 \times 10^3\,J \times \frac{1\,kJ}{1000\,J} = 7.8\,kJ$$

Step 2: Vaporization   Now we must use the molar heat of vaporization to calculate the energy required to vaporize the 25 g of water at 100. °C. The heat of vaporization is given *per mole* rather than per gram, so we must first convert the 25 g of water to moles.

$$25\,g\,H_2O \times \frac{1\,mol\,H_2O}{18\,g\,H_2O} = 1.4\,mol\,H_2O$$

We can now calculate the energy required to vaporize the water.

$$\frac{40.6\,kJ}{mol\,H_2O} \times 1.4\,mol\,H_2O = 57\,kJ$$

Molar heat of   Moles of water
vaporization

The total energy is the sum of the two steps.

$$7.8\,kJ + 57\,kJ = 65\,kJ$$

Heat from   Change to
25 °C to   vapor
100. °C

Self-Check EXERCISE 14.1   Calculate the total energy required to melt 15 g of ice at 0 °C, heat the water to 100. °C, and vaporize it to steam at 100. °C.

HINT: Break the process into three steps and then take the sum.

See Problems 14.15 through 14.18. ■

14.3 Intermolecular Forces

OBJECTIVES: To learn about dipole–dipole attraction, hydrogen bonding, and London dispersion forces. • To understand the effect of these forces on the properties of liquids.

go Chemistry Module 17: Intermolecular Forces covers concepts in this section.

We have seen that covalent bonding forces within molecules arise from the sharing of electrons, but how do intermolecular forces arise? Actually several types of intermolecular forces exist. To illustrate one type, we will consider the forces that exist among water molecules.

> The polarity of a molecule was discussed in Section 12.3.

As we saw in Chapter 12, water is a polar molecule—it has a dipole moment. When molecules with dipole moments are put together, they orient themselves to take advantage of their charge distributions. Molecules with dipole moments can attract each other by lining up so that the positive and negative ends are close to each other, as shown in Figure 14.5a. This is called a **dipole–dipole attraction.** In the liquid, the dipoles find the best compromise between attraction and repulsion, as shown in Figure 14.5b.

Dipole–dipole forces are typically only about 1% as strong as covalent or ionic bonds, and they become weaker as the distance between the dipoles increases. In the gas phase, where the molecules are usually very far apart, these forces are relatively unimportant.

> See Section 12.2 for a discussion of electronegativity.

Particularly strong dipole-dipole forces occur between molecules in which hydrogen is bound to a highly electronegative atom, such as nitrogen, oxygen, or fluorine. Two factors account for the strengths of these interactions: the great polarity of the bond and the close approach of the dipoles, which is made possible by the very small size of the hydrogen atom. Because dipole–dipole attractions of this type are so unusually strong, they are given a special name—**hydrogen bonding.** Figure 14.6 illustrates hydrogen bonding among water molecules.

Hydrogen bonding has a very important effect on various physical properties. For example, the boiling points for the covalent compounds of hydrogen with the elements in Group 6 are given in Figure 14.7. Note that the boiling point of water is much higher than would be expected from the trend shown by the other members of the series. Why? Because the especially large electronegativity value of the oxygen atom compared with that of the other group members causes the O—H bonds to be much more polar

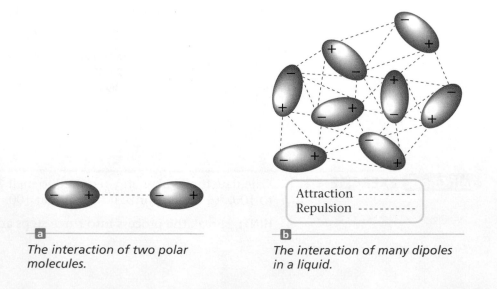

Attraction ----------
Repulsion ----------

a
The interaction of two polar molecules.

b
The interaction of many dipoles in a liquid.

Figure 14.5

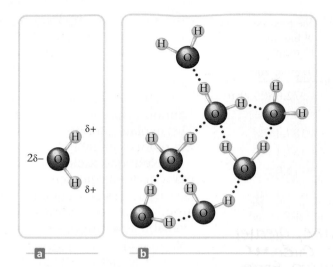

Figure 14.6

a The polar water molecule.
b Hydrogen bonding among water molecules. The small size of the hydrogen atoms allows the molecules to get very close and thus to produce strong interactions.

dramatic contrast

∩ highest

H_2O [δ+][δ−]

Figure 14.7

The boiling points of the covalent hydrides of elements in Group 6.

· *smaller* ↗
· *greater En*
· *decreased atomic size*

than the S—H, Se—H, or Te—H bonds. This leads to very strong hydrogen-bonding forces among the water molecules. An unusually large quantity of energy is required to overcome these interactions and separate the molecules to produce the gaseous state. That is, water molecules tend to remain together in the liquid state even at relatively high temperatures—hence the very high boiling point of water.

However, even molecules without dipole moments must exert forces on each other. We know this because all substances—even the noble gases—exist in the liquid and solid states at very low temperatures. There must be forces to hold the atoms or molecules as close together as they are in these condensed states. The forces that exist among noble gas atoms and nonpolar molecules are called **London dispersion forces.** To understand the origin of these forces, consider a pair of noble gas atoms. Although we usually assume that the electrons of an atom are uniformly distributed about the nucleus (see Figure 14.8a), this is apparently not true at every instant. Atoms

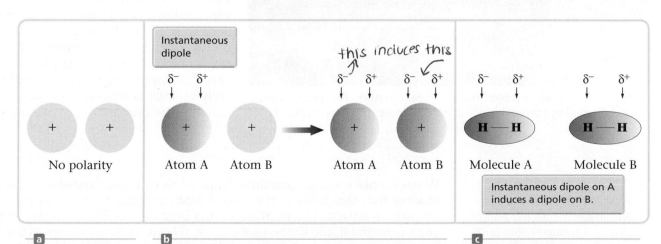

this induces this

a
Two atoms with spherical electron probability. These atoms have no polarity.

b
The atom on the left develops an instantaneous dipole when more electrons happen to congregate on the left than on the right.

c
Nonpolar molecules also interact by developing instantaneous dipoles.

Figure 14.8

Table 14.2	The Freezing Points of the Group 8 Elements
Element	Freezing Point (°C)
helium*	−272.0 (25 atm)
neon	−248.6
argon	−189.4
krypton	−157.3
xenon	−111.9

*Helium will not freeze unless the pressure is increased above 1 atm.

with larger size, greater surface area = Greater London Dispersion Forces

can develop a temporary dipolar arrangement of charge as the electrons move around the nucleus (see Figure 14.8b). This *instantaneous dipole* can then *induce* a similar dipole in a neighboring atom, as shown in Figure 14.8b. The interatomic attraction thus formed is both weak and short-lived, but it can be very significant for large atoms and large molecules, as we will see.

The motions of the atoms must be greatly slowed down before the weak London dispersion forces can lock the atoms into place to produce a solid. This explains, for instance, why the noble gas elements have such low freezing points (see Table 14.2).

Nonpolar molecules such as H_2, N_2, and I_2, none of which has a permanent dipole moment, also attract each other by London dispersion forces (see Figure 14.8c). London forces become more significant as the sizes of atoms or molecules increase. Larger size means there are more electrons available to form the dipoles.

14.4 Evaporation and Vapor Pressure

OBJECTIVE: To understand the relationship among vaporization, condensation, and vapor pressure.

We all know that a liquid can evaporate from an open container. This is clear evidence that the molecules of a liquid can escape the liquid's surface and form a gas. This process, which is called **vaporization** or **evaporation,** requires energy to overcome the relatively strong intermolecular forces in the liquid.

The fact that vaporization requires energy has great practical significance; in fact, one of the most important roles that water plays in our world is to act as a coolant. Because of the strong hydrogen bonding among its molecules in the liquid state, water has an unusually large heat of vaporization (41 kJ/mol). A significant portion of the sun's energy is spent evaporating water from the oceans, lakes, and rivers rather than warming the earth. The vaporization of water is also crucial to our body's temperature-control system, which relies on the evaporation of perspiration.

Water is used to absorb heat from nuclear reactors. The water is then cooled in cooling towers before it is returned to the environment.

▶ Vapor Pressure

Vapor, not *gas*, is the term we customarily use for the gaseous state of a substance that exists naturally as a solid or liquid at 25 °C and 1 atm.

A system at equilibrium is dynamic on the molecular level, but shows no visible changes.

When we place a given amount of liquid in a container and then close it, we observe that the amount of liquid at first decreases slightly but eventually becomes constant. The decrease occurs because there is a transfer of molecules from the liquid to the vapor phase (Figure 14.9). However, as the number of vapor molecules increases, it becomes more and more likely that some of them will return to the liquid. The process by which vapor molecules form a liquid is called **condensation.** Eventually, the same number of molecules are leaving the liquid as are returning to it: the rate of condensation equals the rate of evaporation. *At this point no further change occurs in the amounts of liquid or vapor, because the two opposite processes exactly balance each other;* the system is at *equilibrium*. Note that this system is highly *dynamic* on the mo-

a Net evaporation occurs at first, so the amount of liquid decreases slightly.

a As the number of vapor molecules increases, the rate of condensation increases. Finally the rate of condensation equals the rate of evaporation. The system is at equilibrium.

Figure 14.9

Behavior of a liquid in a closed container.

lecular level—molecules are constantly escaping from and entering the liquid. However, there is no *net* change because the two opposite processes just *balance* each other. As an analogy, consider two island cities connected by a bridge. Suppose the traffic flow on the bridge is the same in both directions. There is motion—we can see the cars traveling across the bridge—but the number of cars in each city is not changing because an equal number enter and leave each one. The result is no *net* change in the number of autos in each city: an equilibrium exists.

The pressure of the vapor present at equilibrium with its liquid is called the *equilibrium vapor pressure* or, more commonly, the **vapor pressure** of the liquid. A simple barometer can be used to measure the vapor pressure of a liquid, as shown in Figure 14.10. Because mercury is so dense, any common liq-

a It is easy to measure the vapor pressure of a liquid by using a simple barometer of the type shown here.

b The water vapor pushed the mercury down 24 mm (760 – 736), so the vapor pressure of water is 24 mm Hg at this temperature.

c Diethyl ether is much more volatile than water and thus shows a higher vapor pressure. In this case, the mercury level has been pushed down 545 mm (760 – 215), so the vapor pressure of diethyl ether is 545 mm Hg at this temperature.

Figure 14.10

uid injected at the bottom of the column of mercury floats to the top, where it produces a vapor, and the pressure of this vapor pushes some mercury out of the tube. When the system reaches equilibrium, the vapor pressure can be determined from the change in the height of the mercury column.

In effect, we are using the space above the mercury in the tube as a closed container for each liquid. However, in this case as the liquid vaporizes, the vapor formed creates a pressure that pushes some mercury out of the tube and lowers the mercury level. The mercury level stops changing when the excess liquid floating on the mercury comes to equilibrium with the vapor. The change in the mercury level (in millimeters) from its initial position (before the liquid was injected) to its final position is equal to the vapor pressure of the liquid.

The vapor pressures of liquids vary widely (see Figure 14.10). Liquids with high vapor pressures are said to be *volatile*—they evaporate rapidly.

The vapor pressure of a liquid at a given temperature is determined by the *intermolecular forces* that act among the molecules. Liquids in which the intermolecular forces are large have relatively low vapor pressures, because such molecules need high energies to escape to the vapor phase. For example, although water is a much smaller molecule than diethyl ether, C_2H_5—O—C_2H_5, the strong hydrogen-bonding forces in water cause its vapor pressure to be much lower than that of ether (see Figure 14.10).

EXAMPLE 14.3   Using Knowledge of Intermolecular Forces to Predict Vapor Pressure

Predict which substance in each of the following pairs will show the largest vapor pressure at a given temperature.

    a. $H_2O(l)$, $CH_3OH(l)$

    b. $CH_3OH(l)$, $CH_3CH_2CH_2CH_2OH(l)$

SOLUTION

    a. Water contains two polar O—H bonds; methanol (CH_3OH) has only one. Therefore, the hydrogen bonding among H_2O molecules is expected to be much stronger than that among CH_3OH molecules. This gives water a lower vapor pressure than methanol.

    b. Each of these molecules has one polar O—H bond. However, because $CH_3CH_2CH_2CH_2OH$ is a much larger molecule than CH_3OH, it has much greater London forces and thus is less likely to escape from its liquid. Thus $CH_3CH_2CH_2CH_2OH(l)$ has a lower vapor pressure than $CH_3OH(l)$. ∎

14.5 The Solid State: Types of Solids

OBJECTIVE: To learn about the various types of crystalline solids.

Solids play a very important role in our lives. The concrete we drive on, the trees that shade us, the windows we look through, the paper that holds this print, the diamond in an engagement ring, and the plastic lenses in eyeglasses are all important solids. Most solids, such as wood, paper, and glass, contain mixtures of various components. However, some natural solids, such as diamonds and table salt, are nearly pure substances.

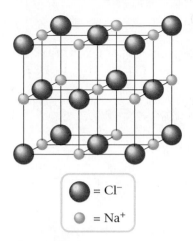

= Cl⁻

= Na⁺

Figure 14.11

The regular arrangement of sodium and chloride ions in sodium chloride, a crystalline solid.

a
Quartz, SiO₂

b
Rock salt, NaCl

c
Iron pyrite, FeS₂

Figure 14.12

Several crystalline solids.

Many substances form **crystalline solids**—those with a regular arrangement of their components. This is illustrated by the partial structure of sodium chloride shown in Figure 14.11. The highly ordered arrangement of the components in a crystalline solid produces beautiful, regularly shaped crystals such as those shown in Figure 14.12.

There are many different types of crystalline solids. For example, both sugar and salt have beautiful crystals that we can easily see. However, although both dissolve readily in water, the properties of the resulting solutions are quite different. The salt solution readily conducts an electric current; the sugar solution does not. This behavior arises from the different natures of the components in these two solids. Common salt, $NaCl$, is an ionic solid that contains Na^+ and Cl^- ions. When solid sodium chloride dissolves in water, sodium ions and chloride ions are distributed throughout the resulting solution. These ions are free to move through the solution to conduct an electric current. Table sugar (sucrose), on the other hand, is composed of neutral molecules that are dispersed throughout the water when the solid dissolves. No ions are present, and the resulting solution does not conduct electricity. These examples illustrate two important types of crystalline solids: **ionic solids,** represented by sodium chloride; and **molecular solids,** represented by sucrose.

A third type of crystalline solid is represented by elements such as graphite and diamond (both pure carbon), boron, silicon, and all metals. These substances, which contain atoms of only one element covalently bonded to each other, are called **atomic solids.**

We have seen that crystalline solids can be grouped conveniently into three classes as shown in Figure 14.13. Notice that the names of the three classes come from the components of the solid. An ionic solid contains ions,

Figure 14.13

The classes of crystalline solids.

Crystalline solids

Ionic solids	Molecular solids	Atomic solids
Components are ions.	Components are molecules.	Components are atoms.

- generally highest mps
- lattice energy (+ - + - / - + - +) ← (less than

- lower but highly variable mps

- lowest mp

note the void spaces created to "line-up" H-bonding interaction. so less dense than H_2O (l)

$\bullet$ = C

Diamond

[a]

An atomic solid. Each sphere represents a carbon atom in diamond.

network covalent solid

$\bullet$ = Cl^-

$\circ$ = Na^+

Sodium chloride

[b]

An ionic solid. The spheres represent alternating Na^+ and Cl^- ions in solid sodium chloride.

= H_2O

Ice

[c]

A molecular solid. Each unit of three spheres represents an H_2O molecule in ice. The dashed lines show the hydrogen bonding among the polar water molecules.

Figure 14.14

Examples of three types of crystalline solids. Only part of the structure is shown in each case. The structures continue in three dimensions with the same patterns.

a molecular solid contains molecules, and an atomic solid contains atoms. Examples of the three types of solids are shown in Figure 14.14.

The properties of a solid are determined primarily by the nature of the forces that hold the solid together. For example, although argon, copper, and diamond are all atomic solids (their components are atoms), they have strikingly different properties. Argon has a very low melting point (-189 °C), whereas diamond and copper melt at high temperatures (about 3500 °C and 1083 °C, respectively). Copper is an excellent conductor of electricity (it is widely used for electrical wires), whereas both argon and diamond are insulators. The shape of copper can easily be changed; it is both malleable (will form thin sheets) and ductile (can be pulled into a wire). Diamond, on the other hand, is the hardest natural substance known. The marked differences in properties among these three atomic solids are due to differences in bonding. We will explore the bonding in solids in the next section.

> The internal forces in a solid determine many of the properties of the solid.

14.6 Bonding in Solids

OBJECTIVES: To understand the interparticle forces in crystalline solids. • To learn about how the bonding in metals determines metallic properties.

We have seen that crystalline solids can be divided into three classes, depending on the fundamental particle or unit of the solid. Ionic solids consist of oppositely charged ions packed together, molecular solids contain molecules, and atomic solids have atoms as their fundamental particles. Examples of the various types of solids are given in Table 14.3.

> Ionic solids were also discussed in Section 12.5.

CO_2
symmetric
zero net dipole; so nonpolar
very weak IMF
very high vapor pressure

$\overset{\longrightarrow}{\ddot{O}=C=\ddot{O}}$
$\overset{\longleftarrow}{}$

= Cl⁻   = Na⁺

Figure 14.15

The packing of Cl⁻ and Na⁺ ions in solid sodium chloride.

Table 14.3	Examples of the Various Types of Solids	
Type of Solid	Examples	Fundamental Unit(s)
ionic	sodium chloride, NaCl(s)	Na⁺, Cl⁻ ions
ionic	ammonium nitrate, $NH_4NO_3(s)$	NH_4^+, NO_3^- ions
molecular	dry ice, $CO_2(s)$	CO_2 molecules
molecular	ice, $H_2O(s)$	H_2O molecules
atomic	diamond, C(s)	C atoms
atomic	iron, Fe(s)	Fe atoms
atomic	argon, Ar(s)	Ar atoms

very different behavior *(for $CO_2(s)$ and ice)*

very different behavior *(for diamond, iron, argon)*

H_2O

H—$\overset{\updownarrow}{\overset{\Phi}{}}$—H

Assymetric
so very polar
- much stronger IMFs
(very low vapor pressure)

▶ Ionic Solids

Ionic solids are stable substances with high melting points that are held together by the strong forces that exist between oppositely charged ions. The structures of ionic solids can be visualized best by thinking of the ions as spheres packed together as efficiently as possible. For example, in NaCl the larger Cl⁻ ions are packed together much like one would pack balls in a box. The smaller Na⁺ ions occupy the small spaces ("holes") left among the spherical Cl⁻ ions, as represented in Figure 14.15.

▶ Molecular Solids

In a molecular solid the fundamental particle is a molecule. Examples of molecular solids include ice (contains H_2O molecules), dry ice (contains CO_2 molecules), sulfur (contains S_8 molecules), and white phosphorus (contains P_4 molecules). The latter two substances are shown in Figure 14.16.

a
Sulfur crystals contain S_8 molecules.

b
White phosphorus contains P_4 molecules. It is so reactive with the oxygen in air that it must be stored under water.

Ken O'Donoghue

Richard Megna/Fundamental Photographs

Figure 14.16

Molecular solids tend to melt at relatively low temperatures because the intermolecular forces that exist among the molecules are relatively weak. If the molecule has a dipole moment, dipole–dipole forces hold the solid together. In solids with nonpolar molecules, London dispersion forces hold the solid together.

Part of the structure of solid phosphorus is represented in Figure 14.17. Note that the distances between P atoms in a given molecule are much shorter than the distances between the P_4 molecules. This is because the covalent bonds *between atoms* in the molecule are so much stronger than the London dispersion forces *between molecules*.

▶ Atomic Solids

The properties of atomic solids vary greatly because of the different ways in which the fundamental particles, the atoms, can interact with each other. For example, the solids of the Group 8 elements have very low melting points (see Table 14.2), because these atoms, having filled valence orbitals, cannot form covalent bonds with each other. So the forces in these solids are the relatively weak London dispersion forces.

On the other hand, diamond, a form of solid carbon, is one of the hardest substances known and has an extremely high melting point (about 3500 °C). The incredible hardness of diamond arises from the very strong covalent carbon–carbon bonds in the crystal, which lead to a giant

Figure 14.17

A representation of part of the structure of solid phosphorus, a molecular solid that contains P_4 molecules.

molecule. In fact, the entire crystal can be viewed as one huge molecule. A small part of the diamond structure is represented in Figure 14.14a. In diamond each carbon atom is bound covalently to four other carbon atoms to produce a very stable solid. Several other elements also form solids whereby the atoms join together covalently to form giant molecules. Silicon and boron are examples.

At this point you might be asking yourself, "Why aren't solids such as a crystal of diamond, which is a 'giant molecule,' classified as molecular solids?" The answer is that, by convention, a solid is classified as a molecular solid only if (like ice, dry ice, sulfur, and phosphorus) it contains small molecules. Substances like diamond that contain giant molecules are called network solids.

▶ Bonding in Metals

Metals represent another type of atomic solid. Metals have familiar physical properties: they can be pulled into wires, can be hammered into sheets, and are efficient conductors of heat and electricity. However, although the shapes of most pure metals can be changed relatively easily, metals are also durable and have high melting points. These facts indicate that it is difficult to separate metal atoms but relatively easy to slide them past each other. In other words, the bonding in most metals is *strong* but *nondirectional*.

The simplest picture that explains these observations is the **electron sea model,** which pictures a regular array of metal atoms in a "sea" of valence electrons that are shared among the atoms in a nondirectional way and that are quite mobile in the metal crystal. The mobile electrons can conduct heat and electricity, and the atoms can be moved rather easily, as, for example, when the metal is hammered into a sheet or pulled into a wire.

Because of the nature of the metallic crystal, other elements can be introduced relatively easily to produce substances called alloys. An **alloy** is best defined as *a substance that contains a mixture of elements and has metallic properties.* There are two common types of alloys.

In a **substitutional alloy** some of the host metal atoms are *replaced* by other metal atoms of similar sizes. For example, in brass approximately one-third of the atoms in the host copper metal have been replaced by zinc atoms, as shown in Figure 14.18a. Sterling silver (93% silver and 7% copper) and pewter (85% tin, 7% copper, 6% bismuth, and 2% antimony) are other examples of substitutional alloys.

An **interstitial alloy** is formed when some of the interstices (holes) among the closely packed metal atoms are occupied by atoms much smaller than the host atoms, as shown in Figure 14.18b. Steel, the

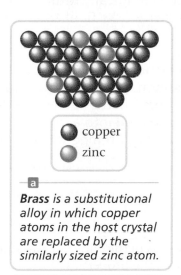

● copper
● zinc

a

Brass *is a substitutional alloy in which copper atoms in the host crystal are replaced by the similarly sized zinc atom.*

● iron
• carbon

b

Steel *is an interstitial alloy in which carbon atoms occupy interstices (holes) among the closely packed iron atoms.*

Figure 14.18

Two types of alloys.

A steel sculpture in Chicago.

Metal with a Memory

A distraught mother walks into the optical shop carrying her mangled pair of $400 eyeglasses. Her child had gotten into her purse, found her glasses, and twisted them into a pretzel. She hands them to the optometrist with little hope that they can be salvaged. The optometrist says not to worry and drops the glasses into a dish of warm water where the glasses magically spring back to their original shape. The optometrist hands the restored glasses to the woman and says there is no charge for repairing them.

How can the frames "remember" their original shape when placed in warm water? The answer is a nickel–titanium alloy called Nitinol that was developed in the late 1950s and early 1960s at the Naval Ordnance Laboratory in White Oak, Maryland, by William J. Buehler. (The name Nitinol comes from *Ni*ckel *Ti*tanium *N*aval *O*rdnance *L*aboratory.)

Nitinol has the amazing ability to remember a shape originally impressed in it. For example,

note the accompanying photos. What causes Nitinol to behave this way? Although the details are too complicated to describe here, this phenomenon results from two different forms of solid Nitinol. When Nitinol is heated to a sufficiently high temperature, the Ni and Ti atoms arrange themselves in a way that leads to the most compact and regular pattern of the atoms—a form called austenite (A). When the alloy is cooled, its atoms rearrange slightly to a form called martensite (M). The shape desired (for example, the word *ICE*) is set into the alloy at a high temperature (A form), then the metal is cooled, causing it to assume the M form. In this process no visible change is noted. Then, if the image is deformed, it will magically return if the alloy is heated (hot water works fine) to a temperature that changes it back to the A form.

Nitinol has many medical applications, including hooks used by orthopedic surgeons to attach ligaments and tendons to bone and "baskets" to catch blood clots. In the latter case a length of Nitinol wire is shaped into a tiny basket and this shape is set at a high temperature. The

best-known interstitial alloy, contains carbon atoms in the "holes" of an iron crystal. The presence of interstitial atoms changes the properties of the host metal. Pure iron is relatively soft, ductile, and malleable because of the absence of strong directional bonding. The spherical metal atoms can be moved rather easily with respect to each other. However, when carbon, which forms strong directional bonds, is introduced into an iron crystal, the presence of the directional carbon–iron bonds makes the resulting alloy harder, stronger, and less ductile than pure iron. The amount of carbon directly affects the properties of steel. *Mild steels* (containing less than 0.2% carbon) are still ductile and malleable and are used for nails, cables, and chains. *Medium steels* (containing 0.2–0.6% carbon) are harder than mild steels and are used in rails and structural steel beams. *High-carbon steels* (containing 0.6–1.5% carbon) are tough and hard and are used for springs, tools, and cutlery.

Many types of steel contain other elements in addition to iron and carbon. Such steels are often called *alloy steels* and can be viewed as being mixed interstitial (carbon) and substitutional (other metals) alloys. An example is stainless steel, which has chromium and nickel atoms substituted for some of the iron atoms. The addition of these metals greatly increases the steel's resistance to corrosion.

wires forming the basket are then straightened so they can be inserted as a small bundle through a catheter. When the wires warm up in the blood, the basket shape springs back and acts as a filter to stop blood clots from moving to the heart.

One of the most promising consumer uses of Nitinol is for eyeglass frames. Nitinol is also now being used for braces to straighten crooked teeth.

The word ICE is formed from Nitinol wire.

The wire is stretched to obliterate the word ICE.

The wire pops back to ICE when immersed in warm water.

© Cengage Learning

EXAMPLE 14.4 | Identifying Types of Crystalline Solids

Name the type of crystalline solid formed by each of the following substances:

  a. ammonia

  b. iron

  c. cesium fluoride

  d. argon

  e. sulfur

SOLUTION

  a. Solid ammonia contains NH_3 molecules, so it is a molecular solid.

  b. Solid iron contains iron atoms as the fundamental particles. It is an atomic solid.

  c. Solid cesium fluoride contains the Cs^+ and F^- ions. It is an ionic solid.

387

d. Solid argon contains argon atoms, which cannot form covalent bonds to each other. It is an atomic solid.

e. Sulfur contains S_8 molecules, so it is a molecular solid.

Self-Check EXERCISE 14.2 Name the type of crystalline solid formed by each of the following substances:

    a. sulfur trioxide

    b. barium oxide

    c. gold

See Problems 14.41 and 14.42. ■

CHAPTER 14 REVIEW

Key Terms

normal boiling point (14.1)

heating/cooling curve (14.1)

normal freezing point (14.1)

intramolecular forces (14.2)

intermolecular forces (14.2)

molar heat of fusion (14.2)

molar heat of vaporization (14.2)

dipole–dipole attraction (14.3)

hydrogen bonding (14.3)

London dispersion forces (14.3)

vaporization (evaporation) (14.4)

condensation (14.4)

vapor pressure (14.4)

crystalline solid (14.5)

ionic solid (14.5)

molecular solid (14.5)

atomic solid (14.5)

electron sea model (14.6)

alloy (14.6)

substitutional alloy (14.6)

interstitial alloy (14.6)

Summary

1. Liquids and solids exhibit some similarities and are very different from the gaseous state.

2. The temperature at which a liquid changes its state to a gas (at 1 atm pressure) is called the normal boiling point of that liquid. Similarly, the temperature at which a liquid freezes (at 1 atm pressure) is the normal freezing point. Changes of state are physical changes, not chemical changes.

3. To convert a substance from the solid to the liquid and then to the gaseous state requires the addition of energy. Forces among the molecules in a solid or a liquid must be overcome by the input of energy. The energy required to melt 1 mole of a substance is called the molar heat of fusion, and the energy required to change 1 mole of liquid to the gaseous state is called the molar heat of vaporization.

4. There are several types of intermolecular forces. Dipole–dipole interactions occur when molecules with dipole moments attract each other. A particularly strong dipole–dipole interaction called hydrogen bonding occurs in molecules that contain hydrogen bonded to a very electronegative element such as N, O, or F. London dispersion forces occur when instantaneous dipoles in atoms or nonpolar molecules lead to relatively weak attractions.

5. The change of a liquid to its vapor is called vaporization or evaporation. The process whereby vapor molecules form a liquid is called condensation. In a closed container, the pressure of the vapor over its liquid reaches a constant value called the vapor pressure of the liquid.

6. Many solids are crystalline (contain highly regular arrangements of their components). The three types of crystalline solids are ionic, molecular, and atomic solids. In ionic solids, the ions are packed together in a way that maximizes the attractions of oppositely charged ions and minimizes the repulsions among identically charged ions. Molecular solids are held together by dipole–dipole attractions if the molecules are polar and by London dispersion forces if the molecules are nonpolar. Atomic solids are held together by covalent bonding forces or London dispersion forces, depending on the atoms present.

Active Learning Questions

These questions are designed to be considered by groups of students in class. Often these questions work well for introducing a particular topic in class.

1. You seal a container half-filled with water. Which best describes what occurs in the container?

   a. Water evaporates until the air becomes saturated with water vapor; at this point, no more water evaporates.
   b. Water evaporates until the air becomes overly saturated (supersaturated) with water, and most of this water recondenses; this cycle continues until a certain amount of water vapor is present, and then the cycle ceases.
   c. The water does not evaporate because the container is sealed.
   d. Water evaporates, and then water evaporates and recondenses simultaneously and continuously.
   e. The water evaporates until it is eventually all in vapor form.

   Justify your choice and for choices you did not pick, explain what is wrong with them.

2. Explain the following: You add 100 mL of water to a 500-mL round-bottomed flask and heat the water until it is boiling. You remove the heat and stopper the flask, and the boiling stops. You then run cool water over the neck of the flask, and the boiling begins again. It seems as though you are boiling water by cooling it.

3. Is it possible for the dispersion forces in a particular substance to be stronger than hydrogen-bonding forces in another substance? Explain your answer.

4. Does the nature of intermolecular forces change when a substance goes from a solid to a liquid, or from a liquid to a gas? What causes a substance to undergo a phase change?

5. How does vapor pressure change with changing temperature? Explain.

6. What occurs when the vapor pressure of a liquid is equal to atmospheric pressure? Explain.

7. What is the vapor pressure of water at 100 °C? How do you know?

8. How do the following physical properties depend on the strength of intermolecular forces? Explain.

   a. melting point
   b. boiling point
   c. vapor pressure

9. Look at Figure 14.2. Why doesn't temperature increase continuously over time? That is, why does the temperature stay constant for periods of time?

10. Which are stronger, intermolecular or intramolecular forces for a given molecule? What observation(s) have you made that supports this position? Explain.

11. Why does water evaporate at all?

12. Sketch a microscopic picture of water and distinguish between *intramolecular bonds* and *intermolecular forces*. Which correspond to the bonds we draw in Lewis structures?

13. Which has the stronger intermolecular forces: N_2 or H_2O? Explain.

14. Which gas would behave more ideally at the same conditions of pressure and temperature: CO or N_2? Why?

15. You have seen that the water molecule has a bent shape and therefore is a polar molecule. This accounts for many of water's interesting properties.

   What if the water molecule were linear? How would this affect the properties of water? How would life be different?

16. True or false? Methane (CH_4) is more likely to form stronger hydrogen bonding than is water because each methane molecule has twice as many hydrogen atoms. Provide a concise explanation of hydrogen bonding to go with your answer.

17. Why should it make sense that N_2 exists as a gas? Given your answer, how is it possible to make liquid nitrogen? Explain why lowering the temperature works.

18. White phosphorus and sulfur both are called molecular solids even though each is made of only phosphorus and sulfur, respectively. How can they be considered molecular solids? If this is true, why isn't diamond (which is made up only of carbon) a molecular solid?

19. Why is it incorrect to use the term "molecule of NaCl" but correct to use the term "molecule of H_2O"? Is the term "molecule of diamond" correct? Explain.

20. Which would you predict should be larger for a given substance: ΔH_{vap} or ΔH_{fus}? Explain why.

21. In the diagram below, which lines represent the hydrogen bonds?

   a. The dotted lines between the hydrogen atoms of one water molecule and the oxygen atoms of a different water molecule.

b. The solid lines between a hydrogen atom and oxygen atom in the same water molecule.
c. Both the solid lines and dotted lines represent the hydrogen bonds.
d. There are no hydrogen bonds represented in the diagram.

22. Use the heating/cooling curve below to answer the following questions.

a. What is the freezing point of the liquid?
b. What is the boiling point of the liquid?
c. Which is greater: the head of fusion or the heat of vaporization? Explain.

23. Assume the two-dimensional structure of an ionic compound, M_xA_y, is

What is the empirical formula of this ionic compound?

15

Solutions

● Seawater is an aqueous solution. *(Georgette Douwma/Getty Images)*

Most of the important chemistry that keeps plants, animals, and humans functioning occurs in aqueous solutions. Even the water that comes out of a tap is not pure water but a solution of various materials in water. For example, tap water may contain dissolved chlorine to disinfect it, dissolved minerals that make it "hard," and traces of many other substances that result from natural and human-initiated pollution. We encounter many other chemical solutions in our daily lives: air, shampoo, orange soda, coffee, gasoline, cough syrup, and many others.

A **solution** is a homogeneous mixture, a mixture in which the components are uniformly intermingled. This means that a sample from one part is the same as a sample from any other part. For example, the first sip of coffee is the same as the last sip.

The atmosphere that surrounds us is a gaseous solution containing $O_2(g)$, $N_2(g)$, and other gases randomly dispersed. Solutions can also be solids. For example, brass is a homogeneous mixture—a solution—of copper and zinc.

These examples illustrate that a solution can be a gas, a liquid, or a solid (see Table 15.1). The substance present in the largest amount is called the **solvent**, and the other substance or substances are called **solutes**. For example, when we dissolve a teaspoon of sugar in a glass of water, the sugar is the solute and the water is the solvent. **Aqueous solutions** are solutions with water as the solvent. Because they are so important, in this chapter we will concentrate on the properties of aqueous solutions.

Brass, a solid solution of copper and zinc, is used to make musical instruments and many other objects.

15.1 Solubility

OBJECTIVES: To understand the process of dissolving. • To learn why certain components dissolve in water.

What happens when you put a teaspoon of sugar in your iced tea and stir it, or when you add salt to water for cooking vegetables? Why do the sugar and salt "disappear" into the water? What does it mean when something dissolves—that is, when a solution forms?

Table 15.1 Various Types of Solutions

Example	State of Solution	Original State of Solute	State of Solvent
air, natural gas	gas	gas	gas
vodka in water, antifreeze in water	liquid	liquid	liquid
brass	solid	solid	solid
carbonated water (soda)	liquid	gas	liquid
seawater, sugar solution	liquid	solid	liquid

Figure 15.1

When solid sodium chloride dissolves, the ions are dispersed randomly throughout the solution.

NaCl   +

H₂O

○ = Na⁺
● = Cl⁻

We saw in Chapter 7 that when sodium chloride dissolves in water, the resulting solution conducts an electric current. This convinces us that the solution contains *ions* that can move (this is how the electric current is conducted). The dissolving of solid sodium chloride in water is represented in Figure 15.1. Notice that in the solid state the ions are packed closely together. However, when the solid dissolves, the ions are separated and dispersed throughout the solution. The strong ionic forces that hold the sodium chloride crystal together are overcome by the strong attractions between the ions and the polar water molecules. This process is represented in Figure 15.2. Notice that each polar water molecule orients itself in a way to maximize its attraction with a Cl⁻ or Na⁺ ion. The negative end of a water molecule is attracted to a Na⁺ ion, while the positive end is attracted to a Cl⁻ ion. The strong forces holding the positive and negative ions in the solid are replaced by strong water–ion interactions, and the solid dissolves (the ions disperse).

It is important to remember that when an ionic substance (such as a salt) dissolves in water, it breaks up into *individual* cations and anions, which are dispersed in the water. For instance, when ammonium nitrate, NH_4NO_3, dissolves in water, the resulting solution contains NH_4^+ and NO_3^- ions, which move around independently. This process can be represented as

Cations are positive ions.
Anions are negative ions.

Figure 15.2

Polar water molecules interact with the positive and negative ions of a salt. These interactions replace the strong ionic forces holding the ions together in the undissolved solid, thus assisting in the dissolving process.

$$NH_4NO_3(s) \xrightarrow{H_2O(l)} NH_4^+(aq) + NO_3^-(aq)$$

where (*aq*) indicates that the ions are surrounded by water molecules.

Anion

δ^+ $2\delta^-$
δ^+

$2\delta^-$
δ^+ δ^+

Cation

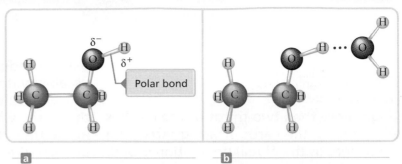

Figure 15.3

a

The ethanol molecule contains a polar O—H bond similar to those in the water molecule.

b

The polar water molecule interacts strongly with the polar O—H bond in ethanol.

A satellite photo of an oil spill in Tokyo Bay.

Water also dissolves many nonionic substances. Sugar is one example of a nonionic solute that is very soluble in water. Another example is ethanol, C_2H_5OH. Wine, beer, and mixed drinks are aqueous solutions of ethanol (and other substances). Why is ethanol so soluble in water? The answer lies in the structure of the ethanol molecule (Figure 15.3a). The molecule contains a polar O—H bond like those in water, which makes it very compatible with water. Just as hydrogen bonds form among water molecules in pure water (see Figure 14.6), ethanol molecules can form hydrogen bonds with water molecules in a solution of the two. This is shown in Figure 15.3b.

The sugar molecule (common table sugar has the chemical name sucrose) is shown in Figure 15.4. Notice that this molecule has many polar O—H groups, each of which can hydrogen-bond to a water molecule. Because of the attractions between sucrose and water molecules, solid sucrose is quite soluble in water.

Many substances do not dissolve in water. For example, when petroleum leaks from a damaged tanker, it does not disperse uniformly in the water (does not dissolve) but rather floats on the surface because its density is less than that of water. Petroleum is a mixture of molecules like the one shown in Figure 15.5. Since carbon and hydrogen have very similar electronegativities, the bonding electrons are shared almost equally and the

Figure 15.4

The structure of common table sugar (called sucrose). The large number of polar O—H groups in the molecule causes sucrose to be very soluble in water.

Figure 15.5

A molecule typical of those found in petroleum. The bonds are not polar.

Water, Water, Everywhere, But . . .

Although more than two thirds of the earth is covered by water, the earth is facing increasing water shortages as the global population grows. Why is that? The problem is that most of the earth's water is ocean water, which contains such high concentrations of dissolved minerals that the ocean water can't be consumed by humans and it kills crops. Humans need "fresh water" to sustain their lives. This fresh water is ultimately derived from rain, which supplies water to our lakes, rivers, and underground aquifers. However, as our populations are expanding, our supply of useable water is running short.

Because the earth has so much "salt water," the obvious answer to the water problem would seem to be removing minerals from ocean water, a process called "desalination." We know how to desalinate seawater. By forcing seawater through special membranes that trap the dissolved ions but allow water molecules to pass through, we can produce useable water. This is the most common method for producing drinking water in the Middle East and other arid regions. Worldwide more than 13,000 desalination plants produce more than 12 billion gallons of useable water every day.

Given the widespread use of desalination in the world, why is the process not used very much in the United States? The answer is simple: cost. Because the desalination process requires high-pressure pumps to force seawater through the special membranes, a great deal of electricity is needed. Also, the special membranes are very expensive. Thus, desalination costs currently are 30% higher than the costs of traditional sources of water. However, as water supplies are becoming tighter, major users of water are more willing to bear the higher costs of desalination. Water-challenged California is a good example. Poseidon Resources Corporation recently signed a contract to build a $300 million desalination plant in Carlsbad, California, a city just north of San Diego. This facility will be the largest in the Western Hemisphere, producing enough water for 100,000 homes (50 million gallons per day). A schematic diagram of this plant is shown on the next page.

Poseidon is planning a second plant in Huntington Beach, California, and as many as 20 other similar projects are in the planning stages. This technology could go a long way toward satisfying our thirst in the future.

---

bonds are essentially nonpolar. The resulting molecule with its nonpolar bonds cannot form attractions to the polar water molecules and this prevents it from being soluble in water. This situation is represented in Figure 15.6.

Notice in Figure 15.6 that the water molecules in liquid water are associated with each other by hydrogen-bonding interactions. For a solute to dissolve in water, a "hole" must be made in the water structure for each solute

Figure 15.6

An oil layer floating on water. For a substance to dissolve, the water–water hydrogen bonds must be broken to make a "hole" for each solute particle. However, the water–water interactions will break only if they are replaced by similar strong interactions with the solute.

Oil layer

Water

Fresh water

Seawater

Membrane

2. The pretreated seawater is forced through a dense, semipermeable membrane at extremely high pressure, which separates salt and minerals from fresh water. The concentrated brine residue is discharged back into the sea at the end of the cycle.

Discharge

1. Seawater enters the plant through a pretreatment filtering system that removes coarser particles, sand, sediment, and dirt.

Fresh

3. Fresh water is stored in a reservoir for later use by the municipal water system.

A schematic diagram of the desalination plant in Carlsbad, California.

particle. This will occur only if the lost water–water interactions are replaced by similar water–solute interactions. In the case of sodium chloride, strong interactions occur between the polar water molecules and the Na^+ and Cl^- ions. This allows the sodium chloride to dissolve. In the case of ethanol or sucrose, hydrogen-bonding interactions can occur between the O—H groups on these molecules and water molecules, making these substances soluble as well. But oil molecules are not soluble in water, because the many water–water interactions that would have to be broken to make "holes" for these large molecules are not replaced by favorable water–solute interactions.

These considerations account for the observed behavior that *"like dissolves like."* In other words, we observe that a given solvent usually dissolves solutes that have polarities similar to its own. For example, water dissolves most polar solutes, because the solute–solvent interactions formed in the solution are similar to the water–water interactions present in the pure solvent. Likewise, nonpolar solvents dissolve nonpolar solutes. For example, dry-cleaning solvents used for removing grease stains from clothes are nonpolar liquids. "Grease" is composed of nonpolar molecules, so a nonpolar solvent is needed to remove a grease stain.

Green Chemistry

Although some chemical industries have been culprits in the past for fouling the earth's environment, that situation is rapidly changing. In fact, a quiet revolution is sweeping through chemistry from academic labs to *Fortune* 500 companies. Chemistry is going green. *Green chemistry* means minimizing hazardous wastes, substituting water and other environmentally friendlier substances for traditional organic solvents, and manufacturing products out of recyclable materials.

A good example of green chemistry is the increasing use of carbon dioxide, one of the by-products of the combustion of fossil fuels. For example, the Dow Chemical Company is now using CO_2 rather than chlorofluorocarbons (CFCs; substances known to catalyze the decomposition of protective stratospheric ozone) to put the "sponginess" into polystyrene egg cartons, meat trays, and burger boxes. Dow does not generate CO_2 for this process but instead uses waste gases captured from its various manufacturing processes.

Another very promising use of carbon dioxide is as a replacement for the solvent perchloroethylene (PERC; Cl₂C=CCl₂), now used by about 80% of dry cleaners in the United States. Chronic exposure to PERC has been linked to kidney and liver damage and cancer. Although PERC is not a hazard to the general public (little PERC adheres to dry-cleaned garments), it represents a major concern for employees in the dry-cleaning industry. At high pressures CO_2 is a liquid that, when used with appropriate detergents, is a very effective solvent for the soil found on dry-clean-only fabrics. When the pressure is lowered, the CO_2 immediately changes to its gaseous form, quickly drying the clothes without the need for added heat. The gas can then be condensed and reused for the next batch of clothes.

The good news is that green chemistry makes sense economically. When all of the costs are taken into account, green chemistry is usually cheaper chemistry as well. Everybody wins.

The dry-cleaning agent PERC is a health concern for workers in the dry-cleaning industry.

Michael Newman/PhotoEdit

15.2 Solution Composition: An Introduction

OBJECTIVE: To learn qualitative terms associated with the concentration of a solution.

Even for very soluble substances, there is a limit to how much solute can be dissolved in a given amount of solvent. For example, when you add sugar to a glass of water, the sugar rapidly disappears at first. However, as you continue to add more sugar, at some point the solid no longer dissolves but collects at the bottom of the glass. When a solution contains as much solute as will dissolve at that temperature, we say it is **saturated**. If a solid solute is added to a solution already saturated with that solute, the added solid does not dissolve. A solution that has *not* reached the limit of solute that will dissolve in it is said to be **unsaturated.** When more solute is added to an unsaturated solution, it dissolves.

Although a chemical compound always has the same composition, a solution is a mixture and the amounts of the substances present can vary in different solutions. For example, coffee can be strong or weak. Strong coffee has more coffee dissolved in a given amount of water than weak coffee. To describe a solution completely, we must specify the amounts of solvent and solute. We sometimes use the qualitative terms *concentrated* and *dilute* to describe a solution. A relatively large amount of solute is dissolved in a **concentrated** solution (strong coffee is concentrated). A relatively small amount of solute is dissolved in a **dilute** solution (weak coffee is dilute).

Although these qualitative terms serve a useful purpose, we often need to know the exact amount of solute present in a given amount of solution. In the next several sections, we will consider various ways to describe the composition of a solution.

15.3 Solution Composition: Mass Percent

OBJECTIVE: To understand the concentration term *mass percent* and learn how to calculate it.

Describing the composition of a solution means specifying the amount of solute present in a given quantity of the solution. We typically give the amount of solute in terms of mass (number of grams) or in terms of moles. The quantity of solution is defined in terms of mass or volume.

One common way of describing a solution's composition is **mass percent** (sometimes called *weight percent*), which expresses the mass of solute present in a given mass of solution. The definition of mass percent is

> The mass of the solution is the sum of the masses of the solute and the solvent.

$$\text{Mass percent} = \frac{\text{mass of solute}}{\text{mass of solution}} \times 100\%$$

$$= \frac{\text{grams of solute}}{\text{grams of solute} + \text{grams of solvent}} \times 100\%$$

For example, suppose a solution is prepared by dissolving 1.0 g of sodium chloride in 48 g of water. The solution has a mass of 49 g (48 g of H_2O plus 1.0 g of NaCl), and there is 1.0 g of solute (NaCl) present. The mass percent of solute, then, is

$$\frac{1.0 \text{ g solute}}{49 \text{ g solution}} \times 100\% = 0.020 \times 100\% = 2.0\% \text{ NaCl}$$

EXAMPLE 15.1 | Solution Composition: Calculating Mass Percent

A solution is prepared by mixing 1.00 g of ethanol, C_2H_5OH, with 100.0 g of water. Calculate the mass percent of ethanol in this solution.

SOLUTION

Where Are We Going?

We want to determine the mass percent of a given ethanol solution.

What Do We Know?

- We have 1.00 g ethanol (C_2H_5OH) in 100.0 g water (H_2O).

- $\text{Mass percent} = \dfrac{\text{mass of solute}}{\text{mass of solution}} \times 100\%.$

How Do We Get There?

In this case we have 1.00 g of solute (ethanol) and 100.0 g of solvent (water). We now apply the definition of mass percent.

$$
\begin{aligned}
\text{Mass percent } C_2H_5OH &= \left(\frac{\text{grams of } C_2H_5OH}{\text{grams of solution}}\right) \times 100\% \\
&= \left(\frac{1.00 \text{ g } C_2H_5OH}{100.0 \text{ g } H_2O + 1.00 \text{ g } C_2H_5OH}\right) \times 100\% \\
&= \frac{1.00 \text{ g}}{101.0 \text{ g}} \times 100\% \\
&= 0.990\% \; C_2H_5OH
\end{aligned}
$$

REALITY CHECK The percent mass is just under 1%, which makes sense because we have 1.00 g of ethanol in a bit more than 100.0 g of solution.

Self-Check **EXERCISE 15.1** A 135-g sample of seawater is evaporated to dryness, leaving 4.73 g of solid residue (the salts formerly dissolved in the seawater). Calculate the mass percent of solute present in the original seawater.

See Problems 15.15 and 15.16. ■

EXAMPLE 15.2 | **Solution Composition: Determining Mass of Solute**

Although milk is not a true solution (it is really a suspension of tiny globules of fat, protein, and other substrates in water), it does contain a dissolved sugar called lactose. Cow's milk typically contains 4.5% by mass of lactose, $C_{12}H_{22}O_{11}$. Calculate the mass of lactose present in 175 g of milk.

SOLUTION

Where Are We Going?

We want to determine the mass of lactose present in 175 g of milk.

What Do We Know?

- We have 175 g of milk.

- Milk contains 4.5% by mass of lactose, $C_{12}H_{22}O_{11}$.

- $\text{Mass percent} = \dfrac{\text{mass of solute}}{\text{mass of solution}} \times 100\%.$

How Do We Get There?

Using the definition of mass percent, we have

$$
\text{Mass percent} = \frac{\text{grams of solute}}{\text{grams of solution}} \times 100\%
$$

We now substitute the quantities we know:

$$\text{Mass percent} = \frac{\overbrace{\text{grams of solute}}^{\text{Mass of lactose}}}{\underbrace{175 \text{ g}}_{\text{Mass of milk}}} \times 100\% = \overset{\text{Mass percent}}{4.5\%}$$

We now solve for grams of solute by multiplying both sides by 175 g,

$$\cancel{175 \text{ g}} \times \frac{\text{grams of solute}}{\cancel{175 \text{ g}}} \times 100\% = 4.5\% \times 175 \text{ g}$$

and then dividing both sides by 100%,

$$\text{Grams of solute} \times \frac{\cancel{100\%}}{\cancel{100\%}} = \frac{4.5\%}{100\%} \times 175 \text{ g}$$

to give

$$\text{Grams of solute} = 0.045 \times 175 \text{ g} = 7.9 \text{ g lactose}$$

Self-Check EXERCISE 15.2   What mass of water must be added to 425 g of formaldehyde to prepare a 40.0% (by mass) solution of formaldehyde? This solution, called formalin, is used to preserve biological specimens.

HINT: Substitute the known quantities into the definition for mass percent, and then solve for the unknown quantity (mass of solvent).

See Problems 15.17 and 15.18. ■

15.4   Solution Composition: Molarity

OBJECTIVES: To understand molarity. • To learn to use molarity to calculate the number of moles of solute present.

When a solution is described in terms of mass percent, the amount of solution is given in terms of its mass. However, it is often more convenient to measure the volume of a solution than to measure its mass. Because of this, chemists often describe a solution in terms of concentration. We define the *concentration* of a solution as the amount of solute in a *given volume* of solution. The most commonly used expression of concentration is **molarity (M).** Molarity describes the amount of solute in moles and the volume of the solution in liters. Molarity is *the number of moles of solute per volume of solution in liters.* That is

$$M = \text{molarity} = \frac{\text{moles of solute}}{\text{liters of solution}} = \frac{\text{mol}}{\text{L}}$$

A solution that is 1.0 molar (written as 1.0 *M*) contains 1.0 mole of solute per liter of solution.

EXAMPLE 15.3 | Solution Composition: Calculating Molarity, I

Calculate the molarity of a solution prepared by dissolving 11.5 g of solid NaOH in enough water to make 1.50 L of solution.

SOLUTION

Where Are We Going?

We want to determine the concentration (M) of a solution of NaOH.

What Do We Know?

- 11.5 g of NaOH is dissolved in 1.50 L of solution.

- $M = \dfrac{\text{moles of solute}}{\text{liters of solution}}$.

What Information Do We Need?

- We need to know the number of moles of NaOH in 11.5 g NaOH.

How Do We Get There?

We have the mass (in grams) of solute, so we need to convert the mass of solute to moles (using the molar mass of NaOH). Then we can divide the number of moles by the volume in liters.

We compute the number of moles of solute, using the molar mass of NaOH (40.0 g).

$$11.5 \text{ g NaOH} \times \frac{1 \text{ mol NaOH}}{40.0 \text{ g NaOH}} = 0.288 \text{ mol NaOH}$$

Then we divide by the volume of the solution in liters.

$$\text{Molarity} = \frac{\text{moles of solute}}{\text{liters of solution}} = \frac{0.288 \text{ mol NaOH}}{1.50 \text{ L solution}} = 0.192 \, M \text{ NaOH} \blacksquare$$

EXAMPLE 15.4 | Solution Composition: Calculating Molarity, II

Calculate the molarity of a solution prepared by dissolving 1.56 g of gaseous HCl into enough water to make 26.8 mL of solution.

SOLUTION

Where Are We Going?

We want to determine the concentration (M) of a solution of HCl.

What Do We Know?

- 1.56 g of HCl is dissolved in 26.8 mL of solution.

- $M = \dfrac{\text{moles of solute}}{\text{liters of solution}}$.

What Information Do We Need?

- We need to know the number of moles of HCl in 1.56 g.

- We need to know the volume of the solution in liters.

How Do We Get There?

We must change 1.56 g of HCl to moles of HCl, and then we must change 26.8 mL to liters (because molarity is defined in terms of liters). First we calculate the number of moles of HCl (molar mass = 36.5 g).

$$1.56 \text{ g HCl} \times \frac{1 \text{ mol HCl}}{36.5 \text{ g HCl}} = 0.0427 \text{ mol HCl}$$
$$= 4.27 \times 10^{-2} \text{ mol HCl}$$

Next we change the volume of the solution from milliliters to liters, using the equivalence statement 1 L = 1000 mL, which gives the appropriate conversion factor.

$$26.8 \text{ mL} \times \frac{1 \text{ L}}{1000 \text{ mL}} = 0.0268 \text{ L}$$
$$= 2.68 \times 10^{-2} \text{ L}$$

Finally, we divide the moles of solute by the liters of solution.

$$\text{Molarity} = \frac{4.27 \times 10^{-2} \text{ mol HCl}}{2.68 \times 10^{-2} \text{ L solution}} = 1.59 \, M \text{ HCl}$$

Self-Check EXERCISE 15.3   Calculate the molarity of a solution prepared by dissolving 1.00 g of ethanol, C_2H_5OH, in enough water to give a final volume of 101 mL.

See Problems 15.37 through 15.42. ■

It is important to realize that the description of a solution's composition may not accurately reflect the true chemical nature of the solute as it is present in the dissolved state. Solute concentration is always written in terms of the form of the solute *before* it dissolves. For example, describing a solution as 1.0 M NaCl means that the solution was prepared by dissolving 1.0 mole of solid NaCl in enough water to make 1.0 L of solution; it does not mean that the solution contains 1.0 mole of NaCl units. Actually the solution contains 1.0 mole of Na^+ ions and 1.0 mole of Cl^- ions. That is, it contains 1.0 M Na^+ and 1.0 M Cl^-.

EXAMPLE 15.5 | **Solution Composition: Calculating Ion Concentration from Molarity**

Remember, ionic compounds separate into the component ions when they dissolve in water.

Give the concentrations of all the ions in each of the following solutions:

    a. 0.50 M $Co(NO_3)_2$

    b. 1 M $FeCl_3$

SOLUTION

    a. When solid $Co(NO_3)_2$ dissolves, it produces ions as follows:

$$Co(NO_3)_2(s) \xrightarrow{H_2O(l)} Co^{2+}(aq) + 2NO_3^-(aq)$$

which we can represent as

$$1 \text{ mol } Co(NO_3)_2(s) \xrightarrow{H_2O(l)} 1 \text{ mol } Co^{2+}(aq) + 2 \text{ mol } NO_3^-(aq)$$

Therefore, a solution that is 0.50 M $Co(NO_3)_2$ contains 0.50 M Co^{2+} and (2×0.50) M NO_3^-, or 1.0 M NO_3^-.

b. When solid $FeCl_3$ dissolves, it produces ions as follows:

$$FeCl_3(s) \xrightarrow{H_2O(l)} Fe^{3+}(aq) + 3Cl^-(aq)$$

or

$$1 \text{ mol } FeCl_3(s) \xrightarrow{H_2O(l)} 1 \text{ mol } Fe^{3+}(aq) + 3 \text{ mol } Cl^-(aq)$$

A solution that is 1 M $FeCl_3$ contains 1 M Fe^{3+} ions and 3 M Cl^- ions.

Self-Check **EXERCISE 15.4** Give the concentrations of the ions in each of the following solutions:

    a. 0.10 M Na_2CO_3

    b. 0.010 M $Al_2(SO_4)_3$

See Problems 15.49 and 15.50. ■

MATH SKILL BUILDER

$M = \dfrac{\text{moles of solute}}{\text{liters of solution}}$

Liters $\times$ M ➡ Moles of solute

Often we need to determine the number of moles of solute present in a given volume of a solution of known molarity. To do this, we use the definition of molarity. When we multiply the molarity of a solution by the volume (in liters), we get the moles of solute present in that sample:

$$\text{Liters of solution} \times \text{molarity} = \cancel{\text{liters of solution}} \times \frac{\text{moles of solute}}{\cancel{\text{liters of solution}}}$$

$$= \text{moles of solute}$$

EXAMPLE 15.6

Solution Composition: Calculating Number of Moles from Molarity

How many moles of Ag^+ ions are present in 25 mL of a 0.75 M $AgNO_3$ solution?

SOLUTION

Where Are We Going?

We want to determine the number of moles of Ag^+ in a solution.

What Do We Know?

- We have 25 mL of 0.75 M $AgNO_3$.

- $M = \dfrac{\text{moles of solute}}{\text{liters of solution}}$.

How Do We Get There?

A 0.75 M $AgNO_3$ solution contains 0.75 M Ag^+ ions and 0.75 M NO_3^- ions. Next we must express the volume in liters. That is, we must convert from mL to L.

A solution of cobalt(II) nitrate.

Tom Pantages

$$25 \text{ mL} \times \frac{1 \text{ L}}{1000 \text{ mL}} = 0.025 \text{ L} = 2.5 \times 10^{-2} \text{ L}$$

Now we multiply the volume times the molarity.

$$2.5 \times 10^{-2} \text{ L solution} \times \frac{0.75 \text{ mol Ag}^+}{\text{L solution}} = 1.9 \times 10^{-2} \text{ mol Ag}^+$$

Self-Check EXERCISE 15.5   Calculate the number of moles of Cl^- ions in 1.75 L of $1.0 \times 10^{-3} M$ $AlCl_3$.

See Problems 15.49 and 15.50. ∎

A **standard solution** is a solution whose *concentration is accurately known.* When the appropriate solute is available in pure form, a standard solution can be prepared by weighing out a sample of solute, transferring it completely to a *volumetric flask* (a flask of accurately known volume), and adding enough solvent to bring the volume up to the mark on the neck of the flask. This procedure is illustrated in Figure 15.7.

EXAMPLE 15.7 | **Solution Composition: Calculating Mass from Molarity**

To analyze the alcohol content of a certain wine, a chemist needs 1.00 L of an aqueous $0.200 M$ $K_2Cr_2O_7$ (potassium dichromate) solution. How much solid $K_2Cr_2O_7$ (molar mass = 294.2 g) must be weighed out to make this solution?

SOLUTION

Where Are We Going?

We want to determine the mass of $K_2Cr_2O_7$ needed to make a given solution.

What Do We Know?

- We want 1.00 L of $0.200 M$ $K_2Cr_2O_7$.

Volume marker (calibration mark)

Wash bottle

Weighed amount of solute

a Put a weighed amount of a substance (the solute) into the volumetric flask, and add a small quantity of water.

b Dissolve the solid in the water by gently swirling the flask (with the stopper in place).

c Add more water (with gentle swirling) until the level of the solution just reaches the mark etched on the neck of the flask. Then mix the solution thoroughly by inverting the flask several times.

Figure 15.7
Steps involved in the preparation of a standard aqueous solution.

• The molar mass of $K_2Cr_2O_7$ is 294.2 g/mol.

• $M = \dfrac{\text{moles of solute}}{\text{liters of solution}}$.

How Do We Get There?

We need to calculate the number of grams of solute ($K_2Cr_2O_7$) present (and thus the mass needed to make the solution). First we determine the number of moles of $K_2Cr_2O_7$ present by multiplying the volume (in liters) by the molarity.

Liters × *M* ➡ Moles of solute

$$1.00 \text{ L solution} \times \frac{0.200 \text{ mol } K_2Cr_2O_7}{\text{L solution}} = 0.200 \text{ mol } K_2Cr_2O_7$$

Then we convert the moles of $K_2Cr_2O_7$ to grams, using the molar mass of $K_2Cr_2O_7$ (294.2 g).

$$0.200 \text{ mol } K_2Cr_2O_7 \times \frac{294.2 \text{ g } K_2Cr_2O_7}{\text{mol } K_2Cr_2O_7} = 58.8 \text{ g } K_2Cr_2O_7$$

Therefore, to make 1.00 L of 0.200 *M* $K_2Cr_2O_7$, the chemist must weigh out 58.8 g of $K_2Cr_2O_7$ and dissolve it in enough water to make 1.00 L of solution. This is most easily done by using a 1.00-L volumetric flask (see Figure 15.7).

Self-Check EXERCISE 15.6   Formalin is an aqueous solution of formaldehyde, HCHO, used as a preservative for biologic specimens. How many grams of formaldehyde must be used to prepare 2.5 L of 12.3 *M* formalin?

See Problems 15.51 and 15.52. ■

15.5 Dilution

OBJECTIVE: To learn to calculate the concentration of a solution made by diluting a stock solution.

The molarities of stock solutions of the common concentrated acids are:

Sulfuric (H_2SO_4)   18 *M*
Nitric (HNO_3)   16 *M*
Hydrochloric (HCl)   12 *M*

To save time and space in the laboratory, solutions that are routinely used are often purchased or prepared in concentrated form (called *stock solutions*). Water (or another solvent) is then added to achieve the molarity desired for a particular solution. The process of adding more solvent to a solution is called **dilution**. For example, the common laboratory acids are purchased as concentrated solutions and diluted with water as they are needed. A typical dilution calculation involves determining how much water must be added to an amount of stock solution to achieve a solution of the desired concentration. The key to doing these calculations is to remember that *only water is added in the dilution.* The amount of solute in the final, more dilute solution is the *same* as the amount of solute in the original, concentrated stock solution. That is,

Moles of solute after dilution = moles of solute before dilution

Dilution with water doesn't alter the number of moles of solute present.

The number of moles of solute stays the same but more water is added, increasing the volume, so the molarity decreases.

For example, suppose we want to prepare 500. mL of 1.00 M acetic acid, $HC_2H_3O_2$, from a 17.5 M stock solution of acetic acid. What volume of the stock solution is required?

The first step is to determine the number of moles of acetic acid needed in the final solution. We do this by multiplying the volume of the solution by its molarity.

$$\text{Volume of dilute solution (liters)} \times \text{molarity of dilute solution} = \text{moles of solute present}$$

The number of moles of solute present in the more dilute solution equals the number of moles of solute that must be present in the more concentrated (stock) solution, because this is the only source of acetic acid.

Because molarity is defined in terms of liters, we must first change 500. mL to liters and then multiply the volume (in liters) by the molarity.

$$\underbrace{500. \text{ mL solution}}_{\substack{V_{\text{dilute solution}} \\ \text{(in mL)}}} \times \underbrace{\frac{1 \text{ L solution}}{1000 \text{ mL solution}}}_{\text{Convert mL to L}} = 0.500 \text{ L solution}$$

$$0.500 \text{ L solution} \times \underbrace{\frac{1.00 \text{ mol } HC_2H_3O_2}{\text{L solution}}}_{M_{\text{dilute solution}}} = 0.500 \text{ mol } HC_2H_3O_2$$

Now we need to find the volume of 17.5 M acetic acid that contains 0.500 mole of $HC_2H_3O_2$. We will call this unknown volume V. Because volume $\times$ molarity = moles, we have

$$V \text{ (in liters)} \times \frac{17.5 \text{ mol } HC_2H_3O_2}{\text{L solution}} = 0.500 \text{ mol } HC_2H_3O_2$$

Solving for V $\left(\text{by dividing both sides by } \dfrac{17.5 \text{ mol}}{\text{L solution}}\right)$ gives

$$V = \frac{0.500 \text{ mol } HC_2H_3O_2}{\dfrac{17.5 \text{ mol } HC_2H_3O_2}{\text{L solution}}} = 0.0286 \text{ L, or } 28.6 \text{ mL, of solution}$$

Therefore, to make 500. mL of a 1.00 M acetic acid solution, we take 28.6 mL of 17.5 M acetic acid and dilute it to a total volume of 500. mL. This process is illustrated in Figure 15.8. Because the moles of solute remain the same before and after dilution, we can write

$$
\begin{array}{ccccc}
\text{Initial Conditions} & & & \text{Final Conditions} & \\
M_1 & \times & V_1 & = \text{moles of solute} = & M_2 & \times & V_2 \\
\text{Molarity} & \text{Volume} & & \text{Molarity} & \text{Volume} \\
\text{before} & \text{before} & & \text{after} & \text{after} \\
\text{dilution} & \text{dilution} & & \text{dilution} & \text{dilution}
\end{array}
$$

We can check our calculations on acetic acid by showing that $M_1 \times V_1 = M_2 \times V_2$. In the above example, $M_1 = 17.5$ M, $V_1 = 0.0286$ L, $V_2 = 0.500$ L, and $M_2 = 1.00$ M, so

$$M_1 \times V_1 = 17.5 \frac{\text{mol}}{\text{L}} \times 0.0286 \text{ L} = 0.500 \text{ mol}$$

$$M_2 \times V_2 = 1.00 \frac{\text{mol}}{\text{L}} \times 0.500 \text{ L} = 0.500 \text{ mol}$$

and therefore

$$M_1 \times V_1 = M_2 \times V_2$$

This shows that the volume (V_2) we calculated is correct.

a	b	c
28.6 mL of 17.5 M acetic acid solution is transferred to a volumetric flask that already contains some water.	Water is added to the flask (with swirling) to bring the volume to the calibration mark, and the solution is mixed by inverting the flask several times.	The resulting solution is 1.00 M acetic acid.

Figure 15.8

EXAMPLE 15.8 | Calculating Concentrations of Diluted Solutions

What volume of 16 M sulfuric acid must be used to prepare 1.5 L of a 0.10 M H_2SO_4 solution?

SOLUTION

Where Are We Going?

We want to determine the volume of sulfuric acid needed to prepare a given volume of a more dilute solution.

What Do We Know?

Initial Conditions (concentrated)	Final Conditions (dilute)
$M_1 = 16 \dfrac{\text{mol}}{\text{L}}$	$M_2 = 0.10 \dfrac{\text{mol}}{\text{L}}$
$V_1 = ?$	$V_2 = 1.5 \text{ L}$

$$\text{Moles of solute} = M_1 \times V_1 = M_2 \times V_2$$

How Do We Get There?

We can solve the equation

$$M_1 \times V_1 = M_2 \times V_2$$

Approximate dilutions can be carried out using a calibrated beaker. Here, concentrated sulfuric acid is being added to water to make a dilute solution.

Tom Pantages

for V_1 by dividing both sides by M_1,

$$\frac{\cancel{M_1} \times V_1}{\cancel{M_1}} = \frac{M_2 \times V_2}{M_1}$$

to give

$$V_1 = \frac{M_2 \times V_2}{M_1}$$

Now we substitute the known values of M_2, V_2, and M_1.

$$V_1 = \frac{\left(0.10\ \frac{\text{mol}}{\text{L}}\right)(1.5\ \text{L})}{16\ \frac{\text{mol}}{\text{L}}} = 9.4 \times 10^{-3}\ \text{L}$$

$$9.4 \times 10^{-3}\ \cancel{\text{L}} \times \frac{1000\ \text{mL}}{1\ \cancel{\text{L}}} = 9.4\ \text{mL}$$

> It is always best to add concentrated acid to water, not water to the acid. That way, if any splashing occurs accidentally, it is dilute acid that splashes.

Therefore, $V_1 = 9.4 \times 10^{-3}$ L, or 9.4 mL. To make 1.5 L of 0.10 M H_2SO_4 using 16 M H_2SO_4, we must take 9.4 mL of the concentrated acid and dilute it with water to a final volume of 1.5 L. The correct way to do this is to add the 9.4 mL of acid to about 1 L of water and then dilute to 1.5 L by adding more water.

Self-Check EXERCISE 15.7   What volume of 12 M HCl must be taken to prepare 0.75 L of 0.25 M HCl?

See Problems 15.57 and 15.58. ■

15.6 Stoichiometry of Solution Reactions

OBJECTIVE: To understand the strategy for solving stoichiometric problems for solution reactions.

Because so many important reactions occur in solution, it is important to be able to do stoichiometric calculations for solution reactions. The principles needed to perform these calculations are very similar to those developed in Chapter 9. It is helpful to think in terms of the following steps:

Steps for Solving Stoichiometric Problems Involving Solutions

Step 1   Write the balanced equation for the reaction. For reactions involving ions, it is best to write the net ionic equation.

> See Section 7.3 for a discussion of net ionic equations.

Step 2   Calculate the moles of reactants.

Step 3   Determine which reactant is limiting.

Step 4   Calculate the moles of other reactants or products, as required.

Step 5   Convert to grams or other units, if required.

EXAMPLE 15.9 | Solution Stoichiometry: Calculating Mass of Reactants and Products

Calculate the mass of solid NaCl that must be added to 1.50 L of a 0.100 M AgNO$_3$ solution to precipitate all of the Ag$^+$ ions in the form of AgCl. Calculate the mass of AgCl formed.

SOLUTION

Where Are We Going?

We want to determine the mass of NaCl.

What Do We Know?

- We have 1.50 L of a 0.100 M AgNO$_3$ solution.

What Information Do We Need?

- We need the balanced equation between AgNO$_3$ and NaCl.
- We need the molar mass of NaCl.

How Do We Get There?

Step 1 *Write the balanced equation for the reaction.*
When added to the AgNO$_3$ solution (which contains Ag$^+$ and NO$_3^-$ ions), the solid NaCl dissolves to yield Na$^+$ and Cl$^-$ ions. Solid AgCl forms according to the following balanced net ionic reaction:

$$Ag^+(aq) + Cl^-(aq) \rightarrow AgCl(s)$$

This reaction was discussed in Section 7.2.

Step 2 *Calculate the moles of reactants.*
In this case we must add enough Cl$^-$ ions to just react with all the Ag$^+$ ions present, so we must calculate the moles of Ag$^+$ ions present in 1.50 L of a 0.100 M AgNO$_3$ solution. (Remember that a 0.100 M AgNO$_3$ solution contains 0.100 M Ag$^+$ ions and 0.100 M NO$_3^-$ ions.)

MATH SKILL BUILDER
Liters $\times$ M ➡ Moles of solute

$$1.50 \; \cancel{L} \times \frac{0.100 \text{ mol Ag}^+}{\cancel{L}} = 0.150 \text{ mol Ag}^+$$

Moles of Ag$^+$ present
in 1.50 L of 0.100 M AgNO$_3$

Step 3 *Determine which reactant is limiting.*
In this situation we want to add just enough Cl$^-$ to react with the Ag$^+$ present. That is, we want to precipitate *all* the Ag$^+$ in the solution. Thus the Ag$^+$ present determines the amount of Cl$^-$ needed.

Step 4 *Calculate the moles of Cl$^-$ required.*
We have 0.150 mole of Ag$^+$ ions and, because one Ag$^+$ ion reacts with one Cl$^-$ ion, we need 0.150 mole of Cl$^-$,

$$0.150 \; \cancel{\text{mol Ag}^+} \times \frac{1 \text{ mol Cl}^-}{1 \; \cancel{\text{mol Ag}^+}} = 0.150 \text{ mol Cl}^-$$

so 0.150 mole of AgCl will be formed.

$$0.150 \text{ mol Ag}^+ + 0.150 \text{ mol Cl}^- \rightarrow 0.150 \text{ mol AgCl}$$

When aqueous sodium chloride is added to a solution of silver nitrate, a white silver chloride precipitate forms.

Step 5   *Convert to grams of NaCl required.*
To produce 0.150 mol Cl⁻, we need 0.150 mol NaCl. We calculate the mass of NaCl required as follows:

$$0.150 \text{ mol NaCl} \times \frac{58.4 \text{ g NaCl}}{\text{mol NaCl}} = 8.76 \text{ g NaCl}$$

Moles   ➡ *times molar mass*   Mass

The mass of AgCl formed is

$$0.150 \text{ mol AgCl} \times \frac{143.3 \text{ g AgCl}}{\text{mol AgCl}} = 21.5 \text{ g AgCl} \blacksquare$$

EXAMPLE 15.10   **Solution Stoichiometry: Determining Limiting Reactants and Calculating Mass of Products**

See Section 7.2 for a discussion of this reaction.

When $Ba(NO_3)_2$ and K_2CrO_4 react in aqueous solution, the yellow solid $BaCrO_4$ is formed. Calculate the mass of $BaCrO_4$ that forms when 3.50×10^{-3} mole of solid $Ba(NO_3)_2$ is dissolved in 265 mL of 0.0100 M K_2CrO_4 solution.

SOLUTION

Where Are We Going?

We want to determine the mass of $BaCrO_4$ that forms in a reaction of known amounts of solutions.

What Do We Know?

- We react 3.50×10^{-3} mol $BaNO_3$ with 265 mL of 0.0100 M K_2CrO_4.

What Information Do We Need?

- We will need the balanced equation between $BaNO_3$ and K_2CrO_4.
- We will need the molar mass of $BaCrO_4$.

How Do We Get There?

Step 1   The original K_2CrO_4 solution contains the ions K^+ and CrO_4^{2-}. When the $Ba(NO_3)_2$ is dissolved in this solution, Ba^{2+} and NO_3^- ions are added. The Ba^{2+} and CrO_4^{2-} ions react to form solid $BaCrO_4$. The balanced net ionic equation is

$$Ba^{2+}(aq) + CrO_4^{2-}(aq) \rightarrow BaCrO_4(s)$$

Step 2   Next we determine the moles of reactants. We are told that 3.50×10^{-3} mole of $Ba(NO_3)_2$ is added to the K_2CrO_4 solution. Each formula unit of $Ba(NO_3)_2$ contains one Ba^{2+} ion, so 3.50×10^{-3} mole of $Ba(NO_3)_2$ gives 3.50×10^{-3} mole of Ba^{2+} ions in solution.

3.50×10^{-3} mol $Ba(NO_3)_2$ ➡ *dissolves to give* 3.50×10^{-3} mol Ba^{2+}

Because $V \times M$ = moles of solute, we can compute the moles of K_2CrO_4 in the solution from the volume and molarity of the original solution. First we must convert the volume of the solution (265 mL) to liters.

$$265 \text{ mL} \times \frac{1 \text{ L}}{1000 \text{ mL}} = 0.265 \text{ L}$$

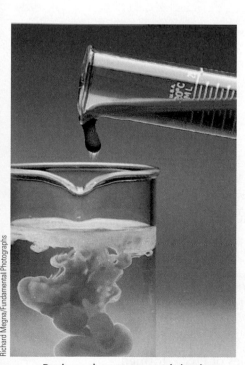

Barium chromate precipitating.

Next we determine the number of moles of K_2CrO_4, using the molarity of the K_2CrO_4 solution (0.0100 M).

$$0.265 \cancel{L} \times \frac{0.0100 \text{ mol } K_2CrO_4}{\cancel{L}} = 2.65 \times 10^{-3} \text{ mol } K_2CrO_4$$

We know that

so the solution contains 2.65×10^{-3} mole of $CrO_4{}^{2-}$ ions.

Step 3   The balanced equation tells us that one Ba^{2+} ion reacts with one $CrO_4{}^{2-}$ ion. Because the number of moles of $CrO_4{}^{2-}$ ions (2.65×10^{-3}) is smaller than the number of moles of Ba^{2+} ions (3.50×10^{-3}), the $CrO_4{}^{2-}$ will run out first.

$$Ba^{2+}(aq) \quad + \quad CrO_4{}^{2-}(aq) \quad \rightarrow \quad BaCrO_4(s)$$

3.50×10^{-3} mol	2.65×10^{-3} mol
	Smaller (runs out first)

Therefore, the $CrO_4{}^{2-}$ is limiting.

Moles of $CrO_4{}^{2-}$	limits	Moles of $BaCrO_4$

Step 4   The 2.65×10^{-3} mole of $CrO_4{}^{2-}$ ions will react with 2.65×10^{-3} mole of Ba^{2+} ions to form 2.65×10^{-3} mole of $BaCrO_4$.

2.65×10^{-3} mol Ba^{2+}	+	2.65×10^{-3} mol $CrO_4{}^{2-}$		2.65×10^{-3} mol $BaCrO_4(s)$

Step 5   The mass of $BaCrO_4$ formed is obtained from its molar mass (253.3 g) as follows:

$$2.65 \times 10^{-3} \cancel{\text{mol BaCrO}_4} \times \frac{253.3 \text{ g BaCrO}_4}{\cancel{\text{mol BaCrO}_4}} = 0.671 \text{ g BaCrO}_4$$

Self-Check EXERCISE 15.8   When aqueous solutions of Na_2SO_4 and $Pb(NO_3)_2$ are mixed, $PbSO_4$ precipitates. Calculate the mass of $PbSO_4$ formed when 1.25 L of 0.0500 M $Pb(NO_3)_2$ and 2.00 L of 0.0250 M Na_2SO_4 are mixed.

HINT: Calculate the moles of Pb^{2+} and $SO_4{}^{2-}$ in the mixed solution, decide which ion is limiting, and calculate the moles of $PbSO_4$ formed.

See Problems 15.65 through 15.68. ■

15.7   Neutralization Reactions

OBJECTIVE: To learn how to do calculations involved in acid–base reactions.

So far we have considered the stoichiometry of reactions in solution that result in the formation of a precipitate. Another common type of solution reaction occurs between an acid and a base. We introduced these reactions in Section 7.4. Recall from that discussion that an acid is a substance that furnishes H^+ ions. A strong acid, such as hydrochloric acid, HCl, dissociates (ionizes) completely in water.

$$HCl(aq) \rightarrow H^+(aq) + Cl^-(aq)$$

Strong bases are water-soluble metal hydroxides, which are completely dissociated in water. An example is NaOH, which dissolves in water to give Na^+ and OH^- ions.

$$NaOH(s) \xrightarrow{H_2O(l)} Na^+(aq) + OH^-(aq)$$

When a strong acid and a strong base react, the net ionic reaction is

$$H^+(aq) + OH^-(aq) \rightarrow H_2O(l)$$

An acid–base reaction is often called a **neutralization reaction.** When just enough strong base is added to react exactly with the strong acid in a solution, we say the acid has been *neutralized.* One product of this reaction is always water. The steps in dealing with the stoichiometry of any neutralization reaction are the same as those we followed in the previous section.

EXAMPLE 15.11 | Solution Stoichiometry: Calculating Volume in Neutralization Reactions

What volume of a 0.100 *M* HCl solution is needed to neutralize 25.0 mL of a 0.350 *M* NaOH solution?

SOLUTION

Where Are We Going?

We want to determine the volume of a given solution of HCl required to react with a known amount of NaOH.

What Do We Know?

- We have 25.0 mL of 0.350 *M* NaOH.
- The concentration of the HCl solution is 0.100 *M*.

What Information Do We Need?

- We need the balanced equation between HCl and NaOH.

How Do We Get There?

Step 1   *Write the balanced equation for the reaction.*
Hydrochloric acid is a strong acid, so all the HCl molecules dissociate to produce H^+ and Cl^- ions. Also, when the strong base NaOH dissolves, the solution contains Na^+ and OH^- ions. When these two solutions are mixed, the H^+ ions from the hydrochloric acid react with the OH^- ions from the sodium

hydroxide solution to form water. The balanced net ionic equation for the reaction is

$$H^+(aq) + OH^-(aq) \rightarrow H_2O(l)$$

Step 2 *Calculate the moles of reactants.*
In this problem we are given a volume (25.0 mL) of 0.350 M NaOH, and we want to add just enough 0.100 M HCl to provide just enough H^+ ions to react with all the OH^-. Therefore, we must calculate the number of moles of OH^- ions in the 25.0-mL sample of 0.350 M NaOH. To do this, we first change the volume to liters and multiply by the molarity.

$$25.0 \text{ mL NaOH} \times \frac{1 \text{ L}}{1000 \text{ mL}} \times \frac{0.350 \text{ mol OH}^-}{\text{L NaOH}} = 8.75 \times 10^{-3} \text{ mol OH}^-$$

Moles of OH^- present
in 25.0 mL of
0.350 M NaOH

Step 3 *Determine which reactant is limiting.*
This problem requires the addition of just enough H^+ ions to react exactly with the OH^- ions present, so the number of moles of OH^- ions present determines the number of moles of H^+ that must be added. The OH^- ions are limiting.

Step 4 *Calculate the moles of H^+ required.*
The balanced equation tells us that the H^+ and OH^- ions react in a 1:1 ratio, so 8.75×10^{-3} mole of H^+ ions is required to neutralize (exactly react with) the 8.75×10^{-3} mole of OH^- ions present.

Step 5 Calculate the volume of 0.100 M *HCl required.*
Next we must find the volume (V) of 0.100 M HCl required to furnish this amount of H^+ ions. Because the volume (in liters) times the molarity gives the number of moles, we have

$$V \times \frac{0.100 \text{ mol H}^+}{L} = 8.75 \times 10^{-3} \text{ mol H}^+$$

Unknown volume (in liters)   Moles of H^+ needed

Now we must solve for V by dividing both sides of the equation by 0.100.

$$V \times \frac{0.100 \text{ mol H}^+}{0.100 \text{ L}} = \frac{8.75 \times 10^{-3} \text{ mol H}^+}{0.100}$$

$$V = 8.75 \times 10^{-2} \text{ L}$$

Changing liters to milliliters, we have

$$V = 8.75 \times 10^{-2} \text{ L} \times \frac{1000 \text{ mL}}{\text{L}} = 87.5 \text{ mL}$$

Therefore, 87.5 mL of 0.100 M HCl is required to neutralize 25.0 mL of 0.350 M NaOH.

Self-Check EXERCISE 15.9   Calculate the volume of 0.10 M HNO_3 needed to neutralize 125 mL of 0.050 M KOH.

See Problems 15.69 through 15.74. ■

15.8   Solution Composition: Normality

OBJECTIVES: To learn about normality and equivalent weight.   • To learn to use these concepts in stoichiometric calculations.

Normality is another unit of concentration that is sometimes used, especially when dealing with acids and bases. The use of normality focuses mainly on the H^+ and OH^- available in an acid–base reaction. Before we discuss normality, however, we need to define some terms. One **equivalent of an acid** is the *amount of that acid that can furnish 1 mole of H^+ ions*. Similarly, one **equivalent of a base** is defined as the *amount of that base that can furnish 1 mole of OH^- ions*. The **equivalent weight** of an acid or a base is the *mass in grams of 1 equivalent (equiv) of that acid or base*.

The common strong acids are HCl, HNO_3, and H_2SO_4. For HCl and HNO_3 each molecule of acid furnishes one H^+ ion, so 1 mole of HCl can furnish 1 mole of H^+ ions. This means that

$$\text{Furnishes 1 mol } H^+$$
$$\downarrow$$
$$1 \text{ mol HCl} = 1 \text{ equiv HCl}$$
$$\text{Molar mass (HCl)} = \text{equivalent weight (HCl)}$$

Likewise, for HNO_3,

$$1 \text{ mol } HNO_3 = 1 \text{ equiv } HNO_3$$
$$\text{Molar mass } (HNO_3) = \text{equivalent weight } (HNO_3)$$

However, H_2SO_4 can furnish *two* H^+ ions per molecule, so 1 mole of H_2SO_4 can furnish *two* moles of H^+. This means that

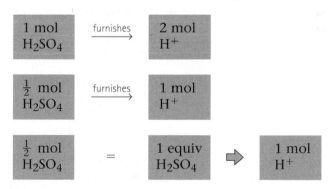

Because each mole of H_2SO_4 can furnish 2 moles of H^+, we need only to take $\frac{1}{2}$ mole of H_2SO_4 to get 1 equiv of H_2SO_4. Therefore,

$$\tfrac{1}{2} \text{ mol } H_2SO_4 = 1 \text{ equiv } H_2SO_4$$

and

$$\text{Equivalent weight } (H_2SO_4) = \tfrac{1}{2} \text{ molar mass } (H_2SO_4)$$
$$= \tfrac{1}{2} (98 \text{ g}) = 49 \text{ g}$$

The equivalent weight of H_2SO_4 is 49 g.

The common strong bases are NaOH and KOH. For NaOH and KOH, each formula unit furnishes one OH^- ion, so we can say

$$1 \text{ mol NaOH} = 1 \text{ equiv NaOH}$$
$$\text{Molar mass (NaOH)} = \text{equivalent weight (NaOH)}$$
$$1 \text{ mol KOH} = 1 \text{ equiv KOH}$$
$$\text{Molar mass (KOH)} = \text{equivalent weight (KOH)}$$

These ideas are summarized in Table 15.2.

Table 15.2 The Molar Masses and Equivalent Weights of the Common Strong Acids and Bases

	Molar Mass (g)	Equivalent Weight (g)
Acid		
HCl	36.5	36.5
HNO_3	63.0	63.0
H_2SO_4	98.0	$49.0 = \dfrac{98.0}{2}$
Base		
NaOH	40.0	40.0
KOH	56.1	56.1

EXAMPLE 15.12

Solution Stoichiometry: Calculating Equivalent Weight

Phosphoric acid, H_3PO_4, can furnish three H^+ ions per molecule. Calculate the equivalent weight of H_3PO_4.

SOLUTION

Where Are We Going?

We want to determine the equivalent weight of phosphoric acid.

What Do We Know?

- The formula for phosphoric acid is H_3PO_4.

- The equivalent weight of an acid is the amount of acid that can furnish 1 mole of H^+ ions.

What Information Do We Need?

- We need to know the molar mass of H_3PO_4.

How Do We Get There?

The key point here involves how many protons (H^+ ions) each molecule of H_3PO_4 can furnish.

Because each H_3PO_4 can furnish three H^+ ions, 1 mole of H_3PO_4 can furnish 3 moles of H^+ ions:

So 1 equiv of H_3PO_4 (the amount that can furnish 1 mole of H^+) is one-third of a mole.

This means the equivalent weight of H_3PO_4 is one-third its molar mass.

$$\boxed{\text{Equivalent weight}} = \boxed{\frac{\text{Molar mass}}{3}}$$

$$\text{Equivalent weight } (H_3PO_4) = \frac{\text{molar mass } (H_3PO_4)}{3}$$

$$= \frac{98.0 \text{ g}}{3} = 32.7 \text{ g} \ \blacksquare$$

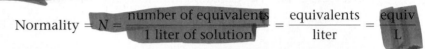

Normality (N) is defined as the number of equivalents of solute per liter of solution.

$$\text{Normality} = N = \frac{\text{number of equivalents}}{1 \text{ liter of solution}} = \frac{\text{equivalents}}{\text{liter}} = \frac{\text{equiv}}{L}$$

This means that a 1 N solution contains 1 equivalent of solute per liter of solution. Notice that when we multiply the volume of a solution in liters by the normality, we get the number of equivalents.

MATH SKILL BUILDER
Liters × Normality ➡ Equiv

$$N \times V = \frac{\text{equiv}}{\cancel{L}} \times \cancel{L} = \text{equiv}$$

EXAMPLE 15.13 | **Solution Stoichiometry: Calculating Normality**

A solution of sulfuric acid contains 86 g of H_2SO_4 per liter of solution. Calculate the normality of this solution.

SOLUTION

Where Are We Going?

We want to determine the normality of a given solution of H_2SO_4.

Whenever you need to calculate the concentration of a solution, first write the appropriate definition. Then decide how to calculate the quantities shown in the definition.

What Do We Know?

- We have 86 g of H_2SO_4 per liter of solution.

- $N = \dfrac{\text{equivalents}}{L}$.

What Information Do We Need?

- We need to know the molar mass of H_2SO_4.

How Do We Get There?

To find the number of equivalents present, we must calculate the number of equivalents represented by 86 g of H_2SO_4. To do this calculation, we focus on the definition of the equivalent: it is the amount of acid that furnishes 1 mole of H^+. Because H_2SO_4 can furnish two H^+ ions per molecule, 1 equiv of H_2SO_4 is $\frac{1}{2}$ mole of H_2SO_4, so

$$\text{Equivalent weight } (H_2SO_4) = \frac{\text{molar mass } (H_2SO_4)}{2}$$

$$= \frac{98.0 \text{ g}}{2} = 49.0 \text{ g}$$

We have 86 g of H_2SO_4.

$$86 \text{ g } H_2SO_4 \times \frac{1 \text{ equiv } H_2SO_4}{49.0 \text{ g } H_2SO_4} = 1.8 \text{ equiv } H_2SO_4$$

$$N = \frac{\text{equiv}}{\text{L}} = \frac{1.8 \text{ equiv } H_2SO_4}{1.0 \text{ L}} = 1.8 \text{ } N \text{ } H_2SO_4$$

REALITY CHECK We know that 86 g is more than 1 equiv of H_2SO_4 (49 g), so this answer makes sense.

Self-Check EXERCISE 15.10 Calculate the normality of a solution containing 23.6 g of KOH in 755 mL of solution.

See Problems 15.79 and 15.80. ∎

The main advantage of using equivalents is that 1 equiv of acid contains the same number of available H^+ ions as the number of OH^- ions present in 1 equiv of base. That is,

0.75 equiv (base) will react exactly with 0.75 equiv (acid).

0.23 equiv (base) will react exactly with 0.23 equiv (acid).

And so on.

In each of these cases, the *number of* H^+ ions furnished by the sample of acid is the same as the *number of* OH^- ions furnished by the sample of base. The point is that *n equivalents of any acid will exactly neutralize n equivalents of any base*.

$$\boxed{\begin{array}{c} n \text{ equiv} \\ \text{acid} \end{array}} \quad \leftarrow \quad \text{reacts exactly with} \quad \rightarrow \quad \boxed{\begin{array}{c} n \text{ equiv} \\ \text{base} \end{array}}$$

Because we know that equal equivalents of acid and base are required for neutralization, we can say that

$$\text{equiv (acid)} = \text{equiv (base)}$$

That is,

$$N_{acid} \times V_{acid} = \text{equiv (acid)} = \text{equiv (base)} = N_{base} \times V_{base}$$

Therefore, for any neutralization reaction, the following relationship holds:

$$N_{acid} \times V_{acid} = N_{base} \times V_{base}$$

EXAMPLE 15.14 | **Solution Stoichiometry: Using Normality in Calculations**

What volume of a 0.075 N KOH solution is required to react exactly with 0.135 L of 0.45 N H_3PO_4?

SOLUTION

Where Are We Going?

We want to determine the volume of a given solution of KOH required to react with a known solution of H_3PO_4.

What Do We Know?

- We have 0.135 L of 0.45 N H_3PO_4.

- The concentration of the KOH solution is 0.075 N.

- We know equivalents$_{acid}$ = equivalents$_{base}$.

- $N_{acid} \times V_{acid} = N_{base} \times V_{base}$

How Do We Get There?

We know that for neutralization, equiv (acid) = equiv (base), or

$$N_{acid} \times V_{acid} = N_{base} \times V_{base}$$

We want to calculate the volume of base, V_{base}, so we solve for V_{base} by dividing both sides by N_{base}.

$$\frac{N_{acid} \times V_{acid}}{N_{base}} = \frac{\cancel{N_{base}} \times V_{base}}{\cancel{N_{base}}} = V_{base}$$

Now we can substitute the given values N_{acid} = 0.45 N, V_{acid} = 0.135 L, and N_{base} = 0.075 N into the equation.

$$V_{base} = \frac{N_{acid} \times V_{acid}}{N_{base}} = \frac{\left(0.45\ \frac{\cancel{equiv}}{L}\right)(0.135\ L)}{0.075\ \frac{\cancel{equiv}}{L}} = 0.81\ L$$

This gives V_{base} = 0.81 L, so 0.81 L of 0.075 N KOH is required to react exactly with 0.135 L of 0.45 N H_3PO_4.

Self-Check EXERCISE 15.11 What volume of 0.50 N H_2SO_4 is required to react exactly with 0.250 L of 0.80 N KOH?

See Problems 15.85 and 15.86. ■

CHAPTER 15 REVIEW

Key Terms

solution (p. 475)
solvent (p. 475)
solute (p. 475)
aqueous solution (p. 475)
saturated (15.2)
unsaturated (15.2)
concentrated (15.2)
dilute (15.2)
mass percent (15.3)
molarity (M) (15.4)

standard solution (15.4)
dilution (15.5)
neutralization reaction (15.7)
equivalent of an acid (15.8)
equivalent of a base (15.8)
equivalent weight (15.8)
normality (N) (15.8)

Summary

1. A solution is a homogeneous mixture. The solubility of a solute in a given solvent depends on the interactions between the solvent and solute particles. Water dissolves many ionic compounds and compounds with polar molecules, because strong forces occur between the solute and the polar water molecules. Nonpolar solvents tend to dissolve nonpolar solutes. "Like dissolves like."

2. Solution composition can be described in many ways. Two of the most important are in terms of mass percent of solute:

$$\text{Mass percent} = \frac{\text{mass of solute}}{\text{mass of solution}} \times 100\%$$

and molarity:

$$\text{Molarity} = \frac{\text{moles of solute}}{\text{liters of solution}}$$

3. A standard solution is one whose concentration is accurately known. Solutions are often made from a stock solution by dilution. When a solution is diluted, only solvent is added, which means that

Moles of solute after dilution =
moles of solute before dilution

4. Normality is defined as the number of equivalents per liter of solution. One equivalent of acid is the amount of acid that furnishes 1 mole of H^+ ions. One equivalent of base is the amount of base that furnishes 1 mole of OH^- ions.

Active Learning Questions

These questions are designed to be considered by groups of students in class. Often these questions work well for introducing a particular topic in class.

1. You have a solution of table salt in water. What happens to the salt concentration (increases, decreases, or stays the same) as the solution boils? Draw pictures to explain your answer.

2. Consider a sugar solution (solution A) with concentration x. You pour one-third of this solution into a beaker, and add an equivalent volume of water (solution B).

    a. What is the ratio of sugar in solutions A and B?
    b. Compare the volumes of solutions A and B.
    c. What is the ratio of the concentrations of sugar in solutions A and B?

3. You need to make 150.0 mL of a 0.10 M NaCl solution. You have solid NaCl, and your lab partner has a 2.5 M NaCl solution. Explain how you each independently make the solution you need.

4. You have two solutions containing solute A. To determine which solution has the highest concentration of A in molarity, which of the following must you know? (There may be more than one answer.)

    a. the mass in grams of A in each solution
    b. the molar mass of A
    c. the volume of water added to each solution
    d. the total volume of the solution

    Explain your answer.

5. Which of the following do you need to know to calculate the molarity of a salt solution? (There may be more than one answer.)

    a. the mass of salt added
    b. the molar mass of the salt

c. the volume of water added
d. the total volume of the solution

Explain your answer.

6. Consider separate aqueous solutions of HCl and H_2SO_4 with the same concentrations in terms of molarity. You wish to neutralize an aqueous solution of NaOH. For which acid solution would you need to add more volume (in mL) to neutralize the base?

    a. The HCl solution.
    b. The H_2SO_4 solution.
    c. You need to know the acid concentrations to answer this question.
    d. You need to know the volume and concentration of the NaOH solution to answer this question.
    e. c and d

    Explain your answer.

7. Draw molecular-level pictures to differentiate between concentrated and dilute solutions.

8. Can one solution have a greater concentration than another in terms of weight percent, but a lower concentration in terms of molarity? Explain.

9. Explain why the formula $M_1V_1 = M_2V_2$ works when solving dilution problems.

10. You have equal masses of different solutes dissolved in equal volumes of solution. Which of the solutes listed below would make the solution with the highest concentration measured in molarity? Defend your answer.

    NaCl, $MgSO_4$, LiF, KNO_3

11. Which of the following solutions contains the greatest number of particles? Support your answer.

    a. 400.0 mL of 0.10 M sodium chloride
    b. 300.0 mL of 0.10 M calcium chloride
    c. 200.0 mL of 0.10 M iron(III) chloride
    d. 200.0 mL of 0.10 M potassium bromide
    e. 800.0 mL of 0.10 M sucrose (table sugar)

12. As with all quantitative problems in chemistry, make sure not to get "lost in the math." In particular, work on visualizing solutions at a molecular level. For example, consider the following.

    You have two separate beakers with aqueous solutions, one with 4 "units" of potassium sulfate and one with 3 "units" of barium nitrate.

    a. Draw molecular-level diagrams of both solutions.
    b. Draw a molecular-level diagram of the mixture of the two solutions before a reaction has taken place.
    c. Draw a molecular-level diagram of the product and solution formed after the reaction has taken place.

13. The figures below are molecular-level representations of four aqueous solutions of the same solute. Arrange the solutions from most to least concentrated.

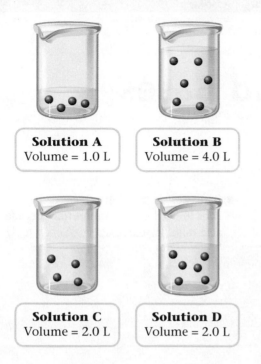

Solution A
Volume = 1.0 L

Solution B
Volume = 4.0 L

Solution C
Volume = 2.0 L

Solution D
Volume = 2.0 L

14. The drawings below represent aqueous solutions. Solution A is 2.00 L of a 2.00 M aqueous solution of cop-

per(II) nitrate. Solution B is 2.00 L of a 3.00 M aqueous solution of potassium hydroxide.

a. Draw a picture of the solution made by mixing solutions A and B together after the precipitation reaction takes place. Make sure this picture shows the correct relative volume compared to solutions A and B and the correct relative number of ions, along with the correct relative amount of solid formed.

b. Determine the concentrations (in M) of all ions left in solution (from part a) and the mass of solid formed.

⊙WL Questions and problems for this chapter are available in OWL. Use the OWL access card that came with this textbook to access assigned questions and problems for this chapter.

16

Acids and Bases

● Gargoyles on the Notre Dame cathedral in Paris in need of restoration from decades of acid rain. *(Witold Skrypczak/SuperStock)*

Acids are very important substances. They cause lemons to be sour, digest food in the stomach (and sometimes cause heartburn), dissolve rock to make fertilizer, dissolve your tooth enamel to form cavities, and clean the deposits out of your coffee maker. Acids are essential industrial chemicals. In fact, the chemical in first place in terms of the amount manufactured in the United States is sulfuric acid, H_2SO_4. Eighty *billion* pounds of this material are used every year in the manufacture of fertilizers, detergents, plastics, pharmaceuticals, storage batteries, and metals. The acid–base properties of substances also can be used to make interesting novelties such as the foaming chewing gum described on page 517.

Royalty-free Corbis

A lemon tastes sour because it contains citric acid.

In this chapter we will consider the most important properties of acids and of their opposites, the bases.

16.1 Acids and Bases

OBJECTIVE: To learn about two models of acids and bases and the relationship of conjugate acid–base pairs.

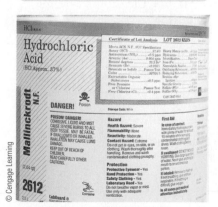

Don't taste chemical reagents!

The label on a bottle of concentrated hydrochloric acid.

© Cengage Learning

Acids were first recognized as substances that taste sour. Vinegar tastes sour because it is a dilute solution of acetic acid; citric acid is responsible for the sour taste of a lemon. Bases, sometimes called *alkalis,* are characterized by their bitter taste and slippery feel. Most hand soaps and commercial preparations for unclogging drains are highly basic.

The first person to recognize the essential nature of acids and bases was Svante Arrhenius. On the basis of his experiments with electrolytes, Arrhenius postulated that **acids** *produce hydrogen ions in aqueous solution,* whereas **bases** *produce hydroxide ions* (review Section 7.4).

For example, when hydrogen chloride gas is dissolved in water, each molecule produces ions as follows:

$$HCl(g) \xrightarrow{H_2O} H^+(aq) + Cl^-(aq)$$

This solution is the strong acid known as hydrochloric acid. On the other hand, when solid sodium hydroxide is dissolved in water, its ions separate producing a solution containing Na^+ and OH^- ions.

$$\text{NaOH}(s) \xrightarrow{\text{H}_2\text{O}} \text{Na}^+(aq) + \text{OH}^-(aq)$$

This solution is called a strong base.

Although the **Arrhenius concept of acids and bases** was a major step forward in understanding acid–base chemistry, this concept is limited because it allows for only one kind of base—the hydroxide ion. A more general definition of acids and bases was suggested by the Danish chemist Johannes Brønsted and the English chemist Thomas Lowry. In the **Brønsted–Lowry model,** *an acid is a proton (H⁺) donor, and a base is a proton acceptor.* According to the Brønsted–Lowry model, the general reaction that occurs when an acid is dissolved in water can best be represented as an acid (HA) donating a proton to a water molecule to form a new acid (the **conjugate acid**) and a new base (the **conjugate base**).

> Recall that (*aq*) means the substance is hydrated—it has water molecules clustered around it.

$$\underset{\text{Acid}}{\text{HA}(aq)} + \underset{\text{Base}}{\text{H}_2\text{O}(l)} \rightarrow \underset{\substack{\text{Conjugate} \\ \text{acid}}}{\text{H}_3\text{O}^+(aq)} + \underset{\substack{\text{Conjugate} \\ \text{base}}}{\text{A}^-(aq)}$$

This model emphasizes the significant role of the polar water molecule in pulling the proton from the acid. Note that the conjugate base is everything that remains of the acid molecule after a proton is lost. The conjugate acid is formed when the proton is transferred to the base. A **conjugate acid–base pair** consists of two substances related to each other by the donating and accepting of a *single proton.* In the above equation there are two conjugate acid–base pairs: HA (acid) and A⁻ (base), and H_2O (base) and H_3O^+ (acid). For example, when hydrogen chloride is dissolved in water it behaves as an acid.

Acid–conjugate base pair

$$\text{HCl}(aq) + \text{H}_2\text{O}(l) \rightarrow \text{H}_3\text{O}^+(aq) + \text{Cl}^-(aq)$$

Base–conjugate acid pair

In this case HCl is the acid that loses an H⁺ ion to form Cl⁻, its conjugate base. On the other hand, H_2O (behaving as a base) gains an H⁺ ion to form H_3O^+ (the conjugate acid).

How can water act as a base? Remember that the oxygen of the water molecule has two unshared electron pairs, either of which can form a covalent bond with an H⁺ ion. When gaseous HCl dissolves in water, the following reaction occurs:

$$\text{H}-\overset{..}{\underset{|}{\text{O}}}: + \underset{\delta^+ \quad \delta^-}{\text{H}-\text{Cl}} \rightarrow \left[\text{H}-\overset{..}{\underset{|}{\text{O}}}-\text{H} \right]^+ + \text{Cl}^-$$

Note that an H⁺ ion is transferred from the HCl molecule to the water molecule to form H_3O^+, which is called the **hydronium ion.**

EXAMPLE 16.1 | Identifying Conjugate Acid–Base Pairs

Which of the following represent conjugate acid–base pairs?

   a. HF, F⁻

   b. NH_4^+, NH_3

   c. HCl, H_2O

Gum That Foams

Mad Dawg chewing gum is a practical joker's dream come true. It is noticeably sour when someone first starts to chew it, but the big surprise comes about ten chews later when brightly colored foam oozes from the person's mouth. Although the effect is dramatic, the cause is simple acid–base chemistry.

The foam consists of sugar and saliva churned into a bubbling mess by carbon dioxide released from the gum. The carbon dioxide is formed when sodium bicarbonate ($NaHCO_3$) present in the gum is mixed with citric acid and malic acid (also present in the gum) in the moist environment of the mouth. As $NaHCO_3$ dissolves in the water of the saliva, it separates into its ions:

$$NaHCO_3(s) \xrightarrow{H_2O} Na^+(aq) + HCO_3^-(aq)$$

The bicarbonate ion, when exposed to H^+ ions from acids, decomposes to carbon dioxide and water:*

$$H^+(aq) + HCO_3^-(aq) \rightarrow H_2O(l) + CO_2(g)$$

The acids present in the gum also cause it to be sour, stimulating extra salivation and thus extra foam.

Although the chemistry behind Mad Dawg is well understood, the development of the gum into a safe, but fun, product was not so easy. In fact, early versions of the gum exploded because the acids and the sodium bicarbonate mixed prematurely. As solids, citric and malic acids and sodium bicarbonate do not react with each other. However, the presence of water frees the ions to move and react. In the manufacture of the gum, colorings and flavorings are applied as aqueous solutions. The water caused the gum to explode in early attempts to manufacture it. The makers of Mad Dawg obviously solved the problem. Buy some Mad Dawg and cut it open to see how they did it.

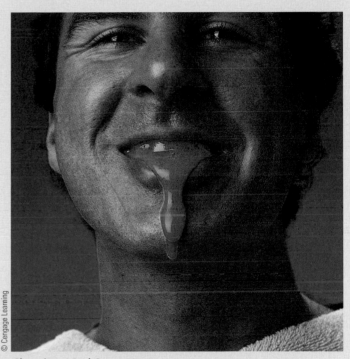

© Cengage Learning

Chewing Mad Dawg gum.

*This reaction is often used to power "bottle rockets" by adding vinegar (dilute acetic acid) to baking soda (sodium bicarbonate).

SOLUTION

a. and b. HF, F^- and NH_4^+, NH_3 are conjugate acid–base pairs because the two species differ by one H^+.

$$HF \rightarrow H^+ + F^-$$
$$NH_4^+ \rightarrow H^+ + NH_3$$

c. HCl and H_2O are not a conjugate acid–base pair because they are not related by the removal or addition of one H^+. The conjugate base of HCl is Cl^-. The conjugate acid of H_2O is H_3O^+. ∎

EXAMPLE 16.2 | Writing Conjugate Bases

Write the conjugate base for each of the following:

   a. $HClO_4$ b. H_3PO_4 c. $CH_3NH_3^+$

SOLUTION

To get the conjugate base for an acid, we must remove an H^+ ion.

   a. $HClO_4 \rightarrow H^+ + ClO_4^-$
      Acid   Conjugate base

   b. $H_3PO_4 \rightarrow H^+ + H_2PO_4^-$
      Acid   Conjugate base

   c. $CH_3NH_3^+ \rightarrow H^+ + CH_3NH_2$
      Acid   Conjugate base

Self-Check EXERCISE 16.1   Which of the following represent conjugate acid–base pairs?

   a. H_2O, H_3O^+

   b. OH^-, HNO_3

   c. H_2SO_4, SO_4^{2-}

   d. $HC_2H_3O_2$, $C_2H_3O_2^-$

See Problems 16.7 through 16.14. ■

16.2 Acid Strength

OBJECTIVES: To understand what acid strength means. • To understand the relationship between acid strength and the strength of the conjugate base.

We have seen that when an acid dissolves in water, a proton is transferred from the acid to water:

$$HA(aq) + H_2O(l) \rightarrow H_3O^+(aq) + A^-(aq)$$

In this reaction a new acid, H_3O^+ (called the conjugate acid), and a new base, A^- (the conjugate base), are formed. The conjugate acid and base can react with one another,

$$H_3O^+(aq) + A^-(aq) \rightarrow HA(aq) + H_2O(l)$$

to re-form the parent acid and a water molecule. Therefore, this reaction can occur "in both directions." The forward reaction is

$$HA(aq) + H_2O(l) \rightarrow H_3O^+(aq) + A^-(aq)$$

and the reverse reaction is

$$H_3O^+(aq) + A^-(aq) \rightarrow HA(aq) + H_2O(l)$$

Note that the products in the forward reaction are the reactants in the reverse reaction. We usually represent the situation in which the reaction can occur in both directions by double arrows:

$$HA(aq) + H_2O(l) \rightleftharpoons H_3O^+(aq) + A^-(aq)$$

This situation represents a competition for the H^+ ion between H_2O (in the forward reaction) and A^- (in the reverse reaction). If H_2O "wins" this competition—that is, if H_2O has a very high attraction for H^+ compared to A^-—then the solution will contain mostly H_3O^+ and A^-. We describe this situation by saying that the H_2O molecule is a much stronger base (more attraction for H^+) than A^-. In this case the forward reaction predominates:

$$HA(aq) + H_2O(l) \Rightarrow H_3O^+(aq) + A^-(aq)$$

We say that the acid HA is **completely ionized** or **completely dissociated.** This situation represents a **strong acid.**

The opposite situation can also occur. Sometimes A^- "wins" the competition for the H^+ ion. In this case A^- is a much stronger base than H_2O and the reverse reaction predominates:

$$HA(aq) + H_2O(l) \Leftarrow H_3O^+(aq) + A^-(aq)$$

Here, A^- has a much larger attraction for H^+ than does H_2O, and most of the HA molecules remain intact. This situation represents a **weak acid.**

We can determine what is actually going on in a solution by measuring its ability to conduct an electric current. Recall from Chapter 7 that a solution can conduct a current in proportion to the number of ions that are present (see Figure 7.2). When 1 mole of solid sodium chloride is dissolved in 1 L of water, the resulting solution is an excellent conductor of an electric current because the Na^+ and Cl^- ions separate completely. We call NaCl a strong electrolyte. Similarly, when 1 mole of hydrogen chloride is dissolved in 1 L of water, the resulting solution is an excellent conductor. Therefore, hydrogen chloride is also a strong electrolyte, which means that each HCl molecule must produce H^+ and Cl^- ions. This tells us that the forward reaction predominates:

$$HCl(aq) + H_2O(l) \rightleftharpoons H_3O^+(aq) + Cl^-(aq)$$

(Accordingly, the arrow pointing right is longer than the arrow pointing left.) In solution there are virtually no HCl molecules, only H^+ and Cl^- ions. This shows that Cl^- is a very poor base compared to the H_2O molecule; it has virtually no ability to attract H^+ ions in water. This aqueous solution of hydrogen chloride (called *hydrochloric acid*) is a strong acid.

In general, the strength of an acid is defined by the position of its ionization (dissociation) reaction:

$$HA(aq) + H_2O(l) \rightleftharpoons H_3O^+(aq) + A^-(aq)$$

A strong acid is one for which *the forward reaction predominates.* This means that almost all the original HA is dissociated (ionized) (see Figure 16.1a). There is an important connection between the strength of an acid and that of its conjugate base. *A strong acid contains a relatively weak conjugate base*—one that has a low attraction for protons. A strong acid can be described as an acid whose conjugate base is a much weaker base than water (Figure 16.2). In this case the water molecules win the competition for the H^+ ions.

In contrast to hydrochloric acid, when acetic acid, $HC_2H_3O_2$, is dissolved in water, the resulting solution conducts an electric current only weakly. That is, acetic acid is a weak electrolyte, which means that only a few ions are present. In other words, for the reaction

$$HC_2H_3O_2(aq) + H_2O(l) \rightleftharpoons H_3O^+(aq) + C_2H_3O_2^-(aq)$$

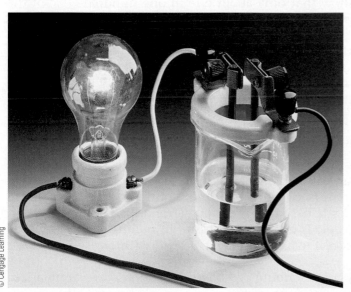

© Cengage Learning

A hydrochloric acid solution readily conducts electric current, as shown by the brightness of the bulb.

A strong acid is completely dissociated in water. No HA molecules remain. Only H_3O^+ and A^- are present.

Figure 16.1

Graphical representation of the behavior of acids of different strengths in aqueous solution.

Relative acid strength

Relative conjugate base strength

the easier HA give up a H+ (handwritten)

the less likely it will take it back (handwritten)

Very strong — Very weak

Strong

Weak

Weak

Strong

less likely HA give up a H+ (handwritten)

Very weak — Very strong *more likely it will take it back* (handwritten)

Figure 16.2

The relationship of acid strength and conjugate base strength for the dissociation reaction

HA(aq) + H$_2$O(l) $\rightleftharpoons$
Acid

H$_3$O$^+$(aq) + A$^-$(aq)
Conjugate base

the reverse reaction predominates (thus the arrow pointing left is longer). In fact, measurements show that only about one in one hundred (1%) of the HC$_2$H$_3$O$_2$ molecules is dissociated (ionized) in a 0.1 M solution of acetic acid. Thus acetic acid is a weak acid. When acetic acid molecules are placed in water, almost all of the molecules remain undissociated. This tells us that the acetate ion, C$_2$H$_3$O$_2^-$, is an effective base—it very successfully attracts H$^+$ ions in water. This means that acetic acid remains largely in the form of HC$_2$H$_3$O$_2$ molecules in solution. A weak acid is one for which the *reverse reaction predominates*.

$$HA(aq) + H_2O(l) \impliedby H_3O^+(aq) + A^-(aq)$$

Most of the acid originally placed in the solution is still present as HA at equilibrium. That is, a weak acid dissociates (ionizes) only to a very small extent in aqueous solution (see Figure 16.1b). In contrast to a strong acid, a weak acid has a conjugate base that is a much stronger base than water. In this case a water molecule is not very successful in pulling an H$^+$ ion away from the conjugate base. *A weak acid contains a relatively strong conjugate base* (Figure 16.2).

The various ways of describing the strength of an acid are summarized in Table 16.1.

Table 16.1 Ways to Describe Acid Strength

Property	Strong Acid	Weak Acid
the acid ionization (dissociation) reaction	forward reaction predominates	reverse reaction predominates
strength of the conjugate base compared with that of water	A$^-$ is a much weaker base than H$_2$O	A$^-$ is a much stronger base than H$_2$O

Carbonation—A Cool Trick

The sensations of taste and smell greatly affect our daily experience. For example, memories are often triggered by an odor that matches one that occurred when an event was originally stored in our memory banks. Likewise, the sense of taste has a powerful effect on our lives. For example, many people crave the intense sensation produced by the compounds found in chili peppers.

One sensation that is quite refreshing for most people is the effect of a chilled, carbonated beverage in the mouth. The sharp, tingling sensation experienced is not directly due to the bubbling of the dissolved carbon dioxide in the beverage. Rather, it arises because protons are produced as the CO_2 interacts with the water in the tissues of the mouth:

$$CO_2 + H_2O \rightleftharpoons H^+ + HCO_3^-$$

This reaction is speeded up by a biologic catalyst—an enzyme—called carbonic anhydrase. The acidification of the fluids in the nerve endings in the mouth leads to the sharp sensation produced by carbonated drinks.

Carbon dioxide also stimulates nerve sites that detect "coolness" in the mouth. In fact, researchers have identified a mutual enhancement between cooling and the presence of CO_2. Studies show that at a given concentration of CO_2, a colder drink feels more "pungent" than a warmer one. When tests were conducted on drinks in which the carbon dioxide concentration was varied, the results showed that a drink felt colder as the CO_2 concentration was increased, even though the drinks were all actually at the same temperature.

Thus a beverage can seem colder if it has a higher concentration of carbon dioxide. At the same time, cooling a carbonated beverage can intensify the tingling sensation caused by the acidity induced by the CO_2. This is truly a happy synergy.

© Cengage Learning

An acetic acid solution conducts only a small amount of current as shown by the dimly lit bulb.

The common strong acids are sulfuric acid, $H_2SO_4(aq)$; hydrochloric acid, $HCl(aq)$; nitric acid, $HNO_3(aq)$; and perchloric acid, $HClO_4(aq)$. Sulfuric acid is actually a **diprotic acid,** an acid that can furnish two protons. The acid H_2SO_4 is a strong acid that is virtually 100% dissociated in water:

$$H_2SO_4(aq) \rightarrow H^+(aq) + HSO_4^-(aq)$$

The HSO_4^- ion is also an acid but it is a weak acid:

$$HSO_4^-(aq) \rightleftharpoons H^+(aq) + SO_4^{2-}(aq)$$

Most of the HSO_4^- ions remain undissociated.

Most acids are **oxyacids,** in which the acidic hydrogen is attached to an oxygen atom (several oxyacids are shown at the bottom of the following page). The strong acids we have mentioned, except hydrochloric acid, are typical examples. **Organic acids,** those with a carbon-atom backbone, commonly contain the **carboxyl group:**

$$-C\begin{matrix} \nearrow O \\ \searrow O-H \end{matrix}$$

Acids of this type are usually weak. An example is acetic acid, CH_3COOH, which is often written as $HC_2H_3O_2$.

Plants Fight Back

Plants sometimes do not seem to get much respect. We often think of them as rather dull life forms. We are used to animals communicating with each other, but we think of plants as mute. However, this perception is now changing. It is now becoming clear that plants communicate with other plants and also with insects. Ilya Roskin and his colleagues at Rutgers University, for example, have found that tobacco plants under attack by disease signal distress using the chemical salicylic acid, a precursor of aspirin. When a tobacco plant is infected with tobacco mosaic virus (TMV), which forms dark blisters on leaves and causes them to pucker and yellow, the sick plant produces large amounts of salicylic acid to alert its immune system to fight the virus. In addition, some of the salicylic acid is converted to methyl salicylate, a volatile compound that evaporates from the sick plant. Neighboring plants absorb this chemical and turn it back to salicylic acid, thus triggering their immune systems to protect them against the impending attack by TMV. Thus, as a tobacco plant gears up to fight an attack by TMV, it also warns its neighbors to be ready for this virus.

In another example of plant communication, a tobacco leaf under attack by a caterpillar emits a chemical signal that attracts a parasitic wasp that stings and kills the insect. Even more impressive is the ability of the plant to customize the emitted signal so that the wasp attracted will be the one that specializes in killing the particular caterpillar involved in the attack. The plant does this by changing the proportions of two chemicals emitted when a caterpillar chews on a leaf. Studies have shown that other plants, such as corn and cotton, also emit wasp-attracting chemicals when they face attack by caterpillars.

This research shows that plants can "speak up" to protect themselves. Scientists hope to learn to help them do this even more effectively.

Agricultural Research Service/USDA

A wasp lays its eggs on a gypsy moth caterpillar on the leaf of a corn plant.

Salicylic acid

Methyl salicylate

Phosphoric acid

Acetic acid

Nitrous acid

Hypochlorous acid

There are some important acids in which the acidic proton is attached to an atom other than oxygen. The most significant of these are the hydrohalic acids HX, where X represents a halogen atom. Examples are HCl(*aq*), a strong acid, and HF(*aq*), a weak acid.

16.3   Water as an Acid and a Base

OBJECTIVE: To learn about the ionization of water.

A substance is said to be *amphoteric* if it can behave either as an acid or as a base. Water is the most common **amphoteric substance.** We can see this clearly in the **ionization of water,** which involves the transfer of a proton from one water molecule to another to produce a hydroxide ion and a hydronium ion.

$$H_2O(l) + H_2O(l) \rightleftharpoons H_3O^+(aq) + OH^-(aq)$$

In this reaction one water molecule acts as an acid by furnishing a proton, and the other acts as a base by accepting the proton. The forward reaction for this process does not occur to a very great extent. That is, in pure water only a tiny amount of H_3O^+ and OH^- exist. At 25 °C the actual concentrations are

$$[H_3O^+] = [OH^-] = 1.0 \times 10^{-7}\ M$$

Notice that in pure water the concentrations of $[H_3O^+]$ and $[OH^-]$ are equal because they are produced in equal numbers in the ionization reaction.

One of the most interesting and important things about water is that the mathematical *product* of the H_3O^+ and OH^- concentrations is always constant. We can find this constant by multiplying the concentrations of H_3O^+ and OH^- at 25 °C:

$$[H_3O^+][OH^-] = (1.0 \times 10^{-7})(1.0 \times 10^{-7}) = 1.0 \times 10^{-14}$$

We call this constant K_w. Thus at 25 °C

$$[H_3O^+][OH^-] = 1.0 \times 10^{-14} = K_w$$

To simplify the notation we often write H_3O^+ as just H^+. Thus we would write the K_w expression as follows:

$$[H^+][OH^-] = 1.0 \times 10^{-14} = K_w$$

K_w is called the **ion-product constant** for water. The units are customarily omitted when the value of the constant is given and used.

It is important to recognize the meaning of K_w. In any aqueous solution at 25 °C, *no matter what it contains*, the product of $[H^+]$ and $[OH^-]$ must always equal 1.0×10^{-14}. This means that if the $[H^+]$ goes up, the $[OH^-]$ must go down so that the product of the two is still 1.0×10^{-14}. For example, if HCl gas is dissolved in water, increasing the $[H^+]$, the $[OH^-]$ must decrease.

> At 25 °C, $K_w = [H^+][OH^-]$
> $= 1.0 \times 10^{-14}$

There are three possible situations we might encounter in an aqueous solution. If we add an acid (an H^+ donor) to water, we get an *acidic solution*. In this case, because we have added a source of H^+, the $[H^+]$ will be greater than the $[OH^-]$. On the other hand, if we add a base (a source of OH^-) to water, the $[OH^-]$ will be greater than the $[H^+]$. This is a *basic solution*. Finally, we might have a situation in which $[H^+] = [OH^-]$. This is called a *neutral solution*. Pure water is automatically neutral but we can also obtain a neutral solution by adding equal amounts of H^+ and OH^-. It is very important that you understand the definitions of neutral, acidic, and basic solutions. In summary:

1. In a **neutral solution,** $[H^+] = [OH^-]$
2. In an **acidic solution,** $[H^+] > [OH^-]$
3. In a **basic solution,** $[OH^-] > [H^+]$

In each case, however, $K_w = [H^+][OH^-] = 1.0 \times 10^{-14}$.

> Remember that H^+ represents H_3O^+.

EXAMPLE 16.3 Calculating Ion Concentrations in Water

Calculate $[H^+]$ or $[OH^-]$ as required for each of the following solutions at 25 °C, and state whether the solution is neutral, acidic, or basic.

a. $1.0 \times 10^{-5}\ M\ OH^-$ b. $1.0 \times 10^{-7}\ M\ OH^-$ c. $10.0\ M\ H^+$

SOLUTION

a. **Where Are We Going?**

We want to determine $[H^+]$ in a solution of given $[OH^-]$ at 25 °C.

What Do We Know?

- At 25 °C, $K_w = [H^+][OH^-] = 1.0 \times 10^{-14}$
- $[OH^-] = 1.0 \times 10^{-5}\ M$

How Do We Get There?

We know that $K_w = [H^+][OH^-] = 1.0 \times 10^{-14}$. We need to calculate the $[H^+]$. However, the $[OH^-]$ is given—it is $1.0 \times 10^{-5}\ M$—so we will solve for $[H^+]$ by dividing both sides by $[OH^-]$.

$$[H^+] = \frac{1.0 \times 10^{-14}}{[OH^-]} = \frac{1.0 \times 10^{-14}}{1.0 \times 10^{-5}} = 1.0 \times 10^{-9}\ M$$

> **MATH SKILL BUILDER**
> $K_w = [H^+][OH^-]$
> $\dfrac{K_w}{[OH^-]} = [H^+]$

Because $[OH^-] = 1.0 \times 10^{-5}\ M$ is greater than $[H^+] = 1.0 \times 10^{-9}\ M$, the solution is basic. (Remember: The more negative the exponent, the smaller the number.)

b. **Where Are We Going?**

We want to determine $[H^+]$ in a solution of given $[OH^-]$ at 25 °C.

What Do We Know?

- At 25 °C, $K_w = [H^+][OH^-] = 1.0 \times 10^{-14}$
- $[OH^-] = 1.0 \times 10^{-7}\ M$

How Do We Get There?

Again the $[OH^-]$ is given, so we solve the K_w expression for $[H^+]$.

$$[H^+] = \frac{1.0 \times 10^{-14}}{[OH^-]} = \frac{1.0 \times 10^{-14}}{1.0 \times 10^{-7}} = 1.0 \times 10^{-7}\ M$$

Here $[H^+] = [OH^-] = 1.0 \times 10^{-7}\ M$, so the solution is neutral.

c. **Where Are We Going?**

We want to determine $[OH^-]$ in a solution of given $[H^+]$ at 25 °C.

What Do We Know?

- At 25 °C, $K_w = [H^+][OH^-] = 1.0 \times 10^{-14}$
- $[H^+] = 10.0\ M$

How Do We Get There?

In this case the $[H^+]$ is given, so we solve for $[OH^-]$.

MATH SKILL BUILDER

$K_w = [H^+][OH^-]$

$\dfrac{K_w}{[H^+]} = [OH^-]$

$$[OH^-] = \frac{1.0 \times 10^{-14}}{[H^+]} = \frac{1.0 \times 10^{-14}}{10.0} = 1.0 \times 10^{-15}\ M$$

Now we compare $[H^+] = 10.0\ M$ with $[OH^-] = 1.0 \times 10^{-15}\ M$. Because $[H^+]$ is greater than $[OH^-]$, the solution is acidic.

Self-Check EXERCISE 16.2   Calculate $[H^+]$ in a solution in which $[OH^-] = 2.0 \times 10^{-2}\ M$. Is this solution acidic, neutral, or basic?

See Problems 16.31 through 16.34. ■

EXAMPLE 16.4 | **Using the Ion-Product Constant in Calculations**

Is it possible for an aqueous solution at 25 °C to have $[H^+] = 0.010\ M$ and $[OH^-] = 0.010\ M$?

SOLUTION

The concentration $0.010\ M$ can also be expressed as $1.0 \times 10^{-2}\ M$. Thus, if $[H^+] = [OH^-] = 1.0 \times 10^{-2}\ M$, the product

$$[H^+][OH^-] = (1.0 \times 10^{-2})(1.0 \times 10^{-2}) = 1.0 \times 10^{-4}$$

This is not possible. The product of $[H^+]$ and $[OH^-]$ must always be 1.0×10^{-14} in water at 25 °C, so a solution could not have $[H^+] = [OH^-] = 0.010\ M$. If H^+ and OH^- are added to water in these amounts, they will react with each other to form H_2O,

$$H^+ + OH^- \rightarrow H_2O$$

until the product $[H^+][OH^-] = 1.0 \times 10^{-14}$.

This is a general result. When H^+ and OH^- are added to water in amounts such that the product of their concentrations is greater than 1.0×10^{-14}, they will react to form water until enough H^+ and OH^- are consumed so that $[H^+][OH^-] = 1.0 \times 10^{-14}$. ■

16.4   The pH Scale

OBJECTIVE: To understand pH and pOH. • To learn to find pOH and pH for various solutions. • To learn to use a calculator in these calculations.

To express small numbers conveniently, chemists often use the "p scale," which is based on common logarithms (base 10 logs). In this system, if N represents some number, then

$$pN = -\log N = (-1) \times \log N$$

Airplane Rash

Because airplanes remain in service for many years, it is important to spot corrosion that might weaken the structure at an early stage. In the past, looking for minute signs of corrosion has been very tedious and labor-intensive, especially for large planes. This situation has changed, however, thanks to the paint system developed by Gerald S. Frankel and Jian Zhang of Ohio State University. The paint they created turns pink in areas that are beginning to corrode, making these areas easy to spot.

The secret to the paint's magic is phenolphthalein, the common acid–base indicator that turns pink in a basic solution. The corrosion of the aluminum skin of the airplane involves a reaction that forms OH$^-$ ions, producing a basic area at the site of the corrosion that turns the phenolphthalein pink. Because this system is highly sensitive, corrosion can be corrected before it damages the plane.

Next time you fly, if the plane has pink spots you might want to wait for a later flight!

---

go Chemistry **Module 9a: pH (Pt. 1)** covers concepts in this section.

> The pH scale provides a compact way to represent solution acidity.

go Chemistry **Module 9b: pH (Pt. 2)** covers concepts in this section.

That is, the p means to take the log of the number that follows and multiply the result by -1. For example, to express the number 1.0×10^{-7} on the p scale, we need to take the negative log of 1.0×10^{-7}.

$$p(1.0 \times 10^{-7}) = -\log(1.0 \times 10^{-7}) = 7.00$$

Because the [H$^+$] in an aqueous solution is typically quite small, using the p scale in the form of the **pH scale** provides a convenient way to represent solution acidity. The pH is defined as

$$pH = -\log[H^+]$$

To obtain the pH value of a solution, we must compute the negative log of the [H$^+$].

In the case where [H$^+$] $= 1.0 \times 10^{-5}$ M, the solution has a pH value of 5.00.

To represent pH to the appropriate number of significant figures, you need to know the following rule for logarithms: *the number of decimal places for a log must be equal to the number of significant figures in the original number.* Thus

2 significant
figures
⌐‾‾⌐

$$[H^+] = 1.0 \times 10^{-5} \, M$$

and

$$pH = 5.00$$
⌐_⌐
2 decimal
places

EXAMPLE 16.5 | Calculating pH

Calculate the pH value for each of the following solutions at 25 °C.

a. A solution in which $[H^+] = 1.0 \times 10^{-9}$ M

b. A solution in which $[OH^-] = 1.0 \times 10^{-6}$ M

SOLUTION

a. For this solution $[H^+] = 1.0 \times 10^{-9}$.

$$-\log 1.0 \times 10^{-9} = 9.00$$
$$pH = 9.00$$

b. In this case we are given the $[OH^-]$. Thus we must first calculate $[H^+]$ from the K_w expression. We solve

$$K_w = [H^+][OH^-] = 1.0 \times 10^{-14}$$

for $[H^+]$ by dividing both sides by $[OH^-]$.

$$[H^+] = \frac{1.0 \times 10^{-14}}{[OH^-]} = \frac{1.0 \times 10^{-14}}{1.0 \times 10^{-6}} = 1.0 \times 10^{-8}$$

Now that we know the $[H^+]$, we can calculate the pH because pH $= -\log[H^+] = -\log[1.0 \times 10^{-8}] = 8.00$.

Self-Check EXERCISE 16.3   Calculate the pH value for each of the following solutions at 25 °C.

a. A solution in which $[H^+] = 1.0 \times 10^{-3}$ M

b. A solution in which $[OH^-] = 5.0 \times 10^{-5}$ M

See Problems 16.41 through 16.44. ■

Table 16.2 The Relationship of the H$^+$ Concentration of a Solution to Its pH

$[H^+]$	pH
1.0×10^{-1}	1.00
1.0×10^{-2}	2.00
1.0×10^{-3}	3.00
1.0×10^{-4}	4.00
1.0×10^{-5}	5.00
1.0×10^{-6}	6.00
1.0×10^{-7}	7.00

The pH decreases as $[H^+]$ increases, and vice versa.

Because the pH scale is a log scale based on 10, *the pH changes by 1 for every power-of-10 change in the [H⁺].* For example, a solution of pH 3 has an H$^+$ concentration of 10^{-3} M, which is 10 times that of a solution of pH 4 ($[H^+] = 10^{-4}$ M) and 100 times that of a solution of pH 5. This is illustrated in Table 16.2. Also note from Table 16.2 that *the pH decreases as the [H⁺] increases.* That is, a lower pH means a more acidic solution. The pH scale and the pH values for several common substances are shown in Figure 16.3.

We often measure the pH of a solution by using a pH meter, an electronic device with a probe that can be inserted into a solution of unknown pH. A pH meter is shown in Figure 16.4. Colored indicator paper is also commonly used to measure the pH of a solution when less accuracy is needed. A drop of the solution to be tested is placed on this special paper, which promptly turns to a color characteristic of a given pH (see Figure 16.5).

Log scales similar to the pH scale are used for representing other quantities. For example,

$$pOH = -\log[OH^-]$$

The symbol p means $-\log$.

Therefore, in a solution in which

$$[OH^-] = 1.0 \times 10^{-12}\ M$$

the pOH is

$$-\log[OH^-] = -\log(1.0 \times 10^{-12}) = 12.00$$

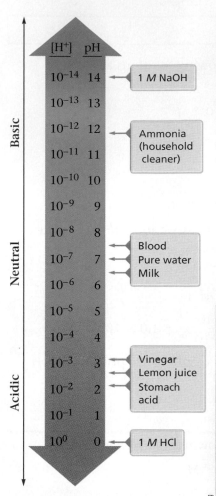

Figure 16.3

The pH scale and pH values of some common substances.

Figure 16.4

A pH meter. The electrodes on the right are placed in the solution with unknown pH. The difference between the [H⁺] in the solution sealed into one of the electrodes and the [H⁺] in the solution being analyzed is translated into an electrical potential and registered on the meter as a pH reading.

Figure 16.5

Indicator paper being used to measure the pH of a solution. The pH is determined by comparing the color that the solution turns the paper to the color chart.

EXAMPLE 16.6 | Calculating pH and pOH

Calculate the pH and pOH for each of the following solutions at 25 °C.

a. $1.0 \times 10^{-3} \, M \, \text{OH}^-$

b. $1.0 \, M \, \text{H}^+$

SOLUTION

a. We are given the [OH⁻], so we can calculate the pOH value by taking $-\log[\text{OH}^-]$.

$$\text{pOH} = -\log[\text{OH}^-] = -\log(1.0 \times 10^{-3}) = 3.00$$

To calculate the pH, we must first solve the K_w expression for [H⁺].

$$[\text{H}^+] = \frac{K_w}{[\text{OH}^-]} = \frac{1.0 \times 10^{-14}}{1.0 \times 10^{-3}} = 1.0 \times 10^{-11} \, M$$

Now we compute the pH.

$$pH = -\log[H^+] = -\log(1.0 \times 10^{-11}) = 11.00$$

b. In this case we are given the $[H^+]$ and we can compute the pH.

$$pH = -\log[H^+] = -\log(1.0) = 0$$

We next solve the K_w expression for $[OH^-]$.

$$[OH^-] = \frac{K_w}{[H^+]} = \frac{1.0 \times 10^{-14}}{1.0} = 1.0 \times 10^{-14} \, M$$

Now we compute the pOH.

$$pOH = -\log[OH^-] = -\log(1.0 \times 10^{-14}) = 14.00 \;\blacksquare$$

We can obtain a convenient relationship between pH and pOH by start-ing with the K_w expression $[H^+][OH^-] = 1.0 \times 10^{-14}$ and taking the negative log of both sides.

$$-\log([H^+][OH^-]) = -\log(1.0 \times 10^{-14})$$

Because the log of a product equals the sum of the logs of the terms—that is, $\log(A \times B) = \log A + \log B$—we have

$$\underbrace{-\log[H^+]}_{pH} \underbrace{-\log[OH^-]}_{pOH} = -\log(1.0 \times 10^{-14}) = 14.00$$

which gives the equation

$$\boxed{pH + pOH = 14.00}$$

This means that once we know either the pH or the pOH for a solution, we can calculate the other. For example, if a solution has a pH of 6.00, the pOH is calculated as follows:

$$pH + pOH = 14.00$$
$$pOH = 14.00 - pH$$
$$pOH = 14.00 - 6.00 = 8.00$$

Red blood cells can exist only over a narrow range of pH.

EXAMPLE 16.7 | Calculating pOH from pH

The pH of blood is about 7.4. What is the pOH of blood?

SOLUTION

$$pH + pOH = 14.00$$
$$pOH = 14.00 - pH$$
$$= 14.00 - 7.4$$
$$= 6.6$$

The pOH of blood is 6.6.

Self-Check EXERCISE 16.4   A sample of rain in an area with severe air pollution has a pH of 3.5. What is the pOH of this rainwater?

See Problems 16.45 and 16.46. ■

It is also possible to find the [H$^+$] or [OH$^-$] from the pH or pOH. To find the [H$^+$] from the pH, we must go back to the definition of pH:

$$pH = -\log[H^+]$$

or

$$-pH = \log[H^+]$$

To arrive at [H$^+$] on the right-hand side of this equation we must "undo" the log operation. This is called taking the *antilog* or the *inverse* log.

> Inverse log ($-$pH) = inverse log (log[H$^+$])
> Inverse log ($-$pH) = [H$^+$]

There are different methods for carrying out the inverse log operation on various calculators. One common method is the two-key $\boxed{\text{inv}}$ $\boxed{\text{log}}$ sequence. (Consult the user's manual for your calculator to find out how to do the antilog or inverse log operation.) The steps in going from pH to [H$^+$] are as follows:

MATH SKILL BUILDER
This operation may involve a 10^x key on some calculators.

Steps for Calculating [H$^+$] from pH

Step 1 Take the inverse log (antilog) of $-$pH to give [H$^+$] by using the $\boxed{\text{inv}}$ $\boxed{\text{log}}$ keys in that order. (Your calculator may require different keys for this operation.)

Step 2 Press the minus [$-$] key.

Step 3 Enter the pH.

For practice, we will convert pH = 7.0 to [H$^+$].

$$pH = 7.0$$
$$-pH = -7.0$$

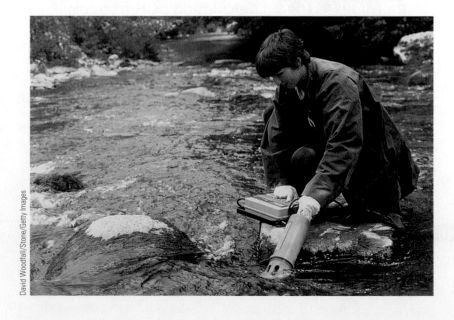

Measuring the pH of the water in a river.

David Woodfall/Stone/Getty Images

The inverse log of -7.0 gives 1×10^{-7}.

$$[H^+] = 1 \times 10^{-7} \, M$$

This process is illustrated further in Example 16.8.

EXAMPLE 16.8 | Calculating [H⁺] from pH

The pH of a human blood sample was measured to be 7.41. What is the $[H^+]$ in this blood?

SOLUTION

$$pH = 7.41$$
$$-pH = -7.41$$
$$[H^+] = \text{inverse log of } -7.41 = 3.9 \times 10^{-8}$$
$$[H^+] = 3.9 \times 10^{-8} \, M$$

Notice that because the pH has two decimal places, we need two significant figures for $[H^+]$.

Self-Check EXERCISE 16.5   The pH of rainwater in a polluted area was found to be 3.50. What is the $[H^+]$ for this rainwater?

See Problems 16.49 and 16.50. ■

A similar procedure is used to change from pOH to $[OH^-]$, as shown in Example 16.9.

EXAMPLE 16.9 | Calculating [OH⁻] from pOH

The pOH of the water in a fish tank is found to be 6.59. What is the $[OH^-]$ for this water?

SOLUTION

We use the same steps as for converting pH to $[H^+]$, except that we use the pOH to calculate the $[OH^-]$.

$$pOH = 6.59$$
$$-pOH = -6.59$$
$$[OH^-] = \text{inverse log of } -6.59 = 2.6 \times 10^{-7}$$
$$[OH^-] = 2.6 \times 10^{-7} \, M$$

Note that two significant figures are required.

Self-Check EXERCISE 16.6   The pOH of a liquid drain cleaner was found to be 10.50. What is the $[OH^-]$ for this cleaner?

See Problems 16.51 and 16.52. ■

Garden-Variety Acid–Base Indicators

What can flowers tell us about acids and bases? Actually, some flowers can tell us whether the soil they are growing in is acidic or basic. For example, in acidic soil, bigleaf hydrangea blossoms will be blue; in basic (alkaline) soil, the flowers will be red. What is the secret? The pigment in the flower is an acid–base indicator.

Generally, acid–base indicators are dyes that are weak acids. Because indicators are usually complex molecules, we often symbolize them as HIn. The reaction of the indicator with water can be written as

$$HIn(aq) + H_2O(l) \rightleftharpoons H_3O^+(aq) + In^-(aq)$$

To work as an acid–base indicator, the conjugate acid–base forms of these dyes must have different colors. The acidity level of the solution will determine whether the indicator is present mainly in its acidic form (HIn) or its basic form (In⁻).

When placed in an acidic solution, most of the basic form of the indicator is converted to the acidic form by the reaction

$$In^-(aq) + H^+(aq) \rightarrow HIn(aq)$$

When placed in a basic solution, most of the acidic form of the indicator is converted to the basic form by the reaction

$$HIn(aq) + OH^-(aq) \rightarrow In^-(aq) + H_2O(l)$$

It turns out that many fruits, vegetables, and flowers can act as acid–base indicators. Red, blue, and purple plants often contain a class of chemicals called anthocyanins, which change color based on the acidity level of the surroundings. Perhaps the most famous of these plants is red cabbage. Red cabbage contains a mixture of anthocyanins and other pigments that allow it to be used as a "universal indicator." Red cabbage juice appears deep red at a pH of 1–2, purple at a pH of 4, blue at a pH of 8, and green at a pH of 11.

16.5 Calculating the pH of Strong Acid Solutions

OBJECTIVE: To learn to calculate the pH of solutions of strong acids.

In this section we will learn to calculate the pH for a solution containing a strong acid of known concentration. For example, if we know a solution contains 1.0 M HCl, how can we find the pH of the solution? To answer this question we must know that when HCl dissolves in water, each molecule dissociates (ionizes) into H⁺ and Cl⁻ ions. That is, we must know that HCl is a strong acid. Thus, although the label on the bottle says 1.0 M HCl, the solution contains virtually no HCl molecules. A 1.0 M HCl solution contains H⁺ and Cl⁻ ions rather than HCl molecules. Typically, container labels indicate the substance(s) used to make up the solution but do not necessarily describe the solution components after dissolution. In this case,

$$1.0\ M\ HCl \rightarrow 1.0\ M\ H^+ \text{ and } 1.0\ M\ Cl^-$$

Therefore, the [H⁺] in the solution is 1.0 M. The pH is then

$$pH = -\log[H^+] = -\log(1.0) = 0$$

Other natural indicators include the skins of beets (which change from red to purple in very basic solutions), blueberries (which change from blue to red in acidic solutions), and a wide variety of flower petals, including delphiniums, geraniums, morning glories, and, of course, hydrangeas.

Royalty-free Corbis

Neil Holmes/Holmes Garden Photos/Alamy

EXAMPLE 16.10 | Calculating the pH of Strong Acid Solutions

Calculate the pH of 0.10 M HNO$_3$.

SOLUTION

HNO$_3$ is a strong acid, so the ions in solution are H$^+$ and NO$_3^-$. In this case,

$$0.10 \ M \ HNO_3 \rightarrow 0.10 \ M \ H^+ \ \text{and} \ 0.10 \ M \ NO_3^-$$

Thus

$$[H^+] = 0.10 \ M \quad \text{and} \quad pH = -\log(0.10) = 1.00$$

Self-Check EXERCISE 16.7 Calculate the pH of a solution of $5.0 \times 10^{-3} \ M$ HCl.

See Problems 16.57 and 16.58. ■

Buffered Solutions

OBJECTIVE: To understand the general characteristics of buffered solutions.

Water: pH = 7
0.01 *M* HCl: pH = 2

A **buffered solution** is one that resists a change in its pH even when a strong acid or base is added to it. For example, when 0.01 mole of HCl is added to 1 L of pure water, the pH changes from its initial value of 7 to 2, a change of 5 pH units. However, when 0.01 mole of HCl is added to a solution containing both 0.1 *M* acetic acid ($HC_2H_3O_2$) and 0.1 *M* sodium acetate ($NaC_2H_3O_2$), the pH changes from an initial value of 4.74 to 4.66, a change of only 0.08 pH unit. The latter solution is buffered—it undergoes only a very slight change in pH when a strong acid or base is added to it.

Buffered solutions are vitally important to living organisms whose cells can survive only in a very narrow pH range. Many goldfish have died because their owners did not realize the importance of buffering the aquarium water at an appropriate pH. For humans to survive, the pH of the blood must be maintained between 7.35 and 7.45. This narrow range is maintained by several different buffering systems.

A solution is **buffered** by the *presence of a weak acid and its conjugate base*. An example of a buffered solution is an aqueous solution that contains acetic acid and sodium acetate. The sodium acetate is a salt that furnishes acetate ions (the conjugate base of acetic acid) when it dissolves. To see how this system acts as a buffer, we must recognize that the species present in this solution are

Hans Reinhard/Bruce Coleman, Inc.

For goldfish to survive, the pH of the water must be carefully controlled.

$$HC_2H_3O_2, \quad \underbrace{Na^+, \quad C_2H_3O_2^-}$$

When $NaC_2H_3O_2$ is
dissolved, it produces
the separated ions

What happens in this solution when a strong acid such as HCl is added? In pure water, the H^+ ions from the HCl would accumulate, thus lowering the pH.

$$HCl \xrightarrow{100\%} H^+ + Cl^-$$

However, this buffered solution contains $C_2H_3O_2^-$ ions, which are basic. That is, $C_2H_3O_2^-$ has a strong affinity for H^+, as evidenced by the fact that $HC_2H_3O_2$ is a weak acid. This means that the $C_2H_3O_2^-$ and H^+ ions do not exist together in large numbers. Because the $C_2H_3O_2^-$ ion has a high affinity for H^+, these two combine to form $HC_2H_3O_2$ molecules. Thus the H^+ from the added HCl does not accumulate in solution but reacts with the $C_2H_3O_2^-$ as follows:

$$H^+(aq) + C_2H_3O_2^-(aq) \rightarrow HC_2H_3O_2(aq)$$

Next consider what happens when a strong base such as sodium hydroxide is added to the buffered solution. If this base were added to pure water, the OH^- ions from the solid would accumulate and greatly change (raise) the pH.

$$NaOH \xrightarrow{100\%} Na^+ + OH^-$$

However, in the buffered solution the OH^- ion, which has a *very strong* affinity for H^+, reacts with $HC_2H_3O_2$ molecules as follows:

$$HC_2H_3O_2(aq) + OH^-(aq) \rightarrow H_2O(l) + C_2H_3O_2^-(aq)$$

Table 16.3   The Characteristics of a Buffer

1. The solution contains a weak acid HA and its conjugate base A⁻.
2. The buffer resists changes in pH by reacting with any added H⁺ or OH⁻ so that these ions do not accumulate.
3. Any added H⁺ reacts with the base A⁻.

$$H^+(aq) + A^-(aq) \rightarrow HA(aq)$$

4. Any added OH⁻ reacts with the weak acid HA.

$$OH^-(aq) + HA(aq) \rightarrow H_2O(l) + A^-(aq)$$

This happens because, although $C_2H_3O_2^-$ has a strong affinity for H⁺, OH⁻ has a much stronger affinity for H⁺ and thus can remove H⁺ ions from acetic acid molecules.

Note that the buffering materials dissolved in the solution prevent added H⁺ or OH⁻ from building up in the solution. Any added H⁺ is trapped by $C_2H_3O_2^-$ to form $HC_2H_3O_2$. Any added OH⁻ reacts with $HC_2H_3O_2$ to form H_2O and $C_2H_3O_2^-$.

The general properties of a buffered solution are summarized in Table 16.3.

CHAPTER 16 REVIEW

Key Terms

acid (16.1)
base (16.1)
Arrhenius concept of
  acids and bases (16.1)
Brønsted–Lowry
  model (16.1)
conjugate acid (16.1)
conjugate base (16.1)
conjugate acid–base
  pair (16.1)
hydronium ion (16.1)
completely ionized
  (dissociated) (16.2)
strong acid (16.2)
weak acid (16.2)

diprotic acid (16.2)
oxyacid (16.2)
organic acid (16.2)
carboxyl group (16.2)
amphoteric
  substance (16.3)
ionization of water (16.3)
ion-product constant,
  K_w (16.3)
neutral solution (16.3)
acidic solution (16.3)
basic solution (16.3)
pH scale (16.4)
buffered solution (16.6)
buffered (16.6)

Summary

1. Acids or bases in water are commonly described by two different models. Arrhenius postulated that acids produce H⁺ ions in aqueous solutions and that bases produce OH⁻ ions. The Brønsted–Lowry model is more general: an acid is a proton donor, and a base is a proton acceptor. Water acts as a Brønsted–Lowry base when it accepts a proton from an acid to form a hydronium ion:

$$HA(aq) + H_2O(l) \rightleftharpoons H_3O^+(aq) + A^-(aq)$$

| Acid | Base | Conjugate acid | Conjugate base |

A conjugate base is everything that remains of the acid molecule after the proton is lost. A conjugate acid is formed when a proton is transferred to the base. Two substances related in this way are called a conjugate acid–base pair.

2. A strong acid or base is one that is completely ionized (dissociated). A weak acid is one that is ionized (dissociated) only to a slight extent. Strong acids have weak conjugate bases. Weak acids have relatively strong conjugate bases.

3. Water is an amphoteric substance—it can behave either as an acid or as a base. The ionization of water reveals this property; one water molecule transfers a proton to another water molecule to produce a hydronium ion and a hydroxide ion.

$$H_2O(l) + H_2O(l) \rightleftharpoons H_3O^+(aq) + OH^-(aq)$$

The expression

$$K_w = [H_3O^+][OH^-] = [H^+][OH^-]$$

is called the ion-product constant. It has been shown experimentally that at 25 °C,

$$[H^+] = [OH^-] = 1.0 \times 10^{-7}\ M$$

so $K_w = 1.0 \times 10^{-14}$.

4. In an acidic solution, $[H^+]$ is greater than $[OH^-]$. In a basic solution, $[OH^-]$ is greater than $[H^+]$. In a neutral solution, $[H^+] = [OH^-]$.

5. To describe $[H^+]$ in aqueous solutions, we use the pH scale.

$$pH = -\log[H^+]$$

Note that the pH decreases as $[H^+]$ (acidity) increases.

6. The pH of strong acid solutions can be calculated directly from the concentration of the acid, because 100% dissociation occurs in aqueous solution.

7. A buffered solution is one that resists a change in its pH even when a strong acid or base is added to it. A buffered solution contains a weak acid and its conjugate base.

Active Learning Questions

These questions are designed to be considered by groups of students in class. Often these questions work well for introducing a particular topic in class.

1. You are asked for the H^+ concentration in a solution of NaOH(*aq*). Because sodium hydroxide is a strong base, can we say there is no H^+, since having H^+ would imply that the solution is acidic?

2. Explain why Cl^- does not affect the pH of an aqueous solution.

3. Write the general reaction for an acid acting in water. What is the base in this case? The conjugate acid? The conjugate base?

4. Differentiate among the terms *concentrated, dilute, weak,* and *strong* in describing acids. Use molecular-level pictures to support your answer.

5. What is meant by "pH"? True or false: A strong acid always has a lower pH than a weak acid does. Explain.

6. Consider two separate solutions: one containing a weak acid, HA, and one containing HCl. Assume that you start with 10 molecules of each.

   a. Draw a molecular-level picture of what each solution looks like.
   b. Arrange the following from strongest to weakest base: Cl^-, H_2O, A^-. Explain.

7. Why is the pH of water at 25 °C equal to 7.00?

8. Can the pH of a solution be negative? Explain.

9. Stanley's grade-point average (GPA) is 3.28. What is Stanley's p(GPA)?

10. A friend asks the following: "Consider a buffered solution made up of the weak acid HA and its salt NaA. If a strong base like NaOH is added, the HA reacts

with the OH^- to make A^-. Thus, the amount of acid (HA) is decreased, and the amount of base (A^-) is increased. Analogously, adding HCl to the buffered solution forms more of the acid (HA) by reacting with the base (A^-). How can we claim that a buffered solution resists changes in the pH of the solution?" How would you explain buffering to your friend?

11. Mixing together aqueous solutions of acetic acid and sodium hydroxide can make a buffered solution. Explain.

12. Could a buffered solution be made by mixing aqueous solutions of HCl and NaOH? Explain.

13. Consider the equation: $HA(aq) + H_2O \rightleftharpoons H_3O^+(aq) + A^-(aq)$.

   a. If water is a better base than A^-, which way will equilibrium lie?
   b. If water is a better base than A^-, does this mean that HA is a strong or a weak acid?
   c. If water is a better base than A^-, is the value for K_a greater or less than 1?

14. Choose the answer that best completes the following statement and defend your answer. When 100.0 mL of water is added to 100.0 mL of 1.00 *M* HCl,

   a. the pH decreases because the solution is diluted.
   b. the pH does not change because water is neutral.
   c. the pH is doubled because the volume is now doubled.
   d. the pH increases because the concentration of H^+ decreases.
   e. the solution is completely neutralized.

15. You mix a solution of a strong acid with a pH of 4 and an equal volume of a strong acid solution with a pH of 6. Is the final pH less than 4, between 4 and 5, 5, between 5 and 6, or greater than 6? Explain.

16. The following figures are molecular-level representations of acid solutions. Label each as a strong acid or a weak acid.

17. Answer the following questions concerning buffered solutions.

   a. Explain what a buffered solution does.
   b. Describe the substances that make up a buffered solution.
   c. Explain how a buffered solution works.

OWL Questions and problems for this chapter are available in OWL. Use the OWL access card that came with this textbook to access assigned questions and problems for this chapter.

17 Equilibrium

● Equilibrium can be analogous to traffic flowing both ways on a bridge, such as San Francisco's Golden Gate Bridge. *(James Martin/Stone/Getty Images)*

not just any collision successful

Chemistry is mostly about reactions—processes in which groups of atoms are reorganized. So far we have learned to describe chemical reactions by using balanced equations and to calculate amounts of reactants and products. However, there are many important characteristics of reactions that we have not yet considered.

Refrigeration prevents food spoilage.

For example, why do refrigerators prevent food from spoiling? That is, why do the chemical reactions that cause food to decompose occur more slowly at lower temperatures? On the other hand, how can a chemical manufacturer speed up a chemical reaction that runs too slowly to be economical?

Another question that arises is why chemical reactions carried out in a closed vessel appear to stop at a certain point. For example, when the reaction of reddish-brown nitrogen dioxide to form colorless dinitrogen tetroxide,

$$2NO_2(g) \quad \rightarrow \quad N_2O_4(g)$$
Reddish-brown   Colorless

is carried out in a closed container, the reddish-brown color at first fades but stops changing after a time and then stays the same color indefinitely if left undisturbed (see Figure 17.1). We will account for all of these important observations about reactions in this chapter.

17.1   How Chemical Reactions Occur

OBJECTIVE: To understand the collision model of how chemical reactions occur.

In writing the equation for a chemical reaction, we put the reactants on the left and the products on the right with an arrow between them. But how do the atoms in the reactants reorganize to form the products?

Figure 17.1

a *A sample containing a large quantity of reddish-brown NO₂ gas.*

b *As the reaction to form colorless N₂O₄ occurs, the sample becomes lighter brown.*

c *After equilibrium is reached [2NO₂(g) ⇌ N₂O₄(g)], the color remains the same.*

a	b	c	d
Two BrNO molecules approach each other at high speeds.	The collision occurs.	The energy of the collision causes the Br—N bonds to break and allows the Br—Br bond to form.	The products: one Br_2 and two NO molecules.

Figure 17.2

Visualizing the reaction $2BrNO(g) \rightarrow 2NO(g) + Br_2(g)$.

Chemists believe that molecules react by colliding with each other. Some collisions are violent enough to break bonds, allowing the reactants to rearrange to form the products. For example, consider the reaction

$$2BrNO(g) \rightleftharpoons 2NO(g) + Br_2(g)$$

which we picture to occur as shown in Figure 17.2. Notice that the Br—N bonds in the two BrNO molecules must be broken and a new Br—Br bond must be formed during a collision for the reactants to become products.

The idea that reactions occur during molecular collisions, which is called the **collision model,** explains many characteristics of chemical reactions. For example, it explains why a reaction proceeds faster if the concentrations of the reacting molecules are increased (higher concentrations lead to more collisions and therefore to more reaction events). The collision model also explains why reactions go faster at higher temperatures, as we will see in the next section.

17.2 Conditions That Affect Reaction Rates

OBJECTIVES: To understand activation energy. • To understand how a catalyst speeds up a reaction.

It is easy to see why reactions speed up when the *concentrations* of reacting molecules are increased: higher concentrations (more molecules per unit volume) lead to more collisions and so to more reaction events. But reactions also speed up when the *temperature* is increased. Why? The answer lies in the fact that not all collisions possess enough energy to break bonds. A minimum energy called the **activation energy (E_a)** is needed for a reaction to occur (see Figure 17.3). If a given collision possesses an energy greater than E_a, that collision can result in a reaction. If a collision has an energy less than E_a, the molecules will bounce apart unchanged.

> Recall from Section 13.8 that the average kinetic energy of a collection of molecules is directly proportional to the temperature (K).

The reason that a reaction occurs faster as the temperature is increased is that the speeds of the molecules increase with temperature. So at higher temperatures, the average collision is more energetic. This makes it more likely that a given collision will possess enough energy to break bonds and to produce the molecular rearrangements needed for a reaction to occur.

Higher temperatures	Higher speeds	More high-energy collisions	More collisions that break bonds	Faster reaction

Figure 17.3

When molecules collide, a certain minimum energy called the activation energy (E_a) is needed for a reaction to occur. If the energy contained in a collision of two BrNO molecules is greater than E_a, the reaction can go "over the hump" to form products. If the collision energy is less than E_a, the colliding molecules bounce apart unchanged.

transition state = maximum energy along reaction pathway

Although O atoms are too reactive to exist near the earth's surface, they do exist in the upper atmosphere.

Cutaways of catalytic converters used in automobiles.

Is it possible to speed up a reaction without changing the temperature or the reactant concentrations? Yes, by using something called a **catalyst,** *a substance that speeds up a reaction without being consumed.* This may sound too good to be true, but it is a very common occurrence. In fact, you would not be alive now if your body did not contain thousands of catalysts called **enzymes.** Enzymes allow our bodies to speed up complicated reactions that would be too slow to sustain life at normal body temperatures. For example, the enzyme carbonic anhydrase speeds up the reaction between carbon dioxide and water

$$CO_2(g) + H_2O(l) \rightleftharpoons H^+(aq) + HCO_3^-(aq)$$

to help prevent an excess accumulation of carbon dioxide in our blood.

Although we cannot consider the details here, a catalyst works because it provides a new pathway for the reaction—a pathway that has a lower activation energy than the original pathway, as illustrated in Figure 17.4. Because of the lower activation energy, more collisions will have enough energy to allow a reaction. This in turn leads to a faster reaction.

A very important example of a reaction involving a catalyst occurs in our atmosphere; it is the breakdown of ozone, O_3, catalyzed by chlorine atoms. Ozone is one constituent of the earth's upper atmosphere that is especially crucial, because it absorbs harmful high-energy radiation from the sun. There are natural processes that result in both the formation and the destruction of ozone in the upper atmosphere. The natural balance of all these opposing processes has resulted in an amount of ozone that has been relatively constant over the years. However, the ozone level now seems to be decreasing, especially over Antarctica (Figure 17.5), apparently because chlorine atoms act as catalysts for the decomposition of ozone to oxygen by the following pair of reactions:

$$\begin{array}{l} Cl + O_3 \rightarrow ClO + O_2 \\ \underline{O + ClO \rightarrow Cl + O_2} \\ \text{Sum: } Cl + O_3 + O + ClO \rightarrow ClO + O_2 + Cl + O_2 \end{array}$$

When species that appear on both sides of the equation are canceled, the end result is the reaction

$$O + O_3 \rightarrow 2O_2$$

Figure 17.4

Comparison of the activation energies for an uncatalyzed reaction (E_a) and for the same reaction with a catalyst present (E_a'). Note that a catalyst works by lowering the activation energy for a reaction.

Protecting the Ozone

Chlorofluorocarbons (CFCs) are ideal compounds for refrigerators and air conditioners because they are nontoxic and noncorrosive. However, the chemical inertness of these substances, once thought to be their major virtue, turns out to be their fatal flaw. When these compounds leak into the atmosphere, as they inevitably do, they are so unreactive they persist there for decades. Eventually these CFCs reach altitudes where ultraviolet light causes them to decompose, producing chlorine atoms that promote the destruction of the ozone in the stratosphere (see discussion in Section 17.2). Because of this problem, the world's industrialized nations have signed an agreement (called the Montreal Protocol) that banned CFCs in 1996 (with a 10-year grace period for developing nations).

Worldwide production of CFCs has already decreased to half of the 1986 level of 1.13 million metric tons. One strategy for replacing the CFCs has been to switch to similar compounds that contain carbon and hydrogen atoms substituted for chlorine atoms. For example, the U.S. appliance industry has switched from Freon-12 (CF_2Cl_2) to the compound CH_2FCH_3 (called HFC-134a) for home refrigerators, and most of the new cars and

Courtesy, Amana

An Amana refrigerator, one of many appliances that now use HFC-134a. This compound is replacing CFCs, which lead to the destruction of atmospheric ozone.

trucks sold in the United States have air conditioners that employ HFC-134a. Converting the 140 million autos currently on the road in the United States that use CF_2Cl_2 will pose a major headache, but experience suggests that replacement of Freon-12 with HFC-134a is less expensive than was originally feared. For example, Volvo Cars of North America estimates that a Volvo can be converted from Freon-12 to HFC-134a for around $300.

A related environmental issue involves replacing the halons for firefighting applications. In particular, scientists are seeking an effective replacement for CF_3Br (halon-1301), the nontoxic "magic gas" used to flood enclosed spaces such as offices, aircraft, race cars, and military tanks in case of fire. The compound CF_3I, which appears to have a lifetime in the atmosphere of only a few days, looks like a promising candidate but much more research on the toxicology and ozone-depleting properties of CF_3I will be required before it receives government approval as a halon substitute.

The chemical industry has responded amazingly fast to the ozone depletion emergency. It is encouraging that we can act rapidly when an environmental crisis occurs. Now we need to get better at keeping the environment at a higher priority as we plan for the future.

Notice that a chlorine atom is used up in the first reaction but a chlorine atom is formed again by the second reaction. Therefore, the amount of chlorine does not change as the overall process occurs. This means that the chlorine atom is a true catalyst: it participates in the process but is not consumed. Estimates show that *one chlorine atom can catalyze the destruction of about one million ozone molecules per second.*

The chlorine atoms that promote this damage to the ozone layer are present because of pollution. Specifically, they come from the decomposition of compounds called Freons, such as CF_2Cl_2, which have been widely used in refrigerators and air conditioners. The Freons have leaked into the at-

Figure 17.5

A photo showing the ozone "hole" over Antarctica.

mosphere, where they are decomposed by light to produce chlorine atoms and other substances. As a result, the manufacture of Freons was banned by agreement among the nations of the world as of the end of 1996. Substitute compounds are now being used in newly manufactured refrigerators and air conditioners.

17.3 The Equilibrium Condition

OBJECTIVE: To learn how equilibrium is established.

Equilibrium is a word that implies balance or steadiness. When we say that someone is maintaining his or her equilibrium, we are describing a state of balance among various opposing forces. The term is used in a similar but more specific way in chemistry. Chemists define **equilibrium** as the *exact balancing of two processes, one of which is the opposite of the other.*

We first encountered the concept of equilibrium in Section 14.4, when we described the way vapor pressure develops over a liquid in a closed container (see Figure 14.9). This equilibrium process is summarized in Figure 17.6. The *equilibrium state* occurs when the rate of evaporation exactly equals the rate of condensation.

So far in this textbook we have usually assumed that reactions proceed to completion—that is, until one of the reactants "runs out." Indeed, many reactions *do* proceed essentially to completion. For such reactions we can assume that the reactants are converted to products until the limiting reactant is completely consumed. On the other hand, there are many chemical reactions that "stop" far short of completion when they are allowed to take place in a closed container. An example is the reaction of nitrogen dioxide to form dinitrogen tetroxide.

$$NO_2(g) + NO_2(g) \rightarrow N_2O_4(g)$$
Reddish-brown   Colorless

The reactant NO_2 is a reddish-brown gas, and the product N_2O_4 is a colorless gas. Imagine an experiment where pure NO_2 is placed in an empty, sealed glass vessel at 25 °C. The initial dark brown color will decrease in intensity as the NO_2 is converted to colorless N_2O_4 (see Figure 17.1). However, even over a long period of time,

Evaporation

Condensation

a At first there is a net transfer of molecules from the liquid state to the vapor state.

b After a while, the amount of the substance in the vapor state becomes constant—both the pressure of the vapor and the level of the liquid stay the same.

c The equilibrium state is very dynamic. The vapor pressure and liquid level remain constant because exactly the same number of molecules escape the liquid as return to it.

Figure 17.6

The establishment of the equilibrium vapor pressure over a liquid in a closed container.

the contents of the reaction vessel do not become colorless. Instead, the intensity of the brown color eventually becomes constant, which means that the concentration of NO_2 is no longer changing. This simple observation is a clear indication that the reaction has "stopped" short of completion. In fact, the reaction has not stopped. Rather, the system has reached **chemical equilibrium,** *a dynamic state where the concentrations of all reactants and products remain constant.*

This situation is similar to the one where a liquid in a closed container develops a constant vapor pressure, except that in this case two opposite chemical reactions are involved. When pure NO_2 is first placed in the closed flask, there is no N_2O_4 present. As collisions between NO_2 molecules occur, N_2O_4 is formed and the concentration of N_2O_4 in the container increases. However, the reverse reaction can also occur. A given N_2O_4 molecule can decompose into two NO_2 molecules.

$$N_2O_4(g) \rightarrow NO_2(g) + NO_2(g)$$

That is, chemical reactions are *reversible;* they can occur in either direction. We usually indicate this fact by using double arrows.

$$2NO_2(g) \underset{\text{Reverse}}{\overset{\text{Forward}}{\rightleftharpoons}} N_2O_4(g)$$

In this case the double arrows mean that either two NO_2 molecules can combine to form an N_2O_4 molecule (the *forward* reaction) or an N_2O_4 molecule can decompose to give two NO_2 molecules (the *reverse* reaction).

Equilibrium is reached whether pure NO_2, pure N_2O_4, or a mixture of NO_2 and N_2O_4 is initially placed in a closed container. In any of these cases, conditions will eventually be reached in the container such that N_2O_4 is being formed and is decomposing at exactly the same rate. This leads to chemical equilibrium, a dynamic situation where the concentrations of reactants and products remain the same indefinitely, as long as the conditions are not changed.

Chemical Equilibrium: A Dynamic Condition

OBJECTIVE: To learn about the characteristics of chemical equilibrium.

> Equilibrium is a dynamic situation.

Because no changes occur in the concentrations of reactants or products in a reaction system at equilibrium, it may appear that everything has stopped. However, this is not the case. On the molecular level there is frantic activity. Equilibrium is not static but is a highly *dynamic* situation. Consider again the analogy between chemical equilibrium and two island cities connected by a single bridge. Suppose the traffic flow on the bridge is the same in both directions. It is obvious that there is motion (we can see the cars traveling across the bridge), but the number of cars in each city does not change because there is an equal flow of cars entering and leaving. The result is no *net* change in the number of cars in each of the two cities.

To see how this concept applies to chemical reactions, let's consider the reaction between steam and carbon monoxide in a closed vessel at a high temperature where the reaction takes place rapidly.

> A double arrow ($\rightleftharpoons$) is used to show that a reaction is occurring in both directions.

$$H_2O(g) + CO(g) \rightleftharpoons H_2(g) + CO_2(g)$$

Assume that the same number of moles of gaseous CO and gaseous H_2O are placed in a closed vessel and allowed to react (Figure 17.7a).

When CO and H_2O, the reactants, are mixed, they immediately begin to react to form the products, H_2 and CO_2. This leads to a decrease in the concentrations of the reactants, but the concentrations of the products, which were initially at zero, are increasing (Figure 17.7b). After a certain

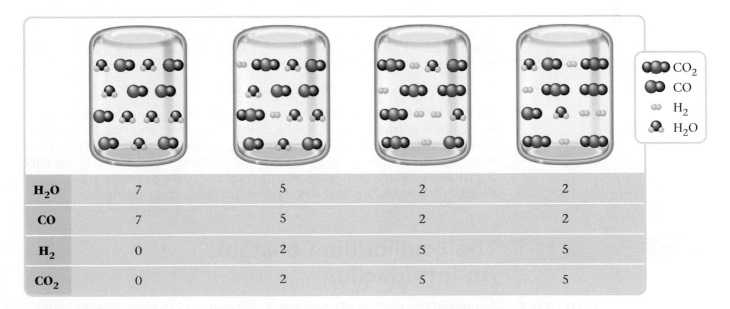

H₂O	7	5	2	2
CO	7	5	2	2
H₂	0	2	5	5
CO₂	0	2	5	5

Figure 17.7

The reaction of H_2O and CO to form CO_2 and H_2 as time passes. **a** Equal numbers of moles of H_2O and CO are mixed in a closed container. **b** The reaction begins to occur, and some products (H_2 and CO_2) are formed. **c** The reaction continues as time passes and more reactants are changed to products. **d** Although time continues to pass, the numbers of reactant and product molecules are the same as in **c**. No further changes are seen as time continues to pass. The system has reached equilibrium.

fastest at beginning
since greatest

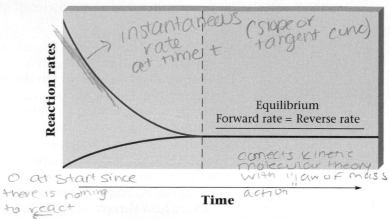

| Rate of forward reaction ($H_2O + CO \rightarrow$) | Rate of reverse reaction ($H_2 + CO_2 \rightarrow$) |

→ instantaneous rate at time t (slope or tangent curve)

Equilibrium
Forward rate = Reverse rate

connects kinetic molecular theory with "law of mass action"

0 at start since there is nothing to react

Reaction rates

Time

Figure 17.8

The changes with time in the rates of the forward and reverse reactions for $H_2O(g) + CO(g) \rightleftharpoons H_2(g) + CO_2(g)$ when equal numbers of moles of $H_2O(g)$ and $CO(g)$ are mixed. At first, the rate of the forward reaction decreases and the rate of the reverse reaction increases. Equilibrium is reached when the forward rate and the reverse rate become the same.

period of time, the concentrations of reactants and products no longer change at all—equilibrium has been reached (Figure 17.7c and d). Unless the system is somehow disturbed, no further changes in the concentrations will occur.

Why does equilibrium occur? We saw earlier in this chapter that molecules react by colliding with one another, and that the more collisions, the faster the reaction. This is why the speed of a reaction depends on concentrations. In this case the concentrations of H_2O and CO are lowered as the forward reaction occurs—that is, as products are formed.

$$H_2O + CO \rightarrow H_2 + CO_2$$

As the concentrations of the reactants decrease, the forward reaction slows down (Figure 17.8). But as in the traffic-on-the-bridge analogy, there is also movement in the reverse direction.

$$H_2 + CO_2 \rightarrow H_2O + CO$$

Initially in this experiment, no H_2 and CO_2 are present, so this reverse reaction cannot occur. However, as the forward reaction proceeds, the concentrations of H_2 and CO_2 build up and the speed (or rate) of the reverse reaction increases (Figure 17.8) as the forward reaction slows down. Eventually the concentrations reach levels at which *the rate of the forward reaction equals the rate of the reverse reaction*. The system has reached equilibrium.

17.5 The Equilibrium Constant: An Introduction

OBJECTIVE: To understand the law of chemical equilibrium and to learn how to calculate values for the equilibrium constant.

Science is based on the results of experiments. The development of the equilibrium concept is typical. On the basis of their observations of many chemical reactions, two Norwegian chemists, Cato Maximilian Guldberg and Peter Waage, proposed in 1864 the **law of chemical equilibrium**

go Chemistry **Module 21: Chemical Equilibrium** covers concepts in this section.

(originally called the *law of mass action*) as a general description of the equilibrium condition. Guldberg and Waage postulated that for a reaction of the type

$$aA + bB \rightleftharpoons cC + dD$$

where A, B, C, and D represent chemical species and *a, b, c,* and *d* are their coefficients in the balanced equation, the law of mass action is represented by the following **equilibrium expression:**

$$K = \frac{[C]^c[D]^d}{[A]^a[B]^b}$$

Square brackets, [ ], indicate concentration units of mol/L.

The square brackets indicate the concentrations of the chemical species *at equilibrium* (in units of mol/L), and *K* is a constant called the **equilibrium constant.** Note that the equilibrium expression is a special ratio of the concentrations of the products to the concentrations of the reactants. Each concentration is raised to a power corresponding to its coefficient in the balanced equation.

The law of chemical equilibrium as proposed by Guldberg and Waage is based on experimental observations. Experiments on many reactions showed that the equilibrium condition could always be described by this special ratio, called the equilibrium expression.

To see how to construct an equilibrium expression, consider the reaction where ozone changes to oxygen:

$$2O_3(g) \rightleftharpoons 3O_2(g)$$

To obtain the equilibrium expression, we place the concentration of the product in the numerator and the concentration of the reactant in the denominator.

$$\frac{[O_2]}{[O_3]} \leftarrow \text{Product} \atop \leftarrow \text{Reactant}$$

Then we use the coefficients as powers.

$$K = \frac{[O_2]^3}{[O_3]^2} \quad \text{Coefficients become powers}$$

EXAMPLE 17.1   **Writing Equilibrium Expressions**

Write the equilibrium expression for the following reactions.

   a. $H_2(g) + F_2(g) \rightleftharpoons 2HF(g)$ b. $N_2(g) + 3H_2(g) \rightleftharpoons 2NH_3(g)$

SOLUTION

Applying the law of chemical equilibrium, we place products over reactants (using square brackets to denote concentrations in units of moles per liter) and raise each concentration to the power that corresponds to the coefficient in the balanced chemical equation.

   a. $K = \dfrac{[HF]^2}{[H_2][F_2]}$ ← Product (coefficient of 2 becomes power of 2) ← Reactants (coefficients of 1 become powers of 1)

   Note that when a coefficient (power) of 1 occurs, it is not written but is understood.

   b. $K = \dfrac{[NH_3]^2}{[N_2][H_2]^3}$

Self-Check EXERCISE 17.1   Write the equilibrium expression for the following reaction.

$$4NH_3(g) + 7O_2(g) \rightleftharpoons 4NO_2(g) + 6H_2O(g)$$

See Problems 17.15 through 17.18. ■

What does the equilibrium expression mean? It means that, for a given reaction at a given temperature, the special ratio of the concentrations of the products to reactants defined by the equilibrium expression will always be equal to the same number—namely, the equilibrium constant K. For example, consider a series of experiments on the ammonia synthesis reaction

$$N_2(g) + 3H_2(g) \rightleftharpoons 2NH_3(g)$$

carried out at 500 °C to measure the concentrations of N_2, H_2, and NH_3 present at equilibrium. The results of these experiments are shown in Table 17.1. In this table, subscript zeros next to square brackets are used to indicate *initial concentrations*: the concentrations of reactants and products originally mixed together before any reaction has occurred.

Consider the results of experiment I. One mole each of N_2 and H_2 were sealed into a 1-L vessel at 500 °C and allowed to reach chemical equilibrium. At equilibrium the concentrations in the flask were found to be $[N_2] = 0.921\ M$, $[H_2] = 0.763\ M$, and $[NH_3] = 0.157\ M$. The equilibrium expression for the reaction

$$N_2(g) + 3H_2(g) \rightleftharpoons 2NH_3(g)$$

is

$$K = \frac{[NH_3]^2}{[N_2][H_2]^3} = \frac{(0.157)^2}{(0.921)(0.763)^3}$$

$$= 0.0602 = 6.02 \times 10^{-2}$$

Similarly, as shown in Table 17.1, we can calculate for experiments II and III that K, the equilibrium constant, has the value 6.02×10^{-2}. In fact, whenever N_2, H_2, and NH_3 are mixed together at this temperature, the system *always* comes to an equilibrium position such that

$$K = 6.02 \times 10^{-2}$$

regardless of the amounts of the reactants and products that are mixed together initially.

Table 17.1   Results of Three Experiments for the Reaction $N_2(g) + 3H_2(g) \rightleftharpoons 2NH_3(g)$ at 500 °C

Experiment	Initial Concentrations			Equilibrium Concentrations			$\dfrac{[NH_3]^2}{[N_2][H_2]^3} = K^*$
	$[N_2]_0$	$[H_2]_0$	$[NH_3]_0$	$[N_2]$	$[H_2]$	$[NH_3]$	
I	1.000 M	1.000 M	0	0.921 M	0.763 M	0.157 M	$\dfrac{(0.157)^2}{(0.921)(0.763)^3} = 0.0602$
II	0	0	1.000 M	0.399 M	1.197 M	0.203 M	$\dfrac{(0.203)^2}{(0.399)(1.197)^3} = 0.0602$
III	2.00 M	1.00 M	3.00 M	2.59 M	2.77 M	1.82 M	$\dfrac{(1.82)^2}{(2.59)(2.77)^3} = 0.0602$

*The units for K are customarily omitted.

It is important to see from Table 17.1 that the *equilibrium concentrations are not always the same.* However, even though the individual sets of equilibrium concentrations are quite different for the different situations, the *equilibrium constant, which depends on the ratio of the concentrations, remains the same.*

Each *set of equilibrium concentrations* is called an **equilibrium position.** It is essential to distinguish between the equilibrium constant and the equilibrium positions for a given reaction system. There is only *one* equilibrium constant for a particular system at a particular temperature, but there are an *infinite* number of equilibrium positions. The specific equilibrium position adopted by a system depends on the initial concentrations; the equilibrium constant does not.

Note that in the above discussion, the equilibrium constant was given without units. In certain cases the units are included when the values of equilibrium constants are given, and in other cases they are omitted. We will not discuss the reasons for this. We will omit the units in this text.

> For a reaction at a given temperature, there are many equilibrium positions but only one value for K.

EXAMPLE 17.2   Calculating Equilibrium Constants

The reaction of sulfur dioxide with oxygen in the atmosphere to form sulfur trioxide has important environmental implications because SO_3 combines with moisture to form sulfuric acid droplets, an important component of acid rain. The following results were collected for two experiments involving the reaction at 600 °C between gaseous sulfur dioxide and oxygen to form gaseous sulfur trioxide:

$$2SO_2(g) + O_2(g) \rightleftharpoons 2SO_3(g)$$

	Initial	Equilibrium
Experiment I	$[SO_2]_0 = 2.00\ M$	$[SO_2] = 1.50\ M$
	$[O_2]_0 = 1.50\ M$	$[O_2] = 1.25\ M$
	$[SO_3]_0 = 3.00\ M$	$[SO_3] = 3.50\ M$
Experiment II	$[SO_2]_0 = 0.500\ M$	$[SO_2] = 0.590\ M$
	$[O_2]_0 = 0$	$[O_2] = 0.045\ M$
	$[SO_3]_0 = 0.350\ M$	$[SO_3] = 0.260\ M$

The law of chemical equilibrium predicts that the value of K should be the same for both experiments. Verify this by calculating the equilibrium constant observed for each experiment.

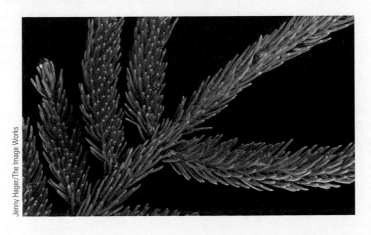

Spruce needles turned brown by acid rain.

Jenny Hager/The Image Works

SOLUTION

The balanced equation for the reaction is

$$2SO_2(g) + O_2(g) \rightleftharpoons 2SO_3(g)$$

From the law of chemical equilibrium, we can write the equilibrium expression

$$K = \frac{[SO_3]^2}{[SO_2]^2[O_2]}$$

For experiment I we calculate the value of K by substituting the observed *equilibrium* concentrations,

$$[SO_3] = 3.50\ M$$
$$[SO_2] = 1.50\ M$$
$$[O_2] = 1.25\ M$$

into the equilibrium expression:

$$K_I = \frac{(3.50)^2}{(1.50)^2(1.25)} = 4.36$$

For experiment II at equilibrium

$$[SO_3] = 0.260\ M$$
$$[SO_2] = 0.590\ M$$
$$[O_2] = 0.045\ M$$

and

$$K_{II} = \frac{(0.260)^2}{(0.590)^2(0.045)} = 4.32$$

Notice that the values calculated for K_I and K_{II} are nearly the same, as we expected. That is, the value of K is constant, within differences due to rounding off and due to experimental error. These experiments show *two different equilibrium positions* for this system, but K, the equilibrium constant, is, indeed, constant. ■

17.6 Heterogeneous Equilibria

OBJECTIVE: To understand the role that liquids and solids play in constructing the equilibrium expression.

So far we have discussed equilibria only for systems in the gaseous state, where all reactants and products are gases. These are examples of **homogeneous equilibria,** in which all substances are in the same state. However, many equilibria involve more than one state and are called **heterogeneous equilibria.** For example, the thermal decomposition of calcium carbonate in the commercial preparation of lime occurs by a reaction involving solids and gases.

$$CaCO_3(s) \rightleftharpoons CaO(s) + CO_2(g)$$
Lime

Straightforward application of the law of equilibrium leads to the equilibrium expression

$$K = \frac{[CO_2][CaO]}{[CaCO_3]}$$

In terms of amount produced, lime is among the top ten chemicals manufactured in the United States.

However, experimental results show that the *position of a heterogeneous equilibrium does not depend on the amounts of pure solids or liquids present.* The fundamental reason for this behavior is that the concentrations of pure solids and liquids cannot change. In other words, we might say that the concentrations of pure solids and liquids are constants. Therefore, we can write the equilibrium expression for the decomposition of solid calcium carbonate as

$$K' = \frac{[CO_2]C_1}{C_2}$$

where C_1 and C_2 are constants representing the concentrations of the solids CaO and $CaCO_3$, respectively. This expression can be rearranged to give

$$\frac{C_2 K'}{C_1} = K = [CO_2]$$

where the constants C_2, K', and C_1 are combined into a single constant K. This leads us to the following general statement: the concentrations of pure solids or pure liquids involved in a chemical reaction *are not included in the equilibrium expression* for the reaction. This applies *only* to pure solids or liquids. It does not apply to solutions or gases, because their concentrations can vary.

For example, consider the decomposition of liquid water to gaseous hydrogen and oxygen:

$$2H_2O(l) \rightleftharpoons 2H_2(g) + O_2(g)$$

where

$$K = [H_2]^2[O_2]$$

Water is not included in the equilibrium expression because it is a pure liquid. However, when the reaction is carried out under conditions where the water is a gas rather than a liquid,

$$2H_2O(g) \rightleftharpoons 2H_2(g) + O_2(g)$$

we have

$$K = \frac{[H_2]^2[O_2]}{[H_2O]^2}$$

because the concentration of water vapor can change.

EXAMPLE 17.3

Writing Equilibrium Expressions for Heterogeneous Equilibria

Write the expressions for K for the following processes.

a. Solid phosphorus pentachloride is decomposed to liquid phosphorus trichloride and chlorine gas.

b. Deep-blue solid copper(II) sulfate pentahydrate is heated to drive off water vapor and form white solid copper(II) sulfate.

SOLUTION

a. The reaction is

$$PCl_5(s) \rightleftharpoons PCl_3(l) + Cl_2(g)$$

In this case, neither the pure solid PCl_5 nor the pure liquid PCl_3 is included in the equilibrium expression. The equilibrium expression is

$$K = [Cl_2]$$

> The concentrations of pure liquids and solids are constant.

As solid copper(II) sulfate pentahydrate, $CuSO_4 \cdot 5H_2O$, is heated, it loses H_2O, eventually forming white $CuSO_4$.

© Cengage Learning

b. The reaction is

$$CuSO_4 \cdot 5H_2O(s) \rightleftharpoons CuSO_4(s) + 5H_2O(g)$$

The two solids are not included. The equilibrium expression is

$$K = [H_2O]^5$$

Self-Check EXERCISE 17.2 Write the equilibrium expression for each of the following reactions.

a. $2KClO_3(s) \rightleftharpoons 2KCl(s) + 3O_2(g)$

This reaction is often used to produce oxygen gas in the laboratory.

b. $NH_4NO_3(s) \rightleftharpoons N_2O(g) + 2H_2O(g)$

c. $CO_2(g) + MgO(s) \rightleftharpoons MgCO_3(s)$

d. $SO_3(g) + H_2O(l) \rightleftharpoons H_2SO_4(l)$

See Problems 17.25 through 17.28. ■

17.7 Le Châtelier's Principle

OBJECTIVE: To learn to predict the changes that occur when a system at equilibrium is disturbed.

It is important to understand the factors that control the *position* of a chemical equilibrium. For example, when a chemical is manufactured, the chemists and chemical engineers in charge of production want to choose conditions that favor the desired product as much as possible. That is, they want the equilibrium to lie far to the right (toward products). When the process for the synthesis of ammonia was being developed, extensive studies were carried out to determine how the equilibrium concentration of ammonia depended on the conditions of temperature and pressure.

In this section we will explore how various changes in conditions affect the equilibrium position of a reaction system. We can predict the effects of changes in concentration, pressure, and temperature on a system at equilibrium by using **Le Châtelier's principle,** which states that when *a change is imposed on a system at equilibrium, the position of the equilibrium shifts in a direction that tends to reduce the effect of that change.*

▶ The Effect of a Change in Concentration

Let us consider the ammonia synthesis reaction. Suppose there is an equilibrium position described by these concentrations:

$$[N_2] = 0.399\ M \qquad [H_2] = 1.197\ M \qquad [NH_3] = 0.203\ M$$

What will happen if 1.000 mol/L of N_2 is suddenly injected into the system? We can begin to answer this question by remembering that for the system at equilibrium, the rates of the forward and reverse reactions exactly balance,

$$N_2(g) + 3H_2(g) \rightleftharpoons 2NH_3(g)$$

as indicated here by arrows of the same length. When N_2 is added, there are suddenly more collisions between N_2 and H_2 molecules. This increases the rate of the forward reaction (shown here by the greater length of the arrow pointing in that direction),

$$N_2(g) + 3H_2(g) \rightleftharpoons 2NH_3(g)$$

and the reaction produces more NH_3. As the concentration of NH_3 increases, the reverse reaction also speeds up (as more collisions between NH_3 molecules occur) and the system again comes to equilibrium. However, the new equilibrium position has more NH_3 than was present in the original position. We say that the equilibrium has shifted to the *right*—toward the products. The original and new equilibrium positions are shown below.

Equilibrium Position I		Equilibrium Position II
$[N_2] = 0.399\ M$		$[N_2] = 1.348\ M$
$[H_2] = 1.197\ M$	1.000 mol/L of N_2 added	$[H_2] = 1.044\ M$
$[NH_3] = 0.203\ M$		$[NH_3] = 0.304\ M$

Note that the equilibrium does, in fact, shift to the right; the concentration of H_2 decreases (from 1.197 *M* to 1.044 *M*), the concentration of NH_3 increases (from 0.203 *M* to 0.304 *M*), and, of course, because nitrogen was added, the concentration of N_2 shows an increase relative to the original amount present.

It is important to note at this point that, although the equilibrium shifted to a new position, the *value of K did not change.* We can demonstrate this by inserting the equilibrium concentrations from positions I and II into the equilibrium expression.

- Position I: $K = \dfrac{[NH_3]^2}{[N_2][H_2]^3} = \dfrac{(0.203)^2}{(0.399)(1.197)^3} = 0.0602$

- Position II: $K = \dfrac{[NH_3]^2}{[N_2][H_2]^3} = \dfrac{(0.304)^2}{(1.348)(1.044)^3} = 0.0602$

These values of *K* are the same. Therefore, although the equilibrium *position* shifted when we added more N_2, the *equilibrium constant K* remained the same.

Could we have predicted this shift by using Le Châtelier's principle? Because the change in this case was to add nitrogen, Le Châtelier's principle

N₂ added

●● N₂ ⁴/₃

●● H₂ /₃

●● NH₃ ³

a The initial equilibrium mixture of N₂, H₂, and NH₃.

b Addition of N₂.
Stress

c The new equilibrium position for the system containing more N₂ (because of the addition of N₂), less H₂, and NH₃ than in **a**.
More

Figure 17.9

predicts that the system will shift in a direction that *consumes* nitrogen. This tends to offset the original change—the addition of N_2. Therefore, Le Châtelier's principle correctly predicts that adding nitrogen will cause the equilibrium to shift to the right (Figure 17.9) as some of the added nitrogen is consumed.

If ammonia had been added instead of nitrogen, the system would have shifted to the left, consuming ammonia. Another way of stating Le Châtelier's principle, then, is to say that *when a reactant or product is added to a system at equilibrium, the system shifts away from the added component.* On the other hand, if *a reactant or product is removed, the system shifts toward the removed component.* For example, if we had removed nitrogen, the system would have shifted to the left and the amount of ammonia present would have been reduced.

A real-life example that shows the importance of Le Châtelier's principle is the effect of high elevations on the oxygen supply to the body. If you have ever traveled to the mountains on vacation, you may have noticed that you felt "light-headed" and especially tired during the first few days of your visit. These feelings resulted from a decreased supply of oxygen to your body because of the lower air pressure that exists at higher elevations. For example, the oxygen supply in Leadville, Colorado (elevation ~ 10,000 ft), is only about two-thirds that found at sea level. We can understand the effects of diminished oxygen supply in terms of the following equilibrium:

$$Hb(aq) + 4O_2(g) \rightleftharpoons Hb(O_2)_4(aq)$$

where Hb represents hemoglobin, the iron-containing protein that transports O_2 from your lungs to your tissues, where it is used to support metabolism. The coefficient 4 in the equation signifies that each hemoglobin molecule picks up four O_2 molecules in the lungs. Note by Le Châtelier's principle that a lower oxygen pressure will cause this equilibrium to shift to the left, away from oxygenated hemoglobin. This leads to an inadequate oxygen supply at the tissues, which in turn results in fatigue and a "woozy" feeling.

This problem can be solved in extreme cases, such as when climbing Mt. Everest or flying in a plane at high altitudes, by supplying extra oxygen from a tank. This extra oxygen pushes the equilibrium to its normal position. However, lugging around an oxygen tank would not be very practical for people who live in the mountains. In fact, nature solves this problem in a very interesting way. The body adapts to living at high elevations by

> A system at equilibrium shifts in the direction that compensates for any imposed change.

producing additional hemoglobin—the other way to shift this equilibrium to the right. Thus, people who live at high elevations have significantly higher hemoglobin levels than those living at sea level. For example, the Sherpas who live in Nepal can function in the rarefied air at the top of Mt. Everest without an auxiliary oxygen supply.

EXAMPLE 17.4 | Using Le Châtelier's Principle: Changes in Concentration

Arsenic, As_4, is obtained from nature by first reacting its ore with oxygen (called *roasting*) to form solid As_4O_6. (As_4O_6, a toxic compound fatal in doses of 0.1 g or more, is the "arsenic" made famous in detective stories.) The As_4O_6 is then reduced using carbon:

$$As_4O_6(s) + 6C(s) \rightleftharpoons As_4(g) + 6CO(g)$$

Predict the direction of the shift in the equilibrium position for this reaction that occurs in response to each of the following changes in conditions.

a. Addition of carbon monoxide

b. Addition or removal of C(s) or $As_4O_6(s)$

c. Removal of $As_4(g)$

SOLUTION

a. Le Châtelier's principle predicts a shift away from the substance whose concentration is increased. The equilibrium position will shift to the left when carbon monoxide is added.

b. Because the amount of a pure solid has no effect on the equilibrium position, changing the amount of carbon or tetraarsenic hexoxide will have no effect.

c. When gaseous arsenic is removed, the equilibrium position will shift to the right to form more products. In industrial processes, the desired product is often continuously removed from the reaction system to increase the yield.

Self-Check EXERCISE 17.3 Novelty devices for predicting rain contain cobalt(II) chloride and are based on the following equilibrium:

$$\underset{\text{Blue}}{CoCl_2(s)} + 6H_2O(g) \rightleftharpoons \underset{\text{Pink}}{CoCl_2 \cdot 6H_2O(s)}$$

What color will this indicator be when rain is likely due to increased water vapor in the air?

See Problems 17.33 through 17.36. ■

When blue anhydrous $CoCl_2$ reacts with water, pink $CoCl_2 \cdot 6H_2O$ is formed.

© Cengage Learning

▶ The Effect of a Change in Volume

When the volume of a gas is decreased (when a gas is compressed), the pressure increases. This occurs because the molecules present are now contained in a smaller space and they hit the walls of their container more often, giving a greater pressure. Therefore, when the volume of a gaseous reaction system at equilibrium is suddenly reduced, leading to a sudden increase in pressure, by Le Châtelier's principle the system will shift in the direction that reduces the pressure.

For example, consider the reaction

$$CaCO_3(s) \rightleftharpoons CaO(s) + CO_2(g)$$

in a container with a movable piston (Figure 17.10). If the volume is suddenly decreased by pushing in the piston, the pressure of the CO_2 gas initially increases. How can the system offset this pressure increase? By shifting to the left—the direction that reduces the amount of gas present. That is, a shift to the left will use up CO_2 molecules, thus lowering the pressure. (There will then be fewer molecules present to hit the walls, because more of the CO_2 molecules have combined with CaO and thus have become part of the solid $CaCO_3$.)

Therefore, when the volume of a gaseous reaction system at equilibrium is decreased (thus increasing the pressure), *the system shifts in the direction that gives the smaller number of gas molecules*. So a decrease in the system volume leads to a shift that decreases the total number of gaseous molecules in the system.

Suppose we are running the reaction

$$N_2(g) + 3H_2(g) \rightleftharpoons 2NH_3(g)$$

and we have a mixture of the gases nitrogen, hydrogen, and ammonia at equilibrium (Figure 17.11a). If we suddenly reduce the volume, what will happen to the equilibrium position? Because the decrease in volume initially increases the pressure, the system moves in the direction that lowers its pressure. The reaction system can reduce its pressure by reducing the number of gas molecules present. This means that the reaction

$$\begin{array}{ccc} N_2(g) + 3H_2(g) & \rightleftharpoons & 2NH_3(g) \\ \text{4 gaseous molecules} & & \text{2 gaseous molecules} \end{array}$$

shifts to the right, because in this direction four molecules (one of nitrogen and three of hydrogen) react to produce two molecules (of ammonia), thus *reducing the total number of gaseous molecules present*. The equilibrium position shifts to the right—toward the side of the reaction that involves the smaller number of gaseous molecules in the balanced equation.

The opposite is also true. When the container volume is increased (which lowers the pressure of the system), the system shifts so as to increase its pressure. An increase in volume in the ammonia synthesis system produces a shift to the left to increase the total number of gaseous molecules present (to increase the pressure).

heterogeneous equilibrium

*response
(absorbs $CO_2(g)$)*

— CO_2

$CaCO_3$ —   — CaO

a The system is initially at equilibrium.

b Then the piston is pushed in, decreasing the volume and increasing the pressure. The system shifts in the direction that consumes CO_2 molecules, thus lowering the pressure again.

Le Châtlier's Principle Applied

Figure 17.10

The reaction system $CaCO_3(s) \rightleftharpoons CaO(s) + CO_2(g)$.

changing amounts of pure solids will not change equilibrium $[CO_2]$ as long as some solid remains

→ increase (P)

Changing amount of NaCl (s) at bottom of beaker does not change $[Na]^+_{sq}$ $[Cl^-]_{sq}$

Pg 470 ✗

*saturated NaCl (aq)
solid NaCl*

	N$_2$
	H$_2$
	NH$_3$

a A mixture of NH$_3$(g), N$_2$(g), and H$_2$(g) at equilibrium.

b The volume is suddenly decreased.

c The new equilibrium position for the system containing more NH$_3$ and less N$_2$ and H$_2$. The reaction N$_2$(g) + 3H$_2$(g) $\rightleftharpoons$ 2NH$_3$(g) shifts to the right (toward the side with fewer molecules) when the container volume is decreased.

Figure 17.11

EXAMPLE 17.5 **Using Le Châtelier's Principle: Changes in Volume**

Predict the shift in equilibrium position that will occur for each of the following processes when the volume is reduced.

    a. The preparation of liquid phosphorus trichloride by the reaction

$$P_4(s) + 6Cl_2(g) \;\rightleftharpoons\; 4PCl_3(l)$$

        6 gaseous molecules   0 gaseous molecules

SOLUTION a

P$_4$ and PCl$_3$ are a pure solid and a pure liquid, respectively, so we need to consider only the effect on Cl$_2$. If the volume is decreased, the Cl$_2$ pressure will initially increase, so the position of the equilibrium will shift to the right, consuming gaseous Cl$_2$ and lowering the pressure (to counteract the original change).

    b. The preparation of gaseous phosphorus pentachloride according to the equation

$$PCl_3(g) + Cl_2(g) \;\rightleftharpoons\; PCl_5(g)$$

        2 gaseous molecules   1 gaseous molecule

SOLUTION b

Decreasing the volume (increasing the pressure) will shift this equilibrium to the right, because the product side contains only one gaseous molecule while the reactant side has two. That is, the system will respond to the decreased volume (increased pressure) by lowering the number of molecules present.

    c. The reaction of phosphorus trichloride with ammonia:

$$PCl_3(g) + 3NH_3(g) \rightleftharpoons P(NH_2)_3(g) + 3HCl(g)$$

SOLUTION c
Both sides of the balanced reaction equation have four gaseous molecules. A change in volume will have no effect on the equilibrium position. There is no shift in this case, because the system cannot change the number of molecules present by shifting in either direction.

Self-Check EXERCISE 17.4   For each of the following reactions, predict the direction the equilibrium will shift when the volume of the container is increased.

   a. $H_2(g) + F_2(g) \rightleftharpoons 2HF(g)$

   b. $CO(g) + 2H_2(g) \rightleftharpoons CH_3OH(g)$

   c. $2SO_3(g) \rightleftharpoons 2SO_2(g) + O_2(g)$

See Problems 17.33 through 17.36. ■

▶ The Effect of a Change in Temperature

It is important to remember that although the changes we have just discussed may alter the equilibrium *position,* they do not alter the equilibrium *constant.* For example, the addition of a reactant shifts the equilibrium position to the right but has no effect on the value of the equilibrium constant; the new equilibrium concentrations satisfy the original equilibrium constant. This was demonstrated earlier in this section for the addition of N_2 to the ammonia synthesis reaction.

The effect of temperature on equilibrium is different, however, because *the value of K changes with temperature.* We can use Le Châtelier's principle to predict the direction of the change in K.

To do this we need to classify reactions according to whether they produce heat or absorb heat. A reaction that produces heat (heat is a "product") is said to be *exothermic.* A reaction that absorbs heat is called *endothermic.* Because heat is needed for an endothermic reaction, energy (heat) can be regarded as a "reactant" in this case.

In an exothermic reaction, heat is treated as a product. For example, the synthesis of ammonia from nitrogen and hydrogen is exothermic (produces heat). We can represent this by treating energy as a product:

$$N_2(g) + 3H_2(g) \rightleftharpoons 2NH_3(g) + 92 \text{ kJ}$$

Energy
released

Le Châtelier's principle predicts that when we add energy to this system at equilibrium by heating it, the shift will be in the direction that consumes energy—that is, to the left.

On the other hand, for an endothermic reaction (one that absorbs energy), such as the decomposition of calcium carbonate,

$$CaCO_3(s) + 556 \text{ kJ} \rightleftharpoons CaO(s) + CO_2(g)$$

Energy
needed

energy is treated as a reactant. In this case an increase in temperature causes the equilibrium to shift to the right.

In summary, to use Le Châtelier's principle to describe the effect of a temperature change on a system at equilibrium, *simply treat energy as a reactant (in an endothermic process) or as a product (in an exothermic process),* and *predict the direction of the shift* in the same way you would if an actual reactant or product were being added or removed.

EXAMPLE 17.6 Using Le Châtelier's Principle: Changes in Temperature

For each of the following reactions, predict how the equilibrium will shift as the temperature is increased.

a. $N_2(g) + O_2(g) \rightleftharpoons 2NO(g)$ (endothermic)

SOLUTION a

This is an endothermic reaction, so energy can be viewed as a reactant.

$$N_2(g) + O_2(g) + energy \rightleftharpoons 2NO(g)$$

Thus the equilibrium will shift to the right as the temperature is increased (energy added).

b. $2SO_2(g) + O_2(g) \rightleftharpoons 2SO_3(g)$ (exothermic)

SOLUTION b

This is an exothermic reaction, so energy can be regarded as a product.

$$2SO_2(g) + O_2(g) \rightleftharpoons 2SO_3(g) + energy$$

As the temperature is increased, the equilibrium will shift to the left.

Self-Check EXERCISE 17.5 For the exothermic reaction

$$2SO_2(g) + O_2(g) \rightleftharpoons 2SO_3(g)$$

predict the equilibrium shift caused by each of the following changes.

a. SO_2 is added.

b. SO_3 is removed.

c. The volume is decreased.

d. The temperature is decreased.

See Problems 17.33 through 17.42. ■

We have seen how Le Châtelier's principle can be used to predict the effects of several types of changes on a system at equilibrium. To summarize these ideas, Table 17.2 shows how various changes affect the equilibrium position of the endothermic reaction $N_2O_4(g) \rightleftharpoons 2NO_2(g)$. The effect of a temperature change on this system is depicted in Figure 17.12.

heat reactant — endothermic

Table 17.2 Shifts in the Equilibrium Position for the Reaction $N_2O_4(g) + Energy \rightleftharpoons 2NO_2(g)$

addition of $N_2O_4(g)$	right
addition of $NO_2(g)$	left
removal of $N_2O_4(g)$	left
removal of $NO_2(g)$	right
decrease in container volume	left
increase in container volume	right
increase in temperature	right
decrease in temperature	left

heat product — exothermic

© Cengage Learning

a — *At 100 °C the flask is definitely reddish-brown due to a large amount of NO_2 present.*

b — *At 0 °C the equilibrium is shifted toward colorless $N_2O_4(g)$.*

Figure 17.12
Shifting the $N_2O_4(g) \rightleftharpoons 2NO_2(g)$ equilibrium by changing the temperature.

17.8 Applications Involving the Equilibrium Constant

OBJECTIVE: To learn to calculate equilibrium concentrations from equilibrium constants.

Knowing the value of the equilibrium constant for a reaction allows us to do many things. For example, the size of K tells us the inherent tendency of the reaction to occur. A value of K much larger than 1 means that at equilibrium, the reaction system will consist of mostly products—the equilibrium lies to the right. For example, consider a general reaction of the type

$$A(g) \rightarrow B(g)$$

where

$$K = \frac{[B]}{[A]}$$

If K for this reaction is 10,000 (10^4), then at equilibrium,

$$\frac{[B]}{[A]} = 10,000 \quad \text{or} \quad \frac{[B]}{[A]} = \frac{10,000}{1}$$

That is, at equilibrium [B] is 10,000 times greater than [A]. This means that the reaction strongly favors the product B. Another way of saying this is that the reaction goes essentially to completion. That is, virtually all of A becomes B.

On the other hand, a small value of K means that the system at equilibrium consists largely of reactants—the equilibrium position is far to the left. The given reaction does not occur to any significant extent.

Another way we use the equilibrium constant is to calculate the equilibrium concentrations of reactants and products. For example, if we know the value of K and the concentrations of all the reactants and products except one, we can calculate the missing concentration. This is illustrated in Example 17.7.

EXAMPLE 17.7 | Calculating Equilibrium Concentration Using Equilibrium Expressions

Gaseous phosphorus pentachloride decomposes to chlorine gas and gaseous phosphorus trichloride. In a certain experiment, at a temperature where

$K = 8.96 \times 10^{-2}$, the equilibrium concentrations of PCl_5 and PCl_3 were found to be 6.70×10^{-3} M and 0.300 M, respectively. Calculate the concentration of Cl_2 present at equilibrium.

SOLUTION

Where Are We Going?

We want to determine $[Cl_2]$ present at equilibrium.

What Do We Know?

For this reaction, the balanced equation is

$$PCl_5(g) \rightleftharpoons PCl_3(g) + Cl_2(g)$$

and the equilibrium expression is

$$K = \frac{[PCl_3][Cl_2]}{[PCl_5]} = 8.96 \times 10^{-2}$$

We know that

$$[PCl_5] = 6.70 \times 10^{-3} \ M$$
$$[PCl_3] = 0.300 \ M$$

How Do We Get There?

We want to calculate $[Cl_2]$. We will rearrange the equilibrium expression to solve for the concentration of Cl_2. First we divide both sides of the expression

$$K = \frac{[PCl_3][Cl_2]}{[PCl_5]}$$

by $[PCl_3]$ to give

$$\frac{K}{[PCl_3]} = \frac{\cancel{[PCl_3]}[Cl_2]}{\cancel{[PCl_3]}[PCl_5]} = \frac{[Cl_2]}{[PCl_5]}$$

Next we multiply both sides by $[PCl_5]$.

$$\frac{K[PCl_5]}{[PCl_3]} = \frac{[Cl_2]\cancel{[PCl_5]}}{\cancel{[PCl_5]}} = [Cl_2]$$

Then we can calculate $[Cl_2]$ by substituting the known information.

$$[Cl_2] = K \times \frac{[PCl_5]}{[PCl_3]} = (8.96 \times 10^{-2}) \frac{(6.70 \times 10^{-3})}{(0.300)}$$
$$[Cl_2] = 2.00 \times 10^{-3}$$

The equilibrium concentration of Cl_2 is 2.00×10^{-3} M. ∎

17.9 Solubility Equilibria

OBJECTIVE: To learn to calculate the solubility product of a salt, given its solubility, and vice versa.

Solubility is a very important phenomenon. Consider the following examples.

- Because sugar and table salt dissolve readily in water, we can flavor foods easily.

Toothpastes containing sodium fluoride, an additive that helps prevent tooth decay.

This X ray of the large intestine has been enhanced by the patient's consumption of barium sulfate.

Pure liquids and pure solids are never included in an equilibrium expression.

- Because calcium sulfate is less soluble in hot water than in cold water, it coats tubes in boilers, reducing thermal efficiency.

- When food lodges between teeth, acids form that dissolve tooth enamel, which contains the mineral hydroxyapatite, $Ca_5(PO_4)_3OH$. Tooth decay can be reduced by adding fluoride to toothpaste. Fluoride replaces the hydroxide in hydroxyapatite to produce the corresponding fluorapatite, $Ca_5(PO_4)_3F$, and calcium fluoride, CaF_2, both of which are less soluble in acids than the original enamel.

- The use of a suspension of barium sulfate improves the clarity of X rays of the digestive tract. Barium sulfate contains the toxic ion Ba^{2+}, but its very low solubility makes ingestion of solid $BaSO_4$ safe.

In this section we will consider the equilibria associated with dissolving solids in water to form aqueous solutions. When a typical ionic solid dissolves in water, it dissociates completely into separate cations and anions. For example, calcium fluoride dissolves in water as follows:

$$CaF_2(s) \xrightarrow{H_2O(l)} Ca^{2+}(aq) + 2F^-(aq)$$

When the solid salt is first added to the water, no Ca^{2+} and F^- ions are present. However, as dissolving occurs, the concentrations of Ca^{2+} and F^- increase, and it becomes more and more likely that these ions will collide and re-form the solid. Thus two opposite (competing) processes are occurring— the dissolving reaction shown above and the reverse reaction to re-form the solid:

$$Ca^{2+}(aq) + 2F^-(aq) \rightarrow CaF_2(s)$$

Ultimately, equilibrium is reached. No more solid dissolves and the solution is said to be saturated.

We can write an equilibrium expression for this process according to the law of chemical equilibrium.

$$K_{sp} = [Ca^{2+}][F^-]^2$$

where $[Ca^{2+}]$ and $[F^-]$ are expressed in mol/L. The constant K_{sp} is called the **solubility product constant,** or simply the **solubility product.**

Because CaF_2 is a pure solid, it is not included in the equilibrium expression. It may seem strange at first that the amount of excess solid present does not affect the position of the solubility equilibrium. Surely more solid means more surface area exposed to the solvent, which would seem to result in greater solubility. This is not the case, however, because both dissolving and re-forming of the solid occur at the surface of the excess solid. When a solid dissolves, it is the ions at the surface that go into solution. And when the ions in solution re-form the solid, they do so on the surface of the solid. So doubling the surface area of the solid doubles not only the rate of dissolving but also the rate of re-formation of the solid. The amount of excess solid present therefore has no effect on the equilibrium position. Similarly, although either increasing the surface area by grinding up the solid or stirring the solution speeds up the attainment of equilibrium, neither procedure changes the *amount* of solid dissolved at equilibrium.

EXAMPLE 17.8 Writing Solubility Product Expressions

Write the balanced equation describing the reaction for dissolving each of the following solids in water. Also write the K_{sp} expression for each solid.

a. $PbCl_2(s)$ b. $Ag_2CrO_4(s)$ c. $Bi_2S_3(s)$

SOLUTION

a. $PbCl_2(s) \rightleftharpoons Pb^{2+}(aq) + 2Cl^-(aq)$; $K_{sp} = [Pb^{2+}][Cl^-]^2$

b. $Ag_2CrO_4(s) \rightleftharpoons 2Ag^+(aq) + CrO_4{}^{2-}(aq)$; $K_{sp} = [Ag^+]^2[CrO_4{}^{2-}]$

c. $Bi_2S_3(s) \rightleftharpoons 2Bi^{3+}(aq) + 3S^{2-}(aq)$; $K_{sp} = [Bi^{3+}]^2[S^{2-}]^3$

Self-Check EXERCISE 17.6   Write the balanced equation for the reaction describing the dissolving of each of the following solids in water. Also write the K_{sp} expression for each solid.

   a. $BaSO_4(s)$ b. $Fe(OH)_3(s)$ c. $Ag_3PO_4(s)$

See Problems 17.57 and 17.58. ∎

EXAMPLE 17.9 | **Calculating Solubility Products** |

Copper(I) bromide, CuBr, has a measured solubility of 2.0×10^{-4} mol/L at 25 °C. That is, when excess CuBr(s) is placed in 1.0 L of water, we can determine that 2.0×10^{-4} mole of the solid dissolves to produce a saturated solution. Calculate the solid's K_{sp} value.

SOLUTION

Where Are We Going?

We want to determine the K_{sp} value for solid CuBr at 25 °C.

What Do We Know?

  • The solubility of CuBr at 25 °C is 2.0×10^{-4} M.

  • $CuBr(s) \rightleftharpoons Cu^+(aq) + Br^-(aq)$

  • $K_{sp} = [Cu^+][Br^-]$

How Do We Get There?

We can calculate the value of K_{sp} if we know $[Cu^+]$ and $[Br^-]$, the equilibrium concentrations of the ions. We know that the measured solubility of CuBr is 2.0×10^{-4} mol/L. This means that 2.0×10^{-4} mole of solid CuBr dissolves per 1.0 L of solution to come to equilibrium. The reaction is

$$CuBr(s) \rightarrow Cu^+(aq) + Br^-(aq)$$

so

2.0×10^{-4} mol/L CuBr(s) →
$\qquad\qquad 2.0 \times 10^{-4}$ mol/L $Cu^+(aq) + 2.0 \times 10^{-4}$ mol/L $Br^-(aq)$

We can now write the equilibrium concentrations

$$[Cu^+] = 2.0 \times 10^{-4} \text{ mol/L}$$

and

$$[Br^-] = 2.0 \times 10^{-4} \text{ mol/L}$$

These equilibrium concentrations allow us to calculate the value of K_{sp} for CuBr.

$$K_{sp} = [Cu^+][Br^-] = (2.0 \times 10^{-4})(2.0 \times 10^{-4})$$
$$= 4.0 \times 10^{-8}$$

The units for K_{sp} values are omitted.

Self-Check EXERCISE 17.7 Calculate the K_{sp} value for barium sulfate, $BaSO_4$, which has a solubility of 3.9×10^{-5} mol/L at 25 °C.

See Problems 17.59 through 17.62. ■

> Solubilities must be expressed in mol/L in K_{sp} calculations.

We have seen that the known solubility of an ionic solid can be used to calculate its K_{sp} value. The reverse is also possible: the solubility of an ionic solid can be calculated if its K_{sp} value is known.

EXAMPLE 17.10 | Calculating Solubility from K_{sp} Values

The K_{sp} value for solid $AgI(s)$ is 1.5×10^{-16} at 25 °C. Calculate the solubility of $AgI(s)$ in water at 25 °C.

SOLUTION

Where Are We Going?

We want to determine the solubility of AgI at 25 °C.

What Do We Know?

- $AgI(s) \rightleftharpoons Ag^+(aq) + I^-(aq)$
- At 25 °C, $K_{sp} = [Ag^+][I^-] = 1.5 \times 10^{-16}$

How Do We Get There?

Because we do not know the solubility of this solid, we will assume that x moles per liter dissolves to reach equilibrium. Therefore,

$$x \frac{mol}{L} AgI(s) \rightarrow x \frac{mol}{L} Ag^+(aq) + x \frac{mol}{L} I^-(aq)$$

and at equilibrium,

$$[Ag^+] = x \frac{mol}{L}$$

$$[I^-] = x \frac{mol}{L}$$

Substituting these concentrations into the equilibrium expression gives

$$K_{sp} = 1.5 \times 10^{-16} = [Ag^+][I^-] = (x)(x) = x^2$$

Thus

$$x^2 = 1.5 \times 10^{-16}$$
$$x = \sqrt{1.5 \times 10^{-16}} = 1.2 \times 10^{-8} \, mol/L$$

The solubility of $AgI(s)$ is 1.2×10^{-8} mol/L.

Self-Check EXERCISE 17.8 The K_{sp} value for lead chromate, $PbCrO_4$, is 2.0×10^{-16} at 25 °C. Calculate its solubility at 25 °C.

See Problems 17.69 and 17.70. ■

CHAPTER 17 REVIEW

Key Terms

collision model (17.1)
activation energy
  (E_a) (17.2)
catalyst (17.2)
enzyme (17.2)
equilibrium (17.3)
chemical
  equilibrium (17.3)
law of chemical
  equilibrium (17.5)
equilibrium
  expression (17.5)

equilibrium
  constant (17.5)
equilibrium
  position (17.5)
homogeneous
  equilibria (17.6)
heterogeneous
  equilibria (17.6)
Le Châtelier's
  principle (17.7)
solubility product
  (K_{sp}) (17.9)

Summary

1. Chemical reactions can be described by the collision model, which assumes that molecules must collide to react. In terms of this model, a certain threshold energy, called the activation energy (E_a), must be overcome for a collision to form products.

2. A catalyst is a substance that speeds up a reaction without being consumed. A catalyst operates by providing a lower-energy pathway for the reaction in question. Enzymes are biological catalysts.

3. When a chemical reaction is carried out in a closed vessel, the system achieves chemical equilibrium, the state where the concentrations of both reactants and products remain constant over time. Equilibrium is a highly dynamic state; reactants are converted continually into products, and vice versa, as molecules collide with each other. At equilibrium, the rates of the forward and reverse reactions are equal.

4. The law of chemical equilibrium is a general description of the equilibrium condition. It states that for a reaction of the type

$$a\text{A} + b\text{B} \rightleftharpoons c\text{C} + d\text{D}$$

the equilibrium expression is given by

$$K = \frac{[\text{C}]^c\,[\text{D}]^d}{[\text{A}]^a\,[\text{B}]^b}$$

where K is the equilibrium constant.

5. For each reaction system at a given temperature, there is only one value for the equilibrium constant, but there are an infinite number of possible equilibrium positions. An equilibrium position is defined as a particular set of equilibrium concentrations that satisfy the equilibrium expression. A specific equilibrium position depends on the initial concentrations. The amount of a pure liquid or a pure solid is never included in the equilibrium expression.

6. Le Châtelier's principle allows us to predict the effects of changes in concentration, volume, and temperature on a system at equilibrium. This principle states that when a change is imposed on a system at equilibrium, the equilibrium position will shift in a direction that tends to compensate for the imposed change.

7. The principle of equilibrium can also be applied when an excess of a solid is added to water to form a saturated solution. The solubility product (K_{sp}) is an equilibrium constant defined by the law of chemical equilibrium. Solubility is an equilibrium position, and the K_{sp} value of a solid can be determined by measuring its solubility. Conversely, the solubility of a solid can be determined if its K_{sp} value is known.

Active Learning Questions

These questions are designed to be considered by groups of students in class. Often these questions work well for introducing a particular topic in class.

1. Consider an equilibrium mixture of four chemicals (A, B, C, and D, all gases) reacting in a closed flask according to the following equation:

$$\text{A} + \text{B} \rightleftharpoons \text{C} + \text{D}$$

a. You add more A to the flask. How does the concentration of each chemical compare to its original concentration after equilibrium is reestablished? Justify your answer.

b. You have the original set-up at equilibrium, and add more D to the flask. How does the concentration of each chemical compare to its original concentration after equilibrium is reestablished? Justify your answer.

2. The boxes shown on the following page represent a set of initial conditions for the reaction:

$$\text{\small(figure)} + \text{\small(figure)} \underset{K = 25}{\rightleftharpoons} \text{\small(figure)} + \text{\small(figure)}$$

Draw a quantitative molecular picture that shows what this system looks like after the reactants are mixed in one of the boxes and the system reaches equilibrium. Support your answer with calculations.

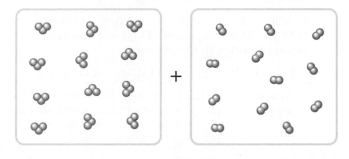

3. For the reaction $H_2 + I_2 \rightleftharpoons 2HI$, consider two possibilities: (a) you add 0.5 mole of each reactant, allow the system to come to equilibrium, and then add 1 mol H_2, and allow the system to reach equilibrium again, or (b) you add 1.5 mol H_2 and 0.5 mol I_2 and allow the system to come to equilibrium. Will the final equilibrium mixture be different for the two procedures? Explain.

4. Given the reaction $A + B \rightleftharpoons C + D$, consider the following situations:

   a. You have 1.3 M A and 0.8 M B initially.
   b. You have 1.3 M A, 0.8 M B, and 0.2 M C initially.
   c. You have 2.0 M A and 0.8 M B initially.

   Order the preceding situations in terms of increasing equilibrium concentration of D and explain your order. Give the order in terms of increasing equilibrium concentration of B and explain.

5. Consider the reaction $A + B \rightleftharpoons C + D$. A friend asks the following: "I know we have been told that if a mixture of A, B, C, and D is in equilibrium and more A is added, more C and D will form. But how can more C and D form if we do not add more B?" What do you tell your friend?

6. Consider the following statements: "Consider the reaction $A(g) + B(g) \rightleftharpoons C(g)$, for which at equilibrium $[A] = 2\ M$, $[B] = 1\ M$, and $[C] = 4\ M$. To a 1-L container of the system at equilibrium you add 3 moles of B. A possible equilibrium condition is $[A] = 1\ M$, $[B] = 3\ M$, and $[C] = 6\ M$, because in both cases, $K = 2$." Indicate everything you think is correct in these statements, and everything that is incorrect. Correct the incorrect statements, and explain.

7. The value of the equilibrium constant, K, is dependent on which of the following? (There may be more than one answer.)

   a. the initial concentrations of the reactants
   b. the initial concentrations of the products
   c. the temperature of the system
   d. the nature of the reactants and products

   Explain.

8. You are browsing through the *Handbook of Hypothetical Chemistry* when you come across a solid that is reported to have a K_{sp} value of zero in water at 25 °C. What does this mean?

9. What do you suppose happens to the K_{sp} value of a solid as the temperature of the solution changes? Consider both increasing and decreasing temperatures, and explain your answer.

10. Consider an equilibrium mixture consisting of $H_2O(g)$, $CO(g)$, $H_2(g)$, and $CO_2(g)$ reacting in a closed vessel according to the equation

   $$H_2O(g) + CO(g) \rightleftharpoons H_2(g) + CO_2(g)$$

   a. You add more H_2O to the flask. How does the new equilibrium concentration of each chemical compare to its original equilibrium concentration after equilibrium is reestablished? Justify your answer.
   b. You add more H_2 to the flask. How does the concentration of each chemical compare to its original concentration after equilibrium is reestablished? Justify your answer.

11. Equilibrium is microscopically dynamic but macroscopically static. Explain what this means.

12. In Section 17.3 of your text, it is mentioned that equilibrium is reached in a "closed system." What is meant by the term "closed system," and why is it necessary for a system to reach equilibrium? Explain why equilibrium is not reached in an open system.

13. Explain why the development of a vapor pressure above a liquid in a closed container represents an equilibrium. What are the opposing processes? How do we recognize when the system has reached a state of equilibrium.

14. Consider the figure below in answering the following questions.

Reaction progress

   a. What does a catalyst do to a chemical reaction?
   b. Which of the pathways in the figure is the catalyzed reaction pathway? How do you know?
   c. What is represented by the double-headed arrow?

18 Oxidation–Reduction Reactions and Electrochemistry

● Workers in China apply paint to a wall to prevent rust. *(China Daily Information Corp/Reuters)*

OWL Sign in to OWL at **www.cengage.com/owl** to view tutorials and simulations, develop problem-solving skills, and complete online homework assigned by your professor.

What do a forest fire, rusting steel, combustion in an automobile engine, and the metabolism of food in a human body have in common? All of these important processes involve oxidation–reduction reactions. In fact, virtually all of the processes that provide energy to heat buildings, power vehicles, and allow people to work and play depend on oxidation–reduction reactions. And every time you start your car, turn on your calculator, look at your digital watch, or listen to a radio at the beach, you are depending on an oxidation–reduction reaction to power the battery in each of these devices. In addition, battery-powered cars have become more common on U.S. roads. This will lead to increased reliance of our society on batteries and will spur the search for new, more efficient batteries.

In this chapter we will explore the properties of oxidation–reduction reactions, and we will see how these reactions are used to power batteries.

© Cengage Learning

The power generated by an alkaline AA battery, a lithium battery, and a mercury battery results from oxidation–reduction reactions.

18.1 Oxidation–Reduction Reactions

OBJECTIVE: To learn about metal–nonmetal oxidation–reduction reactions.

In Section 7.5 we discussed the chemical reactions between metals and nonmetals. For example, sodium chloride is formed by the reaction of elemental sodium and chlorine.

$$2Na(s) + Cl_2(g) \rightarrow 2NaCl(s)$$

Because elemental sodium and chlorine contain uncharged atoms and because sodium chloride is known to contain Na^+ and Cl^- ions, this reaction must involve a transfer of electrons from sodium atoms to chlorine atoms.

Some students use the mnemonic OIL RIG:

Oxidation **I**s **L**oss;
Reduction **I**s **G**ain.

Reactions like this one, in which one or more electrons are transferred, are called **oxidation–reduction reactions,** *or* **redox reactions. Oxidation** is defined as a *loss of electrons.* **Reduction** is defined as a *gain of electrons.* In the reaction of elemental sodium and chlorine, each sodium atom loses one electron, forming a 1+ ion. Therefore, sodium is oxidized. Each chlorine atom gains one electron, forming a negative chloride ion, and is thus reduced. Whenever a metal reacts with a nonmetal to form an ionic compound, electrons are transferred from the metal to the nonmetal. So these reactions are always oxidation–reduction reactions where the metal is oxidized (loses electrons) and the nonmetal is reduced (gains electrons).

EXAMPLE 18.1   Identifying Oxidation and Reduction in a Reaction

In the following reactions, identify which element is oxidized and which element is reduced.

    a. $2Mg(s) + O_2(g) \rightarrow 2MgO(s)$

    b. $2Al(s) + 3I_2(s) \rightarrow 2AlI_3(s)$

SOLUTION

Magnesium burns in air to give a bright, white flame.

    a. We have learned that Group 2 metals form 2+ cations and that Group 6 nonmetals form 2− anions, so we can predict that magnesium oxide contains Mg^{2+} and O^{2-} ions. This means that in the reaction given, each Mg loses two electrons to form Mg^{2+} and so is oxidized. Also each O gains two electrons to form O^{2-} and so is reduced.

    b. Aluminum iodide contains the Al^{3+} and I^- ions. Thus aluminum atoms lose electrons (are oxidized). Iodine atoms gain electrons (are reduced).

Self-Check EXERCISE 18.1   For the following reactions, identify the element oxidized and the element reduced.

    a. $2Cu(s) + O_2(g) \rightarrow 2CuO(s)$　　　　b. $2Cs(s) + F_2(g) \rightarrow 2CsF(s)$

See Problems 18.3 through 18.6. ■

Although we can identify reactions between metals and nonmetals as redox reactions, it is more difficult to decide whether a given reaction between nonmetals is a redox reaction. In fact, many of the most significant redox reactions involve only nonmetals. For example, combustion reactions such as methane burning in oxygen,

$$CH_4(g) + 2O_2(g) \rightarrow CO_2(g) + 2H_2O(g) + \text{energy}$$

are oxidation–reduction reactions. Even though none of the reactants or products in this reaction is ionic, the reaction does involve a transfer of electrons from carbon to oxygen. To explain this, we must introduce the concept of oxidation states.

18.2 Oxidation States

OBJECTIVE: To learn how to assign oxidation states.

The concept of **oxidation states** (sometimes called *oxidation numbers*) lets us keep track of electrons in oxidation–reduction reactions by assigning charges to the various atoms in a compound. Sometimes these charges are quite apparent. For example, in a binary ionic compound the ions have easily identified charges: in sodium chloride, sodium is +1 and chlorine is −1; in magnesium oxide, magnesium is +2 and oxygen is −2; and so on. In such binary ionic compounds the oxidation states are simply the charges of the ions.

Ion	Oxidation State
Na^+	+1
Cl^-	−1
Mg^{2+}	+2
O^{2-}	−2

In an uncombined element, all of the atoms are uncharged (neutral). For example, sodium metal contains neutral sodium atoms, and chlorine gas is made up of Cl_2 molecules, each of which contains two neutral chlorine atoms. Therefore, an atom in a pure element has no charge and is assigned an oxidation state of zero.

In a covalent compound such as water, although no ions are actually present, chemists find it useful to assign imaginary charges to the elements in the compound. The oxidation states of the elements in these compounds are equal to the imaginary charges we determine by assuming that the most electronegative atom (see Section 12.2) in a bond controls or possesses *both* of the shared electrons. For example, in the O—H bonds in water, it is assumed for purposes of assigning oxidation states that the much more electronegative oxygen atom controls both of the shared electrons in each bond. This gives the oxygen eight valence electrons.

$$\begin{array}{c} H \\ \diagdown \\ \ddot{O}\vdots \\ \diagup \\ H \end{array} \begin{array}{l} \longleftarrow 2e^- \\ \\ \longleftarrow 2e^- \end{array}$$

In effect, we say that each hydrogen has lost its single electron to the oxygen. This gives each hydrogen an oxidation state of +1 and the oxygen an oxidation state of −2 (the oxygen atom has formally gained two electrons). In virtually all covalent compounds, oxygen is assigned an oxidation state of −2 and hydrogen is assigned an oxidation state of +1.

Because fluorine is so electronegative, it is always assumed to control any shared electrons. So fluorine is always assumed to have a complete octet of electrons and is assigned an oxidation state of −1. That is, for purposes of assigning oxidation states, fluorine is always imagined to be F^- in its covalent compounds.

The most electronegative elements are F, O, N, and Cl. In general, we give each of these elements an oxidation state equal to its charge as an anion (fluorine is −1, chlorine is −1, oxygen is −2, and nitrogen is −3). When two of these elements are found in the same compound, we assign them in order of electronegativity, starting with the element that has the largest electronegativity.

$$F > O > N > Cl$$

Greatest   Least
electronegativity   electronegativity

For example, in the compound NO_2, because oxygen has a greater electronegativity than nitrogen, we assign each oxygen an oxidation state of −2. This gives a total "charge" of −4 (2×-2) on the two oxygen atoms. Because the NO_2 molecule has zero overall charge, the N must be +4 to exactly balance the −4 on the oxygens. In NO_2, then, the oxidation state of *each* oxygen is −2 and the oxidation state of the nitrogen is +4.

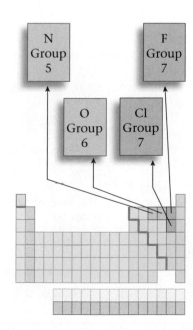

The rules for assigning oxidation states are given below and are illustrated in Table 18.1. Application of these rules allows us to assign oxidation states in most compounds. The principles are illustrated by Example 18.2.

Rules for Assigning Oxidation States

1. The oxidation state of an atom in an uncombined element is 0.
2. The oxidation state of a monatomic ion is the same as its charge.
3. Oxygen is assigned an oxidation state of -2 in most of its covalent compounds. Important exception: peroxides (compounds containing the O_2^{2-} group), in which each oxygen is assigned an oxidation state of -1.
4. In its covalent compounds with nonmetals, hydrogen is assigned an oxidation state of $+1$.
5. In binary compounds, the element with the greater electronegativity is assigned a negative oxidation state equal to its charge as an anion in its ionic compounds.
6. For an electrically neutral compound, the sum of the oxidation states must be zero.
7. For an ionic species, the sum of the oxidation states must equal the overall charge.

Hydrogen peroxide can be used to disinfect a wound.

Table 18.1 Examples of Oxidation States

Substance	Oxidation States	Comments
sodium metal, Na	Na, 0	rule 1
phosphorus, P	P, 0	rule 1
sodium fluoride, NaF	Na, $+1$	rule 2
	F, -1	rule 2
magnesium sulfide, MgS	Mg, $+2$	rule 2
	S, -2	rule 2
carbon monoxide, CO	C, $+2$	
	O, -2	rule 3
sulfur dioxide, SO_2	S, $+4$	
	O, -2	rule 3
hydrogen peroxide, H_2O_2	H, $+1$	
	O, -1	rule 3 (exception)
ammonia, NH_3	H, $+1$	rule 4
	N, -3	rule 5
hydrogen sulfide, H_2S	H, $+1$	rule 4
	S, -2	rule 5
hydrogen iodide, HI	H, $+1$	rule 4
	I, -1	rule 5
sodium carbonate, Na_2CO_3	Na, $+1$	rule 2
	O, -2	rule 3
	C, $+4$	For CO_3^{2-}, the sum of the oxidation states is $+4 + 3(-2) = -2$. rule 7
ammonium chloride, NH_4Cl	N, -3	rule 5
	H, $+1$	rule 4
		For NH_4^+, the sum of the oxidation states is $-3 + 4(+1) = +1$. rule 7
	Cl, -1	rule 2

EXAMPLE 18.2 | Assigning Oxidation States

Assign oxidation states to all atoms in the following molecules or ions.

   a. CO_2

   b. SF_6

   c. NO_3^-

SOLUTION

   a. Rule 3 takes precedence here: oxygen is assigned an oxidation state of -2. We determine the oxidation state for carbon by recognizing that because CO_2 has no charge, the sum of the oxidation states for oxygen and carbon must be 0 (rule 6). Each oxygen is -2 and there are two oxygen atoms, so the carbon atom must be assigned an oxidation state of $+4$.

$$CO_2$$
$$+4 \quad -2 \text{ for } each \text{ oxygen}$$

CHECK: $+4 + 2(-2) = 0$

   b. Because fluorine has the greater electronegativity, we assign its oxidation state first. Its charge as an anion is always -1, so we assign -1 as the oxidation state of each fluorine atom (rule 5). The sulfur must then be assigned an oxidation state of $+6$ to balance the total of -6 from the six fluorine atoms (rule 7).

$$SF_6$$
$$+6 \quad -1 \text{ for } each \text{ fluorine}$$

CHECK: $+6 + 6(-1) = 0$

   c. Oxygen has a greater electronegativity than nitrogen, so we assign its oxidation state of -2 first (rule 5). Because the overall charge on NO_3^- is -1 and because the sum of the oxidation states of the three oxygens is -6, the nitrogen must have an oxidation state of $+5$.

$$NO_3^-$$
$$+5 \quad -2 \text{ for } each \text{ oxygen gives } -6 \text{ total}$$

CHECK: $+5 + 3(-2) = -1$
This is correct; NO_3^- has a -1 charge.

Self-Check EXERCISE 18.2   Assign oxidation states to all atoms in the following molecules or ions.

   a. SO_3

   b. SO_4^{2-}

   c. N_2O_5

   d. PF_3

   e. C_2H_6

See Problems 18.13 through 18.22. ∎

18.3 Oxidation–Reduction Reactions Between Nonmetals

OBJECTIVES: To understand oxidation and reduction in terms of oxidation states.
• To learn to identify oxidizing and reducing agents.

We have seen that oxidation–reduction reactions are characterized by a transfer of electrons. In some cases, the transfer literally occurs to form ions, such as in the reaction

$$2Na(s) + Cl_2(g) \rightarrow 2NaCl(s)$$

We can use oxidation states to verify that electron transfer has occurred.

$$2Na(s) + Cl_2(g) \rightarrow 2NaCl(s)$$

Oxidation state:	0	0	+1 −1
	(element)	(element)	(Na$^+$)(Cl$^-$)

Thus in this reaction, we represent the electron transfer as follows:

$$e^- \Big\langle \begin{matrix} Na \\ Cl \end{matrix} \quad \Rightarrow \quad \begin{matrix} Na^+ \\ Cl^- \end{matrix}$$

In other cases the electron transfer occurs in a different sense, such as in the combustion of methane (the oxidation state for each atom is given below each reactant and product).

$$\underset{\substack{\text{state}: -4 \ +1 \text{ (each H)}}}{\overset{\text{Oxidation}}{CH_4(g)}} \ + \ \underset{0}{2O_2(g)} \ \rightarrow \ \underset{+4 \ -2 \text{ (each O)}}{CO_2(g)} \ + \ \underset{+1 \text{ (each H)} \ -2}{2H_2O(g)}$$

Note that the oxidation state of oxygen in O_2 is 0 because the oxygen is in elemental form. In this reaction there are no ionic compounds, but we can still describe the process in terms of the transfer of electrons. Note that carbon undergoes a change in oxidation state from −4 in CH_4 to +4 in CO_2. Such a change can be accounted for by a loss of eight electrons:

$$\underset{-4}{C \ (\text{in } CH_4)} \quad \overset{}{\Longrightarrow} \quad \underset{+4}{C \ (\text{in } CO_2)}$$
$$\text{Loss of } 8e^-$$

or, in equation form,

$$CH_4 \rightarrow CO_2 + 8e^-$$
$$\underset{-4}{\uparrow} \qquad \underset{+4}{\uparrow}$$

On the other hand, each oxygen changes from an oxidation state of 0 in O_2 to −2 in H_2O and CO_2, signifying a gain of two electrons per atom. Four oxygen atoms are involved, so this is a gain of eight electrons:

$$4O \text{ atoms } (\text{in } 2O_2) \quad \overset{}{\Longrightarrow} \quad 4O^{2-} \ (\text{in } 2H_2O \text{ and } CO_2)$$
$$\text{Gain of } 8e^-$$

or, in equation form,

$$\underset{0}{2O_2} + 8e^- \rightarrow \underset{4(-2) = -8}{CO_2 + 2H_2O}$$

Note that eight electrons are required because four oxygen atoms are going from an oxidation state of 0 to −2. So each oxygen requires two electrons.

No change occurs in the oxidation state of hydrogen, and it is not involved in the electron transfer process.

With this background, we can now define *oxidation* and *reduction* in terms of oxidation states. **Oxidation** is an *increase* in oxidation state (a loss of electrons). **Reduction** is a *decrease* in oxidation state (a gain of electrons). Thus in the reaction

$$2Na(s) + Cl_2(g) \rightarrow 2NaCl(s)$$

sodium is oxidized and chlorine is reduced. Cl_2 is called the **oxidizing agent (electron acceptor)** and Na is called the **reducing agent (electron donor).** We can also define the *oxidizing agent* as the reactant containing the element that is reduced (gains electrons). The *reducing agent* can be defined similarly as the reactant containing the element that is oxidized (loses electrons).

Concerning the reaction

$$CH_4(g) \;+\; 2O_2(g) \;\rightarrow\; CO_2(g) \;+\; 2H_2O(g)$$

with oxidation states: CH_4: C -4, H $+1$; O_2: 0; CO_2: C $+4$, O -2; H_2O: H $+1$, O -2

we can say the following:

1. Carbon is oxidized because there is an increase in its oxidation state (carbon has apparently lost electrons).

2. The reactant CH_4 contains the carbon that is oxidized, so CH_4 is the reducing agent. It is the reactant that furnishes the electrons (those lost by carbon).

3. Oxygen is reduced because there has been a decrease in its oxidation state (oxygen has apparently gained electrons).

4. The reactant that contains the oxygen atoms is O_2, so O_2 is the oxidizing agent. That is, O_2 accepts the electrons.

Note that when the oxidizing or reducing agent is named, the *whole compound* is specified, not just the element that undergoes the change in oxidation state.

Side notes (left margin):

Oxidation:
Loss of electrons
or
Increase in oxidation state

Reduction:
Gain of electrons
or
Decrease in oxidation state

Oxidizing agent:
Accepts electrons
Contains the element reduced

Reducing agent:
Furnishes electrons
Contains the element oxidized

In a redox reaction, an oxidizing agent is reduced (gains electrons) and a reducing agent is oxidized (loses electrons).

EXAMPLE 18.3 | **Identifying Oxidizing and Reducing Agents, I**

When powdered aluminum metal is mixed with pulverized iodine crystals and a drop of water is added, the resulting reaction produces a great deal of energy. The mixture bursts into flames, and a purple smoke of I_2 vapor is produced from the excess iodine. The equation for the reaction is

$$2Al(s) + 3I_2(s) \rightarrow 2AlI_3(s)$$

For this reaction, identify the atoms that are oxidized and those that are reduced, and specify the oxidizing and reducing agents.

Al
Group
3

I
Group
7

SOLUTION

The first step is to assign oxidation states.

$$2Al(s) \;+\; 3I_2(s) \;\rightarrow\; 2AlI_3(s)$$

Al: 0; I_2: 0; AlI_3: Al $+3$, I -1 (each I)

Free elements

$AlI_3(s)$ is a salt that contains Al^{3+} and I^- ions.

Because each aluminum atom changes its oxidation state from 0 to +3 (an increase in oxidation state), aluminum is *oxidized* (loses electrons). On the other hand, the oxidation state of each iodine atom decreases from 0 to −1, and iodine is *reduced* (gains electrons). Because Al furnishes electrons for the reduction of iodine, it is the *reducing agent*. I_2 is the *oxidizing agent* (the reactant that accepts the electrons).

EXAMPLE 18.4 | **Identifying Oxidizing and Reducing Agents, II**

Metallurgy, the process of producing a metal from its ore, always involves oxidation–reduction reactions. In the metallurgy of galena (PbS), the principal lead-containing ore, the first step is the conversion of lead sulfide to its oxide (a process called *roasting*).

$$2PbS(s) + 3O_2(g) \rightarrow 2PbO(s) + 2SO_2(g)$$

The oxide is then treated with carbon monoxide to produce the free metal.

$$PbO(s) + CO(g) \rightarrow Pb(s) + CO_2(g)$$

For each reaction, identify the atoms that are oxidized and those that are reduced, and specify the oxidizing and reducing agents.

SOLUTION

For the first reaction, we can assign the following oxidation states:

$$PbS(s) \ + \ 3O_2(g) \ \rightarrow \ 2PbO(s) \ + \ 2SO_2(g)$$

$$\underset{+2\ \ -2}{} \quad \underset{0}{} \quad \underset{+2\ -2}{} \quad \underset{+4\ -2\ \text{(each O)}}{}$$

The oxidation state for the sulfur atom increases from −2 to +4, so sulfur is oxidized (loses electrons). The oxidation state for each oxygen atom decreases from 0 to −2. Oxygen is reduced (gains electrons). The oxidizing agent (electron acceptor) is O_2, and the reducing agent (electron donor) is PbS.

For the second reaction, we have

$$PbO(s) \ + \ CO(g) \ \rightarrow \ Pb(s) \ + \ CO_2(g)$$

$$\underset{+2\ -2}{} \quad \underset{+2\ -2}{} \quad \underset{0}{} \quad \underset{+4\ -2\ \text{(each O)}}{}$$

Lead is reduced (gains electrons; its oxidation state decreases from +2 to 0), and carbon is oxidized (loses electrons; its oxidation state increases from +2 to +4). PbO is the oxidizing agent (electron acceptor), and CO is the reducing agent (electron donor).

Self-Check EXERCISE 18.3   Ammonia, NH_3, which is widely used as a fertilizer, is prepared by the following reaction:

$$N_2(g) + 3H_2(g) \rightarrow 2NH_3(g)$$

Is this an oxidation–reduction reaction? If so, specify the oxidizing agent and the reducing agent.

See Problems 18.29 through 18.36. ∎

Do We Age by Oxidation?

People (especially those over age 30) seem obsessed about staying young, but the fountain of youth sought since the days of Ponce de Leon has proved elusive. The body inevitably seems to wear out after 70 or 80 years. Is this our destiny or can we find ways to combat aging?

Why do we age? No one knows for certain, but many scientists think that oxidation plays a major role. Although oxygen is essential for life, it can also have a detrimental effect. The oxygen molecule and other oxidizing substances in the body can extract single electrons from the large molecules that make up cell membranes (walls), thus causing them to become very reactive. In fact, these activated molecules can react with each other to change the properties of the cell membranes. If enough of these changes accumulate, the body's immune system comes to view the changed cell as "foreign" and destroys it. This action is particularly harmful to the organism if the cells involved are irreplaceable, such as nerve cells.

Because the human body is so complex, it is very difficult to pinpoint the cause or causes of aging. Scientists are therefore studying simpler life forms. For example, Rajindar Sohal (currently at the University of California) and his coworkers at Southern Methodist University in Dallas are examining aging in common houseflies. Their work indicates that the accumulated damage from oxidation is linked to both the fly's vitality and its life expectancy. One study found that flies that were forced to be sedentary (couldn't fly around) showed much less damage from oxidation (because of their lower oxygen consumption) and lived twice as long as flies that had normal activities.

Accumulated knowledge from various studies indicates that oxidation is probably a major cause of aging. If this is true, how can we protect ourselves? The best way to approach the answer to this question is to study the body's natural defenses against oxidation. A study by Russel J. Reiter of the Texas Health Science Center at San Antonio has shown that melatonin—a chemical secreted by the pineal gland in the brain (but only at night)—protects against oxidation. In addition, it has long been known that vitamin E is an antioxidant. Studies have shown that red blood cells deficient in vitamin E age much faster than cells with normal vitamin E levels. On the basis of this type of evidence many people take daily doses of vitamin E to ward off the effects of aging.

Studies at the Center for Human Nutrition and Aging at Tufts University suggest that a diet rich in antioxidants can reduce the effects of brain aging. Rats that were fed a diet high in antioxidants appeared to have better memory and improved motor skills compared to rats that received a normal diet. Elderly rats fed blueberry diets even recovered some memory and motor skills lost as a result of normal brain aging.

Oxidation is only one possible cause of aging. Research continues on many fronts to find out why we get "older" as time passes.

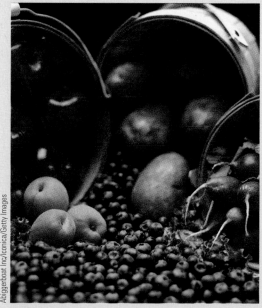

Foods that contain natural antioxidants.

18.4 Balancing Oxidation–Reduction Reactions by the Half-Reaction Method

OBJECTIVE:　To learn to balance oxidation–reduction equations by using half-reactions.

Many oxidation–reduction reactions can be balanced readily by trial and error. That is, we use the procedure described in Chapter 6 to find a set of coefficients that give the same number of each type of atom on both sides of the equation.

However, the oxidation–reduction reactions that occur in aqueous solution are often so complicated that it becomes very tedious to balance them by trial and error. In this section we will develop a systematic approach for balancing the equations for these reactions.

To balance the equations for oxidation–reduction reactions that occur in aqueous solution, we separate the reaction into two half-reactions. **Half-reactions** are equations that have electrons as reactants or products. One half-reaction represents a reduction process and the other half-reaction represents an oxidation process. In a reduction half-reaction, electrons are shown on the reactant side (electrons are gained by a reactant in the equation). In an oxidation half-reaction, the electrons are shown on the product side (electrons are lost by a reactant in the equation).

For example, consider the unbalanced equation for the oxidation–reduction reaction between the cerium(IV) ion and the tin(II) ion.

$$Ce^{4+}(aq) + Sn^{2+}(aq) \rightarrow Ce^{3+}(aq) + Sn^{4+}(aq)$$

This reaction can be separated into a half-reaction involving the substance being *reduced*:

$$e^- + Ce^{4+}(aq) \rightarrow Ce^{3+}(aq) \quad \text{reduction half-reaction}$$

and a half-reaction involving the substance being *oxidized*:

$$Sn^{2+}(aq) \rightarrow Sn^{4+}(aq) + 2e^- \quad \text{oxidation half-reaction}$$

> Ce^{4+} gains $1e^-$ to form Ce^{3+} and is thus reduced.

> Sn^{2+} loses $2e^-$ to form Sn^{4+} and is thus oxidized.

Notice that Ce^{4+} must gain one electron to become Ce^{3+}, so one electron is shown as a reactant along with Ce^{4+} in this half-reaction. On the other hand, for Sn^{2+} to become Sn^{4+}, it must lose two electrons. This means that two electrons must be shown as products in this half-reaction.

The key principle in balancing oxidation–reduction reactions is that the number of electrons lost (from the reactant that is oxidized) must equal the number of electrons gained (from the reactant that is reduced).

$$\text{Number of electrons lost} \quad \xleftarrow[\text{equal to}]{\text{must be}} \quad \text{Number of electrons gained}$$

In the half-reactions shown above, one electron is gained by each Ce^{4+} while two electrons are lost by each Sn^{2+}. We must equalize the number of electrons gained and lost. To do this, we first multiply the reduction half-reaction by 2.

$$2e^- + 2Ce^{4+} \rightarrow 2Ce^{3+}$$

Then we add this half-reaction to the oxidation half-reaction.

$$2e^- + 2Ce^{4+} \rightarrow 2Ce^{3+}$$
$$\underline{Sn^{2+} \rightarrow Sn^{4+} + 2e^-}$$
$$2e^- + 2Ce^{4+} + Sn^{2+} \rightarrow 2Ce^{3+} + Sn^{4+} + 2e^-$$

Finally, we cancel the $2e^-$ on each side to give the overall balanced equation

$$2e^- + 2Ce^{4+} + Sn^{2+} \rightarrow 2Ce^{3+} + Sn^{4+} + 2e^-$$
$$2Ce^{4+} + Sn^{2+} \rightarrow 2Ce^{3+} + Sn^{4+}$$

We can now summarize what we have said about the method for balancing oxidation–reduction reactions in aqueous solution:

1. Separate the reaction into an oxidation half-reaction and a reduction half-reaction.

2. Balance the half-reactions separately.

3. Equalize the number of electrons gained and lost.

4. Add the half-reactions together and cancel electrons to give the overall balanced equation.

It turns out that most oxidation–reduction reactions occur in solutions that are distinctly basic or distinctly acidic. We will cover only the acidic case in this text, because it is the most common. The detailed procedure for balancing the equations for oxidation–reduction reactions that occur in acidic solution is given below, and Example 18.5 illustrates the use of these steps.

The Half-Reaction Method for Balancing Equations for Oxidation–Reduction Reactions Occurring in Acidic Solution

Step 1 Identify and write the equations for the oxidation and reduction half-reactions.

Step 2 For each half-reaction:
   a. Balance all of the elements except hydrogen and oxygen.
   b. Balance oxygen using H_2O.
   c. Balance hydrogen using H^+.
   d. Balance the charge using electrons.

Step 3 If necessary, multiply one or both balanced half-reactions by an integer to equalize the number of electrons transferred in the two half-reactions.

Step 4 Add the half-reactions, and cancel identical species that appear on both sides.

Step 5 Check to be sure the elements and charges balance.

EXAMPLE 18.5   Balancing Oxidation–Reduction Reactions Using the Half-Reaction Method, I

H_2O and H^+ will be added to this equation as we balance it. We do not have to worry about them now.

Balance the equation for the reaction between permanganate and iron(II) ions in acidic solution. The net ionic equation for this reaction is

$$MnO_4^- (aq) + Fe^{2+} (aq) \xrightarrow{\text{Acid}} Fe^{3+} (aq) + Mn^{2+} (aq)$$

This reaction is used to analyze iron ore for its iron content.

SOLUTION

Step 1 *Identify and write equations for the half-reactions.*
The oxidation states for the half-reaction involving the permanganate ion show that manganese is reduced.

$$MnO_4^- \rightarrow Mn^{2+}$$

$$\begin{array}{cc} +7 \quad -2 & +2 \\ \text{(each O)} & \end{array}$$

> Note that the left side contains oxygen but the right side does not. This will be taken care of later when we add water.

Because manganese changes from an oxidation state of $+7$ to $+2$, it is reduced. So this is the *reduction half-reaction*. It will have electrons as reactants, although we will not write them yet. The other half-reaction involves the oxidation of iron(II) to the iron(III) ion and is the *oxidation half-reaction*.

$$Fe^{2+} \rightarrow Fe^{3+}$$

$$\begin{array}{cc} +2 & +3 \end{array}$$

This reaction will have electrons as products, although we will not write them yet.

Step 2 *Balance each half-reaction.*
For the reduction reaction, we have

$$MnO_4^- \rightarrow Mn^{2+}$$

a. The manganese is already balanced.

b. We balance oxygen by adding $4H_2O$ to the right side of the equation.

$$MnO_4^- \rightarrow Mn^{2+} + 4H_2O$$

> The H^+ comes from the acidic solution in which the reaction is taking place.

c. Next we balance hydrogen by adding $8H^+$ to the left side.

$$8H^+ + MnO_4^- \rightarrow Mn^{2+} + 4H_2O$$

d. All of the elements have been balanced, but we need to balance the charge using electrons. At this point we have the following charges for reactants and products in the reduction half-reaction.

$$8H^+ + MnO_4^- \rightarrow Mn^{2+} + 4H_2O$$

$$\begin{array}{cccc} 8+ \quad + \quad 1- & & 2+ \quad + \quad 0 \\ \underbrace{\qquad\qquad}_{7+} & & \underbrace{\qquad\qquad}_{2+} \end{array}$$

A solution containing MnO_4^- ions (*left*) and a solution containing Fe^{2+} ions (*right*).

Ken O'Donoghue

> Always add electrons to the side of the half-reaction with excess positive charge.

We can equalize the charges by adding five electrons to the left side.

$$\underbrace{5e^- + 8H^+ + MnO_4^-}_{2+} \rightarrow \underbrace{Mn^{2+} + 4H_2O}_{2+}$$

Both the *elements* and the *charges* are now balanced, so this represents the balanced reduction half-reaction. The fact that five electrons appear on the reactant side of the equation makes sense, because five electrons are required to reduce MnO_4^- (in which Mn has an oxidation state of $+7$) to Mn^{2+} (in which Mn has an oxidation state of $+2$).

For the oxidation reaction,

$$Fe^{2+} \rightarrow Fe^{3+}$$

the elements are balanced, so all we have to do is balance the charge.

$$\underbrace{Fe^{2+}}_{2+} \rightarrow \underbrace{Fe^{3+}}_{3+}$$

One electron is needed on the right side to give a net $2+$ charge on both sides.

$$\underbrace{Fe^{2+}}_{2+} \rightarrow \underbrace{Fe^{3+} + e^-}_{2+}$$

> The number of electrons gained in the reduction half-reaction must equal the number of electrons lost in the oxidation half-reaction.

Step 3   *Equalize the number of electrons transferred in the two half-reactions.* Because the reduction half-reaction involves a transfer of five electrons and the oxidation half-reaction involves a transfer of only one electron, the oxidation half-reaction must be multiplied by 5.

$$5Fe^{2+} \rightarrow 5Fe^{3+} + 5e^-$$

Step 4   *Add the half-reactions and cancel identical species.*

$$5e^- + 8H^+ + MnO_4^- \rightarrow Mn^{2+} + 4H_2O$$
$$\underline{5Fe^{2+} \rightarrow 5Fe^{3+} + 5e^-}$$
$$\cancel{5e^-} + 8H^+ + MnO_4^- + 5Fe^{2+} \rightarrow Mn^{2+} + 5Fe^{3+} + 4H_2O + \cancel{5e^-}$$

Note that the electrons cancel (as they must) to give the final balanced equation

$$5Fe^{2+}(aq) + MnO_4^-(aq) + 8H^+(aq) \rightarrow 5Fe^{3+}(aq) + Mn^{2+}(aq) + 4H_2O(l)$$

Note that we show the physical states of the reactants and products—(aq) and (l) in this case—only in the final balanced equation.

Step 5   *Check to be sure that elements and charges balance.*

$$\text{Elements}\quad 5Fe, 1Mn, 4O, 8H \rightarrow 5Fe, 1Mn, 4O, 8H$$
$$\text{Charges}\quad 17+ \rightarrow 17+$$

The equation is balanced.

EXAMPLE 18.6   **Balancing Oxidation–Reduction Reactions Using the Half-Reaction Method, II**

When an automobile engine is started, it uses the energy supplied by a lead storage battery. This battery uses an oxidation–reduction reaction between elemental lead (lead metal) and lead(IV) oxide to provide the power to start the engine. The unbalanced equation for a simplified version of the reaction is

$$Pb(s) + PbO_2(s) + H^+(aq) \rightarrow Pb^{2+}(aq) + H_2O(l)$$

Balance this equation using the half-reaction method.

SOLUTION

Step 1 First we identify and write the two half-reactions. One half-reaction must be

$$Pb \rightarrow Pb^{2+}$$

and the other is

$$PbO_2 \rightarrow Pb^{2+}$$

The first reaction involves the oxidation of Pb to Pb^{2+}. The second reaction involves the reduction of Pb^{4+} (in PbO_2) to Pb^{2+}.

> Because Pb^{2+} is the only lead-containing product, it must be the product in both half-reactions.

Step 2 Now we will balance each half-reaction separately.

The oxidation half-reaction

$$Pb \rightarrow Pb^{2+}$$

a–c. All the elements are balanced.

d. The charge on the left is zero and that on the right is +2, so we must add $2e^-$ to the right to give zero overall charge.

$$Pb \rightarrow Pb^{2+} + 2e^-$$

This half-reaction is balanced.

The reduction half-reaction

$$PbO_2 \rightarrow Pb^{2+}$$

a. All elements are balanced except O.

b. The left side has two oxygen atoms and the right side has none, so we add $2H_2O$ to the right side.

$$PbO_2 \rightarrow Pb^{2+} + 2H_2O$$

c. Now we balance hydrogen by adding $4H^+$ to the left.

$$4H^+ + PbO_2 \rightarrow Pb^{2+} + 2H_2O$$

d. Because the left side has a +4 overall charge and the right side has a +2 charge, we must add $2e^-$ to the left side.

$$2e^- + 4H^+ + PbO_2 \rightarrow Pb^{2+} + 2H_2O$$

The half-reaction is balanced.

Step 3 Because each half-reaction involves $2e^-$, we can simply add the half-reactions as they are.

Step 4

$$Pb \rightarrow Pb^{2+} + 2e^-$$
$$\underline{2e^- + 4H^+ + PbO_2 \rightarrow Pb^{2+} + 2H_2O}$$
$$2e^- + 4H^+ + Pb + PbO_2 \rightarrow 2Pb^{2+} + 2H_2O + 2e^-$$

Canceling electrons gives the balanced overall equation

$$Pb(s) + PbO_2(s) + 4H^+(aq) \rightarrow 2Pb^{2+}(aq) + 2H_2O(l)$$

where the appropriate states are also indicated.

Copper metal reacting with nitric acid. The solution is colored by the presence of Cu^{2+} ions. The brown gas is NO_2, which is formed when NO reacts with O_2 in the air.

Step 5 Both the elements and the charges balance.

$$\text{Elements} \quad 2Pb, 2O, 4H \rightarrow 2Pb, 2O, 4H$$
$$\text{Charges} \quad 4+ \rightarrow 4+$$

The equation is correctly balanced.

Self-Check **EXERCISE 18.4** Copper metal reacts with nitric acid, $HNO_3(aq)$, to give aqueous copper(II) nitrate, water, and nitrogen monoxide gas as products. Write and balance the equation for this reaction.

See Problems 18.45 through 18.48. ∎

18.5 Electrochemistry: An Introduction

OBJECTIVES: To understand the term *electrochemistry.* • To learn to identify the components of an electrochemical (galvanic) cell.

Our lives would be very different without batteries. We would have to crank the engines in our cars by hand, wind our watches, and buy very long extension cords if we wanted to listen to a radio on a picnic. Indeed, our society sometimes seems to run on batteries. In this section and the next, we will find out how these important devices produce electrical energy.

A battery uses the energy from an oxidation–reduction reaction to produce an electric current. This is an important illustration of **electrochemistry,** *the study of the interchange of chemical and electrical energy.*

Electrochemistry involves two types of processes:

1. The production of an electric current from a chemical (oxidation–reduction) reaction

2. The use of an electric current to produce a chemical change

To understand how a redox reaction can be used to generate a current, let's reconsider the aqueous reaction between MnO_4^- and Fe^{2+} that we worked with in Example 18.5. We can break this redox reaction into the following half-reactions:

$$8H^+ + MnO_4^- + 5e^- \rightarrow Mn^{2+} + 4H_2O \quad \text{reduction}$$
$$Fe^{2+} \rightarrow Fe^{3+} + e^- \quad \text{oxidation}$$

When the reaction between MnO_4^- and Fe^{2+} occurs in solution, electrons are transferred directly as the reactants collide. No useful work is obtained from the chemical energy involved in the reaction. How can we harness this energy? The key is to *separate the oxidizing agent (electron acceptor) from the reducing agent (electron donor),* thus requiring the electron transfer to occur through a wire. That is, to get from the reducing agent to the oxidizing agent, the electrons must travel through a wire. The current produced in the wire by this electron flow can be directed through a device, such as an electric motor, to do useful work.

For example, consider the system illustrated in Figure 18.1. If our reasoning has been correct, electrons should flow through the wire from Fe^{2+} to MnO_4^-. However, when we construct the apparatus as shown, no flow of electrons occurs. Why? The problem is that if the electrons flowed from the right to the left compartment, the left compartment would become negatively charged and the right compartment would experience a build-up of

> The energy involved in a chemical reaction is customarily not shown in the balanced equation. In the reaction of MnO_4^- with Fe^{2+}, energy is released that can be used to do useful work.

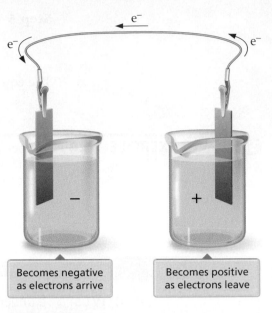

Figure 18.1

Schematic of a method for separating the oxidizing and reducing agents in a redox reaction. (The solutions also contain other ions to balance the charge.) This cell is incomplete at this point.

Figure 18.2

Electron flow under these conditions would lead to a build-up of negative charge on the left and positive charge on the right, which is not feasible without a huge input of energy.

positive charge (Figure 18.2). Creating a charge separation of this type would require large amounts of energy. Therefore, electron flow does not occur under these conditions.

We can, however, solve this problem very simply. The solutions must be connected (without allowing them to mix extensively) so that *ions* can also flow to keep the net charge in each compartment zero (Figure 18.3). This can be accomplished by using a salt bridge (a U-shaped tube filled with a strong electrolyte) or a porous disk in a tube connecting the two solutions (see Figure 18.4). Either of these devices allows ion flow but prevents extensive mixing of the solutions. When we make a provision for ion flow, the circuit is complete. Electrons flow through the wire from reducing agent to ox-

Figure 18.3

Here the ion flow between the two solutions keeps the charge neutral as electrons are transferred. This can be accomplished by having negative ions (anions) flow in the opposite direction to the electrons or by having positive ions (cations) flow in the same direction as the electrons. Both actually occur in a working battery.

a The salt bridge contains a strong electrolyte either as a gel or as a solution; both ends are covered with a membrane that allows only ions to pass.

b The porous disk allows ion flow but does not permit overall mixing of the solutions in the two compartments.

Figure 18.4

A salt bridge or a porous-disk connection allows ions to flow, completing the electric circuit.

The name *galvanic cell* honors Luigi Galvani (1737–1798), an Italian scientist generally credited with the discovery of electricity. These cells are sometimes called *voltaic cells* after Alessandro Volta (1745–1827), another Italian, who first constructed cells of this type around 1800.

Anode: The electrode where oxidation occurs. Cathode: The electrode where reduction occurs.

idizing agent, and ions in the two aqueous solutions flow from one compartment to the other to keep the net charge zero.

Thus an **electrochemical battery,** also called a **galvanic cell,** is a device powered by an oxidation–reduction reaction where the oxidizing agent is separated from the reducing agent so that the electrons must travel through a wire from the reducing agent to the oxidizing agent (Figure 18.5).

Notice that in a battery, the reducing agent loses electrons (which flow through the wire toward the oxidizing agent) and so is oxidized. The electrode where oxidation occurs is called the **anode.** At the other electrode, the oxidizing agent gains electrons and is thus reduced. The electrode where reduction occurs is called the **cathode.**

We have seen that an oxidation–reduction reaction can be used to generate an electric current. In fact, this type of reaction is used to produce elec-

voltaic cell

electron flow can be harnessed externally

consonants

causes oxidation by gaining e⁻

vowels

causes reduction by losing e⁻

Figure 18.5

Schematic of a battery (galvanic cell).

$$Fe_{(s)} + Cu^{2+}_{(aq)} \rightarrow Fe^{2+}_{(aq)} + Cu_{(s)}$$

Alessandro Volta.

tric currents in many space vehicles. An oxidation–reduction reaction that can be used for this purpose is hydrogen and oxygen reacting to form water.

$$2H_2(g) + O_2(g) \rightarrow 2H_2O(l)$$

Oxidation states: 0   0   +1   −2
(each H)

Notice from the changes in oxidation states that in this reaction, hydrogen is oxidized and oxygen reduced. The opposite process can also occur. We can *force* a current through water to produce hydrogen and oxygen gas.

$$2H_2O(l) \xrightarrow[\text{energy}]{\text{Electrical}} 2H_2(g) + O_2(g)$$

This process, where *electrical energy is used to produce a chemical change*, is called **electrolysis.**

In the remainder of this chapter, we will discuss both types of electrochemical processes. In the next section we will concern ourselves with the practical galvanic cells we know as batteries.

18.6 Batteries

OBJECTIVE: To learn about the composition and operation of commonly used batteries.

In the previous section we saw that a galvanic cell is a device that uses an oxidation–reduction reaction to generate an electric current by separating the oxidizing agent from the reducing agent. In this section we will consider several specific galvanic cells and their applications.

▶ Lead Storage Battery

Since about 1915, when self-starters were first used in automobiles, the **lead storage battery** has been a major factor in making the automobile a practical means of transportation. This type of battery can function for several years under temperature extremes from −30 °F to 100 °F and under incessant punishment from rough roads. The fact that this same type of battery has been in use for so many years in the face of all of the changes in science and technology over that span of time attests to how well it does its job.

> Remember: The oxidizing agent accepts electrons and the reducing agent furnishes electrons.

In the lead storage battery, the reducing agent is lead metal, Pb, and the oxidizing agent is lead(IV) oxide, PbO_2. We have already considered a simplified version of this reaction in Example 18.6. In an actual lead storage battery, sulfuric acid, H_2SO_4, furnishes the H^+ needed in the reaction; it also furnishes SO_4^{2-} ions that react with the Pb^{2+} ions to form solid $PbSO_4$. A schematic of one cell of the lead storage battery is shown in Figure 18.6.

In this cell the anode is constructed of lead metal, which is oxidized. In the cell reaction, lead atoms lose two electrons each to form Pb^{2+} ions, which combine with SO_4^{2-} ions present in the solution to give solid $PbSO_4$.

The cathode of this battery has lead(IV) oxide coated onto lead grids. Lead atoms in the +4 oxidation state in PbO_2 accept two electrons each (are reduced) to give Pb^{2+} ions that also form solid $PbSO_4$.

In the cell the anode and cathode are separated (so that the electrons must flow through an external wire) and bathed in sulfuric acid. The half-reactions that occur at the two electrodes and the overall cell reaction are shown on the following page.

Figure 18.6

In a lead storage battery each cell consists of several lead grids that are connected by a metal bar. These lead grids furnish electrons (the lead atoms lose electrons to form Pb^{2+} ions, which combine with SO_4^{2-} ions to give solid $PbSO_4$). Because the lead is oxidized, it functions as the anode of the cell. The substance that gains electrons is PbO_2; it is coated onto lead grids, several of which are hooked together by a metal bar. The PbO_2 formally contains Pb^{4+}, which is reduced to Pb^{2+}, which in turn combines with SO_4^{2-} to form solid $PbSO_4$. The PbO_2 accepts electrons, so it functions as the cathode.

e^- flow

H_2SO_4 electrolyte solution

Pb metal grid (anode)

PbO_2 coated onto a lead grid (cathode)

Anode reaction:

$$Pb + H_2SO_4 \rightarrow PbSO_4 + 2H^+ + 2e^- \quad \text{oxidation}$$

Cathode reaction:

$$PbO_2 + H_2SO_4 + 2e^- + 2H^+ \rightarrow PbSO_4 + 2H_2O \quad \text{reduction}$$

Overall reaction:

$$Pb(s) + PbO_2(s) + 2H_2SO_4(aq) \rightarrow 2PbSO_4(s) + 2H_2O(l)$$

The tendency for electrons to flow from the anode to the cathode in a battery depends on the ability of the reducing agent to release electrons and on the ability of the oxidizing agent to capture electrons. If a battery consists of a reducing agent that releases electrons readily and an oxidizing agent with a high affinity for electrons, the electrons are driven through the connecting wire with great force and can provide much electrical energy. It is useful to think of the analogy of water flowing through a pipe. The greater the pressure on the water, the more vigorously the water flows. The "pressure" on electrons to flow from one electrode to the other in a battery is called the **potential** of the battery and is measured in volts. For example, each cell in a lead storage battery produces about 2 volts of potential. In an actual automobile battery, six of these cells are connected to produce about 12 volts of potential.

▶ **Dry Cell Batteries**

The calculators, electronic watches, CD players, and MP3 players that are so familiar to us are all powered by small, efficient **dry cell batteries.** They are called dry cells because they do not contain a liquid electrolyte. The common dry cell battery was invented more than 100 years ago by George Leclanché (1839–1882), a French chemist. In its *acid version,* the dry cell battery contains a zinc inner case that acts as the anode and a carbon (graphite) rod in contact with a moist paste of solid MnO_2, solid NH_4Cl, and carbon

Figure 18.8

A mercury battery of the type used in small calculators.

Figure 18.7

A common dry cell battery.

that acts as the cathode (Figure 18.7). The half-cell reactions are complex but can be approximated as follows:

Anode reaction: $\quad\quad\quad\quad Zn \rightarrow Zn^{2+} + 2e^-$ oxidation

Cathode reaction:

$$2NH_4^+ + 2MnO_2 + 2e^- \rightarrow Mn_2O_3 + 2NH_3 + H_2O \quad \text{reduction}$$

This cell produces a potential of about 1.5 volts.

In the *alkaline version* of the dry cell battery, the NH_4Cl is replaced with KOH or NaOH. In this case the half-reactions can be approximated as follows:

Anode reaction: $\quad Zn + 2OH^- \rightarrow ZnO(s) + H_2O + 2e^-$ oxidation

Cathode reaction: $\quad 2MnO_2 + H_2O + 2e^- \rightarrow Mn_2O_3 + 2OH^-$ reduction

The alkaline dry cell lasts longer, mainly because the zinc anode corrodes less rapidly under basic conditions than under acidic conditions.

Other types of dry cell batteries include the *silver cell,* which has a Zn anode and a cathode that employs Ag_2O as the oxidizing agent in a basic environment. *Mercury cells,* often used in calculators, have a Zn anode and a cathode involving HgO as the oxidizing agent in a basic medium (see Figure 18.8).

An especially important type of dry cell is the *nickel–cadmium battery,* in which the electrode reactions are

Anode reaction: $\quad Cd + 2OH^- \rightarrow Cd(OH)_2 + 2e^-$ oxidation

Cathode reaction: $\quad NiO_2 + 2H_2O + 2e^- \rightarrow Ni(OH)_2 + 2OH^-$ reduction

In this cell, as in the lead storage battery, the products adhere to the electrodes. Therefore, a nickel–cadmium battery can be recharged an indefinite number of times, because the products can be turned back into reactants by the use of an external source of current.

18.7 Corrosion

OBJECTIVE: To understand the electrochemical nature of corrosion and to learn some ways of preventing it.

Most metals are found in nature in compounds with nonmetals such as oxygen and sulfur. For example, iron exists as iron ore (which contains Fe_2O_3 and other oxides of iron).

Stainless Steel: It's the Pits

One of New York's giants, the Chrysler Building, boasts a much admired art-deco stainless steel pinnacle that has successfully resisted corrosion since it was built in 1929. Stainless steel is the nobility among steels. Consisting of iron, chromium (at least 13%), and nickel (with molybdenum and titanium added to more expensive types), stainless steel is highly resistant to the rusting that consumes regular steel. However, the cheaper grades of stainless steel have an Achilles heel—pit corrosion. In certain environments, pit corrosion can penetrate several millimeters in a matter of weeks.

Metallurgy, the science of producing useful metallic materials, almost always requires some kind of compromise. In the case of stainless steel, inclusions of MnS make the steel easier to machine into useful parts, but such inclusions are also the source of pit corrosion. Recently a group of British researchers analyzed stainless steel using a high-energy beam of ions that blasted atoms loose from the steel surface. Studies of the resultant atom vapor revealed the source of the problem. It turns out that when the stainless steel is cooling, the MnS inclusions "suck" chromium atoms from the surrounding area, leaving behind a

Built in New York in 1929, the Chrysler Building's stainless steel pinnacle has been cleaned only a few times. Despite the urban setting, the material shows few signs of corrosion.

chromium-deficient region. The corrosion occurs in this region, as illustrated in the accompanying diagram. The essential problem is that to resist corrosion steel must contain at least 13% Cr atoms. The low-chromium region around the inclusion is not stainless steel—so it corrodes just like regular steel. This corrosion leads to a pit that causes deterioration of the steel surface.

Now that the reason for the pit corrosion is understood, metallurgists should be able to develop methods of stainless steel formulation that avoid this problem. One British scientist, Mary P. Ryan, suggests that heat treatment of the stainless steel may solve the problem by allowing Cr atoms to diffuse from the inclusion back into the surrounding area. Because corrosion of regular steel is such an important issue, finding ways to make cheaper stainless steel will have a significant economic impact. We need stainless without the pits.

1 *Cr depletion during processing*

2 *Depleted zone dissolves in aqueous solution*

3 *Zone becomes acidic and high in Cl⁻*

4 *Pit develops, undercutting metal surface*

In corrosion, chromium depletion triggers pitting.

Corrosion can be viewed as the process of returning metals to their natural state—the ores from which they were originally obtained. Corrosion involves oxidation of the metal. Because corroded metal often loses its strength and attractiveness, this process causes great economic loss. For example, approximately one-fifth of the iron and steel produced annually is used to replace rusted metal.

Because most metals react with O_2, we might expect them to corrode so fast in air that they wouldn't be useful. It is surprising, therefore, that the problem of corrosion does not virtually prevent the use of metals in air. Part of the explanation is that most metals develop a thin oxide coating, which tends to protect their internal atoms against further oxidation. The best example of this is aluminum. Aluminum readily loses electrons, so it should be very easily oxidized by O_2. Given this fact, why is aluminum so useful for building airplanes, bicycle frames, and so on? Aluminum is such a valuable structural material because it forms a thin adherent layer of aluminum oxide, Al_2O_3, which greatly inhibits further corrosion. Thus aluminum protects itself with this tough oxide coat. Many other metals, such as chromium, nickel, and tin, do the same thing.

Iron can also form a protective oxide coating. However, this oxide is not a very effective shield against corrosion, because it scales off easily, exposing new metal surfaces to oxidation. Under normal atmospheric conditions, copper forms an external layer of greenish copper sulfate or carbonate called *patina*. *Silver tarnish* is silver sulfide, Ag_2S, which in thin layers gives the silver surface a richer appearance. Gold shows no appreciable corrosion in air.

Preventing corrosion is an important way of conserving our natural supplies of metals and energy. The primary means of protection is the application of a coating—most often paint or metal plating—to protect the metal from oxygen and moisture. Chromium and tin are often used to plate steel because they oxidize to form a durable, effective oxide coating.

Alloying is also used to prevent corrosion. *Stainless steel* contains chromium and nickel, both of which form oxide coatings that protect the steel.

Cathodic protection is the method most often employed to protect steel in buried fuel tanks and pipelines. A metal that furnishes electrons more easily than iron, such as magnesium, is connected by a wire to the pipeline or tank that is to be protected (Figure 18.9). Because the magnesium is a better reducing agent than iron, electrons flow through the wire from the magnesium to the iron pipe. Thus the electrons are furnished by the magnesium rather than by the iron, keeping the iron from being oxidized. As oxidation of the magnesium occurs, the magnesium dissolves, so it must be replaced periodically.

> Some metals, such as copper, gold, silver, and platinum, are relatively difficult to oxidize. They are often called noble metals.

> An alloy is a mixture of elements with metallic properties.

Figure 18.9

Cathodic protection of an underground pipe.

18.8 Electrolysis

OBJECTIVE: To understand the process of electrolysis and learn about the commercial preparation of aluminum.

Unless it is recharged, a battery "runs down" because the substances in it that furnish and accept electrons (to produce the electron flow) are consumed. For example, in the lead storage battery (see Section 18.6), PbO_2 and Pb are consumed to form $PbSO_4$ as the battery runs.

$$PbO_2(s) + Pb(s) + 2H_2SO_4(aq) \rightarrow 2PbSO_4(s) + 2H_2O(l)$$

Water-Powered Fireplace

Hydrogen gas is being touted as an environmentally friendly fuel because, unlike fossil fuels, it does not produce the greenhouse gas carbon dioxide. The only product of combustion of H_2 is water. As a result, hydrogen is being investigated as a possible fuel for cars, trucks, and buses. Now comes a manufacturer, Heat & Glo, that is showcasing an in-home fireplace that uses water as the

Courtesy, Hearth & Home Technologies

fuel. The Aqueon fireplace uses electrolysis to decompose the water to $H_2(g)$ and $O_2(g)$; the hydrogen is then burned to furnish heat for the home. The 31,000-Btu fireplace features copper and stainless steel and has a contemporary design (see accompanying photo). To operate, the fireplace is simply hooked up to the water and electrical supplies for the home.

However, one of the most useful characteristics of the lead storage battery is that it can be recharged. *Forcing* current through the battery in the direction opposite to the normal direction reverses the oxidation–reduction reaction. That is, $PbSO_4$ is consumed and PbO_2 and Pb are formed in the charging process. This recharging is done continuously by the automobile's alternator, which is powered by the engine.

> An electrolytic cell uses electrical energy to produce a chemical change that would not otherwise occur.

The process of **electrolysis** involves *forcing a current through a cell to produce a chemical change that would not otherwise occur.*

One important example of this type of process is the electrolysis of water. Water is a very stable substance that can be broken down into its elements by using an electric current (Figure 18.10).

$$2H_2O(l) \xrightarrow[\text{current}]{\substack{\text{Forced} \\ \text{electric}}} 2H_2(g) + O_2(g)$$

The electrolysis of water to produce hydrogen and oxygen occurs whenever a current is forced through an aqueous solution. Thus, when the lead storage battery is charged, or "jumped," potentially explosive mixtures of H_2 and O_2 are produced by the current flow through the solution in the battery. This is why it is very important not to produce a spark near the battery during these operations.

Another important use of electrolysis is in the production of metals from their ores. The metal produced in the greatest quantities by electrolysis is aluminum.

Aluminum is one of the most abundant elements on earth, ranking third behind oxygen and silicon. Because aluminum is a very reactive metal, it is found in nature as its oxide in an ore called *bauxite* (named after Les Baux, France, where it was discovered in 1821). Production of aluminum metal from its ore proved to be more difficult than the production of most other metals. In 1782, Lavoisier, the pioneering French chemist, recognized aluminum as a metal "whose affinity for oxygen is so strong that it cannot be overcome by any known reducing agent." As a result, pure aluminum

Charles D. Winters/Photo Researchers, Inc.

Figure 18.10

The electrolysis of water produces hydrogen gas at the cathode (on the left) and oxygen gas at the anode (on the right). A nonreacting strong electrolyte such as Na_2SO_4 is needed to furnish ions to allow the flow of current.

Figure 18.11

Charles Martin Hall (1863–1914) was a student at Oberlin College in Ohio when he first became interested in aluminum. One of his professors commented that anyone who could find a way to manufacture aluminum cheaply would make a fortune, and Hall decided to give it a try. The 21-year-old worked in a wooden shed near his house with an iron frying pan as a container, a blacksmith's forge as a heat source, and galvanic cells constructed from fruit jars. Using these crude galvanic cells, Hall found that he could produce aluminum by passing a current through a molten mixture of Al_2O_3 and Na_3AlF_6. By a strange coincidence, Paul Heroult, a French chemist who was born and died in the same years as Hall, made the same discovery at about the same time.

The Granger Collection

Table 18.2	The Price of Aluminum, 1855–1990
Date	Price of Aluminum ($/lb)*
1855	$100,000
1885	100
1890	2
1895	0.50
1970	0.30
1980	0.80
1990	0.74

*Note the precipitous drop in price after the discovery of the Hall–Heroult process in 1886.

metal remained unknown. Finally, in 1854, a process was found for producing metallic aluminum by using sodium, but aluminum remained a very expensive rarity. In fact, it is said that Napoleon III served his most honored guests with aluminum forks and spoons, while the others had to settle for gold and silver utensils!

The breakthrough came in 1886 when two men, Charles M. Hall in the United States (Figure 18.11) and Paul Heroult in France, almost simultaneously discovered a practical electrolytic process for producing aluminum, which greatly increased the availability of aluminum for many purposes. Table 18.2 shows how dramatically the price of aluminum dropped after this discovery. The effect of the electrolysis process is to reduce Al^{3+} ions to neutral Al atoms that form aluminum metal. The aluminum produced in this electrolytic process is 99.5% pure. To be useful as a structural material, aluminum is alloyed with metals such as zinc (for trailer and aircraft construction) and manganese (for cooking utensils, storage tanks, and highway signs). The production of aluminum consumes about 4.5% of all electricity used in the United States.

CHAPTER 18 REVIEW

Key Terms

oxidation–reduction (redox) reactions (18.1)
oxidation (18.1, 18.3)
reduction (18.1, 18.3)
oxidation states (18.2)
oxidizing agent (electron acceptor) (18.3)

reducing agent (electron donor) (18.3)
half-reactions (18.4)
electrochemistry (18.5)
electrochemical battery (galvanic cell) (18.5)

anode (18.5)
cathode (18.5)
electrolysis (18.5, 18.8)
lead storage battery (18.6)

potential (18.6)
dry cell battery (18.6)
corrosion (18.7)
cathodic protection (18.7)

Summary

1. Oxidation–reduction reactions involve a transfer of electrons. Oxidation states provide a way to keep track of electrons in these reactions. A set of rules is used to assign oxidation states.

2. Oxidation is an increase in oxidation state (a loss of electrons); reduction is a decrease in oxidation state (a gain of electrons). An oxidizing agent accepts electrons, and a reducing agent donates electrons. Oxidation and reduction always occur together.

3. Oxidation–reduction equations can be balanced by inspection or by the half-reaction method. This method involves splitting a reaction into two parts (the oxidation half-reaction and the reduction half-reaction).

4. Electrochemistry is the study of the interchange of chemical and electrical energy that occurs through oxidation–reduction reactions.

5. When an oxidation–reduction reaction occurs with the reactants in the same solution, the electrons are transferred directly, and no useful work can be obtained. However, when the oxidizing agent is separated from the reducing agent, so that the electrons must flow through a wire from one to the other, chemical energy is transformed into electrical energy. The opposite process, in which electrical energy is used to produce chemical change, is called electrolysis.

6. A galvanic (electrochemical) cell is a device in which chemical energy is transformed into useful electrical energy. Oxidation occurs at the anode of a cell; reduction occurs at the cathode.

7. A battery is a galvanic cell, or group of cells, that serves as a source of electric current. The lead storage battery has a lead anode and a cathode of lead coated with PbO_2, both immersed in a solution of sulfuric acid. Dry cell batteries do not have liquid electrolytes but contain a moist paste instead.

8. Corrosion involves the oxidation of metals to form mainly oxides and sulfides. Some metals, such as aluminum, form a thin protective oxide coating that inhibits their further corrosion. Corrosion of iron can be prevented by a coating (such as paint), by alloying, and by cathodic protection.

Active Learning Questions

These questions are designed to be considered by groups of students in class. Often these questions work well for introducing a particular topic in class.

1. Sketch a galvanic cell, and explain how it works. Look at Figures 18.1 and 18.5. Explain what is occurring in each container and why the cell in Figure 18.5 "works," but the one in Figure 18.1 does not.

2. Make a list of nitrogen compounds with as many different oxidation states for nitrogen as you can.

3. Which of the following are oxidation–reduction reactions? Explain.

   a. $PCl_3 + Cl_2 \rightarrow PCl_5$
   b. $Cu + 2AgNO_3 \rightarrow Cu(NO_3)_2 + 2Ag$
   c. $CO_2 + 2LiOH \rightarrow Li_2CO_3 + H_2O$
   d. $FeCl_2 + 2NaOH \rightarrow Fe(OH)_2 + 2NaCl$
   e. $MnO_2 + 4HCl \rightarrow Cl_2 + 2H_2O + MnCl_2$

4. Which of the following statements is (are) true? Explain. (There may be more than one answer.)

   a. Oxidation and reduction cannot occur independently of each other.
   b. Oxidation and reduction accompany all chemical changes.
   c. Oxidation and reduction describe the loss and gain of electron(s), respectively.

5. Why do we say that when something gains electrons it is reduced? What is being reduced?

6. The equation $Ag^+ + Cu \rightarrow Cu^{2+} + Ag$ has equal numbers of each type of element on each side of the equation. This equation, however, is not balanced. Why is this equation not balanced? Balance the equation.

7. In balancing oxidation–reduction equations, why is it permissible to add water to either side of the equation?

8. What does it mean for a substance to be *oxidized*? The term "oxidation" originally came from substances reacting with oxygen gas. Explain why a substance that reacts with oxygen gas will always be oxidized.

9. Label the following parts of the galvanic cell.

   anode

   cathode

   reducing agent

   oxidizing agent

19 Radioactivity and Nuclear Energy

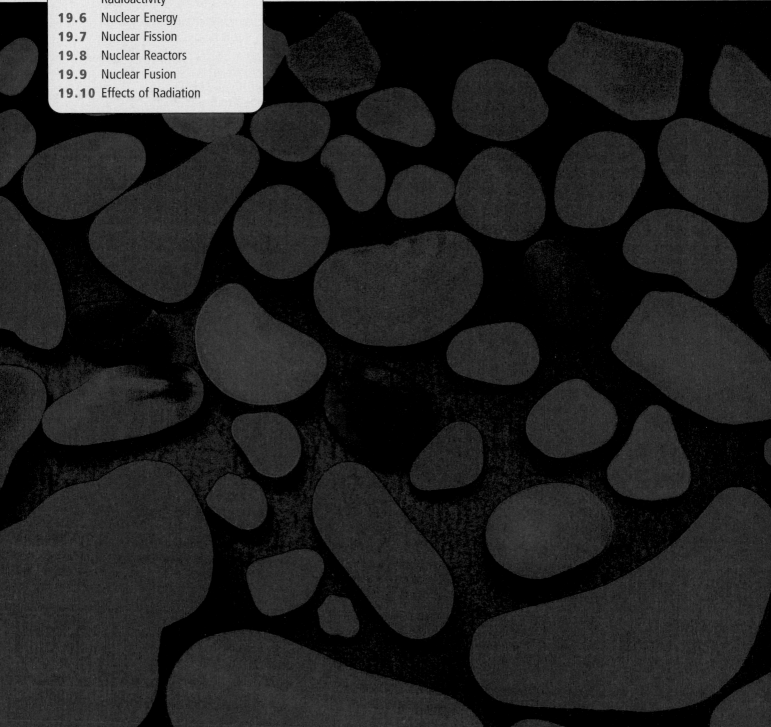

● Radioactive sea glass. *(Courtesy, By the Bay Treasures, photo by Charles Peden)*

Because the chemistry of an atom is determined by the number and arrangement of its electrons, the properties of the nucleus do not strongly affect the chemical behavior of an atom. Therefore, you might be wondering why there is a chapter on the nucleus in a chemistry textbook. The reason for this chapter is that the nucleus is very important to all of us—a quick reading of any daily newspaper will testify to that. Nuclear processes can be used to detect explosives in airline luggage (see "Chemistry in Focus: Measurement: Past, Present, and Future," page 22), to generate electric power, and to establish the ages of very old objects such as human artifacts, rocks, and diamonds (see "Chemistry in Focus: Dating Diamonds," page 624). This chapter considers aspects of the nucleus and its properties that you should know about.

Several facts about the nucleus are immediately impressive: its very small size, its very large density, and the energy that holds it together. The radius of a typical nucleus is about 10^{-13} cm, only a hundred-thousandth the radius of a typical atom. In fact, if the nucleus of the hydrogen atom were the size of a Ping-Pong ball, the electron in the $1s$ orbital would be, on the average, 0.5 km (0.3 mile) away. The density of the nucleus is equally impressive; it is approximately 1.6×10^{14} g/cm^3. A sphere of nuclear material the size of a Ping-Pong ball would have a mass of 2.5 *billion tons!* Finally, the energies involved in nuclear processes are typically millions of times larger than those associated with normal chemical reactions, a fact that makes nuclear processes potentially attractive for generating energy.

The nucleus is believed to be made of particles called **nucleons (neutrons and protons)**.

Leslie Garland/Leslie Garland Picture Library/Alamy Images

Wooden artifacts such as this dragon figurehead from a Viking ship can be dated from their carbon-14 content.

The atomic number (Z) represents the number of protons in a nucleus; the mass number (A) represents the sum of the numbers of protons and neutrons in a nucleus.

Recall from Chapter 4 that the number of protons in a nucleus is called the **atomic number (Z)** and that the sum of the numbers of neutrons and protons is the **mass number (A).** Atoms that have identical atomic numbers but different mass numbers are called **isotopes.** The general term **nuclide** is applied to each unique atom, and we represent it as follows:

$$\text{Mass number} \downarrow$$
$$^{A}_{Z}\text{X} \leftarrow \text{Element symbol}$$
$$\uparrow \text{Atomic number}$$

where X represents the symbol for a particular element. For example, the following nuclides constitute the common isotopes of carbon: carbon-12, $^{12}_{6}$C; carbon-13, $^{13}_{6}$C; and carbon-14, $^{14}_{6}$C. Notice that all the carbon nuclides have six protons ($Z = 6$) and that they have six, seven, and eight neutrons, respectively.

Radioactive Decay

OBJECTIVES: To learn the types of radioactive decay. • To learn to write nuclear equations that describe radioactive decay.

Many nuclei are **radioactive;** that is, they spontaneously decompose, forming a different nucleus and producing one or more particles. An example is carbon-14, which decays as shown in the equation

$$^{14}_{6}C \rightarrow ^{14}_{7}N + ^{0}_{-1}e$$

where $^{0}_{-1}e$ represents an electron, which in nuclear terminology is called a **beta particle,** or **β particle.** This **nuclear equation,** which is typical of those representing radioactive decay, is quite different from the chemical equations we have written before. Recall that in a balanced chemical equation the atoms must be conserved. In a nuclear equation *both the atomic number (Z) and the mass number (A) must be conserved.* That is, the sums of the Z values on both sides of the arrow must be equal, and the same restriction applies to the A values. For example, in the above equation, the sum of the Z values is 6 on both sides of the arrow (6 and $7 - 1$), and the sum of the A values is 14 on both sides of the arrow (14 and $14 + 0$). Notice that the mass number for the β particle is zero; the mass of the electron is so small that it can be neglected here. Of the approximately 2000 known nuclides, only 279 do not undergo radioactive decay. Tin has the largest number of nonradioactive isotopes—ten.

> Over 85% of all known nuclides are radioactive.

▶ Types of Radioactive Decay

There are several different types of radioactive decay. One frequently observed decay process involves production of an **alpha (α) particle,** which is a helium nucleus ($^{4}_{2}He$). **Alpha-particle production** is a very common mode of decay for heavy radioactive nuclides. For example, $^{222}_{88}Ra$, radium-222, decays by α-particle production to give radon-218.

$$^{222}_{88}Ra \rightarrow ^{4}_{2}He + ^{218}_{86}Rn$$

Notice in this equation that the mass number is conserved ($222 = 4 + 218$) and the atomic number is conserved ($88 = 2 + 86$). Another α-particle producer is $^{230}_{90}Th$:

$$^{230}_{90}Th \rightarrow ^{4}_{2}He + ^{226}_{88}Ra$$

Notice that the production of an α particle results in a loss of 4 in mass number (A) and a loss of 2 in atomic number (Z).

    β-particle production is another common decay process. For example, the thorium-234 nuclide produces a β particle as it changes to protactinium-234.

$$^{234}_{90}Th \rightarrow ^{234}_{91}Pa + ^{0}_{-1}e$$

Iodine-131 is also a β-particle producer:

$$^{131}_{53}I \rightarrow ^{0}_{-1}e + ^{131}_{54}Xe$$

> Notice that both Z and A balance in each of these nuclear equations.

Recall that the β particle is assigned a mass number of 0 because its mass is tiny compared with that of a proton or neutron. The value of Z is -1 for the β particle, so the atomic number for the new nuclide is greater by 1 than the atomic number for the original nuclide. Therefore, *the net effect of β-particle production is to change a neutron to a proton.*

    Production of a β particle results in no change in mass number (A) and an increase of 1 in atomic number (Z).

Table 19.1 Various Types of Radioactive Processes

Process	Example
β-particle (electron) production	$^{227}_{89}Ac \rightarrow\ ^{227}_{90}Th +\ ^{0}_{-1}e$
positron production	$^{13}_{7}N \rightarrow\ ^{13}_{6}C +\ ^{0}_{1}e$
electron capture	$^{73}_{33}As +\ ^{0}_{-1}e \rightarrow\ ^{73}_{32}Ge$
α-particle production	$^{210}_{84}Po \rightarrow\ ^{206}_{82}Pb +\ ^{4}_{2}He$
γ-ray production	excited nucleus $\rightarrow$ ground-state nucleus $+\ ^{0}_{0}\gamma$
	excess energy lower energy

To balance nuclear reactions:
balance subscripts and superscripts on
both sides of equation.

A **gamma ray**, or **γ ray**, is a high-energy photon of light. A nuclide in an excited nuclear energy state can release excess energy by producing a gamma ray, and γ-ray production often accompanies nuclear decays of various types. For example, in the α-particle decay of $^{238}_{92}U$,

$$^{238}_{92}U \rightarrow\ ^{4}_{2}He +\ ^{234}_{90}Th + 2\,^{0}_{0}\gamma$$

two γ rays of different energies are produced in addition to the α particle ($^{4}_{2}He$). Gamma rays are photons of light and so have zero charge and zero mass number.

Production of a γ ray results in no change in mass number (A) and no change in atomic number (Z).

The **positron** is a particle with the same mass as the electron but opposite charge. An example of a nuclide that decays by **positron production** is sodium-22:

$$^{22}_{11}Na \rightarrow\ ^{0}_{1}e +\ ^{22}_{10}Ne$$

Note that the production of a positron appears to change a proton into a neutron.

Production of a positron results in no change in mass number (A) and a decrease of 1 in atomic number (Z).

Electron capture is a process in which one of the inner-orbital electrons is captured by the nucleus, as illustrated by the process

$$^{201}_{80}Hg +\ ^{0}_{-1}e \rightarrow\ ^{201}_{79}Au +\ ^{0}_{0}\gamma$$
$$\uparrow$$
Inner-orbital electron

This reaction would have been of great interest to the alchemists, but unfortunately, it does not occur often enough to make it a practical means of changing mercury to gold. Gamma rays are always produced along with electron capture.

Table 19.1 lists the common types of radioactive decay, with examples.

Often a radioactive nucleus cannot achieve a stable (nonradioactive) state through a single decay process. In such a case, a **decay series** occurs until a stable nuclide is formed. A well-known example is the decay series that starts with $^{238}_{92}U$ and ends with $^{206}_{82}Pb$, as shown in Figure 19.1. Similar series exist for $^{235}_{92}U$:

$$^{235}_{92}U \xrightarrow[\text{decays}]{\text{Series of}}\ ^{207}_{82}Pb$$

and for $^{232}_{90}Th$:

$$^{232}_{90}Th \xrightarrow[\text{decays}]{\text{Series of}}\ ^{208}_{82}Pb$$

A gamma ray is a high-energy photon produced in connection with nuclear decay.

The $^{0}_{0}\gamma$ notation indicates $Z = 0$ and $A = 0$ for a γ ray. A gamma ray is often simply indicated by γ.

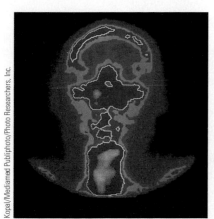

Bone scintigraph of a patient's cranium following administration of the radiopharmaceutical technetium-99.

Figure 19.1

The decay series from $^{238}_{92}$U to $^{206}_{82}$Pb. Each nuclide in the series except $^{206}_{82}$Pb is radioactive, and the successive transformations (shown by the arrows) continue until $^{206}_{82}$Pb is finally formed. The horizontal red arrows indicate β-particle production (Z increases by 1 and A is unchanged). The diagonal blue arrows signify α-particle production (both A and Z decrease).

EXAMPLE 19.1   Writing Nuclear Equations, I

Write balanced nuclear equations for each of the following processes.

a. $^{11}_{6}$C produces a positron.

b. $^{214}_{83}$Bi produces a β particle.

c. $^{237}_{93}$Np produces an α particle.

SOLUTION

a. We must find the product nuclide represented by $^{A}_{Z}$X in the following equation:

$$^{11}_{6}\text{C} \rightarrow \underset{\underset{\text{Positron}}{\uparrow}}{^{0}_{1}\text{e}} + {^{A}_{Z}\text{X}}$$

The key to solving this problem is to recognize that both A and Z must be conserved. That is, we can find the identity of $^{A}_{Z}$X by recognizing that the sums of the Z and A values must be the same on both sides of the equation. Thus, for X, Z must be 5 because $Z + 1 = 6$. A must be 11 because $11 + 0 = 11$. Therefore, $^{A}_{Z}$X is $^{11}_{5}$B. (The fact that Z is 5 tells us that the nuclide is boron. See the periodic table on the inside front cover of the book.) So the balanced equation is $^{11}_{6}$C $\rightarrow$ $^{0}_{1}$e + $^{11}_{5}$B.

CHECK:

Left Side		Right Side
$Z = 6$	$\rightarrow$	$Z = 5 + 1 = 6$
$A = 11$		$A = 11 + 0 = 11$

b. Knowing that a β particle is represented by $_{-1}^{0}e$, we can write

$$^{214}_{83}Bi \rightarrow ^{0}_{-1}e + ^{A}_{Z}X$$

where $Z - 1 = 83$ and $A + 0 = 214$. This means that $Z = 84$ and $A = 214$. We can now write

$$^{214}_{83}Bi \rightarrow ^{0}_{-1}e + ^{214}_{84}X$$

Using the periodic table, we find that $Z = 84$ for the element polonium, so $^{214}_{84}X$ must be $^{214}_{84}Po$.

CHECK:

Left Side		Right Side
$Z = 83$	$\rightarrow$	$Z = 84 - 1 = 83$
$A = 214$		$A = 214 + 0 = 214$

c. Because an α particle is represented by $^{4}_{2}He$, we can write

$$^{237}_{93}Np \rightarrow ^{4}_{2}He + ^{A}_{Z}X$$

where $A + 4 = 237$ or $A = 237 - 4 = 233$, and $Z + 2 = 93$ or $Z = 93 - 2 = 91$. Thus $A = 233$, $Z = 91$, and the balanced equation must be

$$^{237}_{93}Np \rightarrow ^{4}_{2}He + ^{233}_{91}Pa$$

CHECK:

Left Side		Right Side
$Z = 93$	$\rightarrow$	$Z = 91 + 2 = 93$
$A = 237$		$A = 233 + 4 = 237$

Self-Check EXERCISE 19.1   The decay series for $^{238}_{92}U$ is represented in Figure 19.1. Write the balanced nuclear equation for each of the following radioactive decays.

a. Alpha-particle production by $^{226}_{88}Ra$

b. Beta-particle production by $^{214}_{82}Pb$

See Problems 19.25 through 19.28. ■

EXAMPLE 19.2 | Writing Nuclear Equations, II

In each of the following nuclear reactions, supply the missing particle.

a. $^{195}_{79}Au + ? \rightarrow ^{195}_{78}Pt$

b. $^{38}_{19}K \rightarrow ^{38}_{18}Ar + ?$

SOLUTION

a. A does not change and Z for Pt is 1 lower than Z for Au, so the missing particle must be an electron.

$$^{195}_{79}Au + ^{0}_{-1}e \rightarrow ^{195}_{78}Pt$$

CHECK:

Left Side		Right Side
$Z = 79 - 1 = 78$	$\rightarrow$	$Z = 78$
$A = 195 + 0 = 195$		$A = 195$

This is an example of electron capture.

b. For Z and A to be conserved, the missing particle must be a positron.

$$^{38}_{19}K \rightarrow ^{38}_{18}Ar + ^{0}_{1}e$$

CHECK: Left Side   Right Side
$Z = 19$ $\rightarrow$ $Z = 18 + 1 = 19$
$A = 38$ $A = 38 + 0 = 38$

Potassium-38 decays by positron production.

Self-Check **EXERCISE 19.2**   Supply the missing species in each of the following nuclear equations.

a. $^{222}_{86}\text{Rn} \rightarrow \, ^{218}_{84}\text{Po} + \, ?$

b. $^{15}_{8}\text{O} \rightarrow \, ? + \, ^{0}_{1}\text{e}$

See Problems 19.21 through 19.24. ■

19.2   Nuclear Transformations

OBJECTIVE: To learn how one element may be changed into another by particle bombardment.

Irene Curie and Frederick Joliot.

In 1919, Lord Rutherford observed the first **nuclear transformation,** *the change of one element into another.* He found that bombarding $^{14}_{7}\text{N}$ with α particles produced the nuclide $^{17}_{8}\text{O}$:

$$^{14}_{7}\text{N} + \, ^{4}_{2}\text{He} \rightarrow \, ^{17}_{8}\text{O} + \, ^{1}_{1}\text{H}$$

with a proton ($^{1}_{1}\text{H}$) as another product. Fourteen years later, Irene Curie and her husband Frederick Joliot observed a similar transformation from aluminum to phosphorus:

$$^{27}_{13}\text{Al} + \, ^{4}_{2}\text{He} \rightarrow \, ^{30}_{15}\text{P} + \, ^{1}_{0}\text{n}$$

where $^{1}_{0}\text{n}$ represents a neutron that is produced in the process.

Notice that in both these cases the bombarding particle is a helium nucleus (an α particle). Other small nuclei, such as $^{12}_{6}\text{C}$ and $^{15}_{7}\text{N}$, also can be used to bombard heavier nuclei and cause transformations. However, because these positive bombarding ions are repelled by the positive charge of the target nucleus, the bombarding particle must be moving at a very high speed to penetrate the target. These high speeds are achieved in various types of *particle accelerators.*

Neutrons are also employed as bombarding particles to effect nuclear transformations. However, because neutrons are uncharged (and thus not repelled by a target nucleus), they are readily absorbed by many nuclei, producing new nuclides. The most common source of neutrons for this purpose is a fission reactor (see Section 19.8).

By using neutron and positive-ion bombardment, scientists have been able to extend the periodic table—that is, to produce chemical elements that are not present naturally. Prior to 1940, the heaviest known element was uranium ($Z = 92$), but in 1940, neptunium ($Z = 93$) was produced by neutron bombardment of $^{238}_{92}\text{U}$. The process initially gives $^{239}_{92}\text{U}$, which decays to $^{239}_{93}\text{Np}$ by β-particle production:

$$^{238}_{92}\text{U} + \, ^{1}_{0}\text{n} \rightarrow \, ^{239}_{92}\text{U} \rightarrow \, ^{239}_{93}\text{Np} + \, ^{0}_{-1}\text{e}$$

same principle applies to more sophisticated examples

Table 19.2	Syntheses of Some of the Transuranium Elements	
Neutron Bombardment	neptunium ($Z = 93$)	$^{238}_{92}U + ^{1}_{0}n \rightarrow ^{239}_{92}U \rightarrow ^{239}_{93}Np + ^{0}_{-1}e$
	americium ($Z = 95$)	$^{239}_{94}Pu + 2\,^{1}_{0}n \rightarrow ^{241}_{94}Pu \rightarrow ^{241}_{95}Am + ^{0}_{-1}e$
Positive-Ion Bombardment	curium ($Z = 96$)	$^{239}_{94}Pu + ^{4}_{2}He \rightarrow ^{242}_{96}Cm + ^{1}_{0}n$
	californium ($Z = 98$)	$^{242}_{96}Cm + ^{4}_{2}He \rightarrow ^{245}_{98}Cf + ^{1}_{0}n$ or
		$^{238}_{92}U + ^{12}_{6}C \rightarrow ^{246}_{98}Cf + 4\,^{1}_{0}n$
	rutherfordium ($Z = 104$)	$^{249}_{98}Cf + ^{12}_{6}C \rightarrow ^{257}_{104}Rf + 4\,^{1}_{0}n$
	dubnium ($Z = 105$)	$^{249}_{98}Cf + ^{15}_{7}N \rightarrow ^{260}_{105}Db + 4\,^{1}_{0}n$
	seaborgium ($Z = 106$)	$^{249}_{98}Cf + ^{18}_{8}O \rightarrow ^{263}_{106}Sg + 4\,^{1}_{0}n$

In the years since 1940, the elements with atomic numbers 93 through 112, called the **transuranium elements,** have been synthesized. In addition, the production of element 114 (in 1999), elements 113 and 115 (in 2004), and element 118 (in 2006) have been reported. Table 19.2 gives some examples of these processes.

19.3 Detection of Radioactivity and the Concept of Half-life

OBJECTIVES: To learn about radiation detection instruments. • To understand half-life.

> Geiger counters are commonly called *survey meters*.

The most familiar instrument for measuring radioactivity levels is the **Geiger–Müller counter,** or **Geiger counter** (Figure 19.2). High-energy particles from radioactive decay produce ions when they travel through matter. The probe of the Geiger counter contains argon gas. The argon atoms have no charge, but they can be ionized by a rapidly moving particle.

$$Ar(g) \xrightarrow[\text{particle}]{\text{High-energy}} Ar^+(g) + e^-$$

That is, the fast-moving particle "knocks" electrons off some of the argon atoms. Although a sample of uncharged argon atoms does not conduct a current, the ions and electrons formed by the high-energy particle allow a current to flow momentarily, so a "pulse" of current flows every time a particle enters the probe. The Geiger counter detects each pulse of current, and these events are counted.

A **scintillation counter** is another instrument often employed to detect radioactivity. This device uses a substance, such as sodium iodide, that

Figure 19.2

A schematic representation of a Geiger–Müller counter. The high-speed particle knocks electrons off argon atoms to form ions,

$$Ar \xrightarrow{\text{Particle}} Ar^+ + e^-$$

and a pulse of current flows.

gives off light when it is struck by a high-energy particle. A detector senses the flashes of light and thus counts the decay events.

One important characteristic of a given type of radioactive nuclide is its half-life. The **half-life** is the *time required for half the original sample of nuclei to decay*. For example, if a certain radioactive sample contains 1000 nuclei at a given time and 500 nuclei (half of the original number) 7.5 days later, this radioactive nuclide has a half-life of 7.5 days.

A given type of radioactive nuclide always has the same half-life. However, the various radioactive nuclides have half-lives that cover a tremendous range. For example, $^{234}_{91}$Pa, protactinium-234, has a half-life of 1.2 minutes, and $^{238}_{92}$U, uranium-238, has a half-life of 4.5×10^9 (4.5 *billion*) years. This means that a sample containing 100 million $^{234}_{91}$Pa nuclei will have only 50 million $^{234}_{91}$Pa nuclei in it (half of 100 million) after 1.2 minutes have passed. In another 1.2 minutes, the number of nuclei will decrease to half of 50 million, or 25 million nuclei.

$$\text{100 million} ^{234}_{91}\text{Pa} \xrightarrow[\text{minutes}]{1.2} \text{50 million} ^{234}_{91}\text{Pa} \xrightarrow[\text{minutes}]{1.2} \text{25 million} ^{234}_{91}\text{Pa}$$

(50 million decays)   (25 million decays)

This means that a sample of $^{234}_{91}$Pa with 100 million nuclei will show 50 million decay events (50 million $^{234}_{91}$Pa nuclei will decay) over a time of 1.2 minutes. By contrast, a sample containing 100 million $^{238}_{92}$U nuclei will undergo 50 million decay events over 4.5 billion years. Therefore, $^{234}_{91}$Pa shows much greater activity than $^{238}_{92}$U. We sometimes say that $^{234}_{91}$Pa is "hotter" than $^{238}_{92}$U.

Thus, at a given moment, a radioactive nucleus with a short half-life is much more likely to decay than one with a long half-life.

EXAMPLE 19.3   Understanding Half-life

Table 19.3 The Half-lives for Some of the Radioactive Nuclides of Radium

Nuclide	Half-life
$^{223}_{88}$Ra	12 days
$^{224}_{88}$Ra	3.6 days
$^{225}_{88}$Ra	15 days
$^{226}_{88}$Ra	1600 years
$^{228}_{88}$Ra	6.7 years

Table 19.3 lists various radioactive nuclides of radium.

a. Order these nuclides in terms of activity (from most decays per day to least).

b. How long will it take for a sample containing 1.00 mole of $^{223}_{88}$Ra to reach a point where it contains only 0.25 mole of $^{223}_{88}$Ra?

SOLUTION

a. The shortest half-life indicates the greatest activity (the most decays over a given period of time). Therefore, the order is

Most activity   Least activity

(shortest half-life)   (longest half-life)

$^{224}_{88}$Ra $>$ $^{223}_{88}$Ra $>$ $^{225}_{88}$Ra $>$ $^{228}_{88}$Ra $>$ $^{226}_{88}$Ra

3.6 days   12 days   15 days   6.7 years   1600 years

b. In one half-life (12 days), the sample will decay from 1.00 mole of $^{223}_{88}$Ra to 0.50 mole of $^{223}_{88}$Ra. In the next half-life (another 12 days), it will decay from 0.50 mole of $^{223}_{88}$Ra to 0.25 mole of $^{223}_{88}$Ra.

$$\text{1.00 mol} ^{223}_{88}\text{Ra} \xrightarrow{\text{12 days}} \text{0.50 mol} ^{223}_{88}\text{Ra} \xrightarrow{\text{12 days}} \text{0.25 mol} ^{223}_{88}\text{Ra}$$

Therefore, it will take 24 days (two half-lives) for the sample to change from 1.00 mole of $^{223}_{88}$Ra to 0.25 mole of $^{223}_{88}$Ra.

A watch dial with radium paint.

Self-Check EXERCISE 19.3   Watches with numerals that "glow in the dark" formerly were made by including radioactive radium in the paint used to letter the watch faces. Assume that to make the numeral 3 on a given watch, a sample of paint containing 8.0×10^{-7} mole of $^{228}_{88}Ra$ was used. This watch was then put in a drawer and forgotten. Many years later someone finds the watch and wishes to know when it was made. Analyzing the paint, this person finds 1.0×10^{-7} moles of $^{228}_{88}Ra$ in the numeral 3. How much time elapsed between the making of the watch and the finding of the watch?

HINT: Use the half-life of $^{228}_{88}Ra$ from Table 19.3.

See Problems 19.37 through 19.42. ■

19.4 Dating by Radioactivity

OBJECTIVE: To learn how objects can be dated by radioactivity.

Archaeologists, geologists, and others involved in reconstructing the ancient history of the earth rely heavily on the half-lives of radioactive nuclei to provide accurate dates for artifacts and rocks. A method for dating ancient articles made from wood or cloth is **radiocarbon dating**, or **carbon-14 dating**, a technique originated in the 1940s by Willard Libby, an American chemist who received the Nobel Prize for his efforts.

Radiocarbon dating is based on the radioactivity of $^{14}_{6}C$, which decays by β-particle production.

$$^{14}_{6}C \rightarrow {}^{0}_{-1}e + {}^{14}_{7}N$$

Carbon-14 is continuously produced in the atmosphere when high-energy neutrons from space collide with nitrogen-14.

$$^{14}_{7}N + {}^{1}_{0}n \rightarrow {}^{14}_{6}C + {}^{1}_{1}H$$

Just as carbon-14 is produced continuously by this process, it decomposes continuously through β-particle production. Over the years, these two opposing processes have come into balance, causing the amount of $^{14}_{6}C$ present in the atmosphere to remain approximately constant.

Brigham Young researcher Scott Woodward taking a bone sample for carbon-14 dating at an archaeological site in Egypt.

Carbon-14 can be used to date wood and cloth artifacts because the $^{14}_{6}C$, along with the other carbon isotopes in the atmosphere, reacts with oxygen to form carbon dioxide. A living plant consumes this carbon dioxide in the photosynthesis process and incorporates the carbon, including $^{14}_{6}C$, into its molecules. As long as the plant lives, the $^{14}_{6}C$ content in its molecules remains the same as in the atmosphere because of the plant's continuous uptake of carbon. However, as soon as a tree is cut to make a wooden bowl or a flax plant is harvested to make linen, it stops taking in carbon. There is no longer a source of $^{14}_{6}C$ to replace that lost to radioactive decay, so the material's $^{14}_{6}C$ content begins to decrease.

Because the half-life of $^{14}_{6}C$ is known to be 5730 years, a wooden bowl found in an archaeological dig that shows a $^{14}_{6}C$ content of half that found in currently living trees is approximately 5730 years old. That is, because half the $^{14}_{6}C$ present when the tree was cut has disappeared, the tree must have been cut one half-life of $^{14}_{6}C$ ago.

Dating Diamonds

While connoisseurs of gems value the clearest possible diamonds, geologists learn the most from impure diamonds. Diamonds are formed in the earth's crust at depths of about 200 kilometers, where the high pressures and temperatures favor the most dense form of carbon. As the diamond is formed, impurities are sometimes trapped, and these can be used to determine the diamond's date of "birth." One valuable dating impurity is $^{238}_{92}U$, which is radioactive and decays in a series of steps to $^{206}_{82}Pb$, which is stable (nonradioactive). Because the rate at which $^{238}_{92}U$ decays is known, determining how much $^{238}_{92}U$ has been converted to $^{206}_{82}Pb$ tells scientists the amount of time that has elapsed since the $^{238}_{92}U$ was trapped in the diamond as it was formed.

Using these dating techniques, Peter D. Kinney of Curtin University of Technology in Perth, Australia, and Henry O. A. Meyer of Purdue University in West Lafayette, Indiana, have identified the youngest diamond ever found. Discovered in Mbuji Mayi, Zaire, the diamond is 628 million years old, far younger than all previously dated diamonds, which range from 2.4 to 3.2 billion years old.

The great age of all previously dated diamonds had caused some geologists to speculate that all diamond formation occurred billions of years ago. However, this "youngster" suggests that diamonds have formed throughout geologic time and are probably being formed right now in the earth's crust. We won't see these diamonds for a long time, because diamonds typically remain deeply buried in the earth's crust for millions of years until they are brought to the surface by volcanic blasts called kimberlite eruptions.

It's good to know that eons from now there will be plenty of diamonds to mark the engagements of future couples.

Smithsonian Institution, Natural History Museum, Department of Mineral Sciences

The Hope diamond.

19.5 Medical Applications of Radioactivity

OBJECTIVE: To discuss the use of radiotracers in medicine.

> Nuclides used as radiotracers have short half-lives so that they disappear rapidly from the body.

Although we owe the rapid advances of the medical sciences in recent decades to many causes, one of the most important has been the discovery and use of **radiotracers**—radioactive nuclides that can be introduced into organisms in food or drugs and subsequently *traced* by monitoring their radioactivity. For example, the incorporation of nuclides such as $^{14}_{6}C$ and $^{32}_{15}P$ into nutrients has yielded important information about how these nutrients are used to provide energy for the body.

Iodine-131 has proved very useful in the diagnosis and treatment of illnesses of the thyroid gland. Patients drink a solution containing a small amount of NaI that includes[131]I, and the uptake of the iodine by the thyroid gland is monitored with a scanner (Figure 19.3).

Thallium-201 can be used to assess the damage to the heart muscle in a person who has suffered a heart attack because thallium becomes concentrated

PET, the Brain's Best Friend

One of the most valuable applications of radioactivity is in the use of radiotracers for medical diagnosis. Radiotracers are radioactive atoms that are attached to biologically active molecules. The resultant radioactivity is monitored to check on the functioning of organs such as the heart or to trace the path and final destination of a drug.

One particularly valuable radiotracer technique is called positron emission tomography (PET). As its name suggests, PET uses positron producers, such as ^{18}F and ^{11}C, as "labels" on biologic molecules. PET is especially useful for brain scans. For example, a modified form of glucose with ^{18}F attached is commonly used to map glucose metabolism. Areas of the brain where glucose is being rapidly consumed "light up" on the PET screen. The brain of a patient who has a tumor or who has Alzheimer's disease will show a much different picture than will a brain of a patient without Alzheimer's disease. Another application of PET is in seeing how much of a particular labeled drug reaches the intended target. This enables pharmaceutical companies to check the effectiveness of a drug and to set dosages.

One of the challenges of using PET is the speed required to synthesize the labeled molecule. For example, ^{18}F has a half-life of 110 minutes.

Thus, in a little less than 2 hours after the ^{18}F has been produced in a particle accelerator, half of it has already decayed. Also, because of the dangers of handling radioactive ^{18}F, synthesis operations must be carried out by robotic manipulations inside a lead-lined box. The good news is that PET is incredibly sensitive—it can detect amounts of ^{18}F as small as 10^{-12} mol. The use of ^{11}C is even more challenging synthetically than the use of ^{18}F because ^{11}C has a half-life of only 20 minutes.

PET is a rapidly growing technology. In particular, more chemists are needed in this field to improve synthetic methods and to develop new radioactive tracers. If this is of interest to you, do some exploring to see how to prepare yourself for a job in this field.

Pascal Goetgheluck/Photo Researchers, Inc.

A positron emission tomography (PET) scanner.

SIU/Visuals Unlimited

a *Scan of radioactive iodine in a normal thyroid.*

b *Scan of an enlarged thyroid.*

Figure 19.3

After consumption of $Na^{131}I$, the patient's thyroid is scanned for radioactivity levels to determine the efficiency of iodine absorption.

Table 19.4 Some Radioactive Nuclides, Their Half-lives, and Their Medical Applications as Radiotracers*

Nuclide	Half–life	Area of the Body Studied
^{131}I	8.1 days	thyroid
^{59}Fe	45.1 days	red blood cells
^{99}Mo	67 hours	metabolism
^{32}P	14.3 days	eyes, liver, tumors
^{51}Cr	27.8 days	red blood cells
^{87}Sr	2.8 hours	bones
^{99}Tc	6.0 hours	heart, bones, liver, lungs
^{133}Xe	5.3 days	lungs
^{24}Na	14.8 hours	circulatory system

*Z is sometimes not written when listing nuclides.

in healthy muscle tissue. Technetium-99, which is also taken up by normal heart tissue, is used for damage assessment in a similar way.

Radiotracers provide sensitive and nonsurgical methods for learning about biologic systems, for detecting disease, and for monitoring the action and effectiveness of drugs. Some useful radiotracers are listed in Table 19.4.

19.6 Nuclear Energy

OBJECTIVE: To introduce fusion and fission as producers of nuclear energy.

The protons and the neutrons in atomic nuclei are bound together with forces that are much greater than the forces that bind atoms together to form molecules. In fact, the energies associated with nuclear processes are more than a million times those associated with chemical reactions. This potentially makes the nucleus a very attractive source of energy.

Because medium-sized nuclei contain the strongest binding forces ($^{56}_{26}$Fe has the strongest binding forces of all), there are two types of nuclear processes that produce energy:

1. Combining two light nuclei to form a heavier nucleus. This process is called **fusion.**

2. Splitting a heavy nucleus into two nuclei with smaller mass numbers. This process is called **fission.**

As we will see in the next several sections, these two processes can supply amazing quantities of energy with relatively small masses of materials consumed.

19.7 Nuclear Fission

OBJECTIVE: To learn about nuclear fission.

Nuclear fission was discovered in the late 1930s when $^{235}_{92}$U nuclides bombarded with neutrons were observed to split into two lighter elements.

$$^{1}_{0}n + ^{235}_{92}U \rightarrow ^{141}_{56}Ba + ^{92}_{36}Kr + 3\,^{1}_{0}n$$

Upon capturing a neutron, the $^{235}_{92}U$ nucleus undergoes fission to produce two lighter nuclides, more neutrons (typically three), and a large amount of energy.

This process, shown schematically in Figure 19.4, releases 2.1×10^{13} joules of energy per mole of $^{235}_{92}U$. Compared with what we get from typical fuels, this is a huge amount of energy. For example, the fission of 1 mole of $^{235}_{92}U$ produces about *26 million times* as much energy as the combustion of 1 mole of methane.

The process shown in Figure 19.4 is only one of the many fission reactions that $^{235}_{92}U$ can undergo. In fact, over 200 different isotopes of 35 different elements have been observed among the fission products of $^{235}_{92}U$.

In addition to the product nuclides, neutrons are produced in the fission reactions of $^{235}_{92}U$. As these neutrons fly through the solid sample of uranium, they may collide with other $^{235}_{92}U$ nuclei, producing additional fission events. Each of these fission events produces more neutrons that can, in turn, produce the fission of more $^{235}_{92}U$ nuclei. Because each fission event produces neutrons, the process can be self-sustaining. We call it a **chain reaction** (Figure 19.5). For the fission process to be self-sustaining, at least one neutron from each fission event must go on to split another nucleus. If, on average, *less than one* neutron causes another fission event, the process dies out. If *exactly one* neutron from each fission event causes another fission event, the process sustains itself at the same level and is said to be *critical*. If *more than one* neutron from each fission event causes another fission event, the process rapidly escalates and the heat buildup causes a violent explosion.

To achieve the critical state, a certain mass of fissionable material, called the **critical mass,** is needed. If the sample is too small, too many

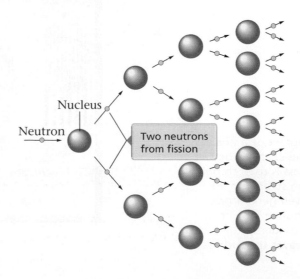

Representation of a fission process in which each event produces two neutrons that can go on to split other nuclei, leading to a self-sustaining chain reaction.

neutrons escape before they have a chance to cause a fission event, and the process stops.

During World War II, the United States carried out an intense research effort called the Manhattan Project to build a bomb based on the principles of nuclear fission. This program produced the fission bomb, which was used with devastating effect on the cities of Hiroshima and Nagasaki in 1945. Basically, a fission bomb operates by suddenly combining two subcritical masses, which results in rapidly escalating fission events that produce an explosion of incredible intensity.

19.8 Nuclear Reactors

OBJECTIVE: To understand how a nuclear reactor works.

> Natural uranium consists mostly of $^{238}_{92}U$.

The core of a nuclear power plant.

Because of the tremendous energies involved, fission has been developed as an energy source to produce electricity in reactors where controlled fission can occur. The resulting energy is used to heat water to produce steam that runs turbine generators, in much the same way that a coal-burning power plant generates energy by heating water to produce steam. A schematic diagram of a nuclear power plant is shown in Figure 19.6.

In the reactor core (Figure 19.7), uranium that has been enriched to approximately 3% $^{235}_{92}U$ (natural uranium contains only 0.7% $^{235}_{92}U$) is housed in metal cylinders. A *moderator* surrounding the cylinders slows the neutrons down so that the uranium fuel can capture them more efficiently. *Control rods,* composed of substances (such as cadmium) that absorb neutrons, are used to regulate the power level of the reactor. The reactor is designed so that if a malfunction occurs, the control rods are automatically inserted into the core to absorb neutrons and stop the reaction. A liquid (usually water) is circulated through the core to extract the heat generated by the energy of fission. This heat energy is then used to change water to steam, which runs turbines that in turn run electrical generators.

A schematic diagram of a nuclear power plant. The energy from the fission process is used to boil water, producing steam for use in a turbine-driven generator. Cooling water from a lake or river is used to condense the steam after it leaves the turbine.

Figure 19.7

A schematic diagram of a reactor core.

the design of the reactor core prevents any possibility of a critical mass

- so no chance of a fusion bomb going off but "meltdowns" can occur

Although the concentration of $^{235}_{92}U$ in the fuel elements is not great enough to allow an explosion such as that which occurs in a fission bomb, a failure of the cooling system can lead to temperatures high enough to melt the reactor core. This means that the building housing the core must be designed to contain the core even in the event of such a "meltdown." A great deal of controversy now exists about the efficiency of the safety systems in nuclear power plants. Accidents such as the one at the Three Mile Island facility in Pennsylvania in 1979 and the one at Chernobyl in the Soviet Union in 1986 have led many people to question the wisdom of continuing to build fission-based power plants.

▶ Breeder Reactors

One potential problem facing the nuclear power industry is the limited supply of $^{235}_{92}U$. Some scientists believe that we have nearly depleted the uranium deposits that are rich enough in $^{235}_{92}U$ to make the production of fissionable fuel economically feasible. Because of this possibility, reactors have been developed in which fissionable fuel is actually produced while the reactor runs. In these **breeder reactors,** the major component of natural uranium, nonfissionable $^{238}_{92}U$, is changed to fissionable $^{239}_{94}Pu$. The reaction involves absorption of a neutron, followed by production of two β particles.

$$^{1}_{0}n + {}^{238}_{92}U \rightarrow {}^{239}_{92}U$$
$$^{239}_{92}U \rightarrow {}^{239}_{93}Np + {}^{0}_{-1}e$$
$$^{239}_{93}Np \rightarrow {}^{239}_{94}Pu + {}^{0}_{-1}e$$

As the reactor runs and $^{235}_{92}U$ is split, some of the excess neutrons are absorbed by $^{238}_{92}U$ to produce $^{239}_{94}Pu$. The $^{239}_{94}Pu$ is then separated out and used to fuel another reactor. Such a reactor thus "breeds" nuclear fuel as it operates.

Although breeder reactors are now used in Europe, the United States is proceeding slowly with their development because much controversy surrounds their use. One problem involves the hazards that arise in handling plutonium, which is very toxic and flames on contact with air.

19.9 Nuclear Fusion

OBJECTIVE: To learn about nuclear fusion.

$^{2}_{1}H$ particles are called *deuterons*.

The process of combining two light nuclei—called **nuclear fusion**—produces even more energy per mole than does nuclear fission. In fact, stars produce their energy through nuclear fusion. Our sun, which presently consists of 73% hydrogen, 26% helium, and 1% other elements, gives off vast quantities of energy from the fusion of protons to form helium. One possible scheme for this process is

$$^{1}_{1}H + {}^{1}_{1}H \rightarrow {}^{2}_{1}H + {}^{0}_{1}e + \text{energy}$$
$$^{1}_{1}H + {}^{2}_{1}H \rightarrow {}^{3}_{2}He + \text{energy}$$
$$^{3}_{2}He + {}^{3}_{2}He \rightarrow {}^{4}_{2}He + 2\,{}^{1}_{1}H + \text{energy}$$
$$^{3}_{2}He + {}^{1}_{1}H \rightarrow {}^{4}_{2}He + {}^{0}_{1}e + \text{energy}$$

Intense efforts are under way to develop a feasible fusion process because of the ready availability of many light nuclides (deuterium, $^{2}_{1}H$, in

Future Nuclear Power

Energy—a crucial commodity in today's world—will become even more important as the pace of world development increases. Because the energy content of the universe is constant, the challenge of energy is not its quantity but rather its quality. We must find economical and environmentally friendly ways to change the energy available in the universe to forms useful to humanity. This process always involves trade-offs. One of the most abundant sources of energy is the energy that binds the atomic nucleus together. We can derive useful energy by assembling small nuclei (fusion) or splitting large nuclei (fission). Although fusion reactors are being studied, a practical fusion reactor appears to be decades away. By contrast, fission reactors have been used since the 1950s. In fact, the production of electricity via fission reactors is widespread. At present, more than 430 nuclear reactors operate in 31 countries, producing over 370 billion watts of electrical power. More than 45 reactors are currently under construction. The 104 reactors currently operating in the United States produce over 100 billion watts of electricity—almost 20% of the country's electrical demands.

Forecasts indicate that the United States will need an additional 355 billion watts of generating capacity in the next 20 years. Where will this energy come from? A significant amount will be derived from coal-fired power plants. However, the major problems with these plants are air pollution and greenhouse gas (CO_2) production (see "Chemistry in Focus: Atmospheric Effects," on page 326).

Another potential source of power is solar energy. It should be an excellent pollution-free energy source, but significant technical problems remain to be solved before it sees widespread use.

Still another important potential power source is nuclear energy. To provide all of the needed 355 billion watts from nuclear energy would require hundreds of new nuclear reactors. However, nuclear power generation is very controversial because of safety, waste disposal, and cost issues.

To address these issues, huge amounts of money are being spent to improve existing reactor designs and to find new types of reactors that will be safer, be more efficient, and generate much less waste. An international consortium—the Generation IV International Forum (GIF)—was formed in 2000 and has decided to study six new reactor designs with the goal of development of one or more of these designs by 2030.

Nuclear power generation is too important to ignore. In the near future we must decide whether its use can be extended safely and economically.

A nuclear reactor.

seawater, for example) that can serve as fuel in fusion reactors. However, initiating the fusion process is much more difficult than initiating fission. The forces that bind nucleons together to form a nucleus become effective only at *very small* distances (approximately 10^{-13} cm), so for two protons to bind together and thereby release energy, they must get very close together. But protons, because they are identically charged, repel each other. This suggests that to get two protons (or two deuterons) close enough to bind together (the strong nuclear binding force is *not* related to charge), they must

A solar flare erupts from the surface of the sun.

be "shot" at each other at speeds high enough to overcome their repulsion from each other. The repulsive forces between two ^2_1H nuclei are so great that temperatures of about 40 million K are thought to be necessary. Only at these temperatures are the nuclei moving fast enough to overcome the repulsions.

Currently, scientists are studying two types of systems to produce the extremely high temperatures required: high-powered lasers and heating by electric currents. At present, many technical problems remain to be solved, and it is not clear whether either method will prove useful.

19.10 Effects of Radiation

OBJECTIVE: To see how radiation damages human tissue.

Everyone knows that being hit by a train is a catastrophic event. The energy transferred in such a collision is very large. In fact, any source of energy is potentially harmful to organisms. Energy transferred to cells can break chemical bonds and cause malfunctioning of the cell systems. This fact is behind our present concern about maintaining the ozone layer in the earth's upper atmosphere, which screens out high-energy ultraviolet radiation arriving from the sun. Radioactive elements, which are sources of high-energy particles, are also potentially hazardous. However, the effects are usually quite subtle, because even though high-energy particles are involved, the quantity of energy actually deposited in tissues *per decay event* is quite small. The resulting damage is no less real, but the effects may not be apparent for years.

Radiation damage to organisms can be classified as somatic or genetic damage. *Somatic damage* is damage to the organism itself, resulting in sickness or death. The effects may appear almost immediately if a massive dose of radiation is received; for smaller doses, damage may appear years later, usually in the form of cancer. *Genetic damage* is damage to the genetic machinery of reproductive cells, creating problems that often afflict the offspring of the organism.

The biologic effects of a particular source of radiation depend on several factors:

1. *The energy of the radiation.* The higher the energy content of the radiation, the more damage it can cause.

2. *The penetrating ability of the radiation.* The particles and rays produced in radioactive processes vary in their ability to penetrate human tissue: γ rays are highly penetrating, β particles can penetrate approximately 1 cm, and α particles are stopped by the skin (Figure 19.8).

3. *The ionizing ability of the radiation.* Because ions behave quite differently from neutral molecules, radiation that removes electrons from molecules in living tissues seriously disturbs their functions. The ionizing ability of radiation varies dramatically. For example, γ rays penetrate very deeply but cause only occasional ionization. On the other hand, α particles, although they are not very penetrating, are very effective at causing ionization and produce serious damage. Therefore, the ingestion of a producer of α particles, such as plutonium, is particularly damaging.

α surface
β in a little more
γ most damaging

Figure 19.8

Radioactive particles and rays vary greatly in penetrating power. Gamma rays are by far the most penetrating.

Nuclear Waste Disposal

Our society does not have a very impressive record for safe disposal of industrial wastes. We have polluted our water and air, and some land areas have become virtually uninhabitable because of the improper burial of chemical wastes. As a result, many people are wary about the radioactive wastes from nuclear reactors. The potential threats of cancer and genetic mutations make these materials especially frightening.

Because of its controversial nature, most of the nuclear waste generated over the past 50 years has been placed in temporary storage. However, in 1982 the U.S. Congress passed the Nuclear Waste Policy Act, which established a timetable for choosing and preparing sites for the deep underground disposal of radioactive materials.

The tentative disposal plan calls for incorporation of the spent nuclear fuel into blocks of glass that will be packed in corrosion-resistant metal containers and then buried in a deep, stable rock formation indicated by the rock layers in Figure 19.9.

There are indications that this method will isolate the waste until the radioactivity decays to safe levels. Some reassuring evidence comes from a natural fission "reactor" that has been discovered at Oklo in Gabon, Africa. Spawned about 2 billion years ago when uranium in ore deposits there formed a critical mass, this "reactor" produced fission and fusion products for several thousand years. Although some of these products have migrated away from the site in the intervening 2 billion years, most have stayed in place.

Finally, more than 25 years after the Nuclear Policy Act, we have begun to store nuclear waste. In 1998, the Waste Isolation Policy Plant (WIPP) in New Mexico was issued a license by the U.S. Environmental Protection Agency to begin receiving nuclear waste, and in March 1999, WIPP received its first waste. This facility employs tunnels carved into the salt beds of an ancient ocean. Once a repository room becomes full, the salt will collapse around the waste, encapsulating it forever.

Another waste depository, under Yucca Mountain in Nevada, may begin to store nuclear waste in the next several years. In June 2008, the Department of Energy submitted a license application to the Nuclear Regulatory Commission (NCR) to construct a repository. This process can take 3 to 4 years for approval and then more time to actually build the repository.

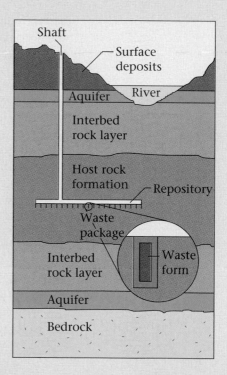

Figure 19.9

A schematic diagram for the tentative plan for deep underground isolation of nuclear waste. The disposal system would consist of a waste package buried in an underground repository. (Reprinted with permission from *Chemical & Engineering News,* July 18, 1983. Copyright © 1983 American Chemical Society.)

4. *The chemical properties of the radiation source.* When a radioactive nuclide is ingested, its capacity to cause damage depends on how long it remains in the body. For example, both $^{85}_{36}Kr$ and $^{90}_{38}Sr$ are β-particle producers. Because krypton, being a noble gas, is chemically inert, it passes through the body quickly and does not have much time to do damage. Strontium, on the other hand, is chemically similar to calcium. It can collect in bones, where it may cause leukemia and bone cancer.

Because of the differences in the behavior of the particles and rays produced by radioactive decay, we have invented a unit called the **rem** that indicates the danger the radiation poses to humans.

Table 19.5 shows the physical effects of short-term exposure to various doses of radiation, and Table 19.6 gives the sources and amounts of the radiation a typical person in the United States is exposed to each year. Note that natural sources contribute about twice as much as human activities do to the total exposure. However, although the nuclear industry contributes only a small percentage of the total exposure, controversy surrounds nuclear power plants because of their *potential* for creating radiation hazards. These hazards arise mainly from two sources: accidents allowing the release of radioactive materials, and improper disposal of the radioactive products in spent fuel elements.

Table 19.5 Effects of Short-Term Exposures to Radiation

Dose (rem)	Clinical Effect
0–25	nondetectable
25–50	temporary decrease in white blood cell counts
100–200	strong decrease in white blood cell counts
500	death of half the exposed population within 30 days after exposure

Table 19.6 Typical Radiation Exposures for a Person Living in the United States (1 millirem = 10^{-3} rem)

Source	Exposure (millirems/year)
cosmic	50
from the earth	47
from building materials	3
in human tissues	21
inhalation of air	5
Total from natural sources	126
X-ray diagnosis	50
radiotherapy X rays, radioisotopes	10
internal diagnosis and therapy	1
nuclear power industry	0.2
luminous watch dials, TV tubes, industrial wastes	2
radioactive fallout	4
Total from human activities	67
Total	193 = 0.193 rem

CHAPTER 19 REVIEW

Key Terms

nucleons (neutrons and protons) (p. 615)

atomic number (Z) (p. 615)

mass number (A) (p. 615)

isotope (p. 615)

nuclide (p. 615)

radioactive (19.1)

beta (β) particle (19.1)

nuclear equation (19.1)

alpha (α) particle (19.1)

alpha (α)-particle production (19.1)

beta (β)-particle production (19.1)

gamma (γ) ray (19.1)

positron (19.1)

positron production (19.1)

electron capture (19.1)

decay series (19.1)

nuclear transformation (19.2)

transuranium elements (19.2)

Geiger–Müller counter (Geiger counter) (19.3)

scintillation counter (19.3)

half-life (19.3)

radiocarbon dating (carbon-14 dating) (19.4)

radiotracers (19.5)

fusion (19.6)

fission (19.6)

chain reaction (19.7)

critical mass (19.7)

breeder reactor (19.8)

nuclear fusion (19.9)

rem (19.10)

Summary

1. Radioactivity is the spontaneous decomposition of a nucleus to form another nucleus and produce one or more particles. We can write a nuclear equation to represent radioactive decay, in which both A (mass number) and Z (atomic number) must be conserved.

2. There are several types of radioactive decay: alpha-particle production, in which an alpha particle (helium nucleus) is produced; beta-particle (or electron) production; the production of gamma rays (high-energy photons of light); and electron capture, in which one of the inner-orbital electrons is captured by the nucleus. Often a series of decays occurs before a radioactive nucleus attains a stable state.

3. The production of new elements by nuclear transformation (the change of one element into another) is carried out by bombarding various nuclei with particles in accelerators. The transuranium elements have been synthesized in this way.

4. The half-life of a radioactive nuclide is the time required for one-half of the original sample to decay. Radiocarbon dating is based on the radioactivity of carbon-14.

5. Radiotracers—radioactive nuclides that can be introduced into organisms in food or drugs and whose pathways can be traced by monitoring their radioactivity—are used diagnostically in medicine.

6. Nuclear fusion is the process of combining two light nuclei to form a heavier, more stable nucleus. Nuclear fission involves the splitting of a heavy nucleus into two (more stable) lighter nuclei. Current nuclear reactors employ controlled fission.

7. Radiation can cause either direct damage to living tissues or damage to reproductive cells that manifests itself in the organism's offspring. The biological effects of radiation depend on the energy of the radiation, the radiation's penetrating ability and ionizing ability, and the chemical properties of the source of the radiation.

Active Learning Questions

These questions are designed to be considered by groups of students in class. Often these questions work well for introducing a particular topic in class.

1. It states in Section 19.1 of your text that approximately 2000 nuclides are known. How can this be possible if there are only about 100 elements?

2. By how many units does the mass number of a nucleus change when the nucleus produces an alpha particle? By how many units does the mass number of a nucleus change when the nucleus produces a beta particle? Is each change an increase or decrease in mass number?

3. Determine the number of half-lives that must pass for only 1% of a particular radioisotope to remain.

4. Hundreds of years ago, alchemists tried to turn lead into gold. Is this transformation possible? If not, why not? If yes, how would you do it?

5. Do radiotracers generally have long or short half-lives? Why?

6. Why is the nucleus of an atom a great source of energy?

7. Discuss what is meant by the term "critical mass" and explain why the ability to achieve a critical mass is essential to sustaining a nuclear reaction.

8. What makes fusion preferable over fission? What makes fusion more complicated?

9. Why is it difficult to determine the effects of radiation on humans?

10. What type of radioactive decay must occur for each of the following transformations?

  a. Transformation 1
  b. Transformation 2
  c. Transformation 3

20

Organic Chemistry

● The National Aquatic Centre (also known as the "Water Cube") used in the 2008 Summer Olympics in Beijing, China. *(Michael Reynolds/ EPA/Corbis)*

The study of carbon-containing compounds and their properties is called **organic chemistry.** The industries based on organic substances have truly revolutionized our lives. In particular, the development of polymers, such as nylon for fabrics; Velcro for fasteners; Kevlar for the composites used in exotic cars, airplanes, and bicycles; and polyvinyl chloride (PVC) for pipes, siding, and toys has produced a wondrous new world.

Two Group 4 elements, carbon and silicon, form the basis of most natural substances. Silicon, with its great affinity for oxygen, forms chains and rings containing Si—O—Si bonds to produce the silica and silicates that form the basic structures for most rocks, sands, and soils. Therefore, silicon compounds are the fundamental inorganic materials of the earth. What silicon is to the geologic world, carbon is to the organic or biologic world. Carbon has the unusual ability of bonding strongly to itself, forming long chains or rings of carbon atoms. In addition, carbon forms strong bonds to other nonmetals such as hydrogen, nitrogen, oxygen, sulfur, and the halogens. Because of these bonding properties, an extraordinary number of carbon compounds exists; several million are now known, and the number continues to grow rapidly. Among these many compounds are the **biomolecules,** molecules that make possible the maintenance and reproduction of life.

Although a few compounds of carbon, such as the oxides of carbon and carbonates, are considered to be inorganic substances, the vast majority of carbon compounds are designated as organic compounds—compounds that typically contain chains or rings of carbon atoms. Originally, the distinction between inorganic and organic substances was based on whether they were produced by living systems. For example, until the early nineteenth century, it was believed that organic compounds had some kind of "life force" and could be synthesized only by living organisms. This misconception was dispelled in 1828 when the German chemist Friedrich Wöhler (1800–1882) prepared urea from the inorganic salt ammonium cyanate by simple heating.

Dr. Jeremy Burgess/SPL/Photo Researchers, Inc.

A close-up photo of Velcro, a synthetic organic material used for fasteners.

$$NH_4OCN \xrightarrow{\text{Heat}} H_2N-\underset{\underset{O}{\|}}{C}-NH_2$$

Ammonium cyanate   Urea

Urea is a component of urine, so it is clearly an organic material formed by living things, yet here was hard evidence that it could be produced in the laboratory as well.

Organic chemistry plays a vital role in our quest to understand living systems. Beyond that, the synthetic fibers, plastics, artificial sweeteners, and medicines that we take for granted are products of industrial organic chemistry. Finally, the energy on which we rely so heavily to power our civilization is based mostly on the combustion of organic materials found in coal and petroleum.

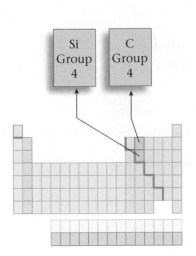

Because organic chemistry is such a vast subject, we can introduce it only briefly in this text. We will begin with the simplest class of organic compounds, the hydrocarbons, and then show how most other organic compounds can be considered to be derived from hydrocarbons.

20.1 Carbon Bonding

OBJECTIVE: To understand the types of bonds formed by the carbon atom.

So many carbon-containing compounds exist because carbon forms strong bonds to itself and to many other elements. A carbon atom can form bonds to a maximum of four other atoms; these can be either carbon atoms or atoms of other elements. One of the hardest, toughest materials known is diamond, a form of pure carbon in which each carbon atom is bound to four other carbon atoms (see Figure 4.18a).

One of the most familiar compounds of carbon is methane, CH_4, the main component of natural gas. The methane molecule consists of a carbon atom with four hydrogen atoms bound to it in a tetrahedral fashion. That is, as predicted by the VSEPR model (see Section 12.9), the four pairs of bonding electrons around the carbon have minimum repulsions when they are located at the corners of a tetrahedron.

 or

This leads to the structure for CH_4 shown in Figure 20.1. *When carbon has four atoms bound to it, these atoms will always have a tetrahedral arrangement about the carbon.*

Carbon can bond to fewer than four elements by forming one or more multiple bonds. Recall that a multiple bond involves the sharing of more than one pair of electrons. For example, a *double bond* involves sharing two pairs of electrons, as in carbon dioxide:

$$\ddot{O}=C=\ddot{O}$$

and a *triple bond* involves sharing three pairs of electrons, as in carbon monoxide:

$$:C\equiv O:$$

Note that carbon is bound to two other atoms in CO_2 and to only one other atom in CO.

Multiple bonding also occurs in organic molecules. Ethylene, C_2H_4, has a double bond:

CO₂ and CO are classed as inorganic substances.

Figure 20.1

Methane is a tetrahedral molecule.

In this case, each carbon is bound to three other atoms (one C atom and two H atoms). A molecule with a triple bond is acetylene, C_2H_2:

$$H-C\equiv C-H$$

Here each carbon is bound to two other atoms (one carbon atom and one hydrogen atom).

More than any other element, carbon has the ability to form chains of atoms, as illustrated by the structures of propane and butane shown in Figure 20.3. In these compounds, each carbon atom is bound to four atoms in a tetrahedral fashion. We will discuss these molecules in detail in the next section.

20.2 Alkanes

OBJECTIVE: To learn about the alkanes—compounds that contain saturated carbon atoms.

[go Chemistry] **Module 15: Naming Organic Compounds** covers concepts in this section.

Hydrocarbons, as the name indicates, are compounds composed of carbon and hydrogen. Those whose carbon–carbon bonds are all single bonds are said to be **saturated** because each carbon is bound to four atoms, the maximum number. Hydrocarbons containing carbon–carbon multiple bonds are described as being **unsaturated** because the carbon atoms involved in a multiple bond can bond to one or more additional atoms. This is shown by the *addition* of hydrogen to ethylene.

Ethylene + H_2 → Ethane

Unsaturated → Saturated

Note that each carbon in ethylene is bonded to three atoms (one carbon and two hydrogens) but can bond to one additional atom after one bond of the carbon–carbon double bond is broken. This leads to ethane, a saturated hydrocarbon (each carbon atom is bonded to four atoms).

Saturated hydrocarbons are called **alkanes.** The simplest alkane is *methane,* CH_4, which has a tetrahedral structure (see Figure 20.1). The next alkane, the one containing two carbon atoms, is *ethane,* C_2H_6, shown in Figure 20.2. Note that each carbon atom in ethane is bonded to four atoms.

The next two members of the series are *propane,* with three carbon atoms and the formula C_3H_8, and *butane,* with four carbon atoms and the

Figure 20.2 **a** The Lewis structure of ethane, C_2H_6. The molecular structure of ethane represented by **b** a space-filling model and **c** a ball-and-stick model.

Figure 20.3

The structures of propane and butane.

Propane   Butane

Computer-generated model of ethane.

formula C_4H_{10}. These molecules are shown in Figure 20.3. Again, these are saturated hydrocarbons (alkanes); each carbon is bonded to four atoms.

Alkanes in which the carbon atoms form long "strings" or chains are called **normal, straight-chain,** or **unbranched hydrocarbons.** As Figure 20.3 illustrates, the chains in normal alkanes are not really straight but zigzag because the tetrahedral C—C—C angle is 109.5°. The normal alkanes can be represented by the structure

$$H-\overset{\displaystyle H}{\underset{\displaystyle H}{C}}-\left(\overset{\displaystyle H}{\underset{\displaystyle H}{C}}\right)_m-\overset{\displaystyle H}{\underset{\displaystyle H}{C}}-H$$

where m is an integer. Note that each member is obtained from the previous one by insertion of a *methylene,* CH_2, group. We can condense the structural formulas by omitting some of the C—H bonds. For example, the general formula for normal alkanes shown above can be condensed to

$$CH_3-(CH_2)_{\overline{m}}\,CH_3$$

EXAMPLE 20.1 | Writing Formulas for Alkanes

Give the formulas for the normal (or straight-chain) alkanes with six and eight carbon atoms.

SOLUTION

The alkane with six carbon atoms can be written as

$$CH_3CH_2CH_2CH_2CH_2CH_3$$

which can be condensed to

$$CH_3-(CH_2)_{\overline{4}}\,CH_3$$

Notice that the molecule contains fourteen hydrogen atoms in addition to the six carbon atoms. Therefore, the formula is C_6H_{14}.

The alkane with eight carbons is

$$CH_3CH_2CH_2CH_2CH_2CH_2CH_2CH_3$$

1   2   3   4   5   6   7   8

which can be written in condensed form as

$$CH_3-(CH_2)_6-CH_3$$

This molecule has eighteen hydrogens. The formula is C_8H_{18}.

Self-Check EXERCISE 20.1   Give the molecular formulas for the alkanes with ten and fifteen carbon atoms.

See Problems 20.11 through 20.14. ■

The first ten straight-chain alkanes are shown in Table 20.1. Note that all alkanes can be represented by the general formula C_nH_{2n+2}, where n represents the number of carbon atoms. For example, nonane, which has nine carbon atoms, is represented by $C_9H_{(2 \times 9)+2}$, or C_9H_{20}. The formula C_nH_{2n+2} reflects the fact that each carbon in the chain has two hydrogen atoms except the two end carbons, which have three each. Thus the number of hydrogen atoms present is twice the number of carbon atoms plus two (for the extra two hydrogen atoms on the ends).

EXAMPLE 20.2 | **Using the General Formula for Alkanes**

Show that the alkane with fifteen carbon atoms can be represented in terms of the general formula C_nH_{2n+2}.

SOLUTION

In this case $n = 15$. The formula is $C_{15}H_{2(15)+2}$, or $C_{15}H_{32}$, as was found in Self-Check Exercise 20.1. ■

Table 20.1 Formulas of the First Ten Straight-Chain Alkanes

Name	Condensed Formula (C_nH_{2n+2})	Extended Formula
methane	CH_4	CH_4
ethane	C_2H_6	CH_3CH_3
propane	C_3H_8	$CH_3CH_2CH_3$
n-butane	C_4H_{10}	$CH_3CH_2CH_2CH_3$
n-pentane	C_5H_{12}	$CH_3CH_2CH_2CH_2CH_3$
n-hexane	C_6H_{14}	$CH_3CH_2CH_2CH_2CH_2CH_3$
n-heptane	C_7H_{16}	$CH_3CH_2CH_2CH_2CH_2CH_2CH_3$
n-octane	C_8H_{18}	$CH_3CH_2CH_2CH_2CH_2CH_2CH_2CH_3$
n-nonane	C_9H_{20}	$CH_3CH_2CH_2CH_2CH_2CH_2CH_2CH_2CH_3$
n-decane	$C_{10}H_{22}$	$CH_3CH_2CH_2CH_2CH_2CH_2CH_2CH_2CH_2CH_3$

Structural Formulas and Isomerism

OBJECTIVE: To learn about structural isomers and how to draw their structural formulas.

Butane and all succeeding alkanes exhibit structural isomerism. **Structural isomerism** occurs when two molecules have the same atoms but different bonds. That is, the molecules have the same formulas but different arrangements of the atoms. For example, butane can exist as a straight-chain molecule (normal butane, or *n*-butane) or as a branched-chain structure (called isobutane), as shown in Figure 20.4. Because of their different structures, these structural isomers have different properties.

EXAMPLE 20.3 | **Drawing Structural Isomers of Alkanes**

Draw the structural isomers of pentane, C_5H_{12}.

SOLUTION

To find the isomeric structures for pentane, C_5H_{12}, we first write the straight carbon chain and then add the hydrogen atoms.

1. The straight-chain structure has the five carbon atoms in a row.

C—C—C—C—C

Figure 20.4

The structural isomers of C_4H_{10}. Each molecule is represented in three ways: a ball-and-stick structure, a space-filling structure, and a structure that shows the shared electrons as lines (a Lewis structure). (top) Normal butane (abbreviated *n*-butane). (bottom) The branched isomer of butane (called isobutane).

We can now add the H atoms.

$$H-\underset{\underset{H}{|}}{\overset{\overset{H}{|}}{C}}-\underset{\underset{H}{|}}{\overset{\overset{H}{|}}{C}}-\underset{\underset{H}{|}}{\overset{\overset{H}{|}}{C}}-\underset{\underset{H}{|}}{\overset{\overset{H}{|}}{C}}-\underset{\underset{H}{|}}{\overset{\overset{H}{|}}{C}}-H$$

This can be written in shorthand form as

$$CH_3-CH_2-CH_2-CH_2-CH_3 \quad \text{or} \quad CH_3-(CH_2)_3-CH_3$$

and is called *n*-pentane.

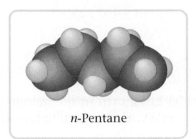

n-Pentane

2. Next we remove one C atom from the main chain and bond it to the second carbon in the chain.

$$\underset{C-\overset{|}{C}-C-C}{\overset{\overset{C}{|}}{}}$$

Then we put on the H atoms so that each carbon has four bonds.

This structure can be represented as

$$CH_3-\underset{}{\overset{\overset{CH_3}{|}}{CH}}-CH_2-CH_3$$

and is called isopentane.

Isopentane

3. Finally, we take two carbons out of the chain to give the arrangement

$$\underset{\underset{C}{|}}{\overset{\overset{C}{|}}{C-\overset{}{C}-C}}$$

Adding the H atoms gives

Neopentane

which can be written in shorthand form as

$$CH_3-\overset{\overset{\displaystyle CH_3}{|}}{\underset{\underset{\displaystyle CH_3}{|}}{C}}-CH_3$$

This molecule is called neopentane.

The space-filling models for these molecules are shown in the margin. Note that all these molecules have the formula C_5H_{12}, as required. Also note that the structures

$$CH_3-CH_2-\overset{\overset{\displaystyle CH_3}{|}}{CH}-CH_3 \qquad CH_3-\overset{\underset{\underset{\displaystyle CH_3}{|}}{CH}}{}-CH_2-CH_3 \qquad CH_3-CH_2-\overset{\underset{\underset{\displaystyle CH_3}{|}}{CH}}{}-CH_3$$

which might at first appear to be additional isomers, are actually identical to structure 2. All three of these structures have exactly the same skeleton of carbons as the structure shown in part 2. All these structures have four carbons in the chain with one carbon on the side:

$$C-C-\overset{\overset{\displaystyle}{|}}{\underset{\underset{\displaystyle C}{|}}{C}}-C$$

20.4 Naming Alkanes

OBJECTIVE: To learn the system for naming alkanes and substituted alkanes.

Because there are literally millions of organic compounds, it would be impossible to remember common names for all of them. Just as we did in Chapter 5 for inorganic compounds, we must learn a systematic method for naming organic compounds. We will first consider the principles applied in naming alkanes and then summarize them as a set of rules.

1. The first four members of the alkane series are called methane, ethane, propane, and butane. The names of the alkanes beyond butane are obtained by adding the suffix -*ane* to the Greek root for the number of carbon atoms.

Number	Greek Root
5	*pent*
6	*hex*
7	*hept*
8	*oct*
9	*non*
10	*dec*

Therefore, the alkane

$$CH_3CH_2CH_2CH_2CH_2CH_2CH_2CH_3$$

which has eight carbons in the chain, is called octane.

(oct   –   ane)

Tells us   Tells us
there are   it is an
eight carbons   alkane

The complete name for this alkane is *n*-octane, the *n* indicating a straight-chain alkane.

2. For a branched hydrocarbon, the longest continuous chain of carbon atoms gives the root name for the hydrocarbon. For example, in the alkane

$$CH_3$$
$$|$$
$$CH_2 \quad \text{Six carbons}$$
$$|$$
$$CH_2$$
$$|$$
$$CH_3-CH_2-CH-CH_2-CH_3$$

Five carbons

the longest continuous chain contains six carbon atoms. The specific name of this compound is not important at this point, but it will be named as a hexane (indicating a six-carbon chain).

3. Alkanes lacking one hydrogen atom can be attached to a hydrocarbon chain in place of one hydrogen atom. For example, the molecule

$$
\begin{array}{ccccc}
H & H & H & H & H \\
| & | & | & | & | \\
H-C-C-C-C&\!\!-\!\!&C-H \\
| & | & | & | & | \\
H & H & H & (CH_3) & H \\
\end{array}
$$

CH_3 substituted for H

can be viewed as a pentane (five-carbon chain) in which one hydrogen atom has been replaced by a —CH_3 group, which is a methane, CH_4, molecule with a hydrogen removed. When a group is substituted for a hydrogen on an alkane chain, we call this group a *substituent*. To name the —CH_3 substituent, we start with the name of its parent alkane, drop the *-ane,* and add *-yl.* Therefore, —CH_3 is called *methyl.* Likewise, when we remove one hydrogen from ethane, CH_3CH_3, we get —CH_3CH_2. Dropping the *-ane* ending and adding *-yl* gives this group the name *ethyl.* Removal of a hydrogen from the end carbon of propane, $CH_3CH_2CH_3$, yields —$CH_3CH_2CH_2$, which is called the *propyl* group.

There are two ways in which the propyl group can be attached as a substituent. A hydrogen can be removed from an end carbon to give the propyl group or from the middle carbon to give the isopropyl group.

$$
\begin{array}{ccc}
| & & \\
CH_2 & & \\
| & & | \\
CH_2 & \text{or} & H_3C-C-CH_3 \\
| & & | \\
CH_3 & & H \\
\end{array}
$$

Name: Propyl   Isopropyl

When a hydrogen is removed from butane, $CH_3CH_2CH_2CH_3$, we get a butyl substituent. In the case of the butyl group, there are four ways in which the atoms can be arranged. These are shown, with their respective names, in Table 20.2.

The general name for an alkane when it functions as a substituent is *alkyl.* All the common alkyl groups are shown in Table 20.2.

4. We specify the positions of substituent groups by numbering sequentially the carbons in the longest chain of carbon atoms, starting at the

Table 20.2 The Most Common Alkyl Substituents and Their Names

Structure*	Name
—CH_3	methyl
—CH_2CH_3	ethyl
—$CH_2CH_2CH_3$	propyl
$CH_3\overset{\mid}{C}HCH_3$	isopropyl
—$CH_2CH_2CH_2CH_3$	butyl
$CH_3\overset{\mid}{C}HCH_2CH_3$	*sec-*butyl
—$CH_2-\overset{\overset{\displaystyle H}{\mid}}{\underset{\underset{\displaystyle CH_3}{\mid}}{C}}-CH_3$	isobutyl
—$\overset{\overset{\displaystyle CH_3}{\mid}}{\underset{\underset{\displaystyle CH_3}{\mid}}{C}}-CH_3$	*tert-*butyl

*The bond with one end open shows the point of attachment of the substituent.

end closest to the branching (the place where the first substituent occurs). For example, the compound

Methyl substituent

$$CH_3 - CH_2 - \underset{\underset{CH_3}{|}}{CH} - CH_2 - CH_2 - CH_3$$

Correct numbering	1	2	3	4	5	6
Incorrect numbering	6	5	4	3	2	1

is called 3-methylhexane. Note that the top set of numbers is correct; the left end of the molecule is closest to the branching, and this gives the smallest number for the position of the substituent. Also note that a hyphen is written between the number and the name of the substituent.

5. When a given type of substituent occurs more than once, we indicate this by using a prefix. The prefix *di-* indicates two identical substituents, and *tri-* indicates three. For example, the compound

$$\overset{1}{C}H_3 - \overset{2}{C}H - \overset{3}{C}H - \overset{4}{C}H_2 - \overset{5}{C}H_3$$
$$\qquad\quad\underset{CH_3}{|} \;\; \underset{CH_3}{|}$$

has the root name pentane (five carbons in the longest chain). We use *di-* to indicate the two methyl substituents and use numbers to locate them on the chain. The name is 2,3-dimethylpentane.

The following rules summarize the principles we have just developed.

Rules for Naming Alkanes

1. Find the longest continuous chain of carbon atoms. This chain (called the *parent chain*) determines the base alkane name.
2. Number the carbons in the parent chain, starting at the end closest to any branching (the first alkyl substituent). When a substituent occurs the same number of carbons from each end, use the next substituent (if any) to determine from which end to start numbering.
3. Using the appropriate name for each alkyl group, specify its position on the parent chain with a number.
4. When a given type of alkyl group occurs more than once, attach the appropriate prefix (*di-* for two, *tri-* for three, and so on) to the alkyl name.
5. The alkyl groups are listed in alphabetical order, *disregarding any prefix*.

EXAMPLE 20.4   Naming Isomers of Alkanes

Draw the structural isomers for the alkane C_6H_{14}, and give the systematic name for each one.

SOLUTION

We proceed systematically, starting with the longest chain and then rearranging the carbons to form the shorter, branched chains.

1. $CH_3CH_2CH_2CH_2CH_2CH_3$
    $\;\;$ 1 $\;\;$ 2 $\;\;$ 3 $\;\;$ 4 $\;\;$ 5 $\;\;$ 6

This alkane has six carbons all in the same continuous chain, so we call it hexane or, more properly, *n*-hexane, indicating that all the carbon atoms are in the same chain.

2. We now take one carbon out of the main chain and make it a methyl substituent. This gives the molecule

$$CH_3CHCH_2CH_2CH_3$$
$$|$$
$$CH_3$$

The carbon skeleton is as follows:

$$
\begin{array}{ccccc}
1 & 2 & 3 & 4 & 5 \\
C-&C-&C-&C-&C \\
 & | & & & \\
 & C & & &
\end{array}
$$

Because the longest chain has five carbons, the base name is pentane. We have numbered the chain from the left, starting closest to the substituent, a methyl group. We indicate the position of the methyl group on the chain by the number 2, the number of the carbon to which it is attached. So the name is 2-methylpentane. Note that if we numbered the chain from the right end, the methyl group would be on carbon 4. We want the smallest possible number, so the numbering shown is correct.

> Note that placing the —CH$_3$ group on carbon 4 gives the same molecule as placing it on carbon 2.

3. The methyl substituent also can be on the number 3 carbon:

$$
\begin{array}{ccccc}
1 & 2 & 3 & 4 & 5 \\
\end{array}
$$
$$CH_3CH_2CHCH_2CH_3$$
$$|$$
$$CH_3$$

The name is 3-methylpentane. We have now exhausted all possibilities for placing a single methyl group on pentane.

4. Next we take two carbons out of the original six-member chain.

$$CH_3CH-CHCH_3$$
$$\quad\ |\quad\ \ |$$
$$\quad CH_3\ \ CH_3$$

The carbon skeleton is

$$
\begin{array}{cccc}
1 & 2 & 3 & 4 \\
C-&C-&C-&C \\
 & | & | & \\
 & C & C &
\end{array}
$$

The longest chain of this molecule has four carbons, so the root name is butane. Because there are two methyl groups (on carbons 2 and 3), we use the prefix *di-*. The name of the molecule is 2,3-dimethylbutane. Note that when two or more numbers are used, they are separated by a comma.

5. Two methyl groups also can be attached to the same carbon atom in the four-carbon chain to give the following molecule:

$$CH_3$$
$$|$$
$$CH_3-C-CH_2CH_3$$
$$|$$
$$CH_3$$

2,2-dimethylbutane

The carbon skeleton is

$$
\begin{array}{c}
\overset{2}{C} \\
\overset{1}{C}-\overset{2}{C}-\overset{3}{C}-\overset{4}{C} \\
C
\end{array}
$$

The root name is butane, and there are two methyl groups on the number 2 carbon. The name is 2,2-dimethylbutane.

6. As we search for more isomers, we might try to place an ethyl substituent on the four-carbon chain to give the molecule

$$
\begin{array}{c}
CH_3-CHCH_2CH_3 \\
| \\
CH_2 \\
| \\
CH_3
\end{array}
$$

The carbon skeleton is

$$
\begin{array}{c}
C-C-C-C \\
| \\
C \\
| \\
C
\end{array}
$$

We might be tempted to name this molecule 2-ethylbutane, but this is incorrect. Notice that there are five carbon atoms in the longest chain.

$$
\begin{array}{c}
C-\overset{}{C}-C-C \\
|\;^{3}^{4}\;^{5} \\
C\;^{2} \\
| \\
C\;^{1}
\end{array}
$$

We can rearrange this carbon skeleton to give

$$
\begin{array}{c}
C \\
| \\
C-C-C-C-C \\
1\;\;2\;\;3\;\;4\;\;5
\end{array}
$$

This molecule is, in fact, a pentane (3-methylpentane), because the longest chain has five carbon atoms, so it is not a new isomer.

In searching for more isomers, we might try a structure such as

$$
\begin{array}{c}
CH_3 \\
| \\
CH_3-C-CH_3 \\
| \\
CH_2 \\
| \\
CH_3
\end{array}
$$

2,2-dimethylbutyne

As we have drawn it, this molecule might appear to be a propane. However, the molecule has a longest chain of four atoms (look vertically), so the correct name is 2,2-dimethylbutane.

Thus there are five distinct structural isomers of C_6H_{14}: n-hexane, 2-methylpentane, 3-methylpentane, 2,3-dimethylbutane, and 2,2-dimethylbutane.

Self-Check EXERCISE 20.2   Name the following molecules.

a. CH$_3$—CH$_2$—CH—CH$_2$—CH—CH$_2$—CH$_2$—CH$_3$
                         | |
                        CH$_3$ CH$_2$
                                        |
                                       CH$_3$

b. CH$_3$—CH$_2$—CH—CH$_2$—CH—CH$_3$
                     | |
                    CH$_2$ CH$_2$
                     | |
                    CH$_3$ CH$_3$

See Problems 20.25 and 20.26. ■

So far we have learned how to name a compound by examining its structural formula. We also must be able to do the reverse: write the structural formula from the name.

EXAMPLE 20.5 | **Writing Structural Isomers from Names**

Write the structural formula for each of the following compounds.

a. 4-ethyl-3,5-dimethylnonane   b. 4-*tert*-butylheptane

SOLUTION

a. The root name nonane signifies a nine-carbon chain. Therefore, we have the following main chain of carbons:

1   2   3   4   5   6   7   8   9
C—C—C—C—C—C—C—C—C

The name indicates an ethyl group attached to carbon 4 and two methyl groups, one on carbon 3 and one on carbon 5. This gives the following carbon skeleton:

1   2   3   4   5   6   7   8   9
C—C—C—C—C—C—C—C—C
        | | |
        C   C   C
            |
            C

When we add the hydrogen atoms, we get the final structure

1   2   3   4   5 6   7   8   9
CH$_3$CH$_2$CH—CH—CHCH$_2$CH$_2$CH$_2$CH$_3$
Methyl (CH$_3$) (CH$_2$) (CH$_3$) Methyl
                  |
                (CH$_3$)
                 Ethyl

b. Heptane signifies a seven-carbon chain, and the *tert*-butyl group (see Table 20.2) is

            |
H$_3$C—C—CH$_3$
            |
           CH$_3$

Thus we have the molecule

$$
\begin{array}{ccccccc}
1 & 2 & 3 & 4 & 5 & 6 & 7
\end{array}
$$

$$CH_3CH_2CH_2CHCH_2CH_2CH_3$$

$$H_3C-\underset{\underset{CH_3}{|}}{\overset{|}{C}}-CH_3$$

Self-Check EXERCISE 20.3 Write the structural formula for 5-isopropyl-4-methyldecane.

See Problems 20.27 and 20.28. ■

20.5 Petroleum

OBJECTIVE: To learn about the composition and uses of petroleum.

Woody plants, coal, petroleum, and natural gas provide a vast resource of energy that originally came from the sun. By the process of photosynthesis, plants store energy that we can claim by burning the plants themselves or, more commonly, burning the decay products that have been converted to fossil fuels. Although the United States currently depends heavily on petroleum for energy, this dependency is a relatively recent phenomenon (see Figure 10.7).

Petroleum and natural gas deposits probably formed from the remains of marine organisms that lived approximately 500 million years ago. **Petroleum** is a thick, dark liquid composed largely of hydrocarbons containing from 5 to more than 25 carbon atoms. **Natural gas,** which is usually associated with petroleum deposits, consists mostly of methane but also contains significant amounts of ethane, propane, and butane.

> Distillation (separation by boiling) was discussed in Section 3.5.

To be used efficiently, petroleum must be separated by boiling into portions called *fractions*. The smaller hydrocarbons can be boiled off at relatively low temperatures; the larger molecules require successively higher temperatures. The major uses of various petroleum fractions are shown in Table 20.3.

The petroleum era began when the demand for lamp oil during the Industrial Revolution outstripped the traditional sources: animal fats and whale oil. In response to this increased demand, Edwin Drake drilled the first oil well in 1859 at Titusville, Pennsylvania. The petroleum from this well was refined to produce *kerosene* (fraction C_{10}–C_{18}), which served as an excellent lamp oil. *Gasoline* (fraction C_5–C_{12}) was of limited use and often was discarded. However, the importance of these fractions was reversed when the development of the electric light reduced the need for kerosene and the advent of the "horseless carriage" signaled the birth of the gasoline age.

As gasoline became more important, new ways were sought to increase the yield of gasoline obtained from each barrel of petroleum. William Burton invented a process called *pyrolytic cracking* at Standard Oil of Indiana. In this process, the heavier molecules of the kerosene fraction are heated to about 700 °C, which causes them to break (crack) into the smaller molecules characteristic of the gasoline fraction. As cars became larger, more efficient internal combustion engines were designed. Because of the uneven burning of the gasoline then available, these engines "knocked," producing unwanted noise and even engine damage. Intensive research to find additives that would promote smoother burning yielded tetraethyl lead, $(C_2H_5)_4Pb$, a very effective "antiknock" agent.

Table 20.3 Uses of the Various Petroleum Fractions

Petroleum Fraction*	Major Uses
C_5–C_{12}	gasoline
C_{10}–C_{18}	kerosene jet fuel
C_{15}–C_{25}	diesel fuel heating oil lubricating oil
>C_{25}	asphalt

*Shows the chain lengths present in each fraction.

Adding tetraethyl lead to gasoline became a common practice, and by 1960, gasoline contained as much as 3 grams of lead per gallon. As we have discovered so often in recent years, technological advances can produce environmental problems. Lead in gasoline causes two important problems. First, it "poisons" the catalytic converters that have been added to exhaust systems to help prevent air pollution. Second, the use of leaded gasoline has greatly increased the amount of lead in the environment, where it can be ingested by animals and humans. For these reasons, the use of lead in gasoline has been largely discontinued. This has required extensive (and expensive) modifications of engines and of the gasoline-refining process.

20.6 Reactions of Alkanes

OBJECTIVE: To learn various types of chemical reactions that alkanes undergo.

At low temperatures, alkanes are not very reactive because the C—C and C—H bonds in alkanes are relatively strong. For example, at 25 °C, alkanes do not react with acids, bases, or strong oxidizing agents. This chemical inertness makes alkanes valuable as lubricating materials and as the backbone for structural materials such as plastics.

At sufficiently high temperatures, however, alkanes *react vigorously with oxygen*. These **combustion reactions** are the basis for the alkanes' widespread use as fuels. For example, the combustion reaction of butane with oxygen is

$$2C_4H_{10}(g) + 13O_2(g) \rightarrow 8CO_2(g) + 10H_2O(g)$$

The alkanes also can undergo **substitution reactions**—reactions in which *one or more hydrogen atoms of the alkane are replaced* (substituted) *by different atoms*. We can represent the substitution reaction of an alkane with a halogen molecule as follows:

$$R—H + X_2 \rightarrow R—X + HX$$

where R represents an alkyl group and X represents a halogen atom. For example, methane can react successively with chlorine as follows:

> The symbol *hv* signifies ultraviolet light used to furnish energy for the reaction.

$$CH_4 + Cl_2 \xrightarrow{hv} CH_3Cl + HCl$$
Chloromethane

$$CH_3Cl + Cl_2 \xrightarrow{hv} CH_2Cl_2 + HCl$$
Dichloromethane

$$CH_2Cl_2 + Cl_2 \xrightarrow{hv} CHCl_3 + HCl$$
Trichloromethane
(chloroform)

$$CHCl_3 + Cl_2 \xrightarrow{hv} CCl_4 + HCl$$
Tetrachloromethane
(carbon tetrachloride)

The *hv* above each arrow signifies that ultraviolet light is needed to furnish the energy to break the Cl—Cl bond to produce chlorine atoms:

$$Cl_2 \rightarrow Cl \cdot + Cl \cdot$$

A chlorine atom has an unpaired electron, indicated by the dot, which makes it very reactive and able to disrupt the C—H bond.

Notice that each step in the process involves replacement of a C—H bond by a C—Cl bond. That is, a chlorine atom *substitutes* for a hydrogen atom. The names of the products of these reactions use the term *chloro* for the chlorine substituents with a prefix that gives the number of chlorine atoms present: *di-* for two, *tri-* for three, and *tetra-* for four. No number is used to describe the chlorine positions in this case because methane has only one carbon atom. Note that the products of the last two reactions have two names, the systematic name and the common name in parentheses.

Besides substitution reactions, alkanes also can undergo **dehydrogenation reactions** in which *hydrogen atoms are removed* and the product is an unsaturated hydrocarbon. For example, in the presence of a catalyst [chromium(III) oxide] at high temperatures, ethane can be dehydrogenated, yielding ethylene, C_2H_4.

$$CH_3CH_3 \xrightarrow[500\,°C]{Cr_2O_3} CH_2{=}CH_2 + H_2$$
Ethylene

20.7 Alkenes and Alkynes

OBJECTIVES: To learn to name hydrocarbons with double bonds (alkenes) and triple bonds (alkynes). • To understand addition reactions.

We have seen that alkanes are saturated hydrocarbons—each of the carbon atoms is bound to four atoms by single bonds. Hydrocarbons that contain carbon–carbon *double bonds*

are called **alkenes.** Hydrocarbons with carbon–carbon *triple bonds* are called **alkynes.** Alkenes and alkynes are unsaturated hydrocarbons.

Multiple carbon–carbon bonds result when hydrogen atoms are removed from alkanes. Alkenes that contain a carbon–carbon double bond have the general formula C_nH_{2n}. The simplest alkene, C_2H_4, commonly known as *ethylene,* has the Lewis structure

Acetylene gas burning. The acetylene is formed by the reaction of calcium carbide, CaC_2, with water in the flask.

The ball-and-stick model of ethylene is shown in Figure 20.5.

The system for naming alkenes and alkynes is similar to the one we have used for alkanes. The following rules are useful.

Figure 20.5
The ball-and-stick model of ethylene (ethene).

Rules for Naming Alkenes and Alkynes

1. Select the longest continuous chain of carbon atoms that contains the double or triple bond.
2. For an alkene, the root name of the carbon chain is the same as for the alkane, except that the *-ane* ending is replaced by *-ene*. For an alkyne, the *-ane* is replaced by *-yne*. For example, for a two-carbon chain we have

CH_3CH_3	$CH_2{=}CH_2$	$CH{\equiv}CH$
Ethane	Ethene	Ethyne

3. Number the parent chain, starting at the end closest to the double or triple bond. The location of the multiple bond is given by the lowest-numbered carbon involved in the bond. For example,

$$\underset{1}{CH_2}=\underset{2}{CH}\underset{3}{CH_2}\underset{4}{CH_3}$$

is called 1-butene and

$$\underset{1}{CH_3}\underset{2}{CH}=\underset{3}{CH}\underset{4}{CH_3}$$

is called 2-butene.

4. Substituents on the parent chain are treated the same way as in naming alkanes. For example, the molecule $ClCH=CHCH_2CH_3$ is called 1-chloro-1-butene.

> In writing shorthand formulas, the hydrogen atoms are often written just after the carbon to which they are attached. For example, the formula for $H-C\equiv C-H$ is often written as $CH\equiv CH$.

EXAMPLE 20.6   Naming Alkenes and Alkynes

Name each of the following molecules.

a. $CH_3CH_2\overset{\displaystyle |}{\underset{\displaystyle CH_3}{CH}}CH=CHCH_3$

b. $CH_3CH_2C\equiv C\overset{\displaystyle |}{\underset{\displaystyle \underset{\displaystyle CH_3}{|}}{\underset{\displaystyle CH_2}{CH}}}CH_2CH_3$

SOLUTION

a. The longest chain contains six carbon atoms, and we number the carbons starting from the end closest to the double bond.

$$\underset{6}{CH_3}\underset{5}{CH_2}\underset{4}{\overset{\displaystyle |}{\underset{\displaystyle CH_3}{CH}}}\underset{3}{CH}=\underset{2}{CH}\underset{1}{CH_3}$$

Thus the root name for the hydrocarbon is 2-hexene. (Remember to use the lower number of the two carbon atoms involved in the double bond.) There is a methyl group attached to the number 4 carbon. Therefore, the name of the compound is 4-methyl-2-hexene.

b. The longest chain of carbon atoms is seven carbons long, and the chain is numbered as shown (starting from the end closest to the triple bond).

$$\underset{1}{CH_3}\underset{2}{CH_2}\underset{3}{C}\equiv \underset{4}{C}\underset{5}{\overset{\displaystyle |}{\underset{\displaystyle \underset{\displaystyle CH_3}{|}}{\underset{\displaystyle CH_2}{CH}}}}\underset{6}{CH_2}\underset{7}{CH_3}$$

The hydrocarbon is a 3-heptyne (we use the lowest-numbered carbon in the triple bond). Because there is an ethyl group on carbon number 5, the full name is 5-ethyl-3-heptyne.

Self-Check EXERCISE 20.4 Name the following molecules.

a. $CH_3CH_2CH_2CH_2CH=CHCHCH_3$
$\qquad\qquad\qquad\qquad\qquad\qquad\;\; |$
$\qquad\qquad\qquad\qquad\qquad\qquad\; CH_3$

b. $CH_3CH_2CH_2C\equiv CH$

See Problems 20.45 and 20.46. ■

▶ Reactions of Alkenes

Because alkenes and alkynes are unsaturated, their most important reactions are **addition reactions,** in which *new atoms form single bonds to the carbons formerly involved in the double or triple bonds.* An addition reaction for an alkene changes the carbon–carbon double bond to a single bond, giving a saturated hydrocarbon (each carbon bonded to four atoms). For example, **hydrogenation reactions,** which use H_2 as a reactant, lead to the addition of a hydrogen atom to each carbon formerly involved in the double bond.

$$CH_2=CHCH_3 + H_2 \xrightarrow{\text{Catalyst}} CH_3CH_2CH_3$$
$$\text{1-Propene} \qquad\qquad\qquad\qquad \text{Propane}$$

Hydrogenation of molecules with double bonds is an important industrial process, particularly in the manufacture of solid shortenings. Unsaturated fats (fats containing double bonds) are generally liquids at room temperature, whereas saturated fats (those containing C—C single bonds) are solids. The liquid unsaturated fats are converted to solid saturated fats by hydrogenation.

Halogenation of unsaturated hydrocarbons involves the addition of halogen atoms. Here is an example:

$$CH_2=CHCH_2CH_2CH_3 + Br_2 \longrightarrow CH_2BrCHBrCH_2CH_2CH_3$$
$$\text{1-Pentene} \qquad\qquad\qquad\qquad \text{1,2-Dibromopentane}$$

Another important reaction of certain unsaturated hydrocarbons is **polymerization,** a process in which many small molecules are joined together to form a large molecule. Polymerization will be discussed in Section 20.16.

20.8 Aromatic Hydrocarbons

OBJECTIVE: To learn about the aromatic hydrocarbons.

When mixtures of hydrocarbons from natural sources, such as petroleum or coal, are separated, certain of the compounds that emerge have pleasant odors and are thus known as **aromatic hydrocarbons.** When these substances, which include wintergreen, cinnamon, and vanillin, are examined, they are all found to contain a common feature: a six-membered ring of carbon atoms called the *benzene ring.* **Benzene** has the formula C_6H_6 and a planar (flat) structure in which all the bond angles are 120° (Figure 20.6).

When we examine the bonding in the benzene ring, we find that more than one Lewis structure can be drawn. That is, the double bonds can be located in different positions, as shown in Figure 20.7. Because the actual bonding is a combi-

Cinnamon is an aromatic hydrocarbon.

Figure 20.6

Benzene, C_6H_6, consists of six carbon atoms bonded together to form a ring. Each carbon has one hydrogen atom bonded to it. All the atoms in benzene lie in the same plane. This representation does not show all the bonds between the carbon atoms in the ring.

Figure 20.7

(Top) Two Lewis structures for the benzene ring. (Bottom) As a shorthand notation, the rings are usually represented without labeling the carbon and hydrogen atoms.

Figure 20.8

To show that the bonding in the benzene ring is a combination of different Lewis structures, the ring is drawn with a circle inside.

nation of the structures represented in Figure 20.7, the benzene ring is usually shown with a circle (Figure 20.8).

20.9 Naming Aromatic Compounds

OBJECTIVE: To learn the system for naming aromatic compounds.

Substituted benzene molecules are formed by replacing one or more of the H atoms on the benzene ring with other atoms or groups of atoms. We will consider benzene rings with one substituent (called *monosubstituted benzenes*) first.

▶ Monosubstituted Benzenes

The systematic method for naming monosubstituted benzenes uses the substituent name as a prefix of benzene. For example, the molecule

Cl

is called chlorobenzene, and the molecule

CH_2CH_3

is called ethylbenzene.

Sometimes, monosubstituted benzene compounds have special names. For example, the molecule

$$CH_3$$

has the systematic name methylbenzene. However, for convenience, it is given the name toluene. Likewise, the molecule

$$OH$$

which might be called hydroxybenzene, has the special name phenol. Several examples of monosubstituted benzenes are shown in Figure 20.9.

Sometimes it is more convenient to name compounds if we view the benzene ring itself as a substituent. For example, the compound

$$\overset{4}{C}H_3\overset{3}{C}H\overset{2}{C}H=\overset{1}{C}H_2$$

is most easily named as a 1-butene with a benzene ring as a substituent on the number 3 carbon. When the benzene ring is used as a substituent, it is called the **phenyl** (pronounced "fen'-ill") **group.** So the name of the preceding compound is 3-phenyl-1-butene. As another example, the compound

$$\overset{1}{C}H_3\overset{2}{C}H\overset{3}{C}H_2\overset{4}{C}H\overset{5}{C}H_2\overset{6}{C}H_3$$
$$\qquad\qquad\qquad\; Cl$$

is named 4-chloro-2-phenylhexane. Remember, we start to number the chain from the end closest to the first substituent and name the substituents in alphabetical order (chloro before phenyl).

Figure 20.9

Names of some common monosubstituted benzenes.

Chlorobenzene

Toluene

Bromobenzene

Phenol

Nitrobenzene

Styrene

▶ Disubstituted Benzenes

When there is more than one substituent on the benzene ring, numbers are used to indicate substituent position. For example, the compound

$$\begin{array}{c} 6 \\ 5 \quad 1 \;Cl \\ 4 \quad 2 \;Cl \\ 3 \end{array}$$

is named 1,2-dichlorobenzene. Another naming system uses the prefix *ortho-* (*o-*) for two adjacent substituents, *meta-* (*m-*) for two substituents with one carbon between them, and *para-* (*p-*) for two substituents opposite each

Ortho- (*o-*) means two adjacent substituents.

Para- (p-) means two substituents directly across the ring from each other.

Meta- (m-) means two substituents with one carbon between them.

other. This means that 1,2-dichlorobenzene also can be called *ortho*-dichlorobenzene or *o*-dichlorobenzene. Likewise, the compound

can be called 1,3-dichlorobenzene or *m*-dichlorobenzene. The compound

is named 1,4-dichlorobenzene or *p*-dichlorobenzene.

Benzenes that have two methyl substituents have the special name xylene. So the compound

which might be called 1,3-dimethylbenzene, is instead called *m*-xylene (*meta*-xylene).

When two different substituents are present on the benzene ring, one is always assumed to be at carbon number 1, and this number is not specified in the name. For example, the compound

is named 2-bromochlorobenzene, not 2-bromo-1-chlorobenzene. Various examples of disubstituted benzenes are shown in Figure 20.10 on page 662.

Benzene is the simplest aromatic molecule. More complex aromatic systems can be viewed as consisting of a number of "fused" benzene rings. Some examples are given in Table 20.4.

Mothballs used to contain naphthalene, composed of "fused" benzene rings, but now contain *p*-dichlorobenzene.

EXAMPLE 20.7   Naming Aromatic Compounds

Name the following compounds.

a. CH_2CH_3

CH_2CH_3

b. CH_3

Br

c. $CH_3—CH—C\equiv CH$

d. CH_3

O_2N NO_2

NO_2

Dyed nylon rope.

Figure 20.10

Some selected disubstituted benzenes and their names. Common names are given in parentheses.

1,2-Dibromobenzene (*o*-dibromobenzene)	1,3-Dibromobenzene (*m*-dibromobenzene)	1,4-Dibromobenzene (*p*-dibromobenzene)
1,4-Dimethylbenzene (*p*-xylene)	1,2-Dimethylbenzene (*o*-xylene)	1,3-Dimethylbenzene (*m*-xylene)
2-Nitrotoluene (*o*-nitrotoluene)	3-Bromonitrobenzene (*m*-bromonitrobenzene)	3-Chlorotoluene (*m*-chlorotoluene)

Table 20.4 More Complex Aromatic Molecules

Structural Formula	Name	Use
	naphthalene	Formerly used in mothballs
	anthracene	dyes
	phenanthrene	dyes, explosives, and synthesis of drugs

SOLUTION

a. There are ethyl groups in the 1 and 3 (or *meta-*) positions, so the name is 1,3-diethylbenzene or *m*-diethylbenzene.

b. The

group is called toluene. The bromine is in the 4 (or *para-*) position. The name is 4-bromotoluene or *p*-bromotoluene.

Termite Mothballing

Termites typically do not get a lot of respect. They are regarded as lowly, destructive insects. However, termites are the first insects known to fumigate their nests with naphthalene, a chemical long used by humans to prevent moths from damaging wool garments. Although termites are not concerned about holes in their sweaters, they may use naphthalene to ward off microbes and predatory ants, among other pests.

Gregg Henderson and Jian Chen, of the Louisiana State University Agricultural Center in Baton Rouge, have observed that Formosan termites are unusually resistant to naphthalene. In fact, these insects build their underground galleries from chewed wood glued together with saliva and excrement. This "glue" (called *carton*) contains significant amounts of naphthalene, which evaporates and permeates the air in the underground tunnels. The source of the naphthalene is unknown—it might be a metabolite

from a food source of the termites or it might be produced from the carton by organisms present in the nest. Whatever the source of the naphthalene, this interesting example shows how organisms use chemistry to protect themselves.

Naphthalene

Agricultural Research/USDA. Photo by Scott Bauer

Formosan subterranean termites.

---

c. In this case, we name the compound as a butyne with a phenyl substituent. The name is 3-phenyl-1-butyne.

d. We name this compound as a substituted toluene (with the —CH_3 group assumed to be on carbon number 1). So the name is 2,4,6-trinitrotoluene. This compound is more commonly known as TNT, a component of high explosives.

Self-Check EXERCISE 20.5   Name the following compounds.

a. NO_2
   Cl

b. $CH_3CH_2CH—CH=CHCH_3$

See Problems 20.55 and 20.56. ∎

20.10 Functional Groups

OBJECTIVE: To learn the common functional groups in organic molecules.

The vast majority of organic molecules contain elements in addition to carbon and hydrogen. However, most of these substances can be viewed as

Table 20.5 The Common Functional Groups

Class	Functional Group	General Formula*	Example
halohydrocarbons[†]	—X(F, Cl, Br, I)	R—X	CH_3I
alcohols	—OH	R—OH	CH_3OH
ethers	—O—	R—O—R'	CH_3—O—CH_3
aldehydes	$-\overset{\overset{O}{\|\|}}{C}-H$	$R-\overset{\overset{O}{\|\|}}{C}-H$	$H-\overset{\overset{O}{\|\|}}{C}-H$
ketones	$-\overset{\overset{O}{\|\|}}{C}-$	$R-\overset{\overset{O}{\|\|}}{C}-R'$	$CH_3-\overset{\overset{O}{\|\|}}{C}-CH_3$
carboxylic acids	$-\overset{\overset{O}{\|\|}}{C}-OH$	$R-\overset{\overset{O}{\|\|}}{C}-OH$	$CH_3-\overset{\overset{O}{\|\|}}{C}-OH$
esters	$-\overset{\overset{O}{\|\|}}{C}-O-$	$R-\overset{\overset{O}{\|\|}}{C}-O-R'$	$CH_3-\overset{\overset{O}{\|\|}}{C}-OCH_2CH_3$
amines	—NH_2	R—NH_2	CH_3NH_2

*R and R' represent hydrocarbon fragments, which may be the same or different.   †These substances are also called alkyl halides.

hydrocarbon derivatives, molecules that are fundamentally hydrocarbons but that have additional atoms or groups of atoms called **functional groups.** The common functional groups are listed in Table 20.5; one example of a compound that contains that functional group is given for each. We will briefly describe some of these functional groups in the next few sections and learn to name the compounds that contain them.

20.11   Alcohols

OBJECTIVE: To learn about simple alcohols and explain how to name them.

Alcohols are characterized by the presence of the —OH group. Some common alcohols are listed in Table 20.6. The systematic name for an alcohol is obtained by replacing the final -*e* of the parent hydrocarbon name with -*ol*. The position of the —OH group is specified by a number (where necessary) chosen such that it is the smallest of the substituent numbers.

The rules for naming alcohols follow.

Rules for Naming Alcohols

1. Select the longest chain of carbon atoms containing the —OH group.
2. Number the chain such that the carbon with the —OH group gets the lowest possible number.
3. Obtain the root name from the name of the parent hydrocarbon chain by replacing the final -*e* with -*ol*.
4. Name any other substituents as usual.

Table 20.6	Some Common Alcohols	
Formula	Systematic Name	Common Name
CH_3OH	methanol	methyl alcohol
CH_3CH_2OH	ethanol	ethyl alcohol
$CH_3CH_2CH_2OH$	1-propanol	n-propyl alcohol
CH_3CHCH_3 $\quad\mid$ $\quad OH$	2-propanol	isopropyl alcohol

For example, the compound

$$CH_3CHCH_2CH_2CH_3$$
$$\mid$$
$$OH$$

is called 2-pentanol because the parent carbon chain is pentane. The compound

$$CH_3CH_2CHCH_2CH_2CH_3$$
$$\mid$$
$$OH$$

is called 3-hexanol.

Alcohols are classified according to the number of hydrocarbon fragments (alkyl groups) bonded to the carbon where the —OH group is attached. Thus we have

$$R-CH_2OH \qquad \underset{R'}{\overset{R}{\diagdown}}CHOH \qquad \underset{R''}{\overset{R}{\mid}}R'-COH$$

Primary alcohol (one R group)   *Secondary* alcohol (two R groups)   *Tertiary* alcohol (three R groups)

where R, R', and R" (which may be the same or different) represent hydrocarbon fragments (alkyl groups).

EXAMPLE 20.8 | Naming Alcohols

Give the systematic name for each of the following alcohols, and specify whether the alcohol is primary, secondary, or tertiary.

a. $CH_3CHCH_2CH_3$
$\quad\quad\mid$
$\quad\quad OH$

b. $ClCH_2CH_2CH_2OH$

c. CH_3
$\quad\mid$
$CH_3CCH_2CH_2CH_2CH_2Br$
$\quad\mid$
$\quad OH$

SOLUTION

a. The chain is numbered as follows:

$$\overset{1}{C}H_3\overset{2}{C}H\overset{3}{C}H_2\overset{4}{C}H_3$$
$$\mid$$
$$OH$$

The compound is called 2-butanol because the —OH group is located at the number 2 position of a four-carbon chain. Note that the carbon to which the —OH is attached also has two R groups (—CH_3 and —CH_2CH_3) attached. Therefore, this is a *secondary* alcohol.

$$
\begin{array}{ccc}
 & \overset{\displaystyle H}{|} & \\
(CH_3)\!-\!&C&\!-\!(CH_2CH_3) \\
 & | & \\
R & OH & R'
\end{array}
$$

b. The chain is numbered as follows:

$$
\overset{3}{Cl}\!-\!\overset{}{CH_2}\!-\!\overset{2}{CH_2}\!-\!\overset{1}{CH_2}\!-\!OH
$$

Remember that in naming an alcohol we give the carbon with the —OH attached the lowest possible number. The name is 3-chloro-1-propanol. This is a *primary* alcohol.

$$
\begin{array}{c}
\overset{\displaystyle H}{|} \\
(Cl\!-\!CH_2CH_2)\!-\!C\!-\!OH \\
| \\
H
\end{array}
$$

One R group
attached to the
carbon with the
—OH group

c. The chain is numbered as follows:

$$
\begin{array}{c}
\overset{2}{}\overset{\displaystyle (CH_3)}{|} \\
\overset{1}{(CH_3)}\!-\!\overset{}{C}\!-\!(\overset{3}{CH_2}\!-\!\overset{4}{CH_2}\!-\!\overset{5}{CH_2}\!-\!\overset{6}{CH_2Br}) \\
| \\
OH
\end{array}
$$

The name is 6-bromo-2-methyl-2-hexanol. This is a tertiary alcohol because the carbon where the —OH is attached also has three R groups attached.

Self-Check EXERCISE 20.6   Name each of the following alcohols, and specify whether it is primary, secondary, or tertiary.

a. $CH_3CH_2CH_2CH_2CH_2OH$

b.
$$
\begin{array}{c}
CH_3 \\
| \\
CH_3\!-\!C\!-\!CH_3 \\
| \\
OH
\end{array}
$$

c.
$$
\begin{array}{c}
Br \\
| \\
CH_3\!-\!CH\!-\!CH_2CH_2CHCH_3 \\
| \\
OH
\end{array}
$$

See Problems 20.61 and 20.62. ∎

20.12 Properties and Uses of Alcohols

OBJECTIVE: To learn about how some alcohols are made and used.

Although there are many important alcohols, the simplest ones, methanol and ethanol, have the greatest commercial value. Methanol, also known as *wood alcohol* because it was formerly obtained by heating wood in the absence of air, is prepared industrially (over 20 million tons per year in the United States) by the hydrogenation of carbon monoxide (catalyzed by a ZnO–Cr$_2$O$_3$ mixture).

$$CO + 2H_2 \xrightarrow[\text{ZnO–Cr}_2\text{O}_3]{400\ °C} CH_3OH$$

Methanol is used as a starting material for the synthesis of acetic acid and many types of adhesives, fibers, and plastics. It also can be used as a motor fuel. In fact, pure methanol has been used for many years in the engines of the cars that are driven in the Indianapolis 500 and similar races. Methanol is especially useful in racing engines because of its resistance to knock. It is advantageous for regular cars because it produces less carbon monoxide (a toxic gas) in the exhaust than does gasoline. Methanol is highly toxic to humans, and swallowing it can lead to blindness and death.

Ethanol is the alcohol found in beverages such as beer, wine, and whiskey; it is produced by the fermentation of the sugar glucose in corn, barley, grapes, and so on.

$$C_6H_{12}O_6 \xrightarrow{\text{Yeast}} 2CH_3CH_2OH + 2CO_2$$
$$\text{Glucose} \qquad\qquad\qquad \text{Ethanol}$$

This reaction is catalyzed by the enzymes (biologic catalysts) found in yeast, and it can proceed only until the alcohol content reaches approximately 13% (that found in most wines), at which point the yeast can no longer survive. Beverages with higher alcohol content are made by distilling the fermentation mixture.

Ethanol, like methanol, can be burned in the internal combustion engines of automobiles and is now commonly added to gasoline to form *gasohol*. It is also used in industry as a solvent and for the preparation of acetic acid. The commercial production of ethanol (half a million tons per year in the United States) is carried out by reaction of water with ethylene.

$$CH_2{=}CH_2 + H_2O \xrightarrow[\text{Catalyst}]{\text{Acid}} CH_3CH_2OH$$

Many alcohols are known that contain more than one —OH group. The one that is the most important commercially is ethylene glycol,

$$\begin{array}{l} H_2C-OH \\ \quad | \\ H_2C-OH \end{array}$$

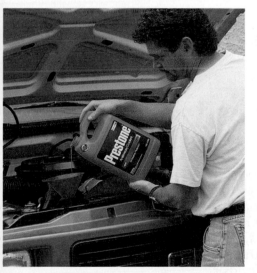

Ethylene glycol is a component of antifreeze, which is used to protect the cooling systems of automobiles.

a toxic substance that is the major constituent of most automobile antifreezes.

The simplest aromatic alcohol is

commonly called **phenol.** Most of the 1 million tons of phenol produced annually in the United States is used to produce polymers for adhesives and plastics.

20.13 Aldehydes and Ketones

OBJECTIVE: To learn the general formulas for aldehydes and ketones and some of their uses.

Aldehydes and ketones contain the **carbonyl group**

$$-\underset{\underset{O}{\|}}{C}-$$

In **ketones,** this group is bonded to two carbon atoms; an example is acetone:

$$CH_3-\underset{\underset{O}{\|}}{C}-CH_3$$

The general formula for a ketone is

$$R-\underset{\underset{O}{\|}}{C}-R'$$

where R and R' are alkyl groups that may or may not be the same. In a ketone, the carbonyl group is never at the end of the hydrocarbon chain. (If it were, it would be an aldehyde.)

In **aldehydes,** the carbonyl group always appears at the end of the hydrocarbon chain. There is always at least one hydrogen bonded to the carbonyl carbon atom. An example of an aldehyde is acetaldehyde:

$$CH_3-\underset{\underset{O}{\|}}{C}-H$$

The general formula for an aldehyde is

$$R-\underset{\underset{O}{\|}}{C}-H$$

We often use compact formulas for aldehydes and ketones. For example, formaldehyde (where R = H) and acetaldehyde (where R = CH$_3$) are usually represented as HCHO and CH$_3$CHO, respectively. Acetone is often written as CH$_3$COCH$_3$ or (CH$_3$)$_2$CO.

Many ketones have useful solvent properties (acetone is often found in nail polish remover, for example) and are used frequently in industry for this purpose. Aldehydes typically have strong odors. Vanillin is responsible for the pleasant odor of vanilla beans; cinnamaldehyde produces the characteristic odor of cinnamon. On the other hand, the unpleasant odor of rancid butter arises from the presence of butyraldehyde (and butyric acid). (See Figure 20.11 for the structures of these compounds.)

Aldehydes and ketones are most often produced commercially by the oxidation of alcohols. Oxidation of a *primary* alcohol gives the corresponding aldehyde, for example,

$$CH_3CH_2OH \xrightarrow{\text{Oxidation}} CH_3C\overset{\displaystyle O}{\underset{\displaystyle H}{{<}}}$$

Oxidation of a *secondary* alcohol results in a ketone:

$$CH_3\underset{\underset{OH}{|}}{C}HCH_3 \xrightarrow{\text{Oxidation}} CH_3\underset{\underset{O}{\|}}{C}CH_3$$

Vanillin

Cinnamaldehyde

Butyraldehyde

Figure 20.11

Some common aldehydes.

$$R-OH \longrightarrow R-C\overset{\displaystyle O}{\underset{\displaystyle H}{{<}}}$$

$$R-\underset{\underset{OH}{|}}{\overset{\overset{H}{|}}{C}}-R' \longrightarrow R-\underset{\underset{O}{\|}}{C}-R'$$

20.14 Naming Aldehydes and Ketones

OBJECTIVE: To learn the systems for naming aldehydes and ketones.

We obtain the systematic name for an aldehyde from the parent alkane by removing the final -*e* and adding -*al*. For ketones, the final -*e* is replaced by -*one*, and a number indicates the position of the carbonyl group where necessary. The carbon chain in ketones is numbered such that the

$$-\overset{\displaystyle\|}{\underset{\displaystyle O}{C}}-$$

carbon gets the lowest possible number. In aldehydes, the

$$-\overset{\displaystyle\|}{\underset{\displaystyle O}{C}}-H$$

is always at the end of the chain and is always assumed to be carbon number 1. The positions of other substituents are specified by numbers, as usual. The following examples illustrate these principles:

$$H-\overset{O}{\overset{\|}{C}}-H$$	$$CH_3-\overset{O}{\overset{\|}{C}}-H$$	$$CH_3-\overset{O}{\overset{\|}{C}}-CH_3$$	$$CH_3-\overset{O}{\overset{\|}{C}}-CH_2CH_2CH_3$$	$$CH_3\overset{}{\underset{\underset{\displaystyle Cl}{\|}}{C}H}CH_2\overset{O}{\overset{\|}{C}}-H$$
Methanal (formaldehyde)	Ethanal (acetaldehyde)	Propanone (acetone)	2-Pentanone	3-Chlorobutanal

The names in parentheses are common names that are used much more often than the systematic names.

Another common aldehyde is benzaldehyde (an aromatic aldehyde), which has the structure

Benzaldehyde

An alternative system for naming ketones specifies the substituents attached to the C=C group. For example, the compound

$$CH_3\underset{\underset{\displaystyle O}{\|}}{C}CH_2CH_3$$

is called 2-butanone when we use the system just described. However, this molecule also can be named methyl ethyl ketone and is commonly referred to in industry as MEK (*methyl ethyl ketone*):

Another example is use of the name ketone for the compound

O=C—CH₃ Methyl

Phenyl

which is commonly called methyl phenyl ketone.

EXAMPLE 20.9 | **Naming Aldehydes and Ketones**

Name the following molecules.

a.
CH₃
|
CH₃C—CHCH₃   (Give two names.)
‖
O

b. H   O
    \\ //
     C

NO₂

c. CH₃CH₂   O
          \\ //
           C

d. CH₃CHCH₂CH₂CHO
        |
        Cl

SOLUTION

a. We can name this molecule as a 2-butanone because the longest chain has four carbon atoms (butane root) with the

$$\begin{array}{c}\diagup\\ \diagup \end{array}C=O$$

group in the number 2 position (the lowest possible number). Because the methyl group is in the number 3 position, the name is 3-methyl-2-butanone. We also can name this compound methyl isopropyl ketone.

b. We name this molecule as a substituted benzaldehyde (the nitro group is in the number 3 position): 3-nitrobenzaldehyde. It also might be named *m*-nitrobenzaldehyde.

c. We name this molecule as a ketone: ethyl phenyl ketone.

d. The name is 4-chloropentanal. Note that an aldehyde group is always at the end of the chain and is automatically assigned as the number 1 carbon.

Self-Check EXERCISE 20.7   Name the following molecules.

a.

$$CH_3CH_2\overset{\displaystyle O}{\overset{\|}{C}}CHCH_2CH_3$$
$$\underset{\displaystyle CH_2CH_3}{|}$$

b.

$$CH_3CH_2CH_2CHCH_2CH_2CH_2CH_2CH_2CHO$$
$$\underset{\displaystyle CH_3CHCH_3}{|}$$

See Problems 20.73 and 20.74. ■

20.15 Carboxylic Acids and Esters

OBJECTIVE: To learn the structures and names of the common carboxylic acids.

Carboxylic acids are characterized by the presence of the **carboxyl group,** —COOH, which has the structure

$$-\overset{\displaystyle O}{\overset{\|}{C}}_{\displaystyle O-H}$$

The general formula of a carboxylic acid is RCOOH, where R represents the hydrocarbon fragment. These molecules typically are weak acids in aqueous solution. That is, the dissociation (ionization) equilibrium

$$RCOOH(aq) + H_2O(l) \rightleftharpoons H_3O^+(aq) + RCOO^-(aq)$$

lies far to the left—only a small percentage of the RCOOH molecules are ionized.

We name carboxylic acids by dropping the final -*e* from the parent alkane (the longest chain containing the —COOH group) and adding -*oic*. Carboxylic acids are frequently known by their common names. For example, CH_3COOH, often written $HC_2H_3O_2$ and commonly called acetic acid, has the systematic name ethanoic acid because the parent alkane is ethane. Several carboxylic acids, their systematic names, and their common names are given in Table 20.7. Other examples of carboxylic acids are shown in Figure 20.12. Note that the —COOH group is always assigned the number 1 position in the chain.

Carboxylic acids can be produced by oxidizing primary alcohols with a strong oxidizing agent. For example, we can oxidize ethanol to acetic acid by using potassium permanganate.

$$CH_3CH_2OH \xrightarrow{\text{KMnO}_4(aq)} CH_3COOH$$

COOH

Benzoic acid

COOH

NO_2

***p*-Nitrobenzoic acid**

$$CH_3CHCH_2CH_2COOH$$
$$\underset{\displaystyle Br}{|}$$

4-Bromopentanoic acid

$$CH_2CH_2COOH$$
$$\underset{\displaystyle Cl}{|}$$

3-Chloropropanoic acid

Figure 20.12
Some carboxylic acids.

Table 20.7   Several Carboxylic Acids, Their Systematic Names, and Their Common Names

Formula	Systematic Name	Common Name
HCOOH	methanoic acid	formic acid
CH_3COOH	ethanoic acid	acetic acid
CH_3CH_2COOH	propanoic acid	propionic acid
$CH_3CH_2CH_2COOH$	butanoic acid	butyric acid
$CH_3CH_2CH_2CH_2COOH$	pentanoic acid	valeric acid

A carboxylic acid reacts with an alcohol to form an ester and a water molecule. For example, the reaction of acetic acid and ethanol produces the ester ethyl acetate and water.

$$CH_3\overset{O}{\overset{\|}{C}}\text{—OH}\quad H\text{—}OCH_2CH_3 \rightarrow CH_3\overset{O}{\overset{\|}{C}}\text{—}OCH_2CH_3 + H_2O$$

React to
form water

This reaction can be represented in general as follows:

$$RCOOH + R'OH \rightarrow RCOOR' + H_2O$$

Acid   Alcohol   Ester   Water

An **ester** has the following general formula:

$$R\text{—}\overset{\displaystyle O}{\underset{\displaystyle O\text{—}R'}{C}}$$

From the acid   From the alcohol

Amyl is a common name for
$CH_3CH_2CH_2CH_2CH_2$—.

Esters often have a sweet, fruity odor that contrasts markedly with the often-pungent odors of the parent carboxylic acids. For example, the odor of bananas derives from *n*-amyl acetate:

$$CH_3C\overset{\displaystyle O}{\underset{\displaystyle OCH_2CH_2CH_2CH_2CH_3}{}}$$

and that of oranges from *n*-octyl acetate:

$$CH_3\underset{\displaystyle O}{\overset{\displaystyle \|}{C}}\text{—}OC_8H_{17}$$

Like carboxylic acids, esters are often referred to by their common names. The name consists of the alkyl name from the alcohol followed by the acid name, where the *-ic* ending is replaced by *-ate*. For example, the ester

$$CH_3C\overset{\displaystyle O}{\underset{\displaystyle O\text{—}CH\underset{\displaystyle CH_3}{\overset{\displaystyle CH_3}{}}}{}}$$

is made from acetic acid, CH_3COOH, and isopropyl alcohol,

$$CH_3\underset{\displaystyle OH}{\overset{\displaystyle |}{CH}}CH_3$$

and is called isopropyl acetate. The systematic name for this ester is isopropylethanoate (from ethanoic acid, the systematic name for acetic acid).

Computer-generated space-filled model of acetylsalicylic acid (aspirin).

A very important ester is formed from the reaction of salicylic acid and acetic acid.

$$\text{Salicylic acid} \quad \cdot \quad \text{Acetic acid} \quad \rightarrow \quad \text{Acetylsalicylic acid} \quad + \quad H_2O$$

The product is acetylsalicylic acid, commonly known as *aspirin,* which is manufactured in huge quantities and is widely used as an analgesic (painkiller).

20.16   Polymers

OBJECTIVE: To learn about some common polymers.

Polymers are large, usually chainlike molecules that are built from small molecules called *monomers.* Polymers form the basis for synthetic fibers, rubbers, and plastics and have played a leading role in the revolution brought about in our lives by chemistry during the past 50 years. (Many important biomolecules are also polymers.)

The simplest and one of the best-known synthetic polymers is *polyethylene,* which is constructed from ethylene monomers. Its structure is

$$nCH_2{=}CH_2 \xrightarrow{\text{Catalyst}} \left(\begin{array}{cc} H & H \\ | & | \\ -C & -C- \\ | & | \\ H & H \end{array}\right)_n$$

Ethylene   Polyethylene

where *n* represents a large number (usually several thousand). Polyethylene is a tough, flexible plastic used for piping, bottles, electrical insulation, film for packaging, garbage bags, and many other purposes. Its properties can be varied by using substituted ethylene monomers. For example, when tetrafluoroethylene is the monomer, the polymer Teflon is obtained.

$$n\left(\begin{array}{cc} F & F \\ \diagdown & \diagup \\ C{=}C \\ \diagup & \diagdown \\ F & F \end{array}\right) \longrightarrow \left(\begin{array}{cc} F & F \\ | & | \\ -C & -C- \\ | & | \\ F & F \end{array}\right)_n$$

Tetrafluoroethylene   Teflon

Because of the resistance of the strong C—F bonds to chemical attack, Teflon is an inert, tough, and nonflammable material that is widely used for electrical insulation, nonstick coatings for cooking utensils, and bearings for low-temperature applications.

Other similar polyethylene-type polymers are made from monomers containing chloro, methyl, cyano, and phenyl substituents (Table 20.8). In each case, the carbon–carbon double bond in the substituted ethylene monomer becomes a single bond in the polymer. The different substituents lead to a wide variety of properties.

Table 20.8 Some Common Synthetic Polymers, Their Monomers, and Applications

Monomer Name and Formula	Polymer Name and Formula	Applications
ethylene $H_2C=CH_2$	polyethylene $-(CH_2-CH_2)_{\overline{n}}$	plastic piping, bottles, electrical insulation, toys
propylene $H_2C=C{\overset{\displaystyle H}{\underset{\displaystyle CH_3}{\mid}}}$	polypropylene $-(CH-CH_2CH-CH_2)_{\overline{n}}$ $\quad CH_3 \qquad CH_3$	film for packaging, carpets, lab wares, toys
vinyl chloride $H_2C=C{\overset{\displaystyle H}{\underset{\displaystyle Cl}{\mid}}}$	polyvinyl chloride (PVC) $-(CH_2-CH)_{\overline{n}}$ $\qquad\qquad Cl$	piping, siding, floor tile, clothing, toys
acrylonitrile $H_2C=C{\overset{\displaystyle H}{\underset{\displaystyle CN}{\mid}}}$	polyacrylonitrile (PAN) $-(CH_2-CH)_{\overline{n}}$ $\qquad\qquad CN$	carpets, fabrics
tetrafluoroethylene $F_2C=CF_2$	Teflon $-(CF_2-CF_2)_{\overline{n}}$	coating for cooking utensils, electrical insulation, bearings
styrene $H_2C=C{\overset{\displaystyle H}{\mid}}$ (benzene ring)	polystyrene $-(CH_2CH)_{\overline{n}}$ (benzene ring)	containers, thermal insulation, toys
butadiene $H_2C=C{\overset{\displaystyle H}{\mid}}-C{\overset{\displaystyle H}{\mid}}=CH_2$	polybutadiene $-(CH_2CH=CHCH_2)_{\overline{n}}$	tire tread, coating resin
butadiene and styrene (see above)	styrene-butadiene rubber $(CH-CH_2-CH_2-CH=CH-CH_2)_{\overline{n}}$ (benzene ring)	synthetic rubber

The polyethylene polymers illustrate one of the major types of polymerization reactions, called **addition polymerization,** in which the monomers simply "add together" to form the polymer and there are no other products.

Another common type of polymerization is **condensation polymerization,** in which a small molecule, such as water, is formed for each extension of the polymer chain. The most familiar polymer produced by condensation is *nylon.* Nylon is a **copolymer** because two different types of monomers combine to form the chain (a **homopolymer,** by contrast, results from the polymerizing of a single type of monomer). One common

The Chemistry of Music

Producing great instrumental music requires a combination of musical talent on the part of the performer and a high-quality instrument. The main component of stringed instruments such as acoustic guitars and violins is wood. Wood is a mixture of natural polymers including cellulose (a linear, high-molar-mass polysaccharide) and hemicellulose (a branched polysaccharide with lower molar mass). These components are held together by lignin, a complex branched polymer. In all wooden instruments, much of the sound comes from the body of the instrument. The specific wood used can have a profound effect on the sound by dampening certain frequencies and enhancing others.

One of the great mysteries relating to musical instruments is the legendary quality of the violins made in the seventeenth and early eighteenth centuries in Cremona, Italy, by Antonio Stradivari, Nicolo Amati, and Bartolomeo Giuseppe Guarneri. The instruments crafted by these masters have sounds unmatched by other violin makers before or since.

Joseph Nagyvary, a retired Texas A&M University biochemistry professor, has worked for the last 30 years to understand the unique properties of the Cremonian instruments. Using various scientific analyses, Nagyvary has decided that the secrets lie mainly in the treatment of the wood before the violins were constructed. First the wood was soaked as it floated downriver to Cremona. Then the wood was apparently treated with a chemical soak to prevent mold and to kill pests such as woodworms that might bore holes in the wood. Nagyvary's analyses indicate that this chemical soak was quite complex and was likely prepared by a local chemist for the violin makers. His studies also indicate that the wood may have been chemically treated by an alkaline solution, which reduced the hemicellulose content of the wood.

Nagyvary has used the knowledge he has gained from studying the Cremonian instruments to build more than 150 violins. He has achieved varying degrees of success in matching the sounds of the Cremonian instruments.

Chemistry is important in constructing modern instruments as well. In acoustic guitars, not only is the type of wood important but the coating used on the wood also influences the quality of the sound. A hard coating tends to emphasize high frequencies, whereas a softer coating resonates at lower frequencies.

Likewise, chemistry is important for the speakers of your audio system. To produce sounds, a speaker's components vibrate rapidly, generating a tremendous amount of heat. In fact, only 2% of the speaker's energy is converted to sound. The remaining 98% is heat. The louder the sound, the greater the heat. To withstand temperatures up to 150 °C, the polymers used in the speaker must be high-performance materials.

Great sounds require sound chemistry.

Courtesy, Dr. Joseph Nagyvary

Joseph Nagyvary.

CHEMISTRY *IN* FOCUS

Mother of Invention

Stephanie Kwolek is not very tall (4 feet 11 inches), but she has had a big impact on modern society. Ms. Kwolek is the scientist responsible for the discovery of Kevlar, a strong, light fiber used in "bulletproof" vests that are worn by many law enforcement officers. The DuPont company, which manufactures Kevlar, estimates that more than 3000 officers have survived life-threatening injuries because of the golden-colored fiber discovered by Stephanie Kwolek. In addition, countless soldiers have been saved. Nearly every U.S. service member has worn a helmet reinforced with Kevlar since 1991. Kevlar also has many other uses, including reinforcement for tires, special cables for suspension bridges, as a component of flame-resistant clothing, and in various types of sports equipment.

Ms. Kwolek was 42 years old in 1965 when she discovered the polymer (a long chain made up of many small units) that forms the basis of Kevlar. Tests showed that, pound for pound, Kevlar is five times stronger than steel. The unusual strength of Kevlar is principally due to the way the long chains "stick" to each other because of the strong attractions between atoms on adjacent strands. Also, instead of being random like a plate of spaghetti, the Kevlar chains tend to line up in parallel to maximize the interchain interactions.

Stephanie Kwolek attributes her scientific curiosity to her father, an amateur naturalist, who took her on many hikes to collect biologic samples in her native Pennsylvania. She was a chemistry major in college and initially wanted to become a doctor, but because she couldn't afford to go to medical school, she took a job in the chemical industry with DuPont, and the rest is history.

Ms. Kwolek has received many honors in her career, including induction into the National Inventors Hall of Fame in Akron, Ohio, in 1995.

Stephanie Kwolek wearing Kevlar gloves.

The structure of Kevlar

form of nylon is produced when hexamethylenediamine and adipic acid react by splitting out a water molecule to form a C—N bond:

Hexamethylenediamine   Adipic acid

Figure 20.13

The reaction to form nylon can be carried out at the interface of two immiscible liquid layers in a beaker. The bottom layer contains adipyl chloride,

$$Cl-\underset{\underset{O}{\|}}{C}-(CH_2)_4-\underset{\underset{O}{\|}}{C}-Cl$$

dissolved in CCl_4, and the top layer contains hexamethylenediamine,

$$H_2N-(CH_2)_6-NH_2$$

dissolved in water. A molecule of HCl is formed as each C—N bond forms. This is a variation of the reaction to form nylon discussed in the text.

The molecule formed, which is called a **dimer** (two monomers joined), can undergo further condensation reactions because it has an amino group at one end and a carboxyl group at the other. Thus both ends are free to react with another monomer. Repetition of this process leads to a long chain of the type

$$\left(\underset{\underset{}{\underset{}{}}}{\overset{\overset{H}{|}}{N}}-(CH_2)_6-\underset{\underset{}{}}{\overset{\overset{H}{|}}{N}}-\underset{\underset{O}{\|}}{C}-(CH_2)_4-\underset{\underset{O}{\|}}{C} \right)_n$$

which is the general structure of nylon. The reaction to form nylon occurs quite readily and is often used as a lecture demonstration (Figure 20.13). The properties of nylon can be varied by changing the number of carbon atoms in the chain of the acid or amine monomers.

More than 1 million tons of nylon are produced annually in the United States for use in clothing, carpets, rope, and so on. Many other types of condensation polymers are also produced. For example, Dacron is a copolymer formed from the condensation reaction of ethylene glycol (a dialcohol) and *p*-terephthalic acid (a dicarboxylic acid).

$$HOCH_2CH_2O-H \quad HO-\underset{\underset{O}{\|}}{C}-\bigcirc-\underset{\underset{O}{\|}}{C}-O-H$$

Ethylene glycol $\quad\downarrow\quad$ *p*-Terephthalic
$\quad\quad H_2O \quad$ acid

The repeating unit of Dacron is

$$\left(OCH_2CH_2-O-\underset{\underset{O}{\|}}{C}-\bigcirc-\underset{\underset{O}{\|}}{C} \right)_n$$

Note that this polymerization involves a carboxylic acid and an alcohol to form an ester group,

$$-O-\underset{\underset{O}{\|}}{C}-$$

Thus Dacron is called a **polyester.** By itself or blended with cotton, Dacron is used widely in fibers for the manufacture of clothing.

CHAPTER 20 REVIEW

Key Terms

organic
   chemistry (p. 607)
biomolecules (p. 607)
hydrocarbon (20.2)
saturated (20.2)
unsaturated (20.2)
alkane (20.2)

normal (straight-chain
   or unbranched)
   hydrocarbon (20.2)
structural
   isomerism (20.3)
petroleum (20.5)
natural gas (20.5)

combustion
   reaction (20.6)
substitution
   reaction (20.6)
dehydrogenation
   reaction (20.6)
alkene (20.7)

alkyne (20.7)
addition reaction (20.7)
hydrogenation
   reaction (20.7)
halogenation (20.7)
polymerization (20.7)

aromatic	functional group (20.10)	carboxylic acid (20.15)	condensation
hydrocarbon (20.8)	alcohol (20.11)	carboxyl group (20.15)	polymerization (20.16)
benzene (20.8)	phenol (20.12)	ester (20.15)	copolymer (20.16)
phenyl group (20.9)	carbonyl group (20.13)	polymer (20.16)	homopolymer (20.16)
hydrocarbon	ketone (20.13)	addition	dimer (20.16)
derivative (20.10)	aldehyde (20.13)	polymerization (20.16)	polyester (20.16)

Summary

1. The study of carbon-containing compounds and their properties is called organic chemistry. Most organic compounds contain chains or rings of carbon atoms. The organic molecules responsible for maintaining and reproducing life are called biomolecules.

2. Hydrocarbons are organic compounds composed of carbon and hydrogen. Those that contain only C—C single bonds are saturated and are called alkanes, and those with carbon–carbon multiple bonds are unsaturated. Unsaturated hydrocarbons can become saturated by the addition of hydrogen, halogens, and/or other substituents.

3. All alkanes can be represented by the general formula C_nH_{2n+2}. Methane, CH_4, is the simplest alkane, and the next three in the series are ethane, C_2H_6; propane, C_3H_8; and butane, C_4H_{10}. In a saturated hydrocarbon, each carbon atom is bonded to four other atoms. Alkanes containing long chains of carbon atoms are called normal, or straight-chain, hydrocarbons.

4. Structural isomerism in alkanes involves the formation of branched structures. Specific rules for systematically naming alkanes indicate the point of attachment of any substituent group, the length of the root chain, and so on.

5. Alkanes can undergo combustion reactions to form carbon dioxide and water or substitution reactions in which hydrogen atoms are replaced by other atoms. Alkanes also can undergo dehydrogenation to form unsaturated hydrocarbons.

6. Hydrocarbons with carbon–carbon double bonds are called alkenes. The simplest alkene is ethylene, C_2H_4. Alkynes are unsaturated hydrocarbons with a carbon–carbon triple bond. The simplest in the series is acetylene, C_2H_2.

7. Unsaturated hydrocarbons undergo addition reactions such as hydrogenation (addition of hydrogen atoms) and halogenation (addition of halogen atoms). Ethylene and substituted ethylene molecules can undergo polymerization, a process by which many molecules (monomers) are joined together to form a large chain-like molecule.

8. Organic molecules that contain elements in addition to carbon and hydrogen can be viewed as hydrocarbon derivatives: hydrocarbons with functional groups. Each functional group exhibits characteristic chemical properties.

9. Alcohols contain the —OH functional group. Aldehydes and ketones contain the carbonyl functional group,

$$\diagup\!\!\!\!\!{C}\!=\!O$$

In aldehydes, this group is bonded to at least one hydrogen atom. Carboxylic acids are characterized by the carboxyl functional group,

$$-C\diagdown\!\!\!\!\!{\diagup}^{O}_{OH}$$

They can react with alcohols to form esters.

10. Polymers can be formed by addition polymerization, in which monomers add together, or by condensation polymerization, which involves the splitting out of small molecules (such as water) as the monomers react.

Active Learning Questions

These questions are designed to be considered by groups of students in class. Often these questions work well for introducing a particular topic in class.

1. What is meant by the term "unsaturated hydrocarbon"? What structural feature characterizes unsaturated hydrocarbons?

2. The following are named incorrectly, but a correct structure can be made from the name given. Draw the following incorrectly named compounds and name them correctly.

   a. 2-ethyl-3-methyl-5-isopropyl hexane
   b. 3-methyl-4-isopropylpentane
   c. 2-ethyl-3-butyne

3. When propane undergoes dehydrogenation, what is the product? Name it and draw a structural formula.

4. How many different possible "tetramethylbenzenes" exist? Provide structural formulas and names for each.

5. For the general formula $C_6H_{14}O$, draw the structures of three isomeric alcohols that illustrate primary, secondary, and tertiary structures.

6. Differentiate between an addition polymer and a condensation polymer and provide an example of each. Do the same for copolymer and homopolymer.

21

Biochemistry

● The energy for exercise is furnished by chemical reactions in the body. *(John Blanding/The Boston Globe/Merlin-net.com)*

Biochemistry, the study of the chemistry of living systems, is a vast and exciting field in which important discoveries about how life is maintained and how diseases occur are being made every day. In particular, there has been rapid growth in the understanding of how living cells manufacture and use the molecules necessary for life. This not only has been beneficial for detection and treatment of diseases but also has spawned a new field—**biotechnology,** which uses nature's "machinery" to synthesize desired substances. For example, insulin is a complex biomolecule that is used in the body to regulate the metabolism of sugars. People who have diabetes are deficient in natural insulin and must take insulin by injection or other means. In the past this insulin was obtained from animal tissues (particularly that of cows). However, our increased understanding of the biochemical processes of the cell has allowed us to "farm" insulin. We have learned to insert the "instructions" for making insulin into the cells of bacteria such as *Escherichia coli* so that as the bacteria grow they produce insulin that can then be harvested for use by diabetics. Many other products, including natural pesticides, are also being produced by the techniques of biotechnology.

An understanding of biochemistry also allows our society to produce healthier processed foods. For example, the food industry is making great efforts to reduce the fat levels in food without destroying the good taste that fats bring to food.

We cannot hope to cover all the important aspects of this field in this chapter; we will concentrate here on the major types of biomolecules that support living organisms. First, however, we will survey the elements found in living systems and briefly describe the constitution of a cell.

This sample of *E. coli* bacteria has been genetically altered to produce human insulin. The insulin production sites are colored orange. The bacteria are cultured and pure human insulin is harvested from them.

Volker Steger/SPL/Photo Researchers, Inc.

At present, 30 elements are definitely known to be essential to human life. These **essential elements** are shown in color in Figure 21.1. The most abundant elements are hydrogen, carbon, nitrogen, and oxygen; sodium, magnesium, potassium, calcium, phosphorus, sulfur, and chlorine are also present in relatively large amounts. Although present only in trace amounts, the first-row transition metals are essential for the action of many enzymes (biologic catalysts). For example, zinc, one of the **trace elements,** is found in nearly 200 biologically important molecules. The functions of the essential elements are summarized in Table 21.1. In time, other elements probably will be found to be essential.

Life is organized around the functions of the **cell,** the smallest unit in living things that exhibits the properties normally associated with life, such as reproduction, metabolism, mutation, and sensitivity to external stimuli.

Figure 21.1

The chemical elements essential for life. Those most abundant in living systems are shown as purple. Nineteen elements, called the *trace elements,* are shown as green.

Table 21.1	Some Essential Elements and Their Major Functions	
Element	**Percent by Mass in the Human Body**	**Function**
oxygen	65	in water and many organic compounds
carbon	18	in all organic compounds
hydrogen	10	in water and many inorganic and organic compounds
nitrogen	3	in both inorganic and organic compounds
calcium	1.5	in bone; essential to some enzymes and to muscle action
phosphorus	1.2	essential in cell membranes and to energy transfer in cells
potassium	0.2	cation in cell fluid
chlorine	0.2	anion inside and outside the cells
sulfur	0.2	in proteins
sodium	0.1	cation in cell fluid
magnesium	0.05	essential to some enzymes
iron	<0.05	in molecules that transport and store oxygen
zinc	<0.05	essential to many enzymes
cobalt	<0.05	found in vitamin B_{12}
iodine	<0.05	essential to thyroid hormones
fluorine	<0.01	in teeth and bones

As the fundamental building blocks of all living systems, aggregates of cells form tissues, which in turn are assembled into the organs that make up complex living systems. Therefore, to understand how life is maintained and reproduced, we must learn how cells operate on the molecular level. This is the main thrust of biochemistry.

21.1 Proteins

OBJECTIVE: To learn about proteins.

Silk is a naturally occurring polymer.

In Chapter 20 we saw that many useful synthetic materials are polymers. A great many natural materials are also polymers: starch, hair, silk and cotton fibers, and the cellulose in woody plants, to name only a few.

In this section we introduce a class of natural polymers, the **proteins,** which make up about 15% of our bodies and have molar masses that range from approximately 6000 to over 1,000,000 grams. Proteins have many functions in the human body. **Fibrous proteins** provide structural integrity and strength for many types of tissue and are the main components of muscle, hair, and cartilage. Other proteins, usually called **globular proteins** because of their roughly spherical shape, are the "worker" molecules of the body. These proteins transport and store oxygen and nutrients, act as catalysts for the thousands of reactions that make life possible, fight invasion of the body by foreign objects, participate in the body's many regulatory systems, and transport electrons in the complex process of metabolizing nutrients.

21.2 Primary Structure of Proteins

OBJECTIVE: To understand the primary structure of proteins.

The building blocks of all proteins are the *α*-**amino acids:**

$$\alpha\text{-Carbon} \qquad NH_2-\overset{\overset{\displaystyle H}{|}}{\underset{\underset{\displaystyle R}{|}}{C}}-\overset{\displaystyle O}{\underset{\displaystyle OH}{C}}$$

The R in this structure may represent H, CH_3, or more complex substituents. These molecules are called *α*-amino acids because the amino group ($-NH_2$) is always attached to the *α*-carbon, the one next to the carboxyl group ($-COOH$). The 20 amino acids most commonly found in proteins are shown in Figure 21.2.

Note from Figure 21.2 that the amino acids are grouped into polar and nonpolar classes on the basis of the composition of the R groups, also called the **side chains.** Nonpolar side chains contain mostly carbon and hydrogen atoms; polar side chains contain nitrogen and oxygen atoms. This difference is important because polar side chains are *hydrophilic* (water-loving), but nonpolar side chains are *hydrophobic* (water-fearing). This greatly affects the three-dimensional structure of the resulting proteins, because they exist in aqueous media in living things.

The protein polymer is built by reactions between amino acids. For example, two amino acids can react as follows, forming a C—N bond with the elimination of water.

At the pH in biologic fluids, the amino acids shown in Figure 21.2 exist in a different form, with the proton of the —COOH group transferred to the —NH₂ group. For example, glycine would be in the form $H_3{}^+NCH_2COO^-$.

Peptide linkage

$$\overset{\overset{\displaystyle H}{|}}{\underset{\underset{\displaystyle H}{|}}{N}}-\overset{\overset{\displaystyle H}{|}}{\underset{\underset{\displaystyle R}{|}}{C}}-\overset{\displaystyle O}{C} \quad + \quad \overset{\overset{\displaystyle H}{|}}{\underset{\underset{\displaystyle H}{|}}{N}}-\overset{\overset{\displaystyle H}{|}}{\underset{\underset{\displaystyle R'}{|}}{C}}-\overset{\displaystyle O}{\underset{\displaystyle OH}{C}} \quad \rightarrow \quad \overset{\overset{\displaystyle H}{|}}{\underset{\underset{\displaystyle H}{|}}{N}}-\overset{\overset{\displaystyle H}{|}}{\underset{\underset{\displaystyle R}{|}}{C}}-\overset{\displaystyle O}{C}-\overset{\displaystyle H}{N}-\overset{\overset{\displaystyle H}{|}}{\underset{\underset{\displaystyle R'}{|}}{C}}-\overset{\displaystyle O}{\underset{\displaystyle OH}{C}} \quad + \quad H_2O$$

Nonpolar R Groups (hydrophobic)

Polar R Groups (hydrophilic)

Figure 21.2

The 20 α-amino acids found in most proteins. The R group is shown in color.

The product shown on page 691 is called a **dipeptide.** The term *peptide* comes from the structure

$$\begin{array}{cc} O & H \\ \parallel & \mid \\ -C- & N- \end{array}$$

which chemists call a **peptide linkage** or a peptide bond. The prefix *di-* indicates that two amino acids have been joined. Additional reactions lengthen the chain to produce a **polypeptide** and eventually a protein.

The 20 amino acids can be assembled in any order, which makes possible an enormous number of different proteins. This variety allows for the many types of proteins needed for the functions that organisms carry out.

The *order* or *sequence* of amino acids in the protein chain is called the **primary structure.** We indicate the primary structure by using three-letter codes for the amino acids (see Figure 21.2), where it is understood that the terminal carboxyl group is on the right and the terminal amino group is on the left. For example, one possible sequence for a tripeptide containing the amino acids lysine, alanine, and leucine is

which is represented in the shorthand notation (three-letter codes) as

lys-ala-leu

EXAMPLE 21.1 Understanding Primary Structure

Write the sequences of all possible tripeptides composed of the amino acids tyrosine (tyr), histidine (his), and cysteine (cys).

SOLUTION

There are six possible sequences:

tyr-his-cys	his-tyr-cys	cys-tyr-his
tyr-cys-his	his-cys-tyr	cys-his-tyr ∎

A striking example of the importance of the primary structure of polypeptides can be seen in the differences between *oxytocin* and *vasopressin*. Oxytocin is a hormone that triggers contraction of the uterus and milk secretion. Vasopressin raises blood pressure and regulates kidney function. Both these molecules are nine-unit polypeptides, and they differ by only two amino acids (Figure 21.3); yet they have completely different functions in the human body.

cys-tyr-ile-gin-asn-cys-pro-leu-gly

a

cys-tyr-phe-gin-asn-cys-pro-arg-gly

b

Figure 21.3

The amino acid sequences in **a** oxytocin and **b** vasopressin. The differing amino acids are enclosed in boxes.

21.3   Secondary Structure of Proteins

OBJECTIVE: To understand the secondary structure of proteins.

a

The primary structure of a protein describes the order of the amino acids in the chain. One example is shown.

b

The secondary structure of a protein describes the arrangement of the chain in space. A spiral (helical) arrangement of the protein chain is shown.

Figure 21.4

So far we have considered the primary structure of proteins—the order of amino acids in the chain (Figure 21.4a). A second level of structure in proteins is the arrangement in space of the chain of the long molecule. This is called the **secondary structure** of the protein (Figure 21.4b).

One common type of secondary structure resembles a spiral staircase. This spiral structure is called an **α-helix** (Figure 21.5). A spiral-like secondary structure gives the protein elasticity (springiness) and is found in the fibrous proteins in wool, hair, and tendons. Another type of secondary structure involves joining several different protein chains in an arrangement called a **pleated sheet,** as shown in Figure 21.6. Silk has this arrangement of proteins, making its fibers flexible but very strong and resistant to stretching. The pleated sheet is also found in muscle fibers.

As you might imagine, a molecule as large as a protein has a great deal of flexibility and can assume a variety of overall shapes. The function the protein is to serve influences the specific shape that it will have. For long, thin structures such as hair, wool and silk fibers, and tendons, an elongated shape is required. This may involve an α-helical secondary structure, as found in the protein α-keratin in hair and wool or in the collagen that occurs in tendons (Figure 21.7a), or it may involve a pleated-sheet secondary

Figure 21.6

Two protein chains bound together in the pleated-sheet secondary structure. Each sphere represents an amino acid.

Figure 21.5

One type of secondary protein structure is like a spiral staircase and is called the α-helix. The circles represent individual amino acids.

a   *b*

Figure 21.7

a Collagen, a protein found in tendons, consists of three protein chains (each with an α-helical structure) twisted together to form a superhelix. The result is a long, relatively narrow protein. *b* Many proteins are bound together in the pleated-sheet arrangement to form the elongated protein found in silk fibers.

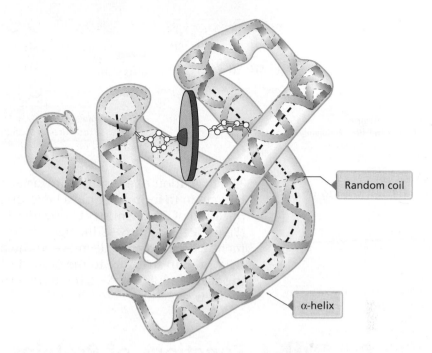

Figure 21.8

The protein myoglobin. Note that the secondary structure is basically α-helical except at the bends. The tertiary structure is globular.

structure, as found in silk (Figure 21.7b). Many of the proteins in the body that have nonstructural functions (such as serving as enzymes) are globular. One is myoglobin (Figure 21.8), which absorbs an O_2 molecule and stores it for use by the cells as it is needed. Note that the secondary structure of myoglobin is basically α-helical. However, in the areas where the chain bends to give the protein its compact globular structure, the α-helix breaks down so that the protein can "turn the corner."

21.4 Tertiary Structure of Proteins

OBJECTIVE: To understand the tertiary structure of proteins.

The overall shape of the protein, long and narrow or globular, is called its **tertiary structure.** To make sure the difference between secondary and tertiary structure is clear, examine Figure 21.8 again. The secondary structure of the myoglobin protein is helical except in the regions where it bends back on itself to give the overall compact (globular) tertiary structure. If the bends did not occur, the tertiary structure would be like a long tube (tubular). In both tertiary arrangements (globular and tubular), the secondary structure of the protein is basically helical.

The amino acid *cysteine* (cys) plays a special role in stabilizing the tertiary structure of many proteins because the —SH groups on two cysteines can react to form an S—S bond called a **disulfide linkage.**

Natural cysteine linkages in hair

Hair set in curlers alters tertiary structure

New cysteine linkages in waved hair

Figure 21.9

The permanent waving of hair.

The formation of a disulfide linkage can fasten together two parts of a protein chain to form and hold a bend in the chain, for example. A practical application of the chemistry of disulfide bonds is the permanent waving of hair (Figure 21.9). The S—S linkages in the protein of hair are broken by treatment with a reducing agent. Next the hair is set in curlers to change the tertiary protein structure to the desired shape. Then treatment with an oxidizing agent causes new S—S bonds to form, which make the hair protein retain the new structure.

21.5   Functions of Proteins

OBJECTIVE: To learn about various functions served by proteins.

Energy

Figure 21.10

A schematic representation of the thermal denaturation of a protein.

The three-dimensional structure of a protein is crucial to its function. The process of breaking down this structure is called **denaturation** (Figure 21.10). For example, heat causes the denaturation of egg proteins when an egg is cooked. Any source of energy can cause denaturation of proteins and is thus potentially dangerous to living organisms. For example, ultraviolet radiation, X-ray radiation, or nuclear radioactivity can disrupt protein structure, which may lead to cancer or genetic damage. The metals lead and mercury, which have a very high affinity for sulfur, cause protein denaturation by disrupting disulfide bonds.

The tremendous variability in the several levels of protein structure allows proteins to be tailored specifically to serve a wide range of functions, some of which are given in Table 21.2.

21.6   Enzymes

OBJECTIVE: To understand how enzymes work.

An enzyme often changes shape slightly as the substrate is bound. This "clamps" the substrate into place.

Enzymes are proteins that catalyze specific biologic reactions. Without the several hundred enzymes now known, life would be impossible. Almost all the critical biochemical reactions would occur far too slowly at the temperatures at which life exists. Enzymes are impressive for their tremendous efficiency (they are typically 1 to 10 million times as efficient as inorganic catalysts) and their incredible selectivity (an enzyme "ignores" the thousands of molecules in body fluids that are not involved in the reaction it catalyzes).

Although the mechanisms of enzyme activity are complex and not fully understood in most cases, a simple theory called the **lock-and-key model** (Figure 21.11) seems to fit many enzymes. This model postulates that the shapes of the reacting molecule (the **substrate**) and the enzyme are such that they fit together much as a key fits a specific lock. The substrate

Table 21.2 Common Functions of Proteins

Function	Comment/Example
Structure	Proteins provide the strength of tendons, bones, and skin. Cartilage, hair, wool, fingernails, and claws are mainly protein. Viruses have an outer layer of protein.
Movement	Proteins are the major components of muscles and are directly responsible for the ability of muscles to contract. The swimming of sperm results from the contraction of protein filaments in their tails.
Catalysis	Nearly all chemical reactions in living organisms are catalyzed by enzymes, which are almost always proteins.
Transport	Oxygen is carried from the lungs to tissues by the protein hemoglobin in red blood cells.
Storage	The protein *ferritin* stores iron in the liver, spleen, and bone marrow.
Energy transformation	Cytochromes are proteins found in all cells. They extract energy from food molecules by transferring electrons in a series of oxidation–reduction reactions.
Protection	*Antibodies* are special proteins that are synthesized in response to foreign substances and cells, such as bacterial cells. They then bind to those substances or cells and provide us with immunity to various diseases. Interferon, a small protein made and released by cells when they are exposed to a virus, protects other cells against viral infection. Blood-clotting proteins protect against bleeding (hemorrhage).
Control	Many *hormones* are proteins produced in the body that have specific effects on the activity of certain organs.
Buffering	Because proteins contain both acidic and basic groups on their side chains, they can neutralize both acids and bases and thus provide buffering for blood and tissues.

and enzyme attach to each other in such a way that the part of the substrate where the reaction is to occur occupies the **active site** of the enzyme. After the reaction occurs, the products are liberated, and the enzyme is ready for a new substrate. We can represent enzyme catalysis by the following steps:

Step 1 The enzyme E and the substrate S come together.

$$E + S \rightleftharpoons E \cdot S$$

Step 2 The reaction occurs to give the product P, which is released from the enzyme.

$$E \cdot S \rightarrow E + P$$

Figure 21.11

Schematic diagram of the lock-and-key model of the functioning of an enzyme.

Urine Farming

Nature is a very good chemist—organisms make hundreds of complex chemicals every second to survive. The rapidly expanding science of biotechnology is learning more every day about how to use natural chemical pathways to make valuable molecules. For example, insulin for diabetics is now made almost exclusively by "farming" genetically altered *Escherichia coli* or yeast.

Turning livestock into four-footed pharmaceutical factories has seemed an attractive option ever since we learned to carry out "genetic engineering." In fact, a great deal of effort has been expended on creating cows, sheep, and goats that are genetically modified to produce useful human proteins that are secreted into the animals' milk, from which they can then be harvested. None of these proteins has yet reached the market—tests are now under way to check their efficacy and safety—but a blood-clotting agent, antithrombin III, developed by Genzyme Transgenics Corporation of Framingham, Massachusetts, is close to commercial application. Creating a transgenic animal is expensive (costing approximately $60,000), but then standard breeding techniques can lead to an entire herd of animals programmed to produce commercially interesting molecules.

As a medium, milk has some disadvantages. It is produced only by mature females and contains a wide variety of proteins that complicate the purification process to obtain the desired product. Urine may be a better alternative. It contains fewer natural proteins and is produced at an early age by both male and female animals. Robert J. Wall and his colleagues at the U.S. Department of Agriculture in Beltsville, Maryland, have developed transgenic mice that produce human growth hormone in the walls of their bladders. For obvious reasons, mice are not ideal for large-scale production of chemicals, but Wall's experiments nevertheless show that the concept works. At present, it is too early to know whether urine farming will prove feasible. Yields from the bladder are about 10,000 times lower than those from the mammary glands. In addition, collecting urine from farm animals could prove to be a very tricky business.

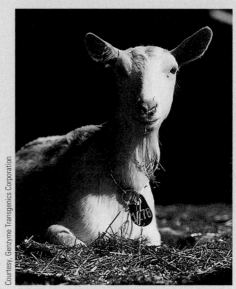

Courtesy, Genzyme Transgenics Corporation

A New Zealand goat, one of the animals used for urine farming.

After the product is released, the enzyme is free to engage another substrate. Because this process occurs so rapidly, only a tiny amount of enzyme is required.

In some cases, a substance other than the substrate can bind to the enzyme's active site. When this occurs, the enzyme is said to be *inhibited*. If the inhibition is permanent, the enzyme is said to be inactivated. Some of the most powerful toxins act by inhibiting, or inactivating, key enzymes.

Because enzymes are crucial to life and because we hope to learn how to mimic their efficiency in our industrial catalysts, the study of enzymes occupies a prominent place in chemical research.

21.7 Carbohydrates

OBJECTIVE: To understand the fundamental properties of carbohydrates.

The **carbohydrates** are another class of biologically important molecules. Carbohydrates serve as food sources for most organisms and as structural materials for plants. Because many carbohydrates have the empirical formula CH_2O, it was believed originally that these substances were hydrates of carbon ($C \cdot H_2O$), which accounts for their name.

Like proteins, carbohydrates occur in almost bewildering varieties. Many of the most important carbohydrates are polymers—large molecules constructed by hooking together many smaller molecules. We have seen that proteins are polymers constructed from amino acids. The polymeric carbohydrates are constructed from molecules called **simple sugars** or, more precisely, **monosaccharides.** Monosaccharides are aldehydes or ketones that contain several hydroxyl (—OH) substituents. An example of a monosaccharide is fructose, with the structure

$$
\begin{array}{c}
CH_2OH \\
| \\
C=O \quad \leftarrow \text{Ketone} \\
| \\
HO-C-H \\
| \\
H-C-OH \\
| \\
H-C-OH \\
| \\
CH_2OH
\end{array}
$$

Fructose

Fructose is a sugar found in honey and fruits. Monosaccharides can have various numbers of carbon atoms, and we name them according to the number of carbon atoms they contain by adding prefixes to the root -*ose*. The general names of monosaccharides are shown in Table 21.3. Notice that fructose is a ketone with six carbon atoms and five —OH substituents. Fructose is a member of the hexose family, where the prefix *hex-* means six.

The most important monosaccharides found in living organisms are pentoses and hexoses. The most common pentoses and hexoses are shown in Table 21.4.

Bread, a carbohydrate food source.

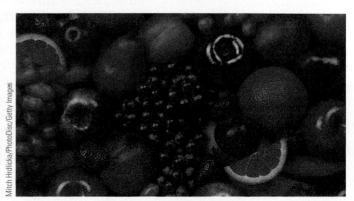

Fruit, a dietary source of fructose.

Table 21.3 The General Names of Monosaccharides

Number of Carbon Atoms	Prefix	General Name of Sugar
3	*tri-*	triose
4	*tetr-*	tetrose
5	*pent-*	pentose
6	*hex-*	hexose
7	*hept-*	heptose
8	*oct-*	octose
9	*non-*	nonose

Table 21.4 Some Important Monosaccharides

Pentoses (five carbon atoms)

Ribose	Arabinose	Ribulose
CHO	CHO	CH_2OH
H—C—OH	HO—C—H	C=O
H—C—OH	H—C—OH	H—C—OH
H—C—OH	H—C—OH	H—C—OH
CH_2OH	CH_2OH	CH_2OH

Hexoses (six carbon atoms)

Ribose	Arabinose	Ribulose
CHO	CHO	CH_2OH
H—C—OH	HO—C—H	C=O
H—C—OH	H—C—OH	H—C—OH
H—C—OH	H—C—OH	H—C—OH
CH_2OH	CH_2OH	CH_2OH

Figure 21.12

The formation of a ring structure for fructose.

Although we have so far represented the monosaccharides as straight-chain molecules, they usually form ring structures in aqueous solution. Figure 21.12 shows how the ring forms for fructose. Note that a new bond is formed between the oxygen of a hydroxyl group and the carbon of the ketone group. In the cyclic form, fructose is a five-membered ring containing a C—O—C bond. Glucose, a hexose that is an aldehyde, forms a six-membered ring structure. The six-membered glucose ring is shown in Figure 21.13a. Note that the ring is nonplanar.

More complex carbohydrates are formed by the combining of monosaccharides. For example, two monosaccharides can combine to form a **disaccharide. Sucrose,** common table sugar, is a disaccharide formed from glucose and fructose by elimination of water to form a C—O—C bond between the rings that is called a **glycoside linkage** (Figure 21.13c). When sucrose is consumed in food, this reaction is reversed (the glycoside bond is broken). An enzyme in saliva catalyzes the breakdown of sucrose into its two monosaccharides.

a Glucose

b Fructose

c Sucrose

Glycoside linkage

$-H_2O \uparrow\downarrow +H_2O$

Figure 21.13

Sucrose **c** is a disaccharide formed from glucose **a** and fructose **b**.

Large polymers containing many monosaccharide units are called **polysaccharides** and can form when each ring forms two glycoside linkages, as shown in Figure 21.14. Three of the most important of these polymers are starch, cellulose, and glycogen. All these substances are polymers of glucose; they differ in the way the glucose rings are linked together.

Starch (Figure 21.14) is the carbohydrate reservoir in plants. It is the form in which glucose is stored by the plant for later use as cellular fuel—both by the plants themselves and by organisms that eat plants.

Cellulose, the major structural component of woody plants and natural fibers such as cotton, is also a polymer of glucose. However, in cellulose the way in which the glucose rings are linked is different from the way they are linked in starch. This difference in linkage has very important consequences. The human digestive system contains enzymes that can catalyze breakage of the glycoside bonds between the glucose molecules in starch. These enzymes are *not* effective on the glycoside bonds of cellulose, however, presumably because the different structure causes a poor fit between the enzyme's active site and the carbohydrate. Interestingly, the enzymes necessary to cleave the glycoside linkages in cellulose are found in bacteria that exist in the digestive tracts of termites, cows, deer, and many other animals. Therefore, unlike humans, these animals can derive nutrition from the cellulose in wood, hay, and other similar substances.

Figure 21.14

Starch is made by hooking together many glucose rings. Only a small part of the chain is shown here.

Great Expectations?
The Chemistry of Placebos

Scientists have been intrigued by the "placebo effect" for more than 50 years. For example, a patient complaining of a pain is given a placebo (an inactive substance such as a sugar pill) but told it is really a drug to relieve the pain. In about one-third of the cases studied over many decades, the patient reports that the pain has gone away. But how does the placebo, which has no inherent medicinal value, relieve the pain? The first explanations had more to do with psychology than with chemistry. The most common explanation is that a patient feels better because the patient *expects* to feel better.

Recent studies, however, suggest there is chemistry involved in the placebo effect—specifically, the actions of endorphins. Endorphins, which were discovered in 1975, are neurotransmitters in the brain that are released when a person encounters stress. These endorphins have a pain-relieving quality much like codeine and morphine.

Researchers from the University of Michigan have found that the placebo effect correlates with an increase in endorphins being released from the brain. In their study, they injected highly concentrated saltwater solutions into the jaws of 14 healthy volunteers to produce pain. As this occurred, the brains of the subjects were scanned. During these scans the men rated their pain on a scale from 0 to 100, and were told when they would be given pain relief.

The brain scans revealed that the endorphin system was activated after the subjects were told that pain relief was coming and the placebo was given. The subjects responded to being told that pain relief was imminent by getting their bodies to release a natural pain reliever. Different studies have supported the link between the placebo effect and endorphin release by showing that when a patient who reports less pain due to a placebo is given naloxone (a drug that blocks endorphins), that person will report the reappearance of pain.

The next time someone tells you that pain is "all in your mind," you can counter by stating that chemistry has a lot to do with pain relief.

The structure of an endorphin molecule.

Glycogen is the main carbohydrate reservoir in animals. For example, it is found in muscles, where it can be broken down into glucose units when energy is required for physical activity.

21.8 Nucleic Acids

OBJECTIVE: To understand the fundamental nucleic acid structures.

Life is possible only because each cell, when it divides, can transmit the vital information about what it is to the next generation of cells. It has been known for a long time that this process involves the chromosomes in the nucleus of the cell. Only since 1953, however, have scientists understood the molecular basis of this intriguing cellular "talent."

Note that the prefix *de-* refers to the absence of the substance that follows. Thus deoxyribose means ribose with a missing oxygen (see Figure 21.15).

Figure 21.15

The structure of the pentoses (top) deoxyribose and (bottom) ribose. Deoxyribose is the sugar molecule present in DNA; ribose is found in RNA. The difference in these sugars is the substitution, in ribose, of an OH for one H in deoxyribose (shown in color).

The substance that stores and transmits the genetic information is a polymer called **deoxyribonucleic acid (DNA),** a huge molecule with a molar mass as high as several billion grams. Together with other similar nucleic acids called the **ribonucleic acids (RNA),** DNA carries the information needed for the synthesis of the various proteins the cell requires to carry out its life functions. The RNA molecules, which are found in the cytoplasm outside the cell nucleus, are much smaller than DNA polymers, with molar mass values of only 20,000 to 40,000 grams.

Both DNA and RNA are polymers—they are constructed by the hooking together of many smaller units. The fundamental unit in these polymers is called a **nucleotide.** Each nucleotide in turn has three parts:

1. *A nitrogen-containing organic base*
2. *A five-carbon sugar*
3. *A phosphate group*

A nucleotide can be represented as follows:

In the DNA polymer, the five-carbon sugar is deoxyribose (Figure 21.15 top); in the RNA polymer, it is ribose (Figure 21.15 bottom). This difference in the sugar molecules present in the polymers is responsible for the names DNA (deoxyribonucleic acid) and RNA (ribonucleic acid).

The organic base molecules in the nucleotides of DNA and RNA are shown in Figure 21.16. Notice that some of these bases are found only in DNA, some are found only in RNA, and some are found in both. The formation of a specific nucleotide containing the base adenine and the sugar ribose is shown in Figure 21.17. To form the DNA and RNA polymers, the nucleotides are hooked together. DNA polymers can contain up to a *billion* nucleotide units. Figure 21.18 shows a small portion of a DNA chain.

The key to DNA's functioning is its *double-helical structure with complementary bases on the two strands.* Complementary bases form hydrogen bonds to each other, as shown in Figure 21.19. Note that the structures of cytosine and guanine make them perfect partners (complementary) for hydrogen bonding and that they are *always* found as pairs on the two strands of DNA. Thymine and adenine form similar hydrogen-bonding pairs.

Uracil (U) RNA   **Cytosine (C)** DNA RNA   **Thymine (T)** DNA   **Adenine (A)** DNA RNA   **Guanine (G)** DNA RNA

Figure 21.16

The organic bases found in DNA and RNA. Note that uracil is found only in RNA and that thymine is found only in DNA.

NH$_2$

Adenine

H_2O

CH_2OH O OH CH_2OH O

C H H C C H H C

H H H H

C—C C—C

OH OH OH OH

Ribose **Adenosine**

Base

NH$_2$

H_2O

Phosphate

OH NH_2 OH **Ribose**

HO—P—O—H H—O—CH$_2$OH O HO—P—O—CH$_2$ O

O C H H C O C H H C

H H H H

Phosphoric acid C—C C—C

OH OH OH OH

Adenosine **Adenosine 5-phosphoric acid**

Figure 21.17

(Top) Adenosine is formed by the reaction of adenine and ribose. (Bottom) The reaction of phosphoric acid with adenosine to form the ester adenosine 5-phosphoric acid, a nucleotide. (At biologic pH, the phosphoric acid would not be fully protonated as is shown here.)

A molecular model of part of the DNA structure.

There is much evidence to suggest that the two strands of DNA unwind during cell division and that new complementary strands are constructed on the unraveled strands (Figure 21.20). Because the bases on the strands always pair in the same way—cytosine with guanine and thymine with adenine— each unraveled strand serves as a template for attaching each complementary base (along with the rest of its nucleotide). This process results in two new double-helix DNA structures that are identical to the original one. Each new double strand contains one strand from the original DNA double helix and one newly synthesized strand. This replication of DNA makes possible the transmission of genetic information when cells divide.

▶ DNA and Protein Synthesis

Besides replication, the other major function of DNA is **protein synthesis.** The proteins consumed by an organism in its food are typically not the specific proteins that organism needs to maintain its existence. The nutrient proteins are broken down into their constituent amino acids, which are then used to construct those proteins that the organism needs. The information for constructing each protein needed by a particular organism is stored in that organism's DNA. A given segment of the DNA, called a **gene,** contains the code for a specific protein. This code for the primary structure of

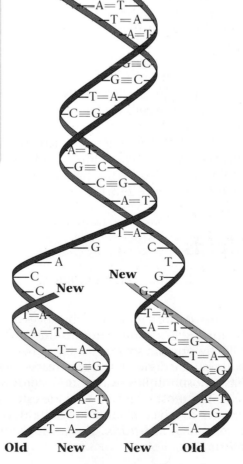

Figure 21.18

A portion of a typical nucleic acid chain, half the DNA double helix shown in Figure 21.19.

Repeating unit along DNA chain

Figure 21.19

a The DNA double helix contains two sugar–phosphate backbones with the bases from the two strands hydrogen-bonded to each other. The **b** thymine–adenine and **c** cytosine–guanine pairs show complementarity. The hydrogen-bonding interactions are shown by dotted lines.

Figure 21.20

During cell division, the original DNA double helix unwinds, and new complementary strands are constructed on each original strand. In this way, the two cells resulting from division have exact copies of the DNA of the original cell.

Nucleus

DNA

RNA

mRNA

Ribosome moves along mRNA

tRNA + Amino acid
(as specified by mRNA)

Growing polypeptide chain

Ribosome + Completed protein

Figure 21.21

The mRNA molecule, constructed from a specific gene on the DNA, is used as the pattern for construction of a given protein with the assistance of ribosomes. The tRNA molecules attach to specific amino acids and put them in place, as dictated by the patterns on the mRNA. This sequence (left to right) shows the protein chain growing.

the protein (the sequence of amino acids) can be transmitted to the construction "machinery" of the cell.

DNA stores the genetic information, and RNA molecules are responsible for transmitting this information to cell components called *ribosomes,* where protein synthesis actually occurs. This process involves, first, the construction of a special RNA molecule called **messenger RNA (mRNA).** The mRNA is built in the cell nucleus, where a specific section of DNA (a gene) is used as the pattern. The mRNA then migrates from the nucleus into the cytoplasm of the cell, where, with the assistance of the ribosomes, the protein is synthesized.

Small RNA fragments, called **transfer RNA (tRNA)** molecules, attach themselves to specific amino acids and bring them to the growing protein chain as dictated by the pattern built into the mRNA. This process is summarized in Figure 21.21.

21.9 Lipids

OBJECTIVE: To learn the four classes of lipids.

The **lipids** are a group of substances defined in terms of their solubility characteristics. They are water-insoluble substances that can be extracted from cells by organic solvents such as benzene. The lipids found in the human body can be divided into four classes according to their molecular structure: fats, phospholipids, waxes, and steroids.

The most common **fats** are esters composed of the trihydroxy alcohol known as glycerol and long-chain carboxylic acids called **fatty acids** (Table 21.5). *Tristearin,* the most common animal fat, is typical of these substances.

Table 21.5 Some Common Fatty Acids and Their Major Sources

Name	Formula	Major Source
Saturated		
arachidic acid	$CH_3(CH_2)_{18}$—COOH	peanut oil
butyric acid	$CH_3(CH_2)_2$—COOH	butter
caproic acid	$CH_3(CH_2)_4$—COOH	butter
lauric acid	$CH_3(CH_2)_{10}$—COOH	coconut oil
stearic acid	$CH_3(CH_2)_{16}$—COOH	animal and vegetable fats
Unsaturated		
oleic acid	$CH_3(CH_2)_7CH{=}CH(CH_2)_7$—COOH	corn oil
linoleic acid	$CH_3(CH_2)_4CH{=}CH$—CH_2—$CH{=}CH(CH_2)_7$—COOH	linseed oil
linolenic acid	$CH_3CH_2CH{=}CH$—$CH_2CH{=}CH$—CH_2—$CH{=}CH$—$(CH_2)_7COOH$	linseed oil

Glycerol   Three stearic acid molecules   Tristearin (fat)

Fats that are esters of glycerol are called **triglycerides** and have the general structure

where the three R groups may be the same or different and may be saturated or unsaturated. Vegetable fats tend to be unsaturated and usually occur as oily liquids; most animal fats are saturated (contain only C—C single bonds) and occur as solids at room temperature.

Unsaturated fats contain one or more C=C bonds.

Triglycerides can be broken down by treatment with aqueous sodium hydroxide. The products are glycerol and the fatty acid salts; the latter are known as *soaps*. This process is called **saponification.**

$$
\begin{matrix}
CH_2-O-\overset{\displaystyle O}{\overset{\|}{C}}-R & & CH_2OH + RCOONa \\
| & & | \\
CH-O-\overset{\displaystyle O}{\overset{\|}{C}}-R' & + \quad 3NaOH & CHOH + R'COONa \\
| & & | \\
CH_2-O-\overset{\displaystyle O}{\overset{\|}{C}}-R'' & & CH_2OH + R''COONa
\end{matrix}
$$

Triglyceride   Sodium   Glycerol   Soaps
                      hydroxide

> Like dissolves like.

Much of what we call greasy dirt is nonpolar. Grease, for example, consists mostly of long-chain hydrocarbons. However, water, the solvent most commonly available to us, is very polar and does not dissolve "greasy dirt." We need to add something to the water that is somehow compatible with both the polar water and the nonpolar grease. Fatty-acid anions are perfect for this role because they have a long nonpolar tail and a polar head. For example, the stearate anion can be represented as

$$
CH_3-CH_2-CH_2-CH_2-CH_2-CH_2-CH_2-CH_2-CH_2-CH_2-CH_2-CH_2-CH_2-CH_2-CH_2-CH_2-CH_2-\overset{\displaystyle O}{\overset{\|}{C}}-O^-
$$

Nonpolar   Polar
tail   head

Such ions can be dispersed in water because they form **micelles** (Figure 21.22). These aggregates of fatty-acid anions have the water-incompatible tails in the interior; the anionic parts (the polar heads) point outward and

Nonpolar   Polar
tail   head

○ Positive ion

◉ Water molecule

Figure 21.22

A two-dimensional "slice" of the structure of a micelle of fatty-acid anions.

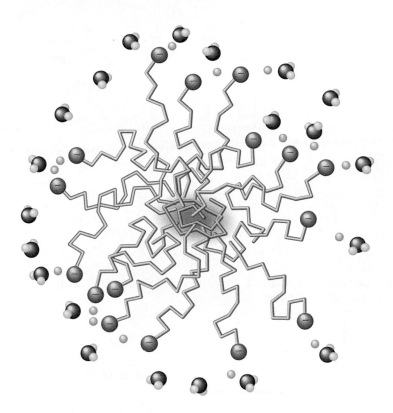

Figure 21.23

Soap micelles absorb grease molecules into their interiors so that the molecules are suspended in the water and can be washed away.

interact with the polar water molecules. A soap solution does not contain *individual* fatty-acid anions dispersed in the water but rather groups of ions (micelles).

Soap dissolves grease by taking the grease molecules into the nonpolar interior of the micelle (Figure 21.23) so that they can be carried away by the water. Soap thus acts to suspend the normally incompatible grease in the water. Because of this ability to assist water in suspending nonpolar materials, soap is also called a *wetting agent,* or **surfactant.**

A major disadvantage is that soap anions form precipitates in hard water (water that contains large concentrations of ions such as Ca^{2+} and Mg^{2+}). These precipitates occur because the Ca^{2+} and Mg^{2+} ions form insoluble solids with soap anions. These precipitates ("soap scum") dull clothes and drastically reduce soap's cleaning efficiency. To help alleviate this problem, a huge industry has developed to produce artificial soaps, called *detergents.* Detergents are similar to natural soaps in that they have a long nonpolar tail and an ionic head. However, detergent anions have the advantage of not forming insoluble solids with Ca^{2+} and Mg^{2+} ions.

Phospholipids are similar in structure to fats in that they are esters of glycerol. Unlike fats, however, they contain only two fatty acids. The third group bound to glycerol is a phosphate group, which gives phospholipids two distinct parts: the long nonpolar "tail" and the polar substituted-phosphate "head" (Figure 21.24).

Waxes are another class of lipids. Like fats and phospholipids, waxes are esters, but unlike these other lipids, they involve monohydroxy alcohols instead of glycerol. For example, *beeswax,* a substance secreted by the wax glands of bees, is mainly myricyl palmitate,

$$CH_3(CH_2)_{14}-\overset{\overset{\displaystyle O}{\|}}{C}-O-(CH_2)_{29}-CH_3$$

Nonpolar tails

Polar head

Figure 21.24

Lecithin, a phospholipid, with its long nonpolar tails and polar substituted-phosphate head.

formed from palmitic acid,

$$CH_3(CH_2)_{14}-C\begin{smallmatrix}O\\\\OH\end{smallmatrix}$$

and carnaubyl alcohol,

$$CH_3(CH_2)_{29}-OH$$

Waxes are solids that furnish waterproof coatings on leaves and fruit and on the skins and feathers of animals. Waxes are also important commercially. For example, whale oil is largely composed of the wax cetyl palmitate. It has been used in so many products, including cosmetics and candles, that the blue whale has been hunted almost to extinction.

Steroids are a class of lipids that have a characteristic carbon ring structure of the type

Steroids comprise four groups: cholesterol, adrenocorticoid hormones, sex hormones, and bile acids.

Cholesterol (Figure 21.25a) is found in virtually all organisms and is the starting material for the formation of the many other steroid-based molecules, such as vitamin D (Figure 21.25b). Although cholesterol is essential for human life, it has been implicated in the formation of plaque on the walls of arteries (a process called *atherosclerosis* or hardening of the arteries), which can lead eventually to clogging. This effect seems especially important in the arteries that supply blood to the heart. Blockage of these arteries leads to heart damage that often results in death from a heart attack.

The **adrenocorticoid hormones,** such as cortisol (Figure 21.25c), are synthesized in the adrenal glands (glands that lie next to each kidney) and are involved in various regulatory functions.

Of the **sex hormones,** the most important male hormone is *testosterone* (Figure 21.25d), a hormone that controls the growth of the reproductive organs and hair and the development of the muscle structure and deep voice that are characteristic of males. There are two types of female sex hormones of particular significance: *progesterone* (Figure 21.25e) and a group of

a cholesterol

b vitamin D$_3$

c cortisol

d testosterone

e progesterone

f estradiol

g ethynodiol diacetate

h cholic acid

Figure 21.25

Several common steroids and steroid derivatives.

estrogens, one of which is *estradiol* (Figure 21.25f). Changes in the concentrations of these hormones cause the periodic changes in the ovaries and uterus that are responsible for the menstrual cycle. During pregnancy, a high level of progesterone is maintained, which prevents ovulation. This effect has led to the use of progesterone-type compounds as birth-control drugs. One of the most common of these is ethynodiol diacetate (Figure 21.25g).

The **bile acids** are produced from cholesterol in the liver and stored in the gallbladder. The primary human bile acid is *cholic acid* (Figure 21.25h), a substance that aids in the digestion of fats by emulsifying them in the intestine. Bile acids also can dissolve cholesterol ingested in food and are therefore important in controlling cholesterol in the body.

CHAPTER 21 REVIEW

Key Terms

biochemistry (p. 689)
biotechnology (p. 689)
essential elements
 (p. 689)
trace elements (p. 689)
cell (p. 689)
protein (21.1)
fibrous protein (21.1)
globular protein (21.1)
α-amino acid (21.2)
side chain (21.2)
dipeptide (21.2)
peptide linkage (21.2)
polypeptide (21.2)
primary structure (21.2)
secondary structure
 (21.3)
α-helix (21.3)
pleated sheet (21.3)
tertiary structure (21.4)
disulfide linkage (21.4)
denaturation (21.5)
enzyme (21.6)
lock-and-key model (21.6)
substrate (21.6)
active site (21.6)
carbohydrate (21.7)
monosaccharide (simple
 sugar) (21.7)
disaccharide (21.7)
sucrose (21.7)

glycoside linkage (21.7)
polysaccharide (21.7)
starch (21.7)
cellulose (21.7)
glycogen (21.7)
deoxyribonucleic acid
 (DNA) (21.8)
ribonucleic acid
 (RNA) (21.8)
nucleotide (21.8)
protein synthesis (21.8)
gene (21.8)
messenger RNA
 (mRNA) (21.8)
transfer RNA
 (tRNA) (21.8)
lipid (21.9)
fat (21.9)
fatty acid (21.9)
triglyceride (21.9)
saponification (21.9)
micelle (21.9)
surfactant (21.9)
phospholipid (21.9)
wax (21.9)
steroid (21.9)
cholesterol (21.9)
adrenocorticoid
 hormone (21.9)
sex hormone (21.9)
bile acid (21.9)

Summary

1. Thirty elements are currently known to be essential for life. The most abundant elements in the human body are hydrogen, carbon, nitrogen, and oxygen, but calcium, phosphorus, sodium, magnesium, potassium, sulfur, and chlorine are also present in large amounts. Other elements found only in trace amounts, such as zinc, are essential for the action of many enzymes.

2. Proteins are a class of natural polymers with molar masses ranging from 6000 to over 1,000,000 grams. Fibrous proteins are employed in the human body for structural purposes in muscle, hair, and cartilage. Globular proteins are molecules that transport and

store oxygen and nutrients, act as catalysts, help regulate the body's systems, fight foreign objects, and so on.

3. The building blocks of proteins are the α-amino acids, which are classified as polar or nonpolar, depending on whether the side chain (R group) attached to the α-carbon is hydrophilic or hydrophobic. A protein polymer is built by successive condensation reactions that produce peptide linkages:

4. The order or sequence of amino acids in the protein chain is called the protein's primary structure. Differences in primary structure are what enable proteins to be tailored for different and very specific functions.

5. The secondary structure is the arrangement of the protein chain in space. The two most common secondary structures are the α-helix and the pleated sheet.

6. The overall shape of the protein is called its tertiary structure. Energy sources and a variety of chemicals can cause the breakdown of tertiary protein structure, which is called denaturation.

7. Enzymes are proteins that act as catalysts in biological reactions.

8. Carbohydrates serve as food sources for most organisms and as structural materials for plants. Simple carbohydrates, called monosaccharides, are most commonly five-carbon and six-carbon polyhydroxy ketones and aldehydes.

9. Monosaccharides combine to form more complex carbohydrates. For example, sucrose is a disaccharide, and starch and cellulose are polymers of glucose (polysaccharides).

10. When a cell divides, the genetic information is transmitted via deoxyribonucleic acid (DNA), which has a double-helical structure. During cell division, the double helix unravels, and a new polymer forms along each strand of the original DNA, creating two double-helical DNA molecules. The DNA contains segments called genes, which store the primary

structures of specific proteins. Various types of ribonucleic acid (RNA) molecules assist in protein synthesis.

11. Lipids are water-insoluble substances found in cells. They can be divided into four classes: fats, phospholipids, waxes, and steroids.

Active Learning Questions

These questions are designed to be considered by groups of students in class. Often these questions work well for introducing a particular topic in class.

1. Differentiate among primary, secondary, and tertiary structure in proteins.

2. Draw the structures of the simple dipeptides *gly-ala* and *ala-gly*.

3. How are proteins able to provide a buffering action?

4. What is meant by inhibition of an enzyme? What happens when an enzyme is irreversibly inhibited?

5. Differentiate between monosaccharide and disaccharide. Provide an example of each.

6. Sketch a representation of sucrose and clearly label the portion that originates from glucose, the portion that originates from fructose, and the glycoside linkage between the rings.

7. What is a polysaccharide? What monomer unit makes up starch and cellulose?

8. Describe the structure of a typical nucleotide.

9. Sketch the structures of the sugars ribose and deoxyribose. Which molecule, RNA or DNA, contains each sugar?

10. Sketch the general structure of a triglyceride. What are the components that go into making a typical triglyceride?

11. Describe the mechanism by which fatty acid salts are able to exert a cleaning action.

OWL Questions and problems for this chapter are available in OWL. Use the OWL access card that came with this textbook to access assigned questions and problems for this chapter.

Using Your Calculator

In this section we will review how to use your calculator to perform common mathematical operations. This discussion assumes that your calculator uses the algebraic operating system, the system used by most brands.

One very important principle to keep in mind as you use your calculator is that it is not a substitute for your brain. Keep thinking as you do the calculations. Keep asking yourself, "Does the answer make sense?"

Addition, Subtraction, Multiplication, and Division

Performing these operations on a pair of numbers always involves the following steps:

1. Enter the first number, using the numbered keys and the decimal (.) key if needed.
2. Enter the operation to be performed.
3. Enter the second number.
4. Press the "equals" key to display the answer.

For example, the operation

$$15.1 + 0.32$$

is carried out as follows:

Press	Display
15.1	15.1
+	15.1
.32	0.32
=	15.42

The answer given by the display is 15.42. If this is the final result of a calculation, you should round it off to the correct number of significant figures (15.4), as discussed in Section 2.5. If this number is to be used in further calculations, use it exactly as it appears on the display. Round off only the final answer in the calculation.

Do the following operations for practice. The detailed procedures are given below.

a. $1.5 + 32.86$ c. 0.33×153

b. $23.5 - 0.41$ d. $\dfrac{9.3}{0.56}$ or $9.3 \div 0.56$

Procedures

a. Press	Display	b. Press	Display
1.5	1.5	23.5	23.5
+	1.5	−	23.5
32.86	32.86	.41	0.41
=	34.36	=	23.09
Rounded:	34.4	Rounded:	23.1

c. Press	Display	d. Press	Display
.33	0.33	9.3	9.3
×	0.33	÷	9.3
153	153	.56	0.56
=	50.49	=	16.607143
Rounded:	50.	Rounded:	17

Squares, Square Roots, Reciprocals, and Logs

Now we will consider four additional operations that we often need to solve chemistry problems.

The *squaring* of a number is done with a key labeled X^2. The *square root* key is usually labeled $\sqrt{X}$. To take the *reciprocal* of a number, you need the 1/X key. The *logarithm* of a number is determined by using a key labeled log or logX.

To perform these operations, take the following steps:

1. Enter the number.
2. Press the appropriate function key.
3. The answer is displayed automatically.

For example, let's calculate the square root of 235.

Press	Display
235	235
$\sqrt{X}$	15.32971
Rounded:	15.3

We can obtain the log of 23 as follows:

Press	Display
23	23
log	1.3617278
Rounded:	1.36

Often a key on a calculator serves two functions. In this case, the first function is listed on the key and the second is shown on the calculator just above the key. For example, on some calculators the top row of keys appears as follows:

$$1/X \qquad X^2$$

$$\boxed{2nd} \ \boxed{R/S} \ \boxed{\sqrt{X}} \ \boxed{off} \ \boxed{on/C}$$

To make the calculator square a number, we must use 2nd and then $\sqrt{X}$; pressing 2nd tells the calculator we want the function that is listed *above* the key. Thus we can obtain the square of 11.56 on this calculator as follows:

Press	Display
11.56	11.56
2nd then $\sqrt{X}$	133.6336
Rounded:	133.6

We obtain the reciprocal of 384 (1/384) on this calculator as follows:

Press	Display
384	384
2nd then R/S	0.0026042
Rounded:	0.00260

Your calculator may be different. See the user's manual if you are having trouble with these operations.

Chain Calculations

In solving problems you often have to perform a series of calculations—a calculation chain. This is generally quite easy if you key in the chain as you read the numbers and operations in order. For example, to perform the calculation

$$\frac{14.68 + 1.58 - 0.87}{0.0850}$$

you should use the appropriate keys as you read it to yourself:

14.68 plus 1.58 equals; minus .87 equals; divided by 0.0850 equals

The details follow.

Press	Display
14.68	14.68
+	14.68
1.58	1.58
=	16.26
−	16.26
.87	0.87
=	15.39
÷	15.39
.0850	0.0850
=	181.05882
Rounded:	181

Note that you must press $\boxed{=}$ after every operation to keep the calculation "up to date."

For more practice, consider the calculation

$$(0.360)(298) + \frac{(14.8)(16.0)}{1.50}$$

Here you are adding two numbers, but each must be obtained by the indicated calculations. One procedure is to calculate each number first and then add them. The first term is

$$(0.360)(298) = 107.28$$

The second term,

$$\frac{(14.8)(16.0)}{1.50}$$

can be computed easily by reading it to yourself. It "reads"

14.8 times 16.0 equals; divided by 1.50 equals

and is summarized as follows:

Press	Display
14.8	14.8
×	14.8
16.0	16.0
=	236.8
÷	236.8
1.50	1.50
=	157.86667

Now we can keep this last number on the calculator and add it to 107.28 from the first calculation.

Press	Display
+	157.86667
107.28	107.28
=	265.14667
Rounded:	265

To summarize,

$$(0.360)(298) + \frac{(14.8)(16.0)}{1.50}$$

becomes

$$107.28 + 157.86667$$

and the sum is 265.14667 or, rounded to the correct number of significant figures, 265. There are other ways to do this calculation, but this is the safest way (assuming you are careful).

A common type of chain calculation involves a number of terms multiplied together in the numerator and the denominator, as in

$$\frac{(323)(.0821)(1.46)}{(4.05)(76)}$$

There are many possible sequences by which this calculation can be carried out, but the following seems the most natural.

323 times .0821 equals; times 1.46 equals; divided by 4.05 equals; divided by 76 equals

This sequence is summarized as follows:

Press	Display
323	323
×	323
.0821	0.0821
=	26.5183
×	26.5183
1.46	1.46
=	38.716718
÷	38.716718
4.05	4.05
=	9.5596835
÷	9.5596835
76	76
=	0.1257853

The answer is 0.1257853, which, when rounded to the correct number of significant figures, is 0.13. Note that when two or more numbers are multiplied in the denominator, you must divide by *each* one.

Here are some additional chain calculations (with solutions) to give you more practice.

a. $15 - (0.750)(243)$

b. $\dfrac{(13.1)(43.5)}{(1.8)(63)}$

c. $\dfrac{(85.8)(0.142)}{(16.46)(18.0)} + \dfrac{(131)(0.0156)}{10.17}$

d. $(18.1)(0.051) - \dfrac{(325)(1.87)}{(14.0)(3.81)} + \dfrac{1.56 - 0.43}{1.33}$

Solutions

a. $15 - 182 = -167$

b. 5.0

c. $0.0411 + 0.201 = 0.242$

d. $0.92 - 11.4 + 0.850 = -9.6$

In performing chain calculations, take the following steps in the order listed.

1. Perform any additions and subtractions that appear inside parentheses.
2. Complete the multiplications and divisions of individual terms.
3. Add and subtract individual terms as required.

Basic Algebra

In solving chemistry problems you will use, over and over again, relatively few mathematical procedures. In this section we review the few algebraic manipulations that you will need.

Solving an Equation

In the course of solving a chemistry problem, we often construct an algebraic equation that includes the unknown quantity (the thing we want to calculate). An example is

$$(1.5)V = (0.23)(0.08206)(298)$$

We need to "solve this equation for V." That is, we need to isolate V on one side of the equals sign with all the numbers on the other side. How can we do this? The key idea in solving an algebraic equation is that *doing the same thing on both sides of the equals sign* does not change the equality. That is, it is always "legal" to do the same thing to both sides of the equation. Here we want to solve for V, so we must get the number 1.5 on the other side of the equals sign. We can do this by dividing *both sides* by 1.5.

$$\frac{(1.5)V}{1.5} = \frac{(0.23)(0.08206)(298)}{1.5}$$

Now the 1.5 in the denominator on the left cancels the 1.5 in the numerator:

$$\frac{(1.5)V}{1.5} = \frac{(0.23)(0.08206)(298)}{1.5}$$

to give

$$V = \frac{(0.23)(0.08206)(298)}{1.5}$$

Using the procedures in "Using Your Calculator" for chain calculations, we can now obtain the value for V with a calculator.

$$V = 3.7$$

Sometimes it is necessary to solve an equation that consists of symbols. For example, consider the equation

$$\frac{P_1V_1}{T_1} = \frac{P_2V_2}{T_2}$$

Let's assume we want to solve for T_2. That is, we want to isolate T_2 on one side of the equation. There are several possible ways to proceed, keeping in mind that we always do the same thing on both sides of the equals sign. First we multiply both sides by T_2.

$$T_2 \times \frac{P_1V_1}{T_1} = \frac{P_2V_2}{T_2} \times T_2$$

This cancels T_2 on the right. Next we multiply both sides by T_1.

$$T_2 \times \frac{P_1V_1}{T_1} \times T_1 = P_2V_2T_1$$

This cancels T_1 on the left. Now we divide both sides by P_1V_1.

$$T_2 \times \frac{P_1V_1}{P_1V_1} = \frac{P_2V_2T_1}{P_1V_1}$$

This yields the desired equation,

$$T_2 = \frac{P_2V_2T_1}{P_1V_1}$$

For practice, solve each of the following equations for the variable indicated.

a. $PV = k$; solve for P

b. $1.5x + 6 = 3$; solve for x

c. $PV = nRT$; solve for n

d. $\dfrac{P_1V_1}{T_1} = \dfrac{P_2V_2}{T_2}$; solve for V_2

e. $\dfrac{°F - 32}{°C} = \dfrac{9}{5}$; solve for $°C$

f. $\dfrac{°F - 32}{°C} = \dfrac{9}{5}$; solve for $°F$

Solutions

a. $\dfrac{P\cancel{V}}{\cancel{V}} = \dfrac{k}{V}$

$P = \dfrac{k}{V}$

b. $1.5x + 6 - 6 = 3 - 6$

$1.5x = -3$

$\dfrac{\cancel{1.5}x}{\cancel{1.5}} = \dfrac{-3}{1.5}$

$x = -\dfrac{3}{1.5} = -2$

c. $\dfrac{PV}{RT} = \dfrac{n\cancel{RT}}{\cancel{RT}}$

$\dfrac{PV}{RT} = n$

d. $\dfrac{P_1V_1}{T_1} \times T_2 = \dfrac{P_2V_2}{\cancel{T_2}} \times \cancel{T_2}$

$\dfrac{P_1V_1T_2}{T_1P_2} = \dfrac{\cancel{P_2}V_2}{\cancel{P_2}}$

$\dfrac{P_1V_1T_2}{T_1P_2} = V_2$

e. $\dfrac{°F - 32}{°\cancel{C}} \times °\cancel{C} = \dfrac{9}{5}°C$

$\dfrac{5}{9}(°F - 32) = \dfrac{\cancel{5}}{\cancel{9}} \times \dfrac{\cancel{9}}{\cancel{5}}°C$

$\dfrac{5}{9}(°F - 32) = °C$

f. $\dfrac{°F - 32}{°\cancel{C}} \times °\cancel{C} = \dfrac{9}{5}°C$

$°F - \cancel{32} + \cancel{32} = \dfrac{9}{5}°C + 32$

$°F = \dfrac{9}{5}°C + 32$

Scientific (Exponential) Notation

The numbers we must work with in scientific measurements are often very large or very small; thus it is convenient to express them using powers of 10. For example, the number 1,300,000 can be expressed as 1.3×10^6, which means multiply 1.3 by 10 six times, or

$$1.3 \times 10^6 = 1.3 \times \underbrace{10 \times 10 \times 10 \times 10 \times 10 \times 10}_{10^6 \,=\, 1\ \text{million}}$$

A number written in scientific notation always has the form:

A number (between 1 and 10) times
the appropriate power of 10

To represent a large number such as 20,500 in scientific notation, we must move the decimal point in such a way as to achieve a number between 1 and 10 and then multiply the result by a power of 10 to compensate for moving the decimal point. In this case, we must move the decimal point four places to the left.

$$2\underset{4\ \ 3\ \ 2\ \ 1}{\underbrace{0\ 5\ 0\ 0}}$$

to give a number between 1 and 10:

2.05

where we retain only the significant figures (the number 20,500 has three significant figures). To compensate for moving the decimal point four places to the left, we must multiply by 10^4. Thus

$$20,500 = 2.05 \times 10^4$$

As another example, the number 1985 can be expressed as 1.985×10^3. To end up with the number 1.985, which is between 1 and 10, we had to move the decimal point three places to the left. To compensate for that, we must multiply by 10^3. Some other examples are given in the accompanying list.

Number	Exponential Notation
5.6	5.6×10^0 or 5.6×1
39	3.9×10^1
943	9.43×10^2
1126	1.126×10^3

So far, we have considered numbers greater than 1. How do we represent a number such as 0.0034 in exponential notation? First, to achieve a number between 1 and 10, we start with 0.0034 and move the decimal point three places to the right.

$$0.0\underset{1\ \ 2\ \ 3}{\underbrace{0\ 3\ }}4$$

This yields 3.4. Then, to compensate for moving the decimal point to the right, we must multiply by a power of 10 with a negative exponent—in this case, 10^{-3}. Thus

$$0.0034 = 3.4 \times 10^{-3}$$

In a similar way, the number 0.00000014 can be written as 1.4×10^{-7}, because going from 0.00000014 to 1.4 requires that we move the decimal point seven places to the right.

Mathematical Operations with Exponentials

We next consider how various mathematical operations are performed using exponential numbers. First we cover the various rules for these operations; then we consider how to perform them on your calculator.

Multiplication and Division

When two numbers expressed in exponential notation are multiplied, the initial numbers are multiplied and the exponents of 10 are *added*.

$$(M \times 10^m)(N \times 10^n) = (MN) \times 10^{m+n}$$

For example (to two significant figures, as required),

$$(3.2 \times 10^4)(2.8 \times 10^3) = 9.0 \times 10^7$$

When the numbers are multiplied, if a result greater than 10 is obtained for the initial number, the decimal point is moved one place to the left and the exponent of 10 is increased by 1.

$$(5.8 \times 10^2)(4.3 \times 10^8) = 24.9 \times 10^{10}$$
$$= 2.49 \times 10^{11}$$
$$= 2.5 \times 10^{11} \text{ (two significant figures)}$$

Division of two numbers expressed in exponential notation involves normal division of the initial numbers and *subtraction* of the exponent of the divisor from that of the dividend. For example,

$$\frac{4.8 \times 10^8}{\underbrace{2.1 \times 10^3}_{\text{Divisor}}} = \frac{4.8}{2.1} \times 10^{(8-3)} = 2.3 \times 10^5$$

If the initial number resulting from the division is less than 1, the decimal point is moved one place to the right and the exponent of 10 is decreased by 1. For example,

$$\frac{6.4 \times 10^3}{8.3 \times 10^5} = \frac{6.4}{8.3} \times 10^{(3-5)} = 0.77 \times 10^{-2}$$
$$= 7.7 \times 10^{-3}$$

Addition and Subtraction

In order for us to add or subtract numbers expressed in exponential notation, *the exponents of the numbers must be the same.* For example, to add 1.31×10^5 and 4.2×10^4, we must rewrite one number so that the exponents of both are the same. The number 1.31×10^5 can be written 13.1×10^4: decreasing the exponent by 1 compensates for moving the decimal point one place to the right. Now we can add the numbers.

$$\begin{array}{r} 13.1 \times 10^4 \\ + \ 4.2 \times 10^4 \\ \hline 17.3 \times 10^4 \end{array}$$

In correct exponential notation, the result is expressed as 1.73×10^5.

To perform addition or subtraction with numbers expressed in exponential notation, we add or subtract only the initial numbers. The exponent of the result is the same as the exponents of the numbers being added or subtracted. To subtract 1.8×10^2 from 8.99×10^3, we first convert 1.8×10^2 to 0.18×10^3 so that both numbers have the same exponent. Then we subtract.

$$\begin{array}{r} 8.99 \times 10^3 \\ -0.18 \times 10^3 \\ \hline 8.81 \times 10^3 \end{array}$$

Powers and Roots

When a number expressed in exponential notation is taken to some power, the initial number is taken to the appropriate power and the exponent of 10 is *multiplied* by that power.

$$(N \times 10^n)^m = N^m \times 10^{m \times n}$$

For example,

$$(7.5 \times 10^2)^2 = (7.5)^2 \times 10^{2 \times 2}$$
$$= 56. \times 10^4$$
$$= 5.6 \times 10^5$$

When a root is taken of a number expressed in exponential notation, the root of the initial number is taken and the exponent of 10 is divided by the number representing the root. For example, we take the square root of a number as follows:

$$\sqrt{N \times 10^n} = (N \times 10^n)^{1/2} = \sqrt{N} \times 10^{n/2}$$

For example,

$$(2.9 \times 10^6)^{1/2} = \sqrt{2.9} \times 10^{6/2}$$
$$= 1.7 \times 10^3$$

Using a Calculator to Perform Mathematical Operations on Exponentials

In dealing with exponential numbers, you must first learn to enter them into your calculator. First the number is keyed in and then the exponent. There is a special key that must be pressed just before the exponent is entered. This key is often labeled $\boxed{\text{EE}}$ or $\boxed{\text{exp}}$. For example, the number 1.56×10^6 is entered as follows:

Press	Display
1.56	1.56
EE or exp	1.56 00
6	1.56 06

To enter a number with a negative exponent, use the change-of-sign key $\boxed{+/-}$ after entering the exponent number. For example, the number 7.54×10^{-3} is entered as follows:

Press	Display
7.54	7.54
EE or exp	7.54 00
3	7.54 03
+/-	7.54 -03

Once a number with an exponent is entered into your calculator, the mathematical operations are performed exactly the same as with a "regular" number. For example, the numbers 1.0×10^3 and 1.0×10^2 are multiplied as follows:

Press	Display
1.0	1.0
EE or exp	1.0 00
3	1.0 03
$\times$	1 03
1.0	1.0
EE or exp	1.0 00
2	1.0 02
=	1 05

The answer is correctly represented as 1.0×10^5.

The numbers 1.50×10^5 and 1.1×10^4 are added as follows:

Press	Display
1.5	1.50
EE or exp	1.50 00
5	1.50 05
+	1.5 05
1.1	1.1
EE or exp	1.1 00
4	1.1 04
=	1.61 05

The answer is correctly represented as 1.61×10^5. Note that when exponential numbers are added, the calculator automatically takes into account any difference in exponents.

To take the power, root, or reciprocal of an exponential number, enter the number first, then press the appropriate key or keys. For example, the square root of 5.6×10^3 is obtained as follows:

Press	Display
5.6	5.6
EE or exp	5.6 00
3	5.6 03
$\sqrt{\mathrm{X}}$	7.4833148 01

The answer is correctly represented as 7.5×10^1.

Practice by performing the following operations that involve exponential numbers. The answers follow the exercises.

a. $7.9 \times 10^2 \times 4.3 \times 10^4$

b. $\dfrac{5.4 \times 10^3}{4.6 \times 10^5}$

c. $1.7 \times 10^2 + 1.63 \times 10^3$

d. $4.3 \times 10^{-3} + 1 \times 10^{-4}$

e. $(8.6 \times 10^{-6})^2$

f. $\dfrac{1}{8.3 \times 10^2}$

g. $\log(1.0 \times 10^{-7})$

h. $-\log(1.3 \times 10^{-5})$

i. $\sqrt{6.7 \times 10^9}$

Solutions

a. 3.4×10^7

b. 1.2×10^{-2}

c. 1.80×10^3

d. 4.4×10^{-3}

e. 7.4×10^{-11}

f. 1.2×10^{-3}

g. -7.00

h. 4.89

i. 8.2×10^4

Graphing Functions

In interpreting the results of a scientific experiment, it is often useful to make a graph. If possible, the function to be graphed should be in a form that gives a straight line. The equation for a straight line (a *linear equation*) can be represented in the general form

$$y = mx + b$$

where y is the *dependent variable*, x is the *independent variable*, m is the *slope*, and b is the *intercept* with the y axis.

To illustrate the characteristics of a linear equation, the function $y = 3x + 4$ is plotted in Figure A.1. For this equation $m = 3$ and $b = 4$. Note that the y intercept occurs when $x = 0$. In this case the y intercept is 4, as can be seen from the equation ($b = 4$).

The slope of a straight line is defined as the ratio of the rate of change in y to that in x:

$$m = \text{slope} = \frac{\Delta y}{\Delta x}$$

For the equation $y = 3x + 4$, y changes three times as fast as x (because x has a coefficient of 3). Thus the slope in

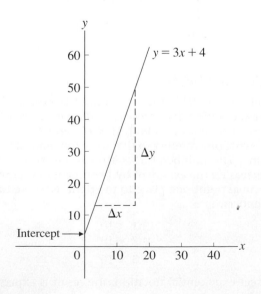

Figure A.1

Graph of the linear equation $y = 3x + 4$.

this case is 3. This can be verified from the graph. For the triangle shown in Figure A.1,

$$\Delta y = 15 - 16 = 36 \quad \text{and} \quad \Delta x = 15 - 3 = 12$$

Thus

$$\text{Slope} = \frac{\Delta y}{\Delta x} = \frac{36}{12} = 3$$

This example illustrates a general method for obtaining the slope of a line from the graph of that line. Simply draw a triangle with one side parallel to the y axis and the other side parallel to the x axis, as shown in Figure A.1. Then determine the lengths of the sides to get Δy and Δx, respectively, and compute the ratio $\Delta y/\Delta x$.

SI Units and Conversion Factors

These conversion factors are given with more significant figures than those typically used in the body of the text.

Length	SI Unit: Meter (m)
1 meter	= 1.0936 yards
1 centimeter	= 0.39370 inch
1 inch	= 2.54 centimeters (exactly)
1 kilometer	= 0.62137 mile
1 mile	= 5280. feet
	= 1.6093 kilometers

Mass	SI Unit: Kilogram (kg)
1 kilogram	= 1000 grams
	= 2.2046 pounds
1 pound	= 453.59 grams
	= 0.45359 kilogram
	= 16 ounces
1 atomic mass unit	= 1.66057×10^{-27} kilograms

Volume	SI Unit: Cubic Meter (m^3)
1 liter	= 10^{-3} m^3
	= 1 dm^3
	= 1.0567 quarts
1 gallon	= 4 quarts
	= 8 pints
	= 3.7854 liters
1 quart	= 32 fluid ounces
	= 0.94635 liter

Pressure	SI Unit: Pascal (Pa)
1 atmosphere	= 101.325 kilopascals
	= 760. torr (mm Hg)
	= 14.70 pounds per square inch

Energy	SI Unit: Joule (J)
1 joule	= 0.23901 calorie
1 calorie	= 4.184 joules

Chapter 2

Self-Check Exercise 2.1

$357 = 3.57 \times 10^2$

$0.0055 = 5.5 \times 10^{-3}$

Self-Check Exercise 2.2

a. Three significant figures. The leading zeros (to the left of the 1) do not count, but the trailing zeros do.

b. Five significant figures. The one captive zero and the two trailing zeros all count.

c. This is an exact number obtained by counting the cars. It has an unlimited number of significant figures.

Self-Check Exercise 2.3

a. $12.6 \times 0.53 = 6.678 = 6.7$
   Limiting

b. $12.6 \times 0.53 = 6.7$; $\begin{array}{r} 6.7 \quad \text{Limiting} \\ -4.59 \\ \hline 2.11 = 2.1 \end{array}$
   Limiting

c. $\begin{array}{r} 25.36 \\ -4.15 \\ \hline 21.21 \end{array}$ $\dfrac{21.21}{2.317} = 9.15408 = 9.154$

Self-Check Exercise 2.4

$$0.750 \ \cancel{L} \times \frac{1.06 \text{ qt}}{1 \ \cancel{L}} = 0.795 \text{ qt}$$

Self-Check Exercise 2.5

$$225 \ \frac{\cancel{mi}}{h} \times \frac{1760 \ \cancel{yd}}{1 \ \cancel{mi}} \times \frac{1 \ \cancel{m}}{1.094 \ \cancel{yd}} \times \frac{1 \text{ km}}{1000 \ \cancel{m}} = 362 \ \frac{\text{km}}{h}$$

Self-Check Exercise 2.6

The best way to solve this problem is to convert 172 K to Celsius degrees. To do this we will use the formula $T_{°C} = T_K - 273$.

In this case

$$T_{°C} = T_K - 273 = 172 - 273 = -101$$

So 172 K = −101 °C, which is a lower temperature than −75 °C. Thus 172 K is colder than −75 °C.

Self-Check Exercise 2.7

The problem is 41 °C = ? °F.

Using the formula

$$T_{°F} = 1.80(T_{°C}) + 32$$

we have

$$T_{°F} = ? \ °F = 1.80(41) + 32 = 74 + 32 = 106$$

That is, 41 °C = 106 °F.

Self-Check Exercise 2.8

This problem can be stated as 239 °F = ? °C.

Using the formula

$$T_{°C} = \frac{T_{°F} - 32}{1.80}$$

we have in this case

$$T_{°C} = ? \ °C = \frac{239 - 32}{1.80} = \frac{207}{1.80} = 115$$

That is, 239 °F = 115 °C.

Self-Check Exercise 2.9

We obtain the density of the cleaner by dividing its mass by its volume.

$$\text{Density} = \frac{\text{mass}}{\text{volume}} = \frac{28.1 \text{ g}}{35.8 \text{ mL}} = 0.785 \text{ g/mL}$$

This density identifies the liquid as isopropyl alcohol.

Chapter 3

Self-Check Exercise 3.1

Items (a) and (c) are physical properties. When the solid gallium melts, it forms liquid gallium. There is no change in composition. Items (b) and (d) reflect the ability to change composition and are thus chemical properties. Statement (b) means that platinum does not react with oxygen to form some new substance. Statement (d) means that copper does react in the air to form a new substance, which is green.

Self-Check Exercise 3.2

a. Milk turns sour because new substances are formed. This is a chemical change.

b. Melting the wax is a physical change (a change of state). When the wax burns, new substances are formed. This is a chemical change.

Self-Check Exercise 3.3

a. Maple syrup is a homogeneous mixture of sugar and other substances dispersed uniformly in water.

b. Helium and oxygen form a homogeneous mixture.

c. Oil and vinegar salad dressing is a heterogeneous mixture. (Note the two distinct layers the next time you look at a bottle of dressing.)

d. Common salt is a pure substance (sodium chloride), so it always has the same composition. (Note that other substances such as iodine are often added to commercial preparations of table salt, which is mostly sodium chloride. Thus commercial table salt is a homogeneous mixture.)

Chapter 4

Self-Check Exercise 4.1

a. P_4O_{10} b. UF_6 c. $AlCl_3$

Self-Check Exercise 4.2

In the symbol $^{90}_{38}\text{Sr}$, the number 38 is the atomic number, which represents the number of protons in the nucleus of a strontium atom. Because the atom is neutral overall, it must also have 38 electrons. The number 90 (the mass number) represents the number of protons plus the number of neutrons. Thus the number of neutrons is $A - Z = 90 - 38 = 52$.

Self-Check Exercise 4.3

The atom $^{201}_{80}\text{Hg}$ has 80 protons, 80 electrons, and $201 - 80 = 121$ neutrons.

Self-Check Exercise 4.4

The atomic number for phosphorus is 15 and the mass number is $15 + 17 = 32$. Thus the symbol for the atom is $^{32}_{15}\text{P}$.

Self-Check Exercise 4.5

Element	Symbol	Atomic Number	Metal or Nonmetal	Family Name
a. argon	Ar	18	nonmetal	noble gas
b. chlorine	Cl	17	nonmetal	halogen
c. barium	Ba	56	metal	alkaline earth metal
d. cesium	Cs	55	metal	alkali metal

Self-Check Exercise 4.6

a. KI $(1+) + (1-) = 0$

b. Mg_3N_2 $3(2+) + 2(3-) = (6+) + (6-) = 0$

c. Al_2O_3 $2(3+) + 3(2-) = 0$

Chapter 5

Self-Check Exercise 5.1

a. rubidium oxide

b. strontium iodide

c. potassium sulfide

Self-Check Exercise 5.2

a. The compound $PbBr_2$ must contain Pb^{2+}—named lead(II)— to balance the charges of the two Br^- ions. Thus the name is lead(II) bromide. The compound $PbBr_4$ must contain Pb^{4+}—named lead(IV)—to balance the charges of the four Br^- ions. The name is therefore lead(IV) bromide.

b. The compound FeS contains the S^{2-} ion (sulfide) and thus the iron cation present must be Fe^{2+}, iron(II). The name is iron(II) sulfide. The compound Fe_2S_3 contains three S^{2-} ions and two iron cations of unknown charge. We can determine the iron charge from the following:

$$2(?+) + 3(2-) = 0$$
$$\quad\uparrow\qquad\quad\uparrow$$
$$\text{Iron}\qquad S^{2-}$$
$$\text{charge}\quad\text{charge}$$

In this case, ? must represent 3 because

$$2(3+) + 3(2-) = 0$$

Thus Fe_2S_3 contains Fe^{3+} and S^{2-}, and its name is iron(III) sulfide.

c. The compound $AlBr_3$ contains Al^{3+} and Br^-. Because aluminum forms only one ion (Al^{3+}), no Roman numeral is required. The name is aluminum bromide.

d. The compound Na_2S contains Na^+ and S^{2-} ions. The name is sodium sulfide. (Because sodium forms only Na^+, no Roman numeral is needed.)

e. The compound $CoCl_3$ contains three Cl^- ions. Thus the cobalt cation must be Co^{3+}, which is named cobalt(III) because cobalt is a transition metal and can form more than one type of cation. Thus the name of $CoCl_3$ is cobalt(III) chloride.

Self-Check Exercise 5.3

Compound	Individual Names	Prefixes	Name
a. CCl_4	carbon chloride	none *tetra-*	carbon tetrachloride
b. NO_2	nitrogen oxide	none *di-*	nitrogen dioxide
c. IF_5	iodine fluoride	none *penta-*	iodine pentafluoride

Self-Check Exercise 5.4

a. silicon dioxide

b. dioxygen difluoride

c. xenon hexafluoride

Self-Check Exercise 5.5

a. chlorine trifluoride

b. vanadium(V) fluoride

c. copper(I) chloride

d. manganese(IV) oxide

e. magnesium oxide

f. water

Self-Check Exercise 5.6

a. calcium hydroxide

b. sodium phosphate

c. potassium permanganate

d. ammonium dichromate

e. cobalt(II) perchlorate (Perchlorate has a 1− charge, so the cation must be Co^{2+} to balance the two ClO_4^- ions.)

f. potassium chlorate

g. copper(II) nitrite (This compound contains two NO_2^- (nitrite) ions and thus must contain a Cu^{2+} cation.)

Self-Check Exercise 5.7

Compound	Name
a. $NaHCO_3$	sodium hydrogen carbonate

Contains Na^+ and HCO_3^-; often called sodium bicarbonate (common name).

b. $BaSO_4$	barium sulfate

Contains Ba^{2+} and SO_4^{2-}.

c. $CsClO_4$	cesium perchlorate

Contains Cs^+ and ClO_4^-.

d. BrF_5	bromine pentafluoride

Both nonmetals (Type III binary).

e. NaBr	sodium bromide

Contains Na^+ and Br^- (Type I binary).

f. KOCl	potassium hypochlorite

Contains K^+ and OCl^-.

g. $Zn_3(PO_4)_2$	zinc(II) phosphate

Contains Zn^{2+} and PO_4^{3-}; Zn is a transition metal and officially requires a Roman numeral. However, because Zn forms only the Zn^{2+} cation, the II is usually left out. Thus the name of the compound is usually given as zinc phosphate.

Self-Check Exercise 5.8

Name	Chemical Formula
a. ammonium sulfate	$(NH_4)_2SO_4$

Two ammonium ions (NH_4^+) are required for each sulfate ion (SO_4^{2-}) to achieve charge balance.

b. vanadium(V) fluoride	VF_5

The compound contains V^{5+} ions and requires five F^- ions for charge balance.

c. disulfur dichloride	S_2Cl_2

The prefix *di-* indicates two of each atom.

d. rubidium peroxide	Rb_2O_2

Because rubidium is in Group 1, it forms only 1+ ions. Thus two Rb^+ ions are needed to balance the 2− charge on the peroxide ion (O_2^{2-}).

e. aluminum oxide	Al_2O_3

Aluminum forms only 3+ ions. Two Al^{3+} ions are required to balance the charge on three O^{2-} ions.

Chapter 6

Self-Check Exercise 6.1

a. $Mg(s) + H_2O(l) \rightarrow Mg(OH)_2(s) + H_2(g)$

Note that magnesium (which is in Group 2) always forms the Mg^{2+} cation and thus requires two OH^- anions for a zero net charge.

b. Ammonium dichromate contains the polyatomic ions NH_4^+ and $Cr_2O_7^{2-}$ (you should have these memorized). Because NH_4^+ has a 1+ charge, two NH_4^+ cations are required for each $Cr_2O_7^{2-}$, with it 2− charge, to give the formula $(NH_4)_2Cr_2O_7$. Chromium(III) oxide contains Cr^{3+} ions—signified by chromium(III)—and O^{2-} (the oxide ion). To achieve a net charge of zero, the solid must contain two Cr^{3+} ions for every three O^{2-} ions, so the formula is Cr_2O_3. Nitrogen gas contains diatomic molecules and is written $N_2(g)$, and gaseous water is written $H_2O(g)$. Thus the unbalanced equation for the decomposition of ammonium dichromate is

$$(NH_4)_2Cr_2O_7(s) \rightarrow Cr_2O_3(s) + N_2(g) + H_2O(g)$$

c. Gaseous ammonia, $NH_3(g)$, and gaseous oxygen, $O_2(g)$, react to form nitrogen monoxide gas, $NO(g)$, plus gaseous water, $H_2O(g)$. The unbalanced equation is

$$NH_3(g) + O_2(g) \rightarrow NO(g) + H_2O(g)$$

Self-Check Exercise 6.2

Step 1 The reactants are propane, $C_3H_8(g)$, and oxygen, $O_2(g)$; the products are carbon dioxide, $CO_2(g)$, and water, $H_2O(g)$. All are in the gaseous state.

Step 2 The unbalanced equation for the reaction is

$$C_3H_8(g) + O_2(g) \rightarrow CO_2(g) + H_2O(g)$$

Step 3 We start with C_3H_8 because it is the most complicated molecule. C_3H_8 contains three carbon atoms per molecule, so a coefficient of 3 is needed for CO_2.

$$C_3H_8(g) + O_2(g) \rightarrow 3CO_2(g) + H_2O(g)$$

Also, each C_3H_8 molecule contains eight hydrogen atoms, so a coefficient of 4 is required for H_2O.

$$C_3H_8(g) + O_2(g) \rightarrow 3CO_2(g) + 4H_2O(g)$$

The final element to be balanced is oxygen. Note that the left side of the equation now has two oxygen atoms, and the right side has ten. We can balance the oxygen by using a coefficient of 5 for O_2.

$$C_3H_8(g) + 5O_2(g) \rightarrow 3CO_2(g) + 4H_2O(g)$$

Step 4 Check:

$$3 \text{ C, } 8 \text{ H, } 10 \text{ O} \rightarrow 3 \text{ C, } 8 \text{ H, } 10 \text{ O}$$

| Reactant atoms | Product atoms |

We cannot divide all coefficients by a given integer to give smaller integer coefficients.

Self-Check Exercise 6.3

a. $NH_4NO_2(s) \rightarrow N_2(g) + H_2O(g)$ (unbalanced)
$NH_4NO_2(s) \rightarrow N_2(g) + 2H_2O(g)$ (balanced)

b. $NO(g) \rightarrow N_2O(g) + NO_2(g)$ (unbalanced)
$3NO(g) \rightarrow N_2O(g) + NO_2(g)$ (balanced)

c. $HNO_3(l) \rightarrow NO_2(g) + H_2O(l) + O_2(g)$ (unbalanced)
$4HNO_3(l) \rightarrow 4NO_2(g) + 2H_2O(l) + O_2(g)$ (balanced)

Chapter 7

Self-Check Exercise 7.1

a. The ions present are

$$Ba^{2+}(aq) + 2NO_3^-(aq) + Na^+(aq) + Cl^-(aq) \rightarrow$$

| Ions in $Ba(NO_3)_2(aq)$ | Ions in $NaCl(aq)$ |

Exchanging the anions gives the possible solid products $BaCl_2$ and $NaNO_3$. Using Table 7.1, we see that both substances are very soluble (rules 1, 2, and 3). Thus no solid forms.

b. The ions present in the mixed solution before any reaction occurs are

$$2Na^+(aq) + S^{2-}(aq) + Cu^{2+}(aq) + 2NO_3^-(aq) \rightarrow$$

| Ions in $Na_2S(aq)$ | Ions in $Cu(NO_3)_2(aq)$ |

Exchanging the anions gives the possible solid products CuS and $NaNO_3$. According to rules 1 and 2 in Table 7.1, $NaNO_3$ is soluble, and by rule 6, CuS should be insoluble. Thus CuS will precipitate. The balanced equation is

$$Na_2S(aq) + Cu(NO_3)_2(aq) \rightarrow CuS(s) + 2NaNO_3(aq)$$

c. The ions present are

$$NH_4^+(aq) + Cl^-(aq) + Pb^{2+}(aq) + 2NO_3^-(aq) \rightarrow$$

| Ions in $NH_4Cl(aq)$ | Ions in $Pb(NO_3)_2(aq)$ |

Exchanging the anions gives the possible solid products NH_4NO_3 and $PbCl_2$. NH_4NO_3 is soluble (rules 1 and 2) and $PbCl_2$ is insoluble (rule 3). Thus $PbCl_2$ will precipitate. The balanced equation is

$$2NH_4Cl(aq) + Pb(NO_3)_2(aq) \rightarrow PbCl_2(s) + 2NH_4NO_3(aq)$$

Self-Check Exercise 7.2

a. *Molecular equation:*
$$Na_2S(aq) + Cu(NO_3)_2(aq) \rightarrow CuS(s) + 2NaNO_3(aq)$$
Complete ionic equation:
$$2Na^+(aq) + S^{2-}(aq) + Cu^{2+}(aq) + 2NO_3^-(aq) \rightarrow$$
$$CuS(s) + 2Na^+(aq) + 2NO_3^-(aq)$$
Net ionic equation:
$$S^{2-}(aq) + Cu^{2+}(aq) \rightarrow CuS(s)$$

b. *Molecular equation:*
$$2NH_4Cl(aq) + Pb(NO_3)_2(aq) \rightarrow PbCl_2(s) + 2NH_4NO_3(aq)$$
Complete ionic equation:
$$2NH_4^+(aq) + 2Cl^-(aq) + Pb^{2+}(aq) + 2NO_3^-(aq) \rightarrow$$
$$PbCl_2(s) + 2NH_4^+(aq) + 2NO_3^-(aq)$$
Net ionic equation:
$$2Cl^-(aq) + Pb^{2+}(aq) \rightarrow PbCl_2(s)$$

Self-Check Exercise 7.3

a. The compound NaBr contains the ions Na^+ and Br^-. Thus each sodium atom loses one electron ($Na \rightarrow Na^+ + e^-$), and each bromine atom gains one electron ($Br + e^- \rightarrow Br^-$).

$$Na + Na + Br - Br \rightarrow (Na^+Br^-) + (Na^+Br^-)$$

b. The compound CaO contains the Ca^{2+} and O^{2-} ions. Thus each calcium atom loses two electrons ($Ca \rightarrow Ca^{2+} + 2e^-$), and each oxygen atom gains two electrons ($O + 2e^- \rightarrow O^{2-}$).

$$Ca + Ca + O - O \rightarrow (Ca^{2+}O^{2-}) + (Ca^{2+}O^{2-})$$

Self-Check Exercise 7.4

a. oxidation–reduction reaction; combustion reaction
b. synthesis reaction; oxidation–reduction reaction; combustion reaction
c. synthesis reaction; oxidation–reduction reaction
d. decomposition reaction; oxidation–reduction reaction
e. precipitation reaction (and double displacement)
f. synthesis reaction; oxidation–reduction reaction
g. acid–base reaction (and double displacement)
h. combustion reaction; oxidation–reduction reaction

Chapter 8

Self-Check Exercise 8.1

The average mass of nitrogen is 14.01 amu. The appropriate equivalence statement is 1 N atom = 14.01 amu, which yields the conversion factor we need:

$$23 \text{ N atoms} \times \frac{14.01 \text{ amu}}{\text{N atom}} = 322.2 \text{ amu}$$

(exact)

Self-Check Exercise 8.2

The average mass of oxygen is 16.00 amu, which gives the equivalence statement 1 O atom = 16.00 amu. The number of oxygen atoms present is

$$288 \text{ amu} \times \frac{1 \text{ O atom}}{16.00 \text{ amu}} = 18.0 \text{ O atoms}$$

Self-Check Exercise 8.3

Note that the sample of 5.00×10^{20} atoms of chromium is less than 1 mole (6.022×10^{23} atoms) of chromium. What fraction of a mole it represents can be determined as follows:

$$5.00 \times 10^{20} \text{ atoms Cr} \times \frac{1 \text{ mol Cr}}{6.022 \times 10^{23} \text{ atoms Cr}} = $$
$$8.30 \times 10^{-4} \text{ mol Cr}$$

Because the mass of 1 mole of chromium atoms is 52.00 g, the mass of 5.00×10^{20} atoms can be determined as follows:

$$8.30 \times 10^{-4} \text{ mol Cr} \times \frac{52.00 \text{ g Cr}}{1 \text{ mol Cr}} = 4.32 \times 10^{-2} \text{ g Cr}$$

Self-Check Exercise 8.4

Each molecule of C_2H_3Cl contains two carbon atoms, three hydrogen atoms, and one chlorine atom, so 1 mole of C_2H_3Cl molecules contains 2 moles of C atoms, 3 moles of H atoms, and 1 mole of Cl atoms.

Mass of 2 mol C atoms: $2 \times 12.01 = 24.02$ g
Mass of 3 mol H atoms: $3 \times 1.008 = 3.024$ g
Mass of 1 mol Cl atoms: $1 \times 35.45 = \underline{35.45}$ g
62.494 g

The molar mass of C_2H_3Cl is 62.49 g (rounding to the correct number of significant figures).

Self-Check Exercise 8.5

The formula for sodium sulfate is Na_2SO_4. One mole of Na_2SO_4 contains 2 moles of sodium ions and 1 mole of sulfate ions.

1 mole of $Na_2SO_4 \rightarrow$ 1 mole of

Mass of 2 mol $Na^+ = 2 \times 22.99 = 45.98$ g
Mass of 1 mol $SO_4{}^{2-} = 32.07 + 4(16.00) = 96.07$ g
Mass of 1 mol $Na_2SO_4 = \underline{142.05}$ g

The molar mass for sodium sulfate is 142.05 g.

A sample of sodium sulfate with a mass of 300.0 g represents more than 1 mol. (Compare 300.0 g to the molar mass of Na_2SO_4.) We calculate the number of moles of Na_2SO_4 present in 300.0 g as follows:

$$300.0 \text{ g Na}_2\text{SO}_4 \times \frac{1 \text{ mol Na}_2\text{SO}_4}{142.05 \text{ g Na}_2\text{SO}_4} = 2.112 \text{ mol Na}_2\text{SO}_4$$

Self-Check Exercise 8.6

First we must compute the mass of 1 mole of C_2F_4 molecules (the molar mass). Because 1 mole of C_2F_4 contains 2 moles of C atoms and 4 moles of F atoms, we have

$$2 \text{ mol C} \times \frac{12.01 \text{ g}}{\text{mol}} = 24.02 \text{ g C}$$

$$4 \text{ mol F} \times \frac{19.00 \text{ g}}{\text{mol}} = 76.00 \text{ g F}$$

Mass of 1 mole of C_2F_4: 100.02 g = molar mass

Using the equivalence statement 100.02 g C_2F_4 = 1 mole C_2F_4, we calculate the moles of C_2F_4 units in 135 g of Teflon.

$$135 \text{ g C}_2\text{F}_4 \text{ units} \times \frac{1 \text{ mol C}_2\text{F}_4}{100.02 \text{ g C}_2\text{F}_4} = 1.35 \text{ mol C}_2\text{F}_4 \text{ units}$$

Next, using the equivalence statement 1 mol = 6.022×10^{23} units, we calculate the number of C_2F_4 units in 135 mol of Teflon.

$$135 \text{ mol C}_2\text{F}_4 \times \frac{6.022 \times 10^{23} \text{ units}}{1 \text{ mol}} = 8.13 \times 10^{23} \text{ C}_2\text{F}_4 \text{ units}$$

Self-Check Exercise 8.7

The molar mass of penicillin F is computed as follows:

$$C: 14 \text{ mol} \times 12.01 \frac{g}{mol} = 168.1 \text{ g}$$

$$H: 20 \text{ mol} \times 1.008 \frac{g}{mol} = 20.16 \text{ g}$$

$$N: 2 \text{ mol} \times 14.01 \frac{g}{mol} = 28.02 \text{ g}$$

$$S: 1 \text{ mol} \times 32.07 \frac{g}{mol} = 32.07 \text{ g}$$

$$O: 4 \text{ mol} \times 16.00 \frac{g}{mol} = 64.00 \text{ g}$$

Mass of 1 mole of $C_{14}H_{20}N_2SO_4$ = 312.39 g = 312.4 g

$$\text{Mass percent of C} = \frac{168.1 \text{ g C}}{312.4 \text{ g C}_{14}\text{H}_{20}\text{N}_2\text{SO}_4} \times 100\%$$
$$= 53.81\%$$

$$\text{Mass percent of H} = \frac{20.16 \text{ g H}}{312.4 \text{ g C}_{14}\text{H}_{20}\text{N}_2\text{SO}_4} \times 100\%$$
$$= 6.453\%$$

$$\text{Mass percent of N} = \frac{28.02 \text{ g N}}{312.4 \text{ g C}_{14}\text{H}_{20}\text{N}_2\text{SO}_4} \times 100\%$$
$$= 8.969\%$$

$$\text{Mass percent of S} = \frac{32.07 \text{ g S}}{312.4 \text{ g C}_{14}\text{H}_{20}\text{N}_2\text{SO}_4} \times 100\%$$
$$= 10.27\%$$

$$\text{Mass percent of O} = \frac{64.00 \text{ g O}}{312.4 \text{ g C}_{14}\text{H}_{20}\text{N}_2\text{SO}_4} \times 100\%$$
$$= 20.49\%$$

Check: The percentages add up to 99.99%.

Self-Check Exercise 8.8

Step 1 0.6884 g lead and 0.2356 g chlorine

Step 2 $0.6884 \text{ g Pb} \times \frac{1 \text{ mol Pb}}{207.2 \text{ g Pb}} = 0.003322 \text{ mol Pb}$

$0.2356 \text{ g Cl} \times \frac{1 \text{ mol Cl}}{35.45 \text{ g Cl}} = 0.006646 \text{ mol Cl}$

Step 3 $\dfrac{0.003322 \text{ mol Pb}}{0.003322} = 1.000 \text{ mol Pb}$

$\dfrac{0.006646 \text{ mol Cl}}{0.003322} = 2.001 \text{ mol Cl}$

These numbers are very close to integers, so step 4 is unnecessary. The empirical formula is $PbCl_2$.

Self-Check Exercise 8.9

Step 1   0.8007 g C, 0.9333 g N, 0.2016 g H, and 2.133 g O

Step 2 $0.8007 \text{ g C} \times \dfrac{1 \text{ mol C}}{12.01 \text{ g C}} = 0.06667 \text{ mol C}$

$0.9333 \text{ g N} \times \dfrac{1 \text{ mol N}}{14.01 \text{ g N}} = 0.06662 \text{ mol N}$

$0.2016 \text{ g H} \times \dfrac{1 \text{ mol H}}{1.008 \text{ g H}} = 0.2000 \text{ mol H}$

$2.133 \text{ g O} \times \dfrac{1 \text{ mol O}}{16.00 \text{ g O}} = 0.1333 \text{ mol O}$

Step 3 $\dfrac{0.06667 \text{ mol C}}{0.06667} = 1.001 \text{ mol C}$

$\dfrac{0.06662 \text{ mol N}}{0.06667} = 1.000 \text{ mol N}$

$\dfrac{0.2000 \text{ mol H}}{0.06662} = 3.002 \text{ mol H}$

$\dfrac{0.1333 \text{ mol O}}{0.06662} = 2.001 \text{ mol O}$

The empirical formula is CNH_3O_2.

Self-Check Exercise 8.10

Step 1   In 100.00 g of Nylon-6 the masses of elements present are 63.68 g C, 12.38 g N, 9.80 g H, and 14.14 g O.

Step 2 $63.68 \text{ g C} \times \dfrac{1 \text{ mol C}}{12.01 \text{ g C}} = 5.302 \text{ mol C}$

$12.38 \text{ g N} \times \dfrac{1 \text{ mol N}}{14.01 \text{ g N}} = 0.8837 \text{ mol N}$

$9.80 \text{ g H} \times \dfrac{1 \text{ mol H}}{1.008 \text{ g H}} = 9.72 \text{ mol H}$

$14.14 \text{ g O} \times \dfrac{1 \text{ mol O}}{16.00 \text{ g O}} = 0.8838 \text{ mol O}$

Step 3 $\dfrac{5.302 \text{ mol C}}{0.8836} = 6.000 \text{ mol C}$

$\dfrac{0.8837 \text{ mol N}}{0.8837} = 1.000 \text{ mol N}$

$\dfrac{9.72 \text{ mol H}}{0.8837} = 11.0 \text{ mol H}$

$\dfrac{0.8838 \text{ mol O}}{0.8837} = 1.000 \text{ mol O}$

The empirical formula for Nylon-6 is $C_6NH_{11}O$.

Self-Check Exercise 8.11

Step 1   First we convert the mass percents to mass in grams. In 100.0 g of the compound, there are 71.65 g of chlorine, 24.27 g of carbon, and 4.07 g of hydrogen.

Step 2   We use these masses to compute the moles of atoms present.

$71.65 \text{ g Cl} \times \dfrac{1 \text{ mol Cl}}{35.45 \text{ g Cl}} = 2.021 \text{ mol Cl}$

$24.27 \text{ g C} \times \dfrac{1 \text{ mol C}}{12.01 \text{ g C}} = 2.021 \text{ mol C}$

$4.07 \text{ g H} \times \dfrac{1 \text{ mol H}}{1.008 \text{ g H}} = 4.04 \text{ mol H}$

Step 3   Dividing each mole value by 2.021 (the smallest number of moles present), we obtain the empirical formula $ClCH_2$.

To determine the molecular formula, we must compare the empirical formula mass to the molar mass. The empirical formula mass is 49.48.

Cl: 35.45
C: 12.01
2 H: $2 \times (1.008)$
$ClCH_2$:   49.48 = empirical formula mass

The molar mass is known to be 98.96. We know that

$$\text{Molar mass} = n \times (\text{empirical formula mass})$$

So we can obtain the value of n as follows:

$$\dfrac{\text{Molar mass}}{\text{Empirical formula mass}} = \dfrac{98.96}{49.48} = 2$$

$$\text{Molecular formula} = (ClCH_2)_2 = Cl_2C_2H_4$$

This substance is composed of molecules with the formula $Cl_2C_2H_4$.

Chapter 9

Self-Check Exercise 9.1

The problem can be stated as follows:

$$4.30 \text{ mol } C_3H_8 \xrightarrow{\text{yields}} ? \text{ mol } CO_2$$

From the balanced equation

$$C_3H_8(g) + 5O_2(g) \rightarrow 3CO_2(g) + 4H_2O(g)$$

we derive the equivalence statement

$$1 \text{ mol } C_3H_8 = 3 \text{ mol } CO_2$$

The appropriate conversion factor (moles of C_3H_8 must cancel) is 3 mol CO_2/1 mol C_3H_8, and the calculation is

$$4.30 \text{ mol } C_3H_8 \times \dfrac{3 \text{ mol } CO_2}{1 \text{ mol } C_3H_8} = 12.9 \text{ mol } CO_2$$

Thus we can say

$$4.30 \text{ mol } C_3H_8 \text{ yields } 12.9 \text{ mol } CO_2$$

Self-Check Exercise 9.2

The problem can be sketched as follows:

$$C_3H_8(g) + 5O_2(g) \rightarrow 3CO_2(g) + 4H_2O(g)$$

We have already done the first step in Example 9.4.

To find out how many moles of CO_2 can be produced from 2.18 moles of C_3H_8, we see from the balanced equation that 3 moles of CO_2 is produced for each mole of C_3H_8 reacted. The mole ratio we need is 3 mol CO_2/1 mol C_3H_8. The conversion is therefore

$$2.18 \text{ mol } C_3H_8 \times \frac{3 \text{ mol } CO_2}{1 \text{ mol } C_3H_8} = 6.54 \text{ mol } CO_2$$

Next, using the molar mass of CO_2, which is $12.01 + 32.00 = 44.01$ g, we calculate the mass of CO_2 produced.

$$6.54 \text{ mol } CO_2 \times \frac{44.01 \text{ g } CO_2}{1 \text{ mol } CO_2} = 288 \text{ g } CO_2$$

The sequence of steps we took to find the mass of carbon dioxide produced from 96.1 g of propane is summarized in the following diagram.

96.1 g C_3H_8	→	$\dfrac{1 \text{ mol } C_3H_8}{44.09 \text{ g } C_3H_8}$	→	2.18 mol C_3H_8
2.18 mol C_3H_8	→	$\dfrac{3 \text{ mol } CO_2}{1 \text{ mol } C_3H_8}$	→	6.54 mol CO_2
6.54 mol CO_2	→	$\dfrac{44.01 \text{ g } CO_2}{1 \text{ mol } CO_2}$	→	288 g CO_2
Mass				Moles

Self–Check Exercise 9.3

We sketch the problem as follows:

$$C_3H_8(g) + 5O_2(g) \rightarrow 3CO_2(g) + 4H_2O(g)$$

Then we do the calculations.

96.1 g C_3H_8	→	$\dfrac{1 \text{ mol } C_3H_8}{44.09 \text{ g}}$	→	2.18 mol C_3H_8
2.18 mol C_3H_8	→	$\dfrac{4 \text{ mol } H_2O}{1 \text{ mol } C_3H_8}$	→	8.72 mol H_2O
8.72 mol H_2O	→	$\dfrac{18.02 \text{ g}}{\text{mol } H_2O}$	→	157 g H_2O

Therefore, 157 g of H_2O is produced from 96.1 g C_3H_8.

Self–Check Exercise 9.4

a. We first write the balanced equation.

$$SiO_2(s) + 4HF(aq) \rightarrow SiF_4(g) + 2H_2O(l)$$

The map of the steps required is

$$SiO_2(s) + 4HF(aq) \rightarrow SiF_4(g) + 2H_2O(l)$$

We convert 5.68 g of SiO_2 to moles as follows:

$$5.68 \text{ g } SiO_2 \times \frac{1 \text{ mol } SiO_2}{60.09 \text{ g } SiO_2} = 9.45 \times 10^{-2} \text{ mol } SiO_2$$

Using the balanced equation, we obtain the appropriate mole ratio and convert to moles of HF.

$$9.45 \times 10^{-2} \text{ mol } SiO_2 \times \frac{4 \text{ mol HF}}{1 \text{ mol } SiO_2} = 3.78 \times 10^{-1} \text{ mol HF}$$

Finally, we calculate the mass of HF by using its molar mass.

$$3.78 \times 10^{-1} \text{ mol HF} \times \frac{20.01 \text{ g HF}}{\text{mol HF}} = 7.56 \text{ g HF}$$

b.   The map for this problem is

$$SiO_2(s) + 4HF(aq) \rightarrow SiF_4(g) + 2H_2O(l)$$

We have already accomplished the first conversion in part a. Using the balanced equation, we obtain moles of H_2O as follows:

$$9.45 \times 10^{-2} \text{ mol } SiO_2 \times \frac{2 \text{ mol } H_2O}{1 \text{ mol } SiO_2} = 1.89 \times 10^{-1} \text{ mol } H_2O$$

The mass of water formed is

$$1.89 \times 10^{-1} \text{ mol } H_2O \times \frac{18.02 \text{ g } H_2O}{\text{mol } H_2O} = 3.41 \text{ g } H_2O$$

Self–Check Exercise 9.5

In this problem, we know the mass of the product to be formed by the reaction

$$CO(g) + 2H_2(g) \rightarrow CH_3OH(l)$$

and we want to find the masses of reactants needed. The procedure is the same one we have been following. We must first convert the mass of CH_3OH to moles, then use the balanced equation to obtain moles of H_2 and CO needed, and then convert these moles to masses. Using the molar mass of CH_3OH (32.04 g/mol), we convert to moles of CH_3OH.

First we convert kilograms to grams.

$$6.0 \text{ kg } CH_3OH \times \frac{1000 \text{ g}}{\text{kg}} = 6.0 \times 10^3 \text{ g } CH_3OH$$

Next we convert 6.0×10^3 g CH_3OH to moles of CH_3OH, using the conversion factor 1 mol CH_3OH/32.04 g CH_3OH.

$$6.0 \times 10^3 \text{ g } CH_3OH \times \frac{1 \text{ mol } CH_3OH}{32.04 \text{ g } CH_3OH} = 1.9 \times 10^2 \text{ mol } CH_3OH$$

Then we have two questions to answer:

To answer these questions, we use the balanced equation

$$CO(g) + 2H_2(g) \rightarrow CH_3OH(l)$$

to obtain mole ratios between the reactants and the products. In the balanced equation the coefficients for both CO and CH_3OH are 1, so we can write the equivalence statement

$$1 \text{ mol CO} = 1 \text{ mol CH}_3\text{OH}$$

Using the mole ratio 1 mol CO/1 mol CH_3OH, we can now convert from moles of CH_3OH to moles of CO.

$$1.9 \times 10^2 \text{ mol CH}_3\text{OH} \times \frac{1 \text{ mol CO}}{1 \text{ mol CH}_3\text{OH}} = 1.9 \times 10^2 \text{ mol CO}$$

To calculate the moles of H_2 required, we construct the equivalence statement between CH_3OH and H_2, using the coefficients in the balanced equation.

$$2 \text{ mol H}_2 = 1 \text{ mol CH}_3\text{OH}$$

Using the mole ratio 2 mol H_2/1 mol CH_3OH, we can convert moles of CH_3OH to moles of H_2.

$$1.9 \times 10^2 \text{ mol CH}_3\text{OH} \times \frac{2 \text{ mol H}_2}{1 \text{ mol CH}_3\text{OH}} = 3.8 \times 10^2 \text{ mol H}_2$$

We now have the moles of reactants required to produce 6.0 kg of CH_3OH. Since we need the masses of reactants, we must use the molar masses to convert from moles to mass.

$$1.9 \times 10^2 \text{ mol CO} \times \frac{28.01 \text{ g CO}}{1 \text{ mol CO}} = 5.3 \times 10^3 \text{ g CO}$$

$$3.8 \times 10^2 \text{ mol H}_2 \times \frac{2.016 \text{ g H}_2}{1 \text{ mol H}_2} = 7.7 \times 10^2 \text{ H}_2$$

Therefore, we need 5.3×10^3 g CO to react with 7.7×10^2 g H_2 to form 6.0×10^3 g (6.0 kg) of CH_3OH. This whole process is mapped in the following diagram.

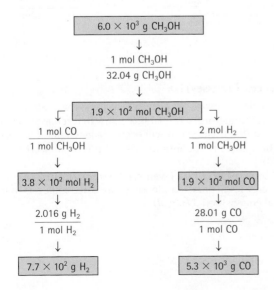

Self-Check Exercise 9.6

Step 1 The balanced equation for the reaction is

$$6\text{Li}(s) + N_2(g) \rightarrow 2\text{Li}_3\text{N}(s)$$

Step 2 To determine the limiting reactant, we must convert the masses of lithium (atomic mass = 6.941 g) and nitrogen (molar mass = 28.02 g) to moles.

$$56.0 \text{ g Li} \times \frac{1 \text{ mol Li}}{6.941 \text{ g Li}} = 8.07 \text{ mol Li}$$

$$56.0 \text{ g N}_2 \times \frac{1 \text{ mol N}_2}{28.02 \text{ g N}_2} = 2.00 \text{ mol N}_2$$

Step 3 Using the mole ratio from the balanced equation, we can calculate the moles of lithium required to react with 2.00 moles of nitrogen.

$$2.00 \text{ mol N}_2 \times \frac{6 \text{ mol Li}}{1 \text{ mol N}_2} = 12.0 \text{ mol Li}$$

Therefore, 12.0 moles of Li is required to react with 2.00 moles of N_2. However, we have only 8.07 mol of Li, so lithium is limiting. It will be consumed before the nitrogen runs out.

Step 4 Because lithium is the limiting reactant, we must use the 8.07 moles of Li to determine how many moles of Li_3N can be formed.

$$8.07 \text{ mol Li} \times \frac{2 \text{ mol Li}_3\text{N}}{6 \text{ mol Li}} = 2.69 \text{ mol Li}_3\text{N}$$

Step 5 We can now use the molar mass of Li_3N (34.83 g) to calculate the mass of Li_3N formed.

$$2.69 \text{ mol Li}_3\text{N} \times \frac{34.83 \text{ g Li}_3\text{N}}{1 \text{ mol Li}_3\text{N}} = 93.7 \text{ g Li}_3\text{N}$$

Self-Check Exercise 9.7

a. **Step 1** The balanced equation is

$$\text{TiCl}_4(g) + O_2(g) \rightarrow \text{TiO}_2(s) + 2\text{Cl}_2(g)$$

Step 2 The numbers of moles of reactants are

$$6.71 \times 10^3 \text{ g TiCl}_4 \times \frac{1 \text{ mol TiCl}_4}{189.68 \text{ g TiCl}_4} = 3.54 \times 10^1 \text{ mol TiCl}_4$$

$$2.45 \times 10^3 \text{ g O}_2 \times \frac{1 \text{ mol O}_2}{32.00 \text{ g O}_2} = 7.66 \times 10^1 \text{ mol O}_2$$

Step 3 In the balanced equation both $TiCl_4$ and O_2 have coefficients of 1, so

$$1 \text{ mol TiCl}_4 = 1 \text{ mol O}_2$$

and

$$3.54 \times 10^1 \text{ mol TiCl}_4 \times \frac{1 \text{ mol O}_2}{1 \text{ mol TiCl}_4}$$

$$= 3.54 \times 10^1 \text{ mol O}_2 \text{ required}$$

We have 7.66×10^1 moles of O_2, so the O_2 is in excess and the $TiCl_4$ is limiting. This makes sense. $TiCl_4$ and O_2 react in a 1:1 mole ratio, so the $TiCl_4$ is limiting because fewer moles of $TiCl_4$ are present than moles of O_2.

Step 4 We will now use the moles of $TiCl_4$ (the limiting reactant) to determine the moles of TiO_2 that would form if the reaction produced 100% of the expected yield (the theoretical yield).

$$3.54 \times 10^1 \text{ mol TiCl}_4 \times \frac{1 \text{ mol TiO}_2}{1 \text{ mol TiCl}_4} = 3.54 \times 10^1 \text{ mol TiO}_2$$

The mass of TiO_2 expected for 100% yield is

$$3.54 \times 10^1 \text{ mol TiO}_2 \times \frac{79.88 \text{ g TiO}_2}{1 \text{ mol TiO}_2} = 2.83 \times 10^3 \text{ g TiO}_2$$

This amount represents the theoretical yield.

b. Because the reaction is said to give only a 75.0% yield of TiO_2, we use the definition of percent yield,

$$\frac{\text{Actual yield}}{\text{Theoretical yield}} \times 100\% = \% \text{ yield}$$

to write the equation

$$\frac{\text{Actual yield}}{2.83 \times 10^3 \text{ g TiO}_2} \times 100\% = 75.0\% \text{ yield}$$

We now want to solve for the actual yield. First we divide both sides by 100%.

$$\frac{\text{Actual yield}}{2.83 \times 10^3 \text{ g TiO}_2} \times \frac{100\%}{100\%} = \frac{75.0}{100} = 0.750$$

Then we multiply both sides by 2.83×10^3 g TiO_2.

$$2.83 \times 10^3 \text{ g TiO}_2 \times \frac{\text{Actual yield}}{2.83 \times 10^3 \text{ g TiO}_2}$$
$$= 0.750 \times 2.83 \times 10^3 \text{ g TiO}_2$$

$$\text{Actual yield} = 0.750 \times 2.83 \times 10^3 \text{ g TiO}_2$$
$$= 2.12 \times 10^3 \text{ g TiO}_2$$

Thus 2.12×10^3 g of $TiO_2(s)$ is actually obtained in this reaction.

Chapter 10

Self-Check Exercise 10.1

The conversion factor needed is $\dfrac{1 \text{ cal}}{4.184 \text{ J}}$, and the conversion is

$$28.4 \text{ J} \times \frac{1 \text{ cal}}{4.184 \text{ J}} = 6.79 \text{ cal}$$

Self-Check Exercise 10.2

We know that it takes 4.184 J of energy to change the temperature of each gram of water by 1 °C, so we must multiply 4.184 by the mass of water (454 g) and the temperature change (98.6 °C − 5.4 °C = 93.2 °C).

$$4.184 \frac{\text{J}}{\text{g} \cdot °\text{C}} \times 454 \text{ g} \times 93.2 \text{ °C} = 1.77 \times 10^5 \text{ J}$$

Self-Check Exercise 10.3

From Table 10.1, the specific heat capacity for solid gold is 0.13 J/g °C. Because it takes 0.13 J to change the temperature of *one* gram of gold by *one* Celsius degree, we must multiply 0.13 by the sample size (5.63 g) and the change in temperature (32 °C − 21 °C = 11 °C).

$$0.13 \frac{\text{J}}{\text{g} \cdot °\text{C}} \times 5.63 \text{ g} \times 11 \text{ °C} = 8.1 \text{ J}$$

We can change this energy to units of calories as follows:

$$8.1 \text{ J} \times \frac{1 \text{ cal}}{4.184 \text{ J}} = 1.9 \text{ cal}$$

Self-Check Exercise 10.4

Table 10.1 lists the specific heat capacities of several metals. We want to calculate the specific heat capacity (s) for this metal and then use Table 10.1 to identify the metal. Using the equation

$$Q = s \times m \times \Delta T$$

we can solve for s by dividing both sides by m (the mass of the sample) and by ΔT:

$$\frac{Q}{m \times \Delta T} = s$$

In this case,

Q = energy (heat) required = 10.1 J
m = 2.8 g
ΔT = temperature change = 36 °C − 21 °C = 15 °C

so

$$s = \frac{Q}{m \times \Delta T} = \frac{10.1 \text{ J}}{(2.8 \text{ g})(15 \text{ °C})} = 0.24 \text{ J/g °C}$$

Table 10.1 shows that silver has a specific heat capacity of 0.24 J/g °C. The metal is silver.

Self-Check Exercise 10.5

We are told that 1652 kJ of energy is *released* when 4 moles of Fe reacts. We first need to determine what number of moles 1.00 g Fe represents.

$$1.00 \text{ g Fe} \times \frac{1 \text{ mol}}{55.85 \text{ g}} = 1.79 \times 10^{-2} \text{ mol Fe}$$

$$1.79 \times 10^{-2} \text{ mol Fe} \times \frac{1652 \text{ kJ}}{4 \text{ mol Fe}} = 7.39 \text{ kJ}$$

Thus 7.39 kJ of energy (as heat) is released when 1.00 g of iron reacts.

Self-Check Exercise 10.6

Noting the reactants and products in the desired reaction

$$S(s) + O_2(g) \rightarrow SO_2(g)$$

We need to reverse the second equation and multiply it by $\frac{1}{2}$. This reverses the sign and cuts the amount of energy by a factor of 2.

$$\tfrac{1}{2}[2SO_3(g) \rightarrow 2SO_2(g) + O_2(g)] \qquad \Delta H = \frac{198.2 \text{ kJ}}{2}$$

or

$$SO_3(g) \rightarrow SO_2(g) + \tfrac{1}{2}O_2(g) \qquad \Delta H = 99.1 \text{ kJ}$$

Now we add this reaction to the first reaction.

$S(s) + \tfrac{3}{2}O_2(g) \rightarrow SO_3(g)$	$\Delta H = -395.2$ kJ
$SO_3(g) \rightarrow SO_2(g) + \tfrac{1}{2}O_2(g)$	$\Delta H = 99.1$ kJ
$S(s) + O_2(g) \rightarrow SO_2(g)$	$\Delta H = -296.1$ kJ

Chapter 11

Self-Check Exercise 11.1

a. Circular pathways for electrons in the Bohr model.
b. Three-dimensional probability maps that represent the likelihood that the electron will occupy a given point in space.
c. The surface that contains 90% of the total electron probability.
d. A set of orbitals of a given type of orbital within a principal energy level. For example, there are three sublevels in principal energy level 3 (s, p, d).

Self-Check Exercise 11.2

Element	Electron Configuration	1s	2s	2p	3s	3p
Al	$1s^2 2s^2 2p^6 3s^2 3p^1$ [Ne]$3s^2 3p^1$	↑↓	↑↓	↑↓ ↑↓ ↑↓	↑↓	↑ _ _
Si	[Ne]$3s^2 3p^2$	↑↓	↑↓	↑↓ ↑↓ ↑↓	↑↓	↑ ↑ _
P	[Ne]$3s^2 3p^3$	↑↓	↑↓	↑↓ ↑↓ ↑↓	↑↓	↑ ↑ ↑
S	[Ne]$3s^2 3p^4$	↑↓	↑↓	↑↓ ↑↓ ↑↓	↑↓	↑↓ ↑ ↑
Cl	[Ne]$3s^2 3p^5$	↑↓	↑↓	↑↓ ↑↓ ↑↓	↑↓	↑↓ ↑↓ ↑
Ar	[Ne]$3s^2 3p^6$	↑↓	↑↓	↑↓ ↑↓ ↑↓	↑↓	↑↓ ↑↓ ↑↓

Self-Check Exercise 11.3

F: $1s^2 2s^2 2p^5$ or [He]$2s^2 2p^5$

Si: $1s^2 2s^2 2p^6 3s^2 3p^2$ or [Ne]$3s^2 3p^2$

Cs: $1s^2 2s^2 2p^6 3s^2 3p^6 4s^2 3d^{10} 4p^6 5s^2 4d^{10} 5p^6 6s^1$ or [Xe]$6s^1$

Pb: $1s^2 2s^2 2p^6 3s^2 3p^6 4s^2 3d^{10} 4p^6 5s^2 4d^{10} 5p^6 6s^2 4f^{14} 5d^{10} 6p^2$ or [Xe]$6s^2 4f^{14} 5d^{10} 6p^2$

I: $1s^2 2s^2 2p^6 3s^2 3p^6 4s^2 3d^{10} 4p^6 5s^2 4d^{10} 5p^5$ or [Kr]$5s^2 4d^{10} 5p^5$

Silicon (Si): In Group 4 and Period 3, it is the second of the "3p elements." The configuration is $1s^2 2s^2 2p^6 3s^2 3p^2$, or [Ne]$3s^2 3p^2$.

Cesium (Cs): In Group 1 and Period 6, it is the first of the "6s elements." The configuration is $1s^2 2s^2 2p^6 3s^2 3p^6 4s^2 3d^{10} 4p^6 5s^2 4d^{10} 5p^6 6s^1$, or [Xe]$6s^1$.

Lead (Pb): In Group 4 and Period 6, it is the second of the "6p elements." The configuration is [Xe]$6s^2 4f^{14} 5d^{10} 6p^2$.

Iodine (I): In Group 7 and Period 5, it is the fifth of the "5p elements." The configuration is [Kr]$5s^2 4d^{10} 5p^5$.

Chapter 12

Self-Check Exercise 12.1

Using the electronegativity values given in Figure 12.3, we choose the bond in which the atoms exhibit the largest difference in electronegativity. (Electronegativity values are shown in parentheses.)

a. H—C > H—P
   (2.1)(2.5)   (2.1)(2.1)

b. O—I > O—F
   (3.5)(2.5)   (3.5)(4.0)

c. S—O > N—O
   (2.5)(3.5)   (3.0)(3.5)

d. N—H > Si—H
   (3.0)(2.1)   (1.8)(2.1)

Self-Check Exercise 12.2

H has one electron, and Cl has seven valence electrons. This gives a total of eight valence electrons. We first draw in the bonding pair:

H—Cl, which could be drawn as H:Cl

We have six electrons yet to place. The H already has two electrons, so we place three lone pairs around the chlorine to satisfy the octet rule.

H—C̈l: or H:C̈l:

Self-Check Exercise 12.3

Step 1 O_3: 3(6) = 18 valence electrons
Step 2 O—O—O
Step 3 Ö=Ö—Ö: and :Ö—Ö=Ö
This molecule shows resonance (it has two valid Lewis structures).

Self-Check Exercise 12.4

See table on top of page A10.

Self-Check Exercise 12.5

a. NH_4^+

The Lewis structure is
$$\left[\begin{array}{c} H \\ | \\ H-N-H \\ | \\ H \end{array} \right]^+$$

(See Self-Check Exercise 12.4.) There are four pairs of electrons around the nitrogen. This requires a tetrahedral arrangement of electron pairs. The NH_4^+ ion has a tetrahedral molecular structure (row 3 in Table 12.4), because all electron pairs are shared.

b. SO_4^{2-}

The Lewis structure is
$$\left[\begin{array}{c} :\ddot{O}: \\ | \\ :\ddot{O}-S-\ddot{O}: \\ | \\ :\ddot{O}: \end{array} \right]^{2-}$$

(See Self-Check Exercise 12.4.) The four electron pairs around the sulfur require a tetrahedral arrangement. The SO_4^{2-} has a tetrahedral molecular structure (row 3 in Table 12.4).

c. NF_3

The Lewis structure is
$$:\ddot{F} \begin{array}{c} \ddot{N} \\ \diagup | \diagdown \end{array} \ddot{F}:$$
$$:\ddot{F}:$$

(See Self-Check Exercise 12.4.) The four pairs of electrons on the nitrogen require a tetrahedral arrangement. In this case, only three of the pairs are shared with the fluorine atoms, leaving one lone pair. Thus the molecular structure is a trigonal pyramid (row 4 in Table 12.4).

d. H_2S

The Lewis structure is H—S̈—H

(See Self-Check Exercise 12.4.) The four pairs of electrons around the sulfur require a tetrahedral arrangement. In this case, two pairs are shared with hydrogen atoms, leaving two lone pairs. Thus the molecular structure is bent or V-shaped (row 5 in Table 12.4).

e. ClO_3^-

The Lewis structure is
$$\left[\begin{array}{c} :\ddot{O}-\ddot{C}l-\ddot{O}: \\ | \\ :\ddot{O}: \end{array} \right]^-$$

(See Self-Check Exercise 12.4.) The four pairs of electrons require a tetrahedral arrangement. In this case, three pairs are shared with oxygen atoms, leaving one lone pair. Thus the molecular structure is a trigonal pyramid (row 4 in Table 12.4).

Molecule or Ion	Total Valence Electrons	Draw Single Bonds	Calculate Number of Electrons Remaining	Use Remaining Electrons to Achieve Noble Gas Configurations	Check	
					Atoms	Electrons
a. NF_3	$5 + 3(7) = 26$	F—N with F above and F below	$26 - 6 = 20$	$:\!\ddot{F}\!-\!N\!-\!\ddot{F}\!:$ with $:\!\ddot{F}\!:$ below	N F	8 8
b. O_2	$2(6) = 12$	O—O	$12 - 2 = 10$	$\ddot{O}\!=\!\ddot{O}$	O	8
c. CO	$4 + 6 = 10$	C—O	$10 - 2 = 8$	$:\!C\!\equiv\!O\!:$	C O	8 8
d. PH_3	$5 + 3(1) = 8$	H H with P and H below	$8 - 6 = 2$	$H\!-\!\ddot{P}\!-\!H$ with H below	P H	8 2
e. H_2S	$2(1) + 6 = 8$	H—S—H	$8 - 4 = 4$	$H\!-\!\ddot{S}\!-\!H$	S H	8 2
f. SO_4^{2-}	$6 + 4(6) + 2 = 32$	O—S—O with O above and O below	$32 - 8 = 24$	$\left[:\!\ddot{O}\!-\!\overset{\ddot{O}:}{\underset{\ddot{O}:}{S}}\!-\!\ddot{O}\!:\right]^{2-}$	S O	8 8
g. NH_4^+	$5 + 4(1) - 1 = 8$	H—N—H with H above and H below	$8 - 8 = 0$	$\left[H\!-\!\overset{H}{\underset{H}{N}}\!-\!H\right]^{+}$	N H	8 2
h. ClO_3^-	$7 + 3(6) + 1 = 26$	Cl with O above and O, O below	$26 - 6 = 20$	$\left[:\!\ddot{O}\!-\!\overset{\ddot{Cl}}{\underset{\ddot{O}:}{}}\!-\!\ddot{O}\!:\right]^{-}$	Cl O	8 8
i. SO_2	$6 + 2(6) = 18$	O—S—O	$18 - 4 = 14$	$\ddot{O}\!=\!\ddot{S}\!-\!\ddot{O}\!:$ and $:\!\ddot{O}\!-\!\ddot{S}\!=\!\ddot{O}$	S O	8 8

Answer to Self-Check Exercise 12.4.

f. BeF_2

The Lewis structure is $:\!\ddot{F}\!-\!Be\!-\!\ddot{F}\!:$

The two electron pairs on beryllium require a linear arrangement. Because both pairs are shared by fluorine atoms, the molecular structure is also linear (row 1 in Table 12.4).

Chapter 13

Self-Check Exercise 13.1

We know that 1.000 atm = 760.0 mm Hg. So

$$525 \text{ mm Hg} \times \frac{1.000 \text{ atm}}{760.0 \text{ mm Hg}} = 0.691 \text{ atm}$$

Self-Check Exercise 13.2

Initial Conditions	**Final Conditions**
$P_1 = 635$ torr	$P_2 = 785$ torr
$V_1 = 1.51$ L	$V_2 = ?$

Solving Boyle's law ($P_1 V_1 = P_2 V_2$) for V_2 gives

$$V_2 = V_1 \times \frac{P_1}{P_2}$$

$$= 1.51 \text{ L} \times \frac{635 \text{ torr}}{785 \text{ torr}} = 1.22 \text{ L}$$

Note that the volume decreased, as the increase in pressure led us to expect.

Self-Check Exercise 13.3

Because the temperature of the gas inside the bubble decreases (at constant pressure), the bubble gets smaller. The conditions are

Initial Conditions
$T_1 = 28\ °C = 28 + 273 = 301$ K
$V_1 = 23$ cm^3

Final Conditions
$T_2 = 18\ °C = 18 + 273 = 291$ K
$V_2 = ?$

Solving Charles's law,

$$\frac{V_1}{T_1} = \frac{V_2}{T_2}$$

for V_2 gives

$$V_2 = V_1 \times \frac{T_2}{T_1} = 23 \text{ cm}^3 \times \frac{291 \text{ K}}{301 \text{ K}} = 22 \text{ cm}^3$$

Self-Check Exercise 13.4

Because the temperature and pressure of the two samples are the same, we can use Avogadro's law in the form

$$\frac{V_1}{n_1} = \frac{V_2}{n_2}$$

The following information is given:

Sample 1	Sample 2
$V_1 = 36.7 \text{ L}$	$V_2 = 16.5 \text{ L}$
$n_1 = 1.5 \text{ mol}$	$n_2 = ?$

We can now solve Avogadro's law for the value of n_2 (the moles of N_2 in sample 2):

$$n_2 = n_1 \times \frac{V_2}{V_1} = 1.5 \text{ mol} \times \frac{16.5 \text{ L}}{36.7 \text{ L}} = 0.67 \text{ mol}$$

Here n_2 is smaller than n_1, which makes sense in view of the fact that V_2 is smaller than V_1.

Note: We isolate n_2 from Avogadro's law as given above by multiplying both sides of the equation by n_2 and then by n_1/V_1,

$$\left(n_2 \times \frac{n_1}{V_1}\right)\frac{V_1}{n_1} = \left(n_2 \times \frac{n_1}{V_1}\right)\frac{V_2}{n_2}$$

to give $n_2 = n_1 \times V_2/V_1$.

Self-Check Exercise 13.5

We are given the following information:

$$P = 1.00 \text{ atm}$$
$$V = 2.70 \times 10^6 \text{ L}$$
$$n = 1.10 \times 10^5 \text{ mol}$$

We solve for T by dividing both sides of the ideal gas law by nR:

$$\frac{PV}{nR} = \frac{nRT}{nR}$$

to give

$$T = \frac{PV}{nR} = \frac{(1.00 \text{ atm})(2.70 \times 10^6 \text{ L})}{(1.10 \times 10^5 \text{ mol})\left(0.08206 \dfrac{\text{L atm}}{\text{K mol}}\right)}$$

$$= 299 \text{ K}$$

The temperature of the helium is 299 K, or $299 - 273 = 26 \text{ °C}$.

Self-Check Exercise 13.6

We are given the following information about the radon sample:

$$n = 1.5 \text{ mol}$$
$$V = 21.0 \text{ L}$$
$$T = 33 \text{ °C} = 33 + 273 = 306 \text{ K}$$
$$P = ?$$

We solve the ideal gas law ($PV = nRT$) for P by dividing both sides of the equation by V:

$$P = \frac{nRT}{V} = \frac{(1.5 \text{ mol})\left(0.08206 \dfrac{\text{L atm}}{\text{K mol}}\right)(306 \text{ K})}{21.0 \text{ L}}$$

$$= 1.8 \text{ atm}$$

Self-Check Exercise 13.7

To solve this problem, we take the ideal gas law and separate those quantities that change from those that remain constant (on opposite sides of the equation). In this case, volume and temperature change, and number of moles and pressure (and, of course, R) remain constant. So $PV = nRT$ becomes $V/T = nR/P$, which leads to

$$\frac{V_1}{T_1} = \frac{nR}{P} \quad \text{and} \quad \frac{V_2}{T_2} = \frac{nR}{P}$$

Combining these gives

$$\frac{V_1}{T_1} = \frac{nR}{P} = \frac{V_2}{T_2} \quad \text{or} \quad \frac{V_1}{T_1} = \frac{V_2}{T_2}$$

We are given

Initial Conditions
$T_1 = 5 \text{ °C} = 5 + 273 = 278 \text{ K}$
$V_1 = 3.8 \text{ L}$
Final Conditions
$T_2 = 86 \text{ °C} = 86 + 273 = 359 \text{ K}$
$V_2 = ?$

Thus

$$V_2 = \frac{T_2 V_1}{T_1} = \frac{(359 \text{ K})(3.8 \text{ L})}{278 \text{ K}} = 4.9 \text{ L}$$

Check: Is the answer sensible? In this case, the temperature was increased (at constant pressure), so the volume should increase. The answer makes sense.

Note that this problem could be described as a "Charles's law problem." The real advantage of using the ideal gas law is that you need to remember only *one* equation to do virtually any problem involving gases.

Self-Check Exercise 13.8

We are given the following information:

Initial Conditions
$P_1 = 0.747 \text{ atm}$
$T_1 = 13 \text{ °C} = 13 + 273 = 286 \text{ K}$
$V_1 = 11.0 \text{ L}$
Final Conditions
$P_2 = 1.18 \text{ atm}$
$T_2 = 56 \text{ °C} = 56 + 273 = 329 \text{ K}$
$V_2 = ?$

In this case, the number of moles remains constant. Thus we can say

$$\frac{P_1 V_1}{T_1} = nR \quad \text{and} \quad \frac{P_2 V_2}{T_2} = nR$$

or

$$\frac{P_1 V_1}{T_1} = \frac{P_2 V_2}{T_2}$$

Solving for V_2 gives

$$V_2 = V_1 \times \frac{T_2}{T_1} \times \frac{P_1}{P_2} = (11.0 \text{ L})\left(\frac{329 \text{ K}}{286 \text{ K}}\right)\left(\frac{0.747 \text{ atm}}{1.18 \text{ atm}}\right)$$

$$= 8.01 \text{ L}$$

Self-Check Exercise 13.9

As usual when dealing with gases, we can use the ideal gas equation $PV = nRT$. First consider the information given:

$$P = 0.91 \text{ atm} = P_{\text{total}}$$
$$V = 2.0 \text{ L}$$
$$T = 25 \text{ °C} = 25 + 273 = 298 \text{ K}$$

Given this information, we can calculate the number of moles of gas in the mixture: $n_{total} = n_{N_2} + n_{O_2}$. Solving for n in the ideal gas equation gives

$$n_{total} = \frac{P_{total}V}{RT} = \frac{(0.91 \text{ atm})(2.0 \text{ L})}{\left(0.08206 \dfrac{\text{L atm}}{\text{K mol}}\right)(298 \text{ K})} = 0.074 \text{ mol}$$

We also know that 0.050 mole of N_2 is present. Because

$$n_{total} = n_{N_2} + n_{O_2} = 0.074 \text{ mol}$$
$$\uparrow$$
$$(0.050 \text{ mol})$$

we can calculate the moles of O_2 present.

$$0.050 \text{ mol} + n_{O_2} = 0.074 \text{ mol}$$
$$n_{O_2} = 0.074 \text{ mol} - 0.050 \text{ mol} = 0.024 \text{ mol}$$

Now that we know the moles of oxygen present, we can calculate the partial pressure of oxygen from the ideal gas equation.

$$P_{O_2} = \frac{n_{O_2}RT}{V} = \frac{(0.024 \text{ mol})\left(0.08206 \dfrac{\text{L atm}}{\text{K mol}}\right)(298 \text{ K})}{2.0 \text{ L}}$$
$$= 0.29 \text{ atm}$$

Although it is not requested, note that the partial pressure of the N_2 must be 0.62 atm, because

$$\underbrace{0.62 \text{ atm}}_{P_{N_2}} + \underbrace{0.29 \text{ atm}}_{P_{O_2}} = \underbrace{0.91 \text{ atm}}_{P_{total}}$$

Self-Check Exercise 13.10

The volume is 0.500 L, the temperature is 25 °C (or 25 + 273 = 298 K), and the total pressure is given as 0.950 atm. Of this total pressure, 24 torr is due to the water vapor. We can calculate the partial pressure of the H_2 because we know that

$$P_{total} = P_{H_2} + P_{H_2O} = 0.950 \text{ atm}$$
$$\uparrow$$
$$24 \text{ torr}$$

Before we carry out the calculation, however, we must convert the pressures to the same units. Converting P_{H_2O} to atmospheres gives

$$24 \text{ torr} \times \frac{1.000 \text{ atm}}{760.0 \text{ torr}} = 0.032 \text{ atm}$$

Thus

$$P_{total} = P_{H_2} + P_{H_2O} = 0.950 \text{ atm} = P_{H_2} + 0.032 \text{ atm}$$

and

$$P_{H_2} = 0.950 \text{ atm} - 0.032 \text{ atm} = 0.918 \text{ atm}$$

Now that we know the partial pressure of the hydrogen gas, we can use the ideal gas equation to calculate the moles of H_2.

$$n_{H_2} = \frac{P_{H_2}V}{RT} = \frac{(0.918 \text{ atm})(0.500 \text{ L})}{\left(0.08206 \dfrac{\text{L atm}}{\text{K mol}}\right)(298 \text{ K})}$$
$$= 0.0188 \text{ mol} = 1.88 \times 10^{-2} \text{ mol}$$

The sample of gas contains 1.88×10^{-2} mole of H_2, which exerts a partial pressure of 0.918 atm.

Self-Check Exercise 13.11

We will solve this problem by taking the following steps:

| Grams of zinc | ⟹ | Moles of zinc | ⟹ | Moles of H_2 | ⟹ | Volume of H_2 |

Step 1   Using the atomic mass of zinc (65.38), we calculate the moles of zinc in 26.5 g.

$$26.5 \text{ g Zn} \times \frac{1 \text{ mol Zn}}{65.38 \text{ g Zn}} = 0.405 \text{ mol Zn}$$

Step 2   Using the balanced equation, we next calculate the moles of H_2 produced.

$$0.405 \text{ mol Zn} \times \frac{1 \text{ mol } H_2}{1 \text{ mol Zn}} = 0.405 \text{ mol } H_2$$

Step 3   Now that we know the moles of H_2, we can compute the volume of H_2 by using the ideal gas law, where

$$P = 1.50 \text{ atm}$$
$$V = ?$$
$$n = 0.405 \text{ mol}$$
$$R = 0.08206 \text{ L atm/K mol}$$
$$T = 19 \text{ °C} = 19 + 273 = 292 \text{ K}$$

$$V = \frac{nRT}{P} = \frac{(0.405 \text{ mol})\left(0.08206 \dfrac{\text{L atm}}{\text{K mol}}\right)(292 \text{ K})}{1.50 \text{ atm}}$$
$$= 6.47 \text{ L of } H_2$$

Self-Check Exercise 13.12

Although there are several possible ways to do this problem, the most convenient method involves using the molar volume at STP. First we use the ideal gas equation to calculate the moles of NH_3 present:

$$n = \frac{PV}{RT}$$

where $P = 15.0$ atm, $V = 5.00$ L, and $T = 25$ °C + 273 = 298 K.

$$n = \frac{(15.0 \text{ atm})(5.00 \text{ L})}{\left(0.08206 \dfrac{\text{L atm}}{\text{K mol}}\right)(298 \text{ K})} = 3.07 \text{ mol}$$

We know that at STP each mole of gas occupies 22.4 L. Therefore, 3.07 mol has the volume

$$3.07 \text{ mol} \times \frac{22.4 \text{ L}}{1 \text{ mol}} = 68.8 \text{ L}$$

The volume of the ammonia at STP is 68.8 L.

Chapter 14

Self-Check Exercise 14.1

Energy to melt the ice:

$$15 \text{ g } H_2O \times \frac{1 \text{ mol } H_2O}{18 \text{ g } H_2O} = 0.83 \text{ mol } H_2O$$

$$0.83 \text{ mol } H_2O \times 6.02 \frac{\text{kJ}}{\text{mol } H_2O} = 5.0 \text{ kJ}$$

Energy to heat the water from 0 °C to 100 °C:

$$4.18 \frac{\text{J}}{\text{g °C}} \times 15 \text{ g} \times 100 \text{ °C} = 6300 \text{ J}$$

$$6300 \text{ J} \times \frac{1 \text{ kJ}}{1000 \text{ J}} = 6.3 \text{ kJ}$$

Energy to vaporize the water at 100 °C:

$$0.83 \text{ mol } H_2O \times 40.6 \frac{\text{kJ}}{\text{mol } H_2O} = 34 \text{ kJ}$$

Total energy required:

$$5.0 \text{ kJ} + 6.3 \text{ kJ} + 34 \text{ kJ} = 45 \text{ kJ}$$

Self-Check Exercise 14.2

a. Contains SO_3 molecules—a molecular solid.
b. Contains Ba^{2+} and O^{2-} ions—an ionic solid.
c. Contains Au atoms—an atomic solid.

Chapter 15

Self-Check Exercise 15.1

$$\text{Mass percent} = \frac{\text{mass of solute}}{\text{mass of solution}} \times 100\%$$

For this sample, the mass of solution is 135 g and the mass of the solute is 4.73 g, so

$$\text{Mass percent} = \frac{4.73 \text{ g solute}}{135 \text{ g solution}} \times 100\%$$
$$= 3.50\%$$

Self-Check Exercise 15.2

Using the definition of mass percent, we have

$$\frac{\text{Mass of solute}}{\text{Mass of solution}} =$$

$$\frac{\text{grams of solute}}{\text{grams of solute} + \text{grams of solvent}} \times 100\% = 40.0\%$$

There are 425 grams of solute (formaldehyde). Substituting, we have

$$\frac{425 \text{ g}}{425 \text{ g} + \text{grams of solvent}} \times 100\% = 40.0\%$$

We must now solve for grams of solvent (water). This will take some patience, but we can do it if we proceed step by step. First we divide both sides by 100%.

$$\frac{425 \text{ g}}{425 \text{ g} + \text{grams of solvent}} \times \frac{100\%}{100\%} = \frac{40.0\%}{100\%} = 0.400$$

Now we have

$$\frac{425 \text{ g}}{425 \text{ g} + \text{grams of solvent}} = 0.400$$

Next we multiply both sides by (425 g + grams of solvent).

$$(425 \text{ g} + \text{grams of solvent}) \times \frac{425 \text{ g}}{425 \text{ g} + \text{grams of solvent}}$$
$$= 0.400 \times (425 \text{ g} + \text{grams of solvent})$$

This gives

$$425 \text{ g} = 0.400 \times (425 \text{ g} + \text{grams of solvent})$$

Carrying out the multiplication gives

$$425 \text{ g} = 170. \text{ g} + 0.400 \text{ (grams of solvent)}$$

Now we subtract 170. g from both sides,

$$425 \text{ g} - 170. \text{ g} = 170. \text{ g} - 170. \text{ g} + 0.400 \text{ (grams of solvent)}$$
$$255 \text{ g} = 0.400 \text{ (grams of solvent)}$$

and divide both sides by 0.400.

$$\frac{255 \text{ g}}{0.400} = \frac{0.400}{0.400} \text{ (grams of solvent)}$$

We finally have the answer:

$$\frac{255 \text{ g}}{0.400} = 638 \text{ g} = \text{grams of solvent}$$
$$= \text{mass of water needed}$$

Self-Check Exercise 15.3

The moles of ethanol can be obtained from its molar mass (46.1).

$$1.00 \text{ g } C_2H_5OH \times \frac{1 \text{ mol } C_2H_5OH}{46.1 \text{ g } C_2H_5OH} = 2.17 \times 10^{-2} \text{ mol } C_2H_5OH$$

$$\text{Volume in liters} = 101 \text{ mL} \times \frac{1 \text{ L}}{1000 \text{ mL}} = 0.101 \text{ L}$$

$$\text{Molarity of } C_2H_5OH = \frac{\text{moles of } C_2H_5OH}{\text{liters of solution}}$$
$$= \frac{2.17 \times 10^{-2} \text{ mol}}{0.101 \text{ L}}$$
$$= 0.215 \text{ M}$$

Self-Check Exercise 15.4

When Na_2CO_3 and $Al_2(SO_4)_3$ dissolve in water, they produce ions as follows:

$$Na_2CO_3(s) \xrightarrow{H_2O(l)} 2Na^+(aq) + CO_3^{2-}(aq)$$

$$Al_2(SO_4)_3(s) \xrightarrow{H_2O(l)} 2Al^{3+}(aq) + 3SO_4^{2-}(aq)$$

Therefore, in a 0.10 M Na_2CO_3 solution, the concentration of Na^+ ions is $2 \times 0.10 \text{ M} = 0.20 \text{ M}$ and the concentration of CO_3^{2-} ions is 0.10 M. In a 0.010 M $Al_2(SO_4)_3$ solution, the concentration of Al^{3+} ions is $2 \times 0.010 \text{ M} = 0.020 \text{ M}$ and the concentration of SO_4^{2-} ions is $3 \times 0.010 \text{ M} = 0.030 \text{ M}$.

Self-Check Exercise 15.5

When solid $AlCl_3$ dissolves, it produces ions as follows:

$$AlCl_3(s) \xrightarrow{H_2O(l)} Al^{3+}(aq) + 3Cl^-(aq)$$

so a 1.0×10^{-3} M $AlCl_3$ solution contains 1.0×10^{-3} M Al^{3+} ions and 3.0×10^{-3} M Cl^- ions.

To calculate the moles of Cl^- ions in 1.75 L of the 1.0×10^{-3} M $AlCl_3$ solution, we must multiply the volume by the molarity.

$$1.75 \text{ L solution} \times 3.0 \times 10^{-3} \text{ M } Cl^-$$

$$= 1.75 \text{ L solution} \times \frac{3.0 \times 10^{-3} \text{ mol } Cl^-}{\text{L solution}}$$

$$= 5.25 \times 10^{-3} \text{ mol } Cl^- = 5.3 \times 10^{-3} \text{ mol } Cl^-$$

Self-Check Exercise 15.6

We must first determine the number of moles of formaldehyde in 2.5 L of 12.3 M formalin. Remember that volume of solution (in liters) times molarity gives moles of solute. In this case, the volume of solution is 2.5 L and the molarity is 12.3 moles of HCHO per liter of solution.

$$2.5 \text{ L solution} \times \frac{12.3 \text{ mol HCHO}}{\text{L solution}} = 31 \text{ mol HCHO}$$

Next, using the molar mass of HCHO (30.0 g), we convert 31 moles of HCHO to grams.

$$31 \text{ mol HCHO} \times \frac{30.0 \text{ g HCHO}}{1 \text{ mol HCHO}} = 9.3 \times 10^2 \text{ g HCHO}$$

Therefore, 2.5 L of 12.3 M formalin contains 9.3×10^2 g of formaldehyde. We must weigh out 930 g of formaldehyde and dissolve it in enough water to make 2.5 L of solution.

Self-Check Exercise 15.7

We are given the following information:

$$M_1 = 12 \frac{\text{mol}}{\text{L}} \qquad\qquad M_2 = 0.25 \frac{\text{mol}}{\text{L}}$$

$$V_1 = ? \text{ (what we need to find)} \qquad V_2 = 0.75 \text{ L}$$

Using the fact that the moles of solute do not change upon dilution, we know that

$$M_1 \times V_1 = M_2 \times V_2$$

Solving for V_1 by dividing both sides by M_1 gives

$$V_1 = \frac{M_2 \times V_2}{M_1} = \frac{0.25 \frac{\text{mol}}{\text{L}} \times 0.75 \text{ L}}{12 \frac{\text{mol}}{\text{L}}}$$

and

$$V_1 = 0.016 \text{ L} = 16 \text{ mL}$$

Self-Check Exercise 15.8

Step 1 When the aqueous solutions of Na_2SO_4 (containing Na^+ and SO_4^{2-} ions) and $Pb(NO_3)_2$ (containing Pb^{2+} and NO_3^- ions) are mixed, solid $PbSO_4$ is formed.

$$Pb^{2+}(aq) + SO_4^{2-}(aq) \rightarrow PbSO_4(s)$$

Step 2 We must first determine whether Pb^{2+} or SO_4^{2-} is the limiting reactant by calculating the moles of Pb^{2+} and SO_4^{2-} ions present. Because $0.0500\ M\ Pb(NO_3)_2$ contains $0.0500\ M\ Pb^{2+}$ ions, we can calculate the moles of Pb^{2+} ions in 1.25 L of this solution as follows:

$$1.25\ \cancel{L} \times \frac{0.0500\ \text{mol}\ Pb^{2+}}{\cancel{L}} = 0.0625\ \text{mol}\ Pb^{2+}$$

The $0.0250\ M\ Na_2SO_4$ solution contains $0.0250\ M\ SO_4^{2-}$ ions, and the number of moles of SO_4^{2-} ions in 2.00 L of this solution is

$$2.00\ \cancel{L} \times \frac{0.0250\ \text{mol}\ SO_4^{2-}}{\cancel{L}} = 0.0500\ \text{mol}\ SO_4^{2-}$$

Step 3 Pb^{2+} and SO_4^{2-} react in a 1:1 ratio, so the amount of SO_4^{2-} ions is limiting because SO_4^{2-} is present in the smaller number of moles.

Step 4 The Pb^{2+} ions are present in excess, and only 0.0500 mole of solid $PbSO_4$ will be formed.

Step 5 We calculate the mass of $PbSO_4$ by using the molar mass of $PbSO_4$ (303.3 g).

$$0.0500\ \cancel{\text{mol}\ PbSO_4} \times \frac{303.3\ \text{g}\ PbSO_4}{1\ \cancel{\text{mol}\ PbSO_4}} = 15.2\ \text{g}\ PbSO_4$$

Self-Check Exercise 15.9

Step 1 Because nitric acid is a strong acid, the nitric acid solution contains H^+ and NO_3^- ions. The KOH solution contains K^+ and OH^- ions. When these solutions are mixed, the H^+ and OH^- react to form water.

$$H^+(aq) + OH^-(aq) \rightarrow H_2O(l)$$

Step 2 The number of moles of OH^- present in 125 mL of $0.050\ M$ KOH is

$$125\ \cancel{\text{mL}} \times \frac{1\ \cancel{L}}{1000\ \cancel{\text{mL}}} \times \frac{0.050\ \text{mol}\ OH^-}{\cancel{L}} = 6.3 \times 10^{-3}\ \text{mol}\ OH^-$$

Step 3 H^+ and OH^- react in a 1:1 ratio, so we need 6.3×10^{-3} mole of H^+ from the $0.100\ M\ HNO_3$.

Step 4 6.3×10^{-3} mole of OH^- requires 6.3×10^{-3} mole of H^+ to form 6.3×10^{-3} mole of H_2O.
Therefore,

$$V \times \frac{0.100\ \text{mol}\ H^+}{L} = 6.3 \times 10^{-3}\ \text{mol}\ H^+$$

where V represents the volume in liters of $0.100\ M\ HNO_3$ required. Solving for V, we have

$$V = \frac{6.3 \times 10^{-3}\ \cancel{\text{mol}\ H^+}}{\dfrac{0.100\ \cancel{\text{mol}\ H^+}}{L}} = 6.3 \times 10^{-2}\ L$$

$$= 6.3 \times 10^{-2}\ \cancel{L} \times \frac{1000\ \text{mL}}{\cancel{L}} = 63\ \text{mL}$$

Self-Check Exercise 15.10

From the definition of normality, N = equiv/L, we need to calculate (1) the equivalents of KOH and (2) the volume of the solution in liters. To find the number of equivalents, we use the equivalent weight of KOH, which is 56.1 g (see Table 15.2).

$$23.6\ \text{g}\ \cancel{KOH} \times \frac{1\ \text{equiv}\ KOH}{56.1\ \text{g}\ \cancel{KOH}} = 0.421\ \text{equiv}\ KOH$$

Next we convert the volume to liters.

$$755\ \cancel{\text{mL}} \times \frac{1\ L}{1000\ \cancel{\text{mL}}} = 0.755\ L$$

Finally, we substitute these values into the equation that defines normality.

$$\text{Normality} = \frac{\text{equiv}}{L} = \frac{0.421\ \text{equiv}}{0.755\ L} = 0.558\ N$$

Self-Check Exercise 15.11

To solve this problem, we use the relationship

$$N_{\text{acid}} \times V_{\text{acid}} = N_{\text{base}} \times V_{\text{base}}$$

where

$$N_{\text{acid}} = 0.50\ \frac{\text{equiv}}{L}$$

$$V_{\text{acid}} = ?$$

$$N_{\text{base}} = 0.80\ \frac{\text{equiv}}{L}$$

$$V_{\text{base}} = 0.250\ L$$

We solve the equation

$$N_{\text{acid}} \times V_{\text{acid}} = N_{\text{base}} \times V_{\text{base}}$$

for V_{acid} by dividing both sides by N_{acid}.

$$\frac{N_{\text{acid}} \times V_{\text{acid}}}{N_{\text{acid}}} = \frac{N_{\text{base}} \times V_{\text{base}}}{N_{\text{acid}}}$$

$$V_{\text{acid}} = \frac{N_{\text{base}} \times V_{\text{base}}}{N_{\text{acid}}} = \frac{(0.80\ \frac{\text{equiv}}{\cancel{L}}) \times (0.250\ L)}{0.50\ \frac{\text{equiv}}{\cancel{L}}}$$

$$V_{\text{acid}} = 0.40\ L$$

Therefore, 0.40 L of 0.50 N H_2SO_4 is required to neutralize 0.250 L of 0.80 N KOH.

Chapter 16

Self-Check Exercise 16.1

The conjugate acid–base pairs are

$$H_2O, \qquad H_3O^+$$
$$\text{Base} \qquad \text{Conjugate acid}$$

and

$$HC_2H_3O_2, \qquad C_2H_3O_2^-$$
$$\text{Acid} \qquad \text{Conjugate base}$$

The members of both pairs differ by one H^+.

Self-Check Exercise 16.2

Because $[H^+][OH^-] = 1.0 \times 10^{-14}$, we can solve for $[H^+]$.

$$[H^-] = \frac{1.0 \times 10^{-14}}{[OH^-]} = \frac{1.0 \times 10^{-14}}{2.0 \times 10^{-2}} = 5.0 \times 10^{-13}\ M$$

This solution is basic: $[OH^-] = 2.0 \times 10^{-2}\ M$ is greater than $[H^+] = 5.0 \times 10^{-13}\ M$.

Self-Check Exercise 16.3

a. Because $[H^+] = 1.0 \times 10^{-3}\ M$, we get pH = 3.00 because pH = $-\log[H^+] = -\log[1.0 \times 10^{-3}] = 3.00$.

b. Because $[OH^-] = 5.0 \times 10^{-5}\ M$, we can find $[H^+]$ from the K_w expression.

$$[H^+] = \frac{K_w}{[OH^-]} = \frac{1.0 \times 10^{-14}}{5.0 \times 10^{-5}} = 2.0 \times 10^{-10}\ M$$

$$\text{pH} = -\log[H^+] = -\log[2.0 \times 10^{-10}] = 9.70$$

Self-Check Exercise 16.4

$$\text{pOH} + \text{pH} = 14.00$$
$$\text{pOH} = 14.00 - \text{pH} = 14.00 - 3.5$$
$$\text{pOH} = 10.5$$

Self-Check Exercise 16.5

Step 1 pH = 3.50
Step 2 $-\text{pH} = -3.50$

Step 3 $\boxed{\text{inv}}\ \boxed{\text{log}}\ -3.50 = 3.2 \times 10^{-4}$

$[H^+] = 3.2 \times 10^{-4}\ M$

Self-Check Exercise 16.6

Step 1 pOH = 10.50

Step 2 $-$pOH = -10.50

Step 3 $\boxed{\text{inv}}\ \boxed{\text{log}}\ -10.50 = 3.2 \times 10^{-11}$

$[OH^-] = 3.2 \times 10^{-11}\ M$

Self-Check Exercise 16.7

Because HCl is a strong acid, it is completely dissociated:

$5.0 \times 10^{-3}\ M\ HCl \rightarrow 5.0 \times 10^{-3}\ M\ H^+$ and $5.0 \times 10^{-3}\ M\ Cl^-$

so $[H^+] = 5.0 \times 10^{-3}\ M$.

pH $= -\log(5.0 \times 10^{-3}) = 2.30$

Chapter 17

Self-Check Exercise 17.1

Applying the law of chemical equilibrium gives

$$K = \frac{[NO_2]^4[H_2O]^6}{[NH_3]^4[O_2]^7}$$

Coefficient of NO_2 → $[NO_2]^4$; Coefficient of H_2O → $[H_2O]^6$; ← Coefficient of O_2; ↑ Coefficient of NH_3

Self-Check Exercise 17.2

a. $K = [O_2]^3$ The solids are not included.

b. $K = [N_2O][H_2O]^2$ The solid is not included. Water is gaseous in this reaction, so it is included.

c. $K = \dfrac{1}{[CO_2]}$ The solids are not included.

d. $K = \dfrac{1}{[SO_3]}$ Water and H_2SO_4 are pure liquids and so are not included.

Self-Check Exercise 17.3

When rain is imminent, the concentration of water vapor in the air increases. This shifts the equilibrium to the right, forming $CoCl_2 \cdot 6H_2O(s)$, which is pink.

Self-Check Exercise 17.4

a. No change. Both sides of the equation contain the same number of gaseous components. The system cannot change its pressure by shifting its equilibrium position.

b. Shifts to the left. The system can increase the number of gaseous components present, and so increase the pressure, by shifting to the left.

c. Shifts to the right to increase the number of gaseous components and thus its pressure.

Self-Check Exercise 17.5

a. Shifts to the right away from added SO_2.

b. Shifts to the right to replace removed SO_3.

c. Shifts to the right to decrease its pressure.

d. Shifts to the right. Energy is a product in this case, so a decrease in temperature favors the forward reaction (which produces energy).

Self-Check Exercise 17.6

a. $BaSO_4(s) \rightleftharpoons Ba^{2+}(aq) + SO_4^{2-}(aq); K_{sp} = [Ba^{2+}][SO_4^{2-}]$

b. $Fe(OH)_3(s) \rightleftharpoons Fe^{3+}(aq) + 3OH^-(aq); K_{sp} = [Fe^{3+}][OH^-]^3$

c. $Ag_3PO_4(s) \rightleftharpoons 3Ag^+(aq) + PO_4^{3-}(aq); K_{sp} = [Ag^+]^3[PO_4^{3-}]$

Self-Check Exercise 17.7

$(3.9 \times 10^{-5})^2 = 1.5 \times 10^{-9} = K_{sp}$

Self-Check Exercise 17.8

$PbCrO_4(s) \rightleftharpoons Pb^{2+}(aq) + CrO_4^{2-}(aq)$

$K_{sp} = [Pb^{2+}][CrO_4^{2-}] = 2.0 \times 10^{-16}$

$[Pb^{2+}] = x$

$[CrO_4^{2-}] = x$

$K_{sp} = 2.0 \times 10^{-16} = x^2$

$x = [Pb^{2+}] = [CrO_4^{2-}] = 1.4 \times 10^{-8}$

Chapter 18

Self-Check Exercise 18.1

a. CuO contains Cu^{2+} and O^{2-} ions, so copper is oxidized $(Cu \rightarrow Cu^{2+} + 2e^-)$ and oxygen is reduced $(O + 2e^- \rightarrow O^{2-})$.

b. CsF contains Cs^+ and F^- ions. Thus cesium is oxidized $(Cs \rightarrow Cs^+ + e^-)$ and fluorine is reduced $(F + e^- \rightarrow F^-)$.

Self-Check Exercise 18.2

a. SO_3

We assign oxygen first. Each O is assigned an oxidation state of -2, giving a total of -6 (3×-2) for the three oxygen atoms. Because the molecule has zero charge overall, the sulfur must have an oxidation state of $+6$.

Check: $+6 + 3(-2) = 0$

b. SO_4^{2-}

As in part a, each oxygen is assigned an oxidation state of -2, giving a total of -8 (4×-2) on the four oxygen atoms. The anion has a net charge of -2, so the sulfur must have an oxidation state of $+6$.

Check: $+6 + 4(-2) = -2$

SO_4^{2-} has a charge of -2, so this is correct.

c. N_2O_5

We assign oxygen before nitrogen because oxygen is more electronegative. Thus each O is assigned an oxidation state of -2, giving a total of -10 (5×-2) on the five oxygen atoms. Therefore, the oxidation states of the *two* nitrogen atoms must total $+10$ because N_2O_5 has no overall charge. Each N is assigned an oxidation state of $+5$.

Check: $2(+5) + 5(-2) = 0$

d. PF_3

First we assign the fluorine an oxidation state of -1, giving a total of -3 (3×-1) on the three fluorine atoms. Thus P must have an oxidation state of $+3$.

Check: $+3 + 3(-1) = 0$

e. C_2H_6

In this case, it is best to recognize that hydrogen is always $+1$ in compounds with nonmetals. Thus each H is assigned an oxidation state of $+1$, which means that the six H atoms account for a total of $+6$ ($6 \times +1$). Therefore, the *two* carbon atoms must account for -6, and each carbon is assigned an oxidation state of -3.

Check: $2(-3) + 6(+1) = 0$

Self-Check Exercise 18.3

We can tell whether this is an oxidation–reduction reaction by comparing the oxidation states of the elements in the reactants and products:

$$N_2 + 3H_2 \rightarrow 2NH_3$$

Oxidation states: 0   0   −3   +1 (each H)

Nitrogen goes from 0 to −3. Thus it gains three electrons and is reduced. Each hydrogen atom goes from 0 to +1 and is thus oxidized, so this is an oxidation–reduction reaction. The oxidizing agent is N_2 (it takes electrons from H_2). The reducing agent is H_2 (it gives electrons to N_2).

$$N_2 + 3H_2 \rightarrow 2NH_3$$
6e⁻

Self-Check Exercise 18.4

The unbalanced equation for this reaction is

$$Cu(s) + HNO_3(aq) \rightarrow Cu(NO_3)_2(aq) + H_2O(l) + NO(g)$$

Copper metal   Nitric acid   Aqueous copper(II) nitrate (contains Cu^{2+})   Water   Nitrogen monoxide

Step 1   The oxidation half-reaction is

$$Cu + HNO_3 \rightarrow Cu(NO_3)_2$$

Oxidation states: 0   +1 +5 −2   +2 +5 −2
                          (each O)   (each O)

The copper goes from 0 to +2 and thus is oxidized. This reduction reaction is

$$HNO_3 \rightarrow NO$$

Oxidation states: +1 +5 −2   +2 −2
                          (each O)

In this case, nitrogen goes from +5 in HNO_3 to +2 in NO and so is reduced. Notice two things about these reactions:

1. The HNO_3 must be included in the oxidation half-reaction to supply NO_3^- in the product $Cu(NO_3)_2$.
2. Although water is a product in the overall reaction, it does not need to be included in either half-reaction at the beginning. It will appear later as we balance the equation.

Step 2   *Balance the oxidation half-reaction.*

$$Cu + HNO_3 \rightarrow Cu(NO_3)_2$$

a. Balance nitrogen first.

$$Cu + 2HNO_3 \rightarrow Cu(NO_3)_2$$

b. Balancing nitrogen also caused oxygen to balance.
c. Balance hydrogen using H^+.

$$Cu + 2HNO_3 \rightarrow Cu(NO_3)_2 + 2H^+$$

d. Balance the charge using e^-.

$$Cu + 2HNO_3 \rightarrow Cu(NO_3)_2 + 2H^+ + 2e^-$$

This is the balanced oxidation half-reaction.
Balance the reduction half-reaction.

$$HNO_3 \rightarrow NO$$

a. All elements are balanced except hydrogen and oxygen.
b. Balance oxygen using H_2O.

$$HNO_3 \rightarrow NO + 2H_2O$$

c. Balance hydrogen using H^+.

$$3H^+ + HNO_3 \rightarrow NO + 2H_2O$$

d. Balance the charge using e^-.

$$3e^- + 3H^+ + HNO_3 \rightarrow NO + 2H_2O$$

This is the balanced reduction half-reaction.

Step 3   We equalize electrons by multiplying the oxidation half-reaction by 3:

$$3 \times [Cu + 2HNO_3 \rightarrow Cu(NO_3)_2 + 2H^+ + 2e^-]$$

gives

$$3Cu + 6HNO_3 \rightarrow 3Cu(NO_3)_2 + 6H^+ + 6e^-$$

Multiplying the reduction half-reaction by 2:

$$2 \times [3e^- + 3H^+ + HNO_3 \rightarrow NO + 2H_2O]$$

gives

$$6e^- + 6H^+ + 2HNO_3 \rightarrow 2NO + 4H_2O$$

Step 4   We can now add the balanced half-reactions, which both involve a six-electron change.

$$3Cu + 6HNO_3 \rightarrow 3Cu(NO_3)_2 + 6H^+ + 6e^-$$
$$6e^- + 6H^+ + 2HNO_3 \rightarrow 2NO + 4H_2O$$
_____
$$6e^- + 6H^+ + 3Cu + 8HNO_3 \rightarrow 3Cu(NO_3)_2 + 2NO + 4H_2O + 6H^+ + 6e^-$$

Canceling species common to both sides gives the balanced overall equation:

$$3Cu(s) + 8HNO_3(aq) \rightarrow 3Cu(NO_3)_2(aq) + 2NO(g) + 4H_2O(l)$$

Step 5   Check the elements and charges.

Elements   3Cu, 8H, 8N, 24O → 3Cu, 8H, 8N, 24O
Charges   0 → 0

Chapter 19

Self-Check Exercise 19.1

a. An alpha particle is a helium nucleus, ^4_2He. We can initially represent the production of an α particle by $^{226}_{88}Ra$ as follows:

$$^{226}_{88}Ra \rightarrow {}^4_2He + {}^A_ZX$$

Because we know that both A and Z are conserved, we can write

$$A + 4 = 226 \quad \text{and} \quad Z + 2 = 88$$

Solving for A gives 222 and for Z gives 86, so ^A_ZX is $^{222}_{86}X$. Because Rn has $Z = 86$, ^A_ZX is $^{222}_{86}Rn$. The overall balanced equation is

$$^{226}_{88}Ra \rightarrow {}^4_2He + {}^{222}_{86}Rn$$

Check: $Z = 88$ $Z = 86 + 2 = 88$
                                →
            $A = 226$ $A = 222 + 4 = 226$

b. Using a similar strategy, we have

$$^{214}_{82}Pb \rightarrow {}^0_{-1}e + {}^A_ZX$$

Because $Z - 1 = 82$, $Z = 83$, and because $A + 0 = 214$, $A = 214$. Therefore, $^A_ZX = {}^{214}_{83}Bi$. The balanced equation is

$$^{214}_{82}Pb \rightarrow {}^0_{-1}e + {}^{214}_{83}Bi$$

Check: $Z = 82$ $Z = 83 - 1 = 82$
                                →
            $A = 214$ $A = 214 + 0 = 214$

Self-Check Exercise 19.2

a. The missing particle must be ^4_2H (an α particle), because

$$^{222}_{86}Rn \rightarrow {}^{218}_{84}Po + {}^4_2He$$

is a balanced equation.

Check: $Z = 86$ $Z = 84 + 2 = 86$
                                →
            $A = 222$ $A = 218 + 4 = 222$

b. The missing species must be $^{15}_7X$ or $^{15}_7N$, because the balanced equation is

$$^{15}_8O \rightarrow {}^{15}_7N + {}^0_1e$$

Check: $Z = 8$ $Z = 7 - 1 = 8$
                                →
            $A = 15$ $A = 15 + 0 = 15$

Self-Check Exercise 19.3

Let's do this problem by thinking about the number of half-lives required to go from 8.0×10^{-7} mole to 1.0×10^{-7} mole of $^{228}_{88}Ra$.

$$8.0 \times 10^{-7} \text{ mol} \longrightarrow 4.0 \times 10^{-7} \text{ mol} \longrightarrow$$

$$\text{First} \qquad\qquad \text{Second}$$
$$\text{half-life} \qquad\qquad \text{half-life}$$

$$2.0 \times 10^{-7} \text{ mol} \longrightarrow 1.0 \times 10^{-7} \text{ mol}$$

$$\text{Third}$$
$$\text{half-life}$$

It takes three half-lives, then, for the sample to go from 8.0×10^{-7} mole of $^{228}_{88}Ra$ to 1.0×10^{-7} mole of $^{228}_{88}Ra$. From Table 19.3, we know that the half-life of $^{228}_{88}Ra$ is 6.7 years. Therefore, the elapsed time is $3(6.7 \text{ years}) = 20.1$ years, or 2.0×10^1 years when we use the correct number of significant figures.

Chapter 20

Self-Check Exercise 20.1

The alkane with ten carbon atoms can be represented as $CH_3-(CH_2)_8-CH_3$ and its formula is $C_{10}H_{22}$. The alkane with fifteen carbons,

$$CH_3-(CH_2)_{13}-CH_3$$

has the formula $C_{15}H_{32}$.

Self-Check Exercise 20.2

a.
$$\overset{1}{C}H_3-\overset{2}{C}H_2-\overset{3}{C}H-\overset{4}{C}H_2-\overset{5}{C}H-\overset{6}{C}H_2-\overset{7}{C}H_2-\overset{8}{C}H_3$$

with a CH$_3$ (Methyl) branch at carbon 3 and a CH$_2$–CH$_3$ (Ethyl) branch at carbon 5.

This molecule is 5-ethyl-3-methyloctane.

b.
$$\overset{7}{C}H_3-\overset{6}{C}H_2-\overset{5}{C}H-\overset{4}{C}H_2-\overset{3}{C}H-(CH_3) \text{ Methyl}$$

with CH$_2$–CH$_3$ (Ethyl) branch and CH$_2$ $_2$ / CH$_3$ $_1$.

This molecule is 5-ethyl-3-methylheptane. Note that this chain could be numbered from the opposite direction to give the name 3-ethyl-5-methylheptane. These two names are equally correct.

Self-Check Exercise 20.3

The root name *decate* indicates a ten-carbon chain. There is a methyl group at the number 4 position and an isopropyl group at the number 5 position. The structural formula is

Self-Check Exercise 20.4

a. The longest chain has eight carbon atoms with a double bond, so the root name is octene. The double bond exists between carbons 3 and 4, so the name is 3-octene. There is a methyl group on the number-2 carbon. The name is 2-methyl-3-octene.

b. The carbon chain has five carbons with a triple bond between carbons 1 and 2. The name is 1-pentyne.

Self-Check Exercise 20.5

a. 2-chloronitrobenzene or *o*-chloronitrobenzene

b. 4-phenyl-2-hexene

Self-Check Exercise 20.6

a. 1-pentanol; primary alcohol

b. 2-methyl-2-propanol (but this alcohol is usually called tertiary butyl alcohol); tertiary alcohol

c. 5-bromo-2-hexanol; secondary alcohol

Self-Check Exercise 20.7

a. 4-ethyl-3-hexanone
   Because the compound is named as a hexanone, the carbonyl group is assigned the lowest possible number.

b. 7-isopropyldecanal

Page numbers followed by *n* refer to margin notes. Page numbers followed by *f* refer to figures. Page numbers followed by *t* refer to tables.